Laser Optics of Condensed Matter

VOLUME 2
THE PHYSICS OF
OPTICAL PHENOMENA AND THEIR
USE AS PROBES OF MATTER

Laser Optics of Condensed Matter

**VOLUME 2
THE PHYSICS OF
OPTICAL PHENOMENA AND THEIR
USE AS PROBES OF MATTER**

Edited by

Elsa Garmire
University of Southern California
Los Angeles, California

Alexei A. Maradudin
University of California
Irvine, California

and

Karl K. Rebane
Estonian Academy of Sciences
Tallinn, Estonia, USSR

Springer Science+Business Media, LLC

Library of Congress Cataloging-in-Publication Data

Binational USA-USSR Symposium on Laser Optics of Condensed Matter (4th
: 1990 : Irvine, Calif.)
    Laser optics of condensed matter. Volume 2, the physics of
optical phenomena and their use as probes of matter / edited by Elsa
Garmire, Alexei A. Maradudin and Karl K. Rebane.
        p.   cm.
    "Proceedings of the fourth Binational USA-USSR Symposium on Laser
Optics of Condesed Matter, held January 23-27, 1990, in Irvine,
California."
    Includes bibliographical references and index.
    ISBN 978-1-4613-6658-4   ISBN 978-1-4615-3726-7 (eBook)
    DOI 10.1007/978-1-4615-3726-7
    1. Condensed matter--Optical properties--Congresses. 2. Lasers-
-Congresses.   I. Garmire, Elsa. II. Maradudin, Alexei A.
III. Rebane, Karl Karlovich, 1926-   . IV. Title.
QC173.4.C65B56   1990
530.4'12--dc20   91-2849
                                                                   CIP

Proceedings of the Fourth Binational USA-USSR Symposium
on Laser Optics of Condensed Matter, held January 23-27, 1990,
in Irvine, California

© 1991 Springer Science+Business Media New York
Originally published by Plenum Press New York in 1991
Softcover reprint of the hardcover 1st edition 1991

PREFACE

The Fourth USA-USSR Symposium. on The Physics of Optical Phenomena and Their Use as Probes of Matter, was held in Irvine, California, January 23-27, 1990.

Participating in the Symposium were 22 scientists from the USSR and 29 from the USA. In addition, to provide an international dimension to this Symposium without, however, compromising significantly its essentially binational character, 7 non-US and non-USSR scientists were invited to take part in it.

The present volume is the proceedings of that Symposium, and contains all manuscripts received prior to August 1, 1990, representing scientific contributions presented. A few manuscripts were not received, but for completeness the corresponding abstract is printed.

Three previous USA/USSR Binational Symposia on related topics have been held, viz. "Theory of Light Scattering in Condensed Matter" (Moscow, 1975), "Light Scattering in Solids" (New York, 1979), and "Laser Optics of Condensed Matter" (Leningrad, 1987). These meetings were evaluated by the participants as highly successful and provided invaluable opportunities for researchers to exchange information and to initiate collaborative work which led to research visits by US physicist to Soviet laboratories, and vice versa, and which continue to the present day.

The suggestion for holding the Fourth Symposium arose during the third Symposium, from participants on both sides who appreciated the opportunity the latter had provided, after a gap of eight years during which formal opportunities for direct interaction did not exist for learning at first hand about the intense research going on in both countries in the general subject area of these Symposia. Discussions with colleagues in both countries after the last symposia revealed that the interest in a fourth symposium was not confined to just those who had participated in the earlier meetings, but was widespread.

A Symposium such as this one cannot take place without the participation and support of many individuals and organizations. The binational organization of this Symposium, and the agencies funding it, are indicated on the following pages.

It is finally our very pleasant duty to thank the following persons who helped in various ways during the Symposium and also in putting together the Proceedings: Dr. L. A. Bureyeva, Ms. Jeannie Brown, and Mrs. M. B. Maradudin. The American co-editors are grateful to their

Soviet co-editor, Professor Karl K. Rebane, for his continued help and assistance.

E. Garmire
A.A. Maradudin

Los Angeles and Irvine
August 1, 1990

CONTENTS

SESSION V
OPTICAL PROPERTIES OF CRITICAL PHENOMENA
RANDOM SYSTEMS, AND COHERENT PHENOMENA

SESSION VI
NONLINEAR OPTICAL PROPERTIES OF SEMICONDUCTORS
ORGANICS AND FIBERS

SESSION IX
ULTRASHORT PHENOMENA

OPENING REMARKS

Ladies and Gentlemen, Dear Friends!

A month ago died one of the best physicists of our time, world-wide known, world-wide respected fighter for the human rights, Academician Andrei Dmitrievich Sakharov. May I ask you to rise for a moment of silence in memory of a great and decent man - Andrei Dmitrievich Sakharov. Thank you!

I would like to point out a remarkable circumstance about our meeting here today: we got together for the Fourth Symposium exactly in the time and at the place as it was agreed three years ago in Leningrad. A considerable delegation of twenty Soviet physicists has arrived to Irvine. It indicates, firstly, the continuing warming of the political climate in the world, between our two countries. On the other hand it shows that some real interest does exist in the topics selected for our Symposia and that there has been progress in our topics, in the work the participants are performing in laser optics studies of condensed matter. And last, but not least, the US organizing committee, especially Professor Alexei Maradudin, have given an example of excellent work.

On behalf of the Soviet participants I wish all of us a successful symposium! Thank you!

Professor K. K. Rebane
Institute of Physics, Tartu

Dear Colleagues!

I am pleased and honored to open the Fourth US/USSR Binational Symposium on "The Physics of Optical Phenomena and Their Use as Probes of Matter." This 1990 Symposium in Irvine follows the Symposia held in Moscow in 1975, in New York in 1979, and in Leningrad in 1987.

On behalf of the US Steering Committee for this Symposium I welcome our colleagues from the Soviet Union and those from the United States. In a break from the tradition of the previous Symposia, we have also invited a small number of colleagues from Western Europe and Japan, and I am happy to welcome them here as well.

It is fitting that we meet today in the Arnold and Mabel Beckman Center for the National Academies of Science and Engineering, in view of the support the US National Academy of Sciences, together with the Academy of Sciences of the USSR, has given this series of Symposia.

The Organizing Committee have worked hard to prepare a high-level program that reflects the current state of the subject of the Symposium. A full program of post-Symposium visits for our Soviet colleagues has also been arranged.

Among the objectives underlying these Symposia have been strengthening and deepening American-Soviet scientific cooperation in the optics of condensed matter; bringing together theorists and experimentalists; and widening the circle of participants. We hope that the present Symposium keeps us on the path toward achieving these objectives. In this connection, I note that more than half of the US group are first-time participants in these binational Symposia, and we also see new faces among the Soviet participants. This augurs well for the continuity of our Symposia.

It is also worth noting that whereas four years elapsed between the first and second Symposia, and eight years between the second and third, we begin the fourth Symposium only two and a half years after the conclusion of the third Symposium. I think these intervals have been a sensitive barometer of the state of international relations, and hope that this has now reached the point where the intervals between future Symposia will be determined only by the rate at which we are able to generate new results.

It is now time to begin the scientific program that will occupy us for the remainder of this week. Therefore, as the Chairman of the US Steering Committee for it, I declare the Fourth USA/USSR Binational Symposium, on The Physics of Optical Phenomena and Their Use as Probes of Matter, open.

Professor A. A. Maradudin
University of California, Irvine

PICOSECOND RESOLVED OPTICALLY DRIVEN PHONON DYNAMICS

W. E. Bron

Department of Physics
University of California, Irvine
Irvine, CA 92717 U.S.A.

INTRODUCTION

New dual synchronously pumped and synchronously amplified variable frequency pulsed dye lasers[1],[2] have been assembled. These lasers have been applied[3]-[6] to observations of phonon and polariton dynamics directly in the subpico- and in the picosecond time domain. We describe below observation of the dephasing of longitudinal optical (LO) phonons, the dephasing of polaritons, both near thermal equilibrium, and also the dynamics of LO phonons when the phonon occupation number is driven far above thermal equilibrium.

EXPERIMENTAL METHODS

The standard technique to determine the "lifetime" of optical phonons is to determine the spectral width of the incoherent Raman scattering cross section. Since spectral width and temporal duration are related by a Fourier transformation, information on the duration of the optical phonon excitation can, in principle, be obtained in this way. More recently it has become apparent that we may circumvent this transformation by simply making measurements of the optical phonon dynamics directly in the time domain. Since, however, optical phonon "lifetimes" are of the order of picoseconds, the experimental techniques must be capable of resolution within that very short temporal domain. Moreover since phonon energies correspond to the far infrared (FIR) region of the electromagnetic spectrum, it would be necessary to construct a picosecond source in this spectral range. Unfortunately, to date no such source has been demonstrated. On the other hand, many pico- and even femtosecond lasers, with energies in the visible spectrum, have been developed. It is, therefore, necessary to use nonlinear difference mixing of two visible laser frequencies to form a FIR excitation source of phonons provided the phonons to be excited are Raman active.[7]

As an example of the application of such laser systems I cite our work on the generation of near-zone-center coherent optical phonons in GaP (and ZnSe) through coherent Raman excitation (CRE), and our study of the subsequent dynamics of the coherent LO phonon state. These investigations were carried out directly in the time domain through time resolved coherent anti-Stokes Raman scattering (TRCARS).[3],[4] Only the results on GaP will be discussed here.

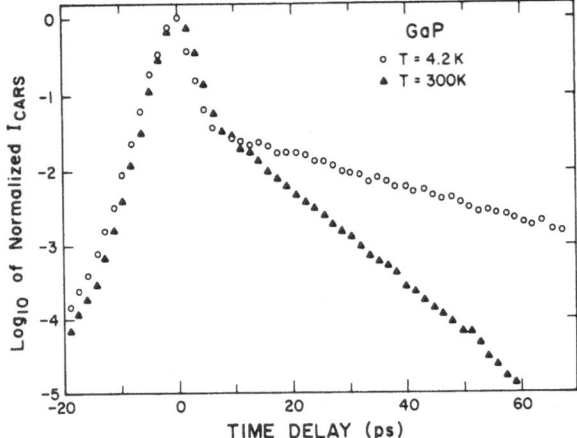

Fig. 1. Temporal evolution of the TRCARS
signal intensity for LO phonons in
GaP for ambient temperatures of 4.2
and 300K.

Coherent Raman excitation (CRE) and coherent anti-Stokes Raman scattering (CARS) are particular forms of nonlinear mixing of laser
fields.[8] Two electromagnetic fields $\vec{E}_\ell$ and $\vec{E}_s$, with photon energies $\hbar\omega_\ell$
and $\hbar\omega_s$ and wavevectors $\vec{k}_\ell$, $\vec{k}_s$, can in general give rise to a series of
nonlinear interactions in a solid. Of particular interest here is the
case in which $\omega_\ell - \omega_s$ corresponds to a Raman active vibrational excitation.
In the resultant CRE, the energy $\hbar(\omega_\ell - \omega_s)$ and wavevector $\vec{q} = \vec{k}_\ell - \vec{k}_s$ of the
active mode are in the ideal case exactly specified, and the phase space
available to the excited vibrations is severely limited. In this sense
it differs markedly from stimulated and spontaneous Raman scattering for
which the available phase space is much larger. Thus, the occupation
probabilities of the components of the coherent state produced through
CRE may be very high as compared to those of incoherent Raman excitation. The coherent excitation may be probed by a third laser field, $\vec{E}_p$,
whose energy $\hbar\omega_p$ is taken for convenience to be $\hbar\omega_\ell$. The nonlinear
interaction of this field, with the coherent lattice vibrations results
in yet another nonlinear polarization[7] which drives an electromagnetic
field with energy $\hbar\omega_c = \hbar(2\omega_\ell - \omega_s)$. The probe technique clearly involves
coherent anti-Stokes Raman scattering, giving rise to the CARS
designation for the entire nonlinear process.

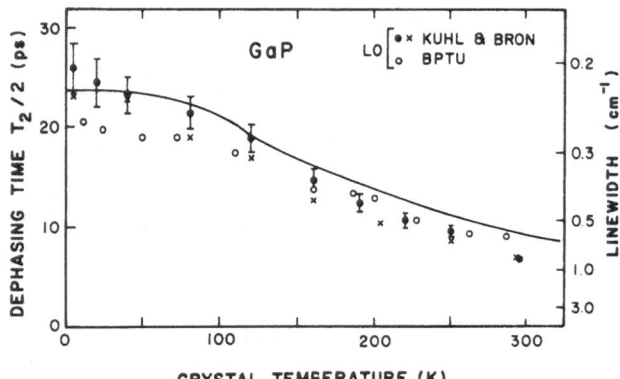

Fig. 2. Temperature dependence of the LO
phonon dephasing time and of the
Raman linewidth. The open circles
are from data taken from Ref. 9.

Phonon Dephasing Near Thermal Equilibrium

The components of the coherent phonon state described above are expected to dephase relative to each other after the excitation fields $\vec{E}_\ell$ and $\vec{E}_s$ cease. In general, phonon dephasing in solids results from scattering at boundaries, lattice imperfections, carriers, or through population decay brought about by anharmonic interactions. Near-zone-center LO phonons possess, however, nearly zero group velocities so that their generation in the bulk of the solid (as is done here) eliminates boundary scattering as an active dephasing mode. Impurity scattering may also be neglected in high quality crystals. Carrier scattering may also be neglected for observations in excess of a few picoseconds after carrier generation. We have demonstrated elsewhere, that a one-component plasma also has little effect on the LO phonon lifetime, providing, as done here, only very high purity ($< 10^{16}$ cm^{-3}) samples are used. Thus, contrary to the assumptions we made in Ref. [3], we now assume that, under our experimental conditions, anharmonic phonon decay dominates the dephasing process.

The experimental details have been discussed elsewhere.[3],[4] Figure 1 illustrates typical normalized TRCARS intensities as a function of the time delay Δt of the two pump beams (E_ℓ, E_s) relative to the probe beam (E_p) for the case of GaP held at 4.2K and 300K, and with $\omega_\ell - \omega_s = \omega_{LO}$, so that the coherent LO phonon state is excited. Two components of the CARS beam intensity $I_c(\Delta t)$, are clearly visible. The first component near $\Delta t = 0$, which can be independently observed when $\omega_\ell - \omega_s \neq \omega_{LO}$, is a slightly asymmetric "bell-shaped" curve. This component of $I_c(\Delta t)$ is attributed to the nonlinear response of bound electrons.[4] The remaining part of $I_c(\Delta t)$ decreases exponentially with delay time for $\Delta t \gg 0$ and can be observed to do so over several orders of magnitude. The temperature dependence of $T_2/2$ has also been determined through linewidth measurements of the spontaneous Raman scattering intensity. The results of the temporally resolved and those from linewidth measurements are displayed in Fig. 2. For completeness we also include similar linewidth data for a different crystal of GaP as reported by Bairamov, et al.[9]

If anharmonic decay is the sole dephasing mechanism, then at low temperatures, i.e., for $\hbar\omega \gg k_B T$, three phonon interactions dominate.[10] Of the possible interactions only LO phonon decay into two longitudinal acoustic phonons (LA) (and their recombination) needs to be considered. The phonon dispersion relations of GaP[11] plus energy and crystal momentum conservation, rule out all interactions except

$$LO(\omega_{LO}, 0) \longleftrightarrow LA(\omega_{LO}/2, \vec{k}') + LA(\omega_{LO}/2, -\vec{k}'). \tag{1}$$

Here, $\vec{k}'$ is the wavevector of the LA phonon at half the LO phonon energy. Three-particle anharmonic interactions lead[12] to a LO phonon decay rate Γ, at a frequency ω, as

$$\Gamma(\omega, T) = \frac{\pi}{2} \sum_{\vec{k}} \left| V_3 \begin{pmatrix} 0 & \vec{k} & \vec{k}' \\ LO & j & j' \end{pmatrix} \right|^2 \delta\left(\omega - \omega_{\vec{k}j} - \omega_{\vec{k}'j'} \right)$$

$$\times \left\{ n(T)_{\vec{k}j} + n(T)_{\vec{k}'j'} + 1 \right\} \tag{2}$$

In Eq. 2, V_3 is the anharmonic coupling coefficient, and $\omega_{\vec{k}'j'}$, $n(T)_{\vec{k}j}$ and $\omega_{\vec{k}'j'}$, $n(T)_{\vec{k}'j'}$ represent, respectively, the frequency and thermal occupation numbers of the two acoustic phonons into which the LO phonon

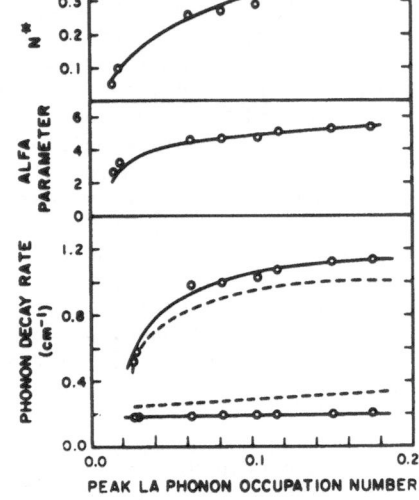

Fig. 3. TRCARS signal intensity as
a function of time delay
between the formation of
LO phonons and the time of
the probe pulse.

decays. The sum over the delta function leads to the phonon joint
density of states. Equation (1) is normally shortened to
$\Gamma(\omega,T) = \Gamma_o(\omega)\{n_1(\omega_1,T) + n_2(\omega_2 + 1\}$. The solid curve given in Fig. 2
indicates the temperature variation of the dephasing time for GaP as
obtained from Eq. (1) and (2). The experimental observation of a more
rapid decay with increasing temperature than that predicted by Eq. 2
implies that higher order phonon-phonon interactions, or thermally
excited carriers, are beginning to play a role.

Nonequilibrium Phonon Dephasing

We have observed a quite different phonon decay mechanism if, the
system described in the section above, is optically driven into a
strongly nonequilibrium state.[5] As noted above, near thermal equilibrium
the lifetime of an LO phonon in GaP, decaying into two acoustic phonons,
decreases as the ambient temperature increases (see Fig. 2). A theoretical analysis by Bulgadaev and Levinson[13] predicts that these effects are
supplanted by others when thermalized phonons are replaced by nonequilibrium phonon distributions. In fact, they predict that a strongly optically pumped LO phonon spectral distribution, in the presence of a highly
excited narrow-bandwidth, acoustic phonon spectral distribution, will
exhibit two (rather than one) decay rates below a critical acoustic
phonon occupation number N^*. We have recently reported the first
observation of two decay rates for the dephasing of LO phonons.[5]

It is clear that Eq. 2 above must be altered so as to replace the
thermal occupation number n(T) by that representing the nonequilibrium
state. Bulgadaev and Levinson have proposed that the distribution of the
occupation numbers of the acoustic phonons is Lorentzian such that

$$n_{\vec{k}j}(\omega) = n_A \frac{(\Delta\omega/2)^2}{(\omega-\omega_0/2)^2 + (\Delta\omega/2)^2} , \qquad (3)$$

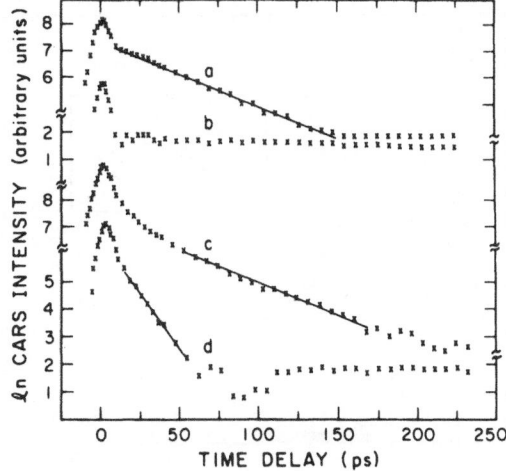

Fig.4. (Lower figure) Compilation of observed (open
circles) and theoretical-
ly predicted curves
(dashed lines) of the two
decay rates (Γ_1, Γ_2) as a
function of n_A. (Middle
figure) Alpha parameter
and (upper figure) N^* as
a function of n_A.

in which n_A is the occupation number at the peak of the distribution and
$\Delta\omega$ is its FWHM. The applicability of this particular distribution is
shown by Levinson[13] to be valid on theoretical grounds in the limits of
very low and very high optical excitation. As shown below, we observe
exponential decay throughout the experiment which is consistent only with
Lorentzian spectral distribution. We assume that the intrinsic lifetime
of the LA phonons toward decay to even lower energy is long compared to
the experimental observation time (~ 200 ps) so that their linewidth is
very small compared to $\Delta\omega$. Thus, in this model of the nonequilibrium
acoustic phonon state a "collective mode"[13] is formed of LA phonon
states, each of very narrow bandwidth, but whose occupation number
distribution follows equation (3). Replacing n(T) in Eq. 2 by
$n_{\vec{kj}}(\omega)$ of Eq. 3, leads to the somewhat unexpected result that the LO
phonon excitation has two, rather than a single, exponential decay
rates. It is further found that [for n_A less than $N^* = (\alpha-1)^2/8\alpha$]

$$\Gamma_{1,2} = \tfrac{1}{2}\,\Gamma_0\left[(\alpha+1) \mp \left\{(\alpha-1)^2 -\beta\right\}^{1/2}\right], \quad (4)$$

with $\alpha = 2\Delta\omega/\Gamma_0$ and $\beta = 8\alpha n_A$ (typical values for GaP are $\alpha \sim 1$, so that
$2\Delta\omega \sim \Gamma_0 \sim 1\ cm^{-1}$ and $\omega_0 \sim 400\ cm^{-1}$). Note that as n_A approaches zero,
Γ_1 and Γ_2 approach Γ_0 and $2\Delta\omega$, respectively. Thus, the solution with
$\Gamma_1 = \Gamma_0$ is the one usually observed in Raman measurements at relatively
low laser intensities. Bulgadaev and Levinson[13] predict that the
strength of the Raman signal associated with Γ_2, for small n_A, is so weak
as to be unobservable. This branch of the solution of Eq. (4), however,
gains intensity with increasing n_A until N^* is reached, beyond which the
intensities are the same and $\Gamma_1 = \Gamma_2$. Thus, Eq. (1) leads directly to a

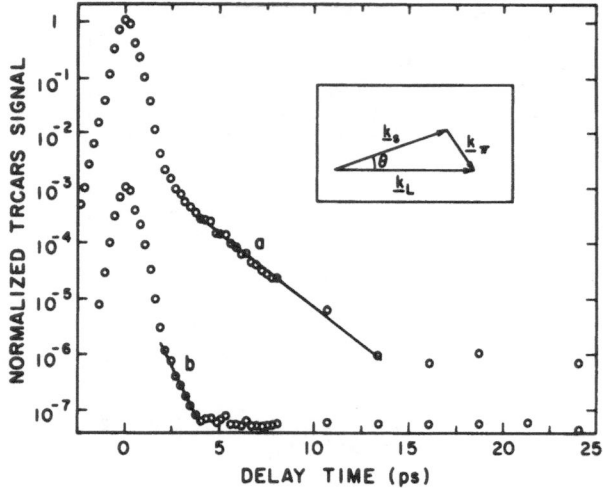

Fig. 5. a) TRCARS signal intensity as a function of probe pulse delay of
b) TRCARS signal after long term component has been subtracted.

single decay rate as observed for LO phonons[3],[14] when the acoustic-phonon occupation probabilities are evaluated at thermal equilibrium in terms of Bose-Einstein distribution. If, on the other hand, $n_{\vec{k}j}(\omega)$ follows a Lorentzian spectral distribution as in Eq. (3), Equation (1) leads directly to two decay rates for the optical phonon. We demonstrate below that, indeed, the two decay rates are also observed experimentally.

In order to search for the nonequilibrium phonon state, we have performed picosecond time resolved anti-Stokes Raman scattering (TRCARS) measurements using a recently developed dual synchronously pumped and synchronously, strongly amplified, laser system.[1],[2] For values of $n_A \lesssim 10^{-2}$ a typical result of the TRCARS measurement is shown in Fig. 3, curves a and b. As noted above, the TRCARS signal intensity has two readily distinguishable components. Subtraction of the long-term, exponentially decreasing phonon part of the total signal, leaves only the response of bound electrons and a time independent residue ascribable to various sources of stray background radiation. For values of $n_A \gtrsim 10^{-12}$, however, a second, additional (shorter) dephasing time is readily observed. The faster decay rate (Fig. 3, curves c and d) is extracted by subtracting the long-term component from the total TRCARS signal. Figures 4 (open circles) is a compilation of observed values of Γ_1 and Γ_2 (in cm^{-1}) for the range of n_A over which two different decay rates are observed. Moreover, separate evaluations of n_A for fixed values of lead to Γ_1 and Γ_2 as predicted through Eq. 4 (dashed lines)'. The difference between the observed and predicted values of Γ_1 and Γ_2, increase roughly linearly with n_A. The origin of this difference is not yet understood, although the neglect of higher order terms in the nonlinear optical excitation, appears to be the likely sources.

We conclude that evidence for the new nonequilibrium phonon state previously proposed by Bulgadaev and Levinson has been obtained, and that its properties, for the most part, agree with the predicted ones. Time-resolved coherent anti-Stokes scattering offers a unique technique for determining the detailed properties of this state. The precise role of the new state as a precursor to laser damage, and in other strongly opti-cally driven phenomena, will need to be investigated further.

6

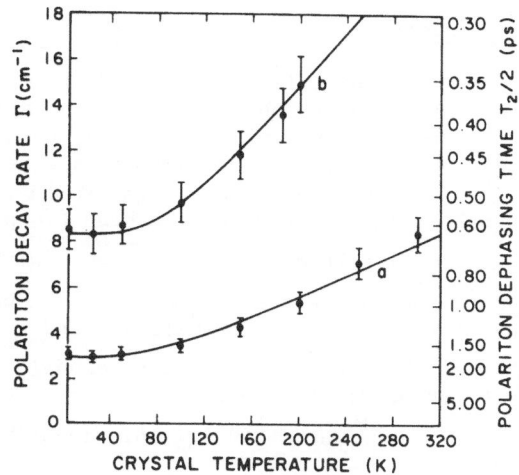

Fig. 6. Temperature dependence of the
polariton dephasing time. a)
"X-point" decay channel, b) "half-
energy" decay channel.

$\hbar\omega_\pi \gg k_B T$, Eq. 2 again applies. We observe two exponential decays of the
TRCARS signal. Thus, there exist two homogeneously broadened states with
bandwidths (at OK) Γ_0^s and Γ_0^f which lead, respectively, to the slow and to
the fast decay rates. In order to obtain ω_1 and ω_2 and Γ_0 for each decay
channel we have performed two-parameter, least-χ^2 fits to Eq. (2) using
the data displayed in Fig. 6 . The two fitting parameters are Γ_0 and ω_1
femtosecond laser system.[2] In this experiment the central polariton wave
vector $k_\pi \sim 2\times10^3$ cm^{-1}, which corresponds to a spectral distribution
peaked at ~ 354 cm^{-1}. However, a 120-mm focal-length lens is used to
focus the l, s, and p beams into the sample, which results in an angular
spread in the wave vector of the laser beam and, hence, in a spread in
polariton wave vector of between ~1.4×10^3 and ~2.6×10^3 cm^{-1}. The
corresponding range of polariton frequencies is from ~330 to ~360 cm^{-1}.
Figure 5 , plot a, illustrates the typical temporal dependence of the
TRCARS signal amplitude as a function of Δt, and at an ambient temperature of 5 K and with $\omega_1 - \omega_s \approx 354$ cm^{-1}. Note that this time the second
component of the TRCARS signal (~1.5 ps < Δt < ~5 ps), which occurs after
the CRE has stopped, indicates a curved structure on a semilogarithmic
plot, i.e., that the decay of the polaritons in this time interval does
not correspond to a single decay constant. The third component (~5 ps
< Δt < 12.5 ps) illustrates a clearly exponential decay corresponding to
a polariton decay time $T_2/2 = 1.72 \pm 0.16$ ps (slower decay rate, Γ^s).
Subtracting this third component from the total signal uncovers a second
exponential decay (see Fig. 5, plot b) with a decay time $T_2/2 = 620 \pm 60$ fs
(faster decay rate, Γ^f)'. If instead $\omega_1 - \omega_s \approx 344$ cm^{-1}, only one decay,
namely, the fast rate, is observed. Thus, we have again observed two
decay rates when only one is expected. We show below that the origin of
the two decay rates differs between the case of polariton decay and the
new nonequilibrium state.

In order to investigate the temperature dependence of the polariton
decay, we have repeated these measurements at 50K intervals from 5 to
300K, with $\omega_1 - \omega_s \approx 354$ cm^{-1}. The temperature dependence of $T_2/2$ is
indicated in Fig. 6. Since the decays are exponential their Fourier
transforms are Lorentzian with bandwidths Γ_π (in cm^{-1}) given by
$(2\pi c T_2/2)^{-1}$. At low ambient temperatures such that the polariton energy

Subpicosecond Polariton decay

We return now to a TRCARS experiment conducted again near thermal equilibrium. However, instead of LO phonons we now investigate the decay of polaritons in GaP.[6] Since the polariton lifetimes turn out to be in the subpicosecond regime, we apply for this purpose our recently reported with $\omega_2 = \omega_T - \omega_1$. The fits to the data appear as the solid lines in Fig. 6. For the slower decay rate (Fig. 6, curve a) the best two-parameter fit yields $\Gamma_0^s = 2.96 \pm 0.4$ cm^{-1}, $\omega_1 = 118 \pm 16$ cm^{-1} $\omega_2 = 236 \pm 16$ cm^{-1} with $\chi^2 = 0.9$. This result, when compared to the phonon frequencies at the X-point as obtained from neutron scattering[15] $\omega_{TA(X)} = 107 \pm 3$cm^{-1} and $\omega_{LA(X)} = 249 \pm 4$ cm^{-1}), identifies this as the slow X-point decay channel. For the faster decay rate (see Fig. 6, curve b), the best fit yields $\Gamma_0^f = 8.2 \pm 0.3$ cm^{-1}, $\omega_1 = 172 \pm 26$ cm^{-1}, $\omega_2 = 1.72 \pm 26$ cm^{-1}, and $\chi^2 = 0.5$. This identifies the fast half-energy decay channel.

We conclude that time resolved coherent anti-Stokes Raman scattering is an ideal technique with which to study the dynamical properties of coherently excited phonons and polaritons both near thermal equilibrium and in a strongly perturbed nonequilibrium state. The technique also offers experiments seeking insights into the dynamics of charge carriers and their interaction with phonons, polaritons and other excitations. Results in this area are to be published elsewhere.

WEB acknowledges funding through NSF Grant No. DMR 86-03888, DMR 89-13289, and NATO Grant No. D34188. He also acknowledges strong contributions by his coworkers Dr. J. Kuhl, Dr. A. Mayer, Dr. T. Juhasz, Dr. S. Mehta, Dr. B. K. Rhee, and K. Harris.

References

1. T. Juhasz, J. Kuhl and W. E. Bron. Opt. Lett. 13, 577 (1988).
2. T. Juhasz, G. O. Smith, S. M. Mehta, K. Harris, and W. E. Bron, IEEE J. Quantum Electronics, 25, 1704 (1989).
3. W. E. Bron, J. Kuhl and B. K. Rhee, Phys. Rev. b34, 6961 (1986).
4. B. K. Rhee and W. E. Bron, Phys. Rev. B34, 7107 (1986).
5. W. E. Bron, T. Juhasz and S. Mehta, Phys. Rev. Lett. 62, 1655 (1989).
6. T. Juhasz and W. E. Bron, Phys. Rev. Lett. 63, 2385 (1989).
7. M. D. Levinson and N. Bloembergen, Phys. Rev. B10, 4447 (1974).
8. W. Demdröter, "Laser Spectroscopy" (Springer: Berlin, Heidelberg and New York) 1981.
9. B. Kh. Bairamov, D. A. Parshin, V. V. Toporov, and Sh. B. Ubaidullav, Sov. Techn. Phys. Lett. 5, 466 (1979).
10. R. Orbach and L. A. Vredevoe, Physics 1, 91 (1964).
11. H. Bilz and W. Kress, "Phonon Dispersion Relations in Insulators," (Springer: Berlin) 1979.
12. W. E. Bron, Rep. Prog. Phys. 43, 301 (1980).
13. I. B. Levinson, Sov. Phys. JETP 38, 162 (1973); S. E. Bulgadaev and I. B. Levinson, JETP Lett. 19, 304 (1974); S. E. Bulgadaev and I. B. Levinson, Sov. Phys JETP 40, 1161 (1974).
14. A. A. Maradudin and A. E. Fein, Phys. Rev. 128, 2589 (1962).
15. J. L. Yarell, J. L. Warren, R. G. Wenzel, and P. A. Dean, in "Neutron Inelastic Scattering" (International Atomic Energy Agency: Vienna) 1968, Vol. 1, pg. 301.

RELAXATION AND PROPAGATION OF HIGH FREQUENCY PHONONS IN THIN CRYSTALLINE PLATES AFTER INTENSE LASER PUMPING

D. V. Kazakovtsev, A. A. Maksimov, D. A. Pronin, and I. I. Tartakovskii

Institute of Solid State Physics
USSR Academy of Sciences
142432, Chernogolovka, Moscow Region, USSR

One of the most interesting and important problems in phonon physics is that of the relaxation of nonequilibrium phonons in the course of the establishment of an equilibrium temperature formation in the crystal. The initial stage of such relaxation in the case of weak excitation of the phonon system, when spontaneous phonon decay dominates, was investigated, in particular, in Refs. 1-2. To describe the whole course of phonon relaxation, including the temperature formation stage, one must take into account the phonon-phonon coalescence processes along with the decay processes[3]. At sufficiently high excitation levels, when the temperature increase $\Delta T = T_f - T_o \simeq T_o$, phonon coalescence can be significant from the very beginning of the relaxation process. The latter situation is rather common for intense laser excitation, and can also occur in the case of current pulse excitation of phonons.

We measured the nonequilibrium phonon occupation numbers in the crystal as functions of the time elapsed after the appliction of an intense laser pulse. These were compared with the results of model numerical calculations obtained in the quasicontinuous phonon spectrum approximation. Different regimes of nonequilibrium phonon propagation in a thin crystalline plate are also investigated experimentally.

EXPERIMENT

We have used an optical method with frequency selectivity and high spatial and temporal resolution, proposed in Ref. 4, to detect nonequilibrium acoustic phonons. It is based on the fact the change in the occupation numbers of high frequency acoustic phonons Δn with frequency Ω. in the crystal volume can be determined from the increase of the light absorption coefficient $\Delta k = k(\omega) - k_o(\omega)$ at the frequency $\omega < \omega_T$ near the bottom ω_T of the lowest exciton b-band of an anthracene crystal, namely $\Delta n(\Omega) \propto \Delta k$, $\Omega \approx \omega_T - \omega$. We changed the lasing frequency ω of a tunable dye laser, and the value of the time delay between the pumping and probe pulses, thus obtaining the changes in the occupation numbers $\Delta n(\Omega)$ of nonequilibrium phonons of various frequencies at various time intervals after the pumping pulse.

The experiments have been performed in thin ($d \le 2\mu m$) single crystal anthracene plates placed in a helium bath at $T = 4.3K$. Their front surface (the developed ab-plane) was uniformly excited by pulses from a

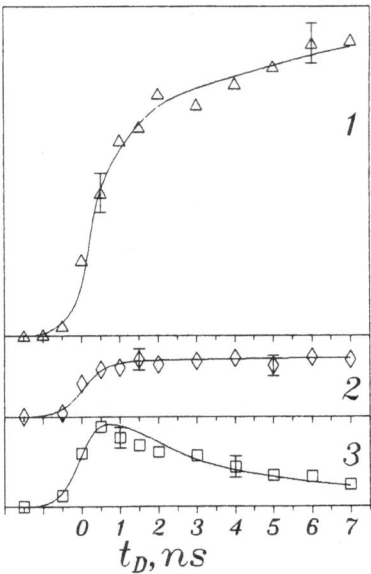

Fig. 1. The light absorption coefficients $\Delta k(\omega)$ at different frequencies $\Omega = \omega_T - \omega_{ex}$: 1 - 14 cm^{-1}, 2 - 25 cm^{-1}, 3 - 32 cm^{-1} as functions of time. The excitation by a laser pulse is at $t_D = 0$.

nitrogen laser (planar experimental geometry). The results of measurements at several frequencies are presented in Fig. 1. They reveal that the kinetics of the phonon occupation numbers $n(\Omega)$ depend significantly on phonon frequency Ω. One can see also that pulsed laser pumping results in acoustic phonon generation in a wide frequency range, with most of the energy at the initial moments being concentrated in the high frequency domain of the phonon spectrum, i.e. the initial energy distribution of the acoustic phonons differs significantly from the equilibrium one. In the course of the subsequent relaxation the occupation numbers of high frequency phonons with $\Omega > 25$ cm^{-1} decrease, while those of the lower frequency phonons increase. Note, that the total energy of the phonon distribution is conserved, since adiabatic conditions are rigorously maintained in the experiment up to a time $t_D \simeq 1$ μs.

MODEL CALCULATIONS

To describe the relaxation of the nonequilibrium phonon system theoretically we calculated numerically the kinetics of the phonon occupation numbers for the whole range of acoustic phonon frequencies on the basis of the following model. We consider two phonon branches with an isotropic Debye spectrum. The dispersion of the longitudinal ℓ-mode is $\Omega = u_\ell|k|$; that of the transverse t-mode is $\Omega = u_t|k|$. The latter mode is assumed to be twofold degenerate according to two transverse phonon polarizations. Here u_ℓ and u_t are the phase velocities for the longitudinal and transverse phonons, respectively. For calculational convenience, all velocities were measured in units of u_t i.e., in dimensionless form u_t was unity, while the speed of longitudinal phonons was $s = u_\ell/u_t > 1$. All frequencies and energies were measured in the frequency of the boundary Debye acoustic phonon Ω_D; thus, $h = 1$ also.

We have taken into account all normal three-phonon anharmonic processes, consistent with the energy and quasimomentum conservation laws. The Umklapp processes were neglected, which is reasonable for $T \ll 1$. The intermode conversion processes $\ell \leftrightarrow t$ were assumed to be rapid as compared to the anharmonic ones considered in the model.

We followed the time evolution of the phonon occupation numbers $n(\Omega,t)$ for various frequencies Ω resulting from a given spatially homogeneous initial energy distribution $n(\Omega,0) = n_o(\Omega)$ at $t = 0$. We investigated the entire process of the transformation of the nonequilibrium function $n_o(\Omega)$ into an equilibrium one $n_T(\Omega,\infty)$ at $t \to \infty$, this being the Planck distribution with temperature T. This temperature was determined by the total energy of the initial phonon distribution $n_o(\Omega)$.

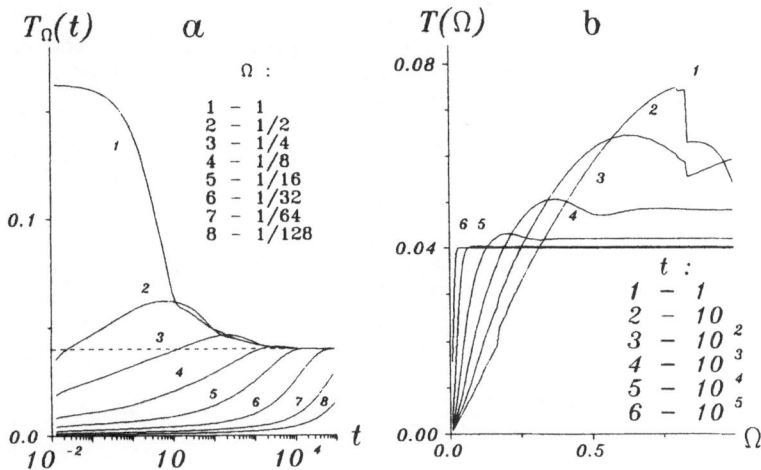

Fig. 2. The calculated mode temeprature: a) as a function of time; b) as a function of phonon frequency.

The model phonon spectrum is a set of delta-function-like spheres in $\vec{k}$-spa with radii $r_\ell(i) = \frac{1}{Ns}$ i for the longitudinal modes and $r_t(i) = \frac{1}{N}$ i for the transverse modes. Here i is the level number labelling the phonon energy, $1 \le i \le N$, where N = 128 is the total number of levels in the set. For this set to correspond the Debye spectrum the density of states of its levels must be taken as:

$$\rho_\alpha(i,\vec{k}) = \rho_\alpha \delta\left(|\vec{k}| - r_\alpha(i)\right),\tag{1}$$

where α is the mode type, ℓ or t, and $\rho_\ell = 1/s$, $\rho_t = 2$.

The kinetic equation for this model, which takes into account three-phonon anharmonic processes $i \leftrightarrow j + k$, yields a system of N nonlinear differential equations, where the m-th has the following form

$$\dot{n}_m(t) = \frac{1}{(\rho_\ell+\rho_t)m^2R} \left(-\frac{1}{2} \sum_{j=1}^{N-m} R_{m \leftrightarrow j+(m-j)} F_1 + \sum_{j=1}^{N-m} R_{(m+j) \leftrightarrow m+j} F_2 \right), \qquad (2)$$

$$F_1 = n_m(t)\left[1+n_j(t)\right]\left[1+n_{(m-j)}(t)\right] - \left[1+n_m(t)\right]n_j(t)n_{(m-j)}(t),$$

$$F_2 = n_{(m+j)}(t)\left[1+n_j(t)\right]\left[1+n_m(t)\right] - \left[1+n_{(m+j)}(t)\right]n_j(t)n_m(t).$$

Here R is a normalization constant, chosen so that the time is measured in units of the lifetime of the boundary Debye phonon $\tau(\Omega_D)$, calculated in the spontaneous decay limit. $R_{i \leftrightarrow j+k}$ is the sum of the probabilities $W_{i \leftrightarrow j+k}$ of the three types of interlevel processes:

$$i_\ell \rightarrow j_t + k_t, \quad i_\ell \rightarrow j_\ell + k_t, \quad i_\ell \leftrightarrow j_t + k_\ell.$$

The matrix elements were assumed to be proportional to the product of the module of the wave vectors of the three phonons involved, e.g. for the first of the processes mentioned above

$$M_{\ell \leftrightarrow t+t}(k_i, k_j, k_k) = C_{\ell \leftrightarrow t+t} r_\ell(i) r_t(j) r_t(k),$$

the total probability of the process being

$$W_{\ell \leftrightarrow t+t} = C^2_{\ell \leftrightarrow t+t}\rho_\ell \rho_t \rho_t \frac{\left[r_\ell(i) r_t(j) r_t(k)\right]^3}{ijk} \theta\left(r_\ell(i) - |r_t(j) - r_t(k)|\right) \times$$

$$\times \theta\left(|r_t(j) + r_t(k)| - r_\ell(i)\right) \delta_{i,(j+k)}. \qquad (3)$$

The numerical calculations were carried out in the fourth order Runge-Kutta scheme with the error controlled at every step.

The quality of the model can be illustrated by the results of a calculation for an initial δ-function-like phonon distribution $n_o(i) = 0$, $i \neq N$, $n_o(i=N) \neq 0$. The excitation energy corresponds to $T_f = 0.04$. Figure 2a presents time evolution of the mode temperature for phonons with energies 1, 1/2, 1/4, ..., 1/128 (which means (i=128, 64, 32, ..., 1, respectively). One can see how the mode temperature equalizes first for energies 1 and 1/2, then 1, 1/2 and 1/4, and so on, and the maximum of the phonon distribution gradually shifts to the low frequency region. The pattern may be visualized by depicting the same mode temperature as a function of the corresponding phonon frequency for several moments of time. This is done in Fig. 2b. We see a gradual expansion of the thermalized domain, which first appears at the highest phonon frequencies, into the region of lower frequencies, the rate of this expansion slowing down significantly wit decreasing frequency, and at $t \rightarrow \infty$ the equilibrium temperature $T(\Omega) = T_f = 0.04$ is achieved for all Ω.

One of the tests for the reliability of the model was calculating the spontaneous decay lifetime $\tau_{decay}(\Omega)$ for the ℓ-mode for a low level of excitation. The obtained Ω-dependence appeared to be consistent with the well-known relation $\tau_{decay}(\Omega) \propto (\Omega/\Omega_D)^{-5}$.

12

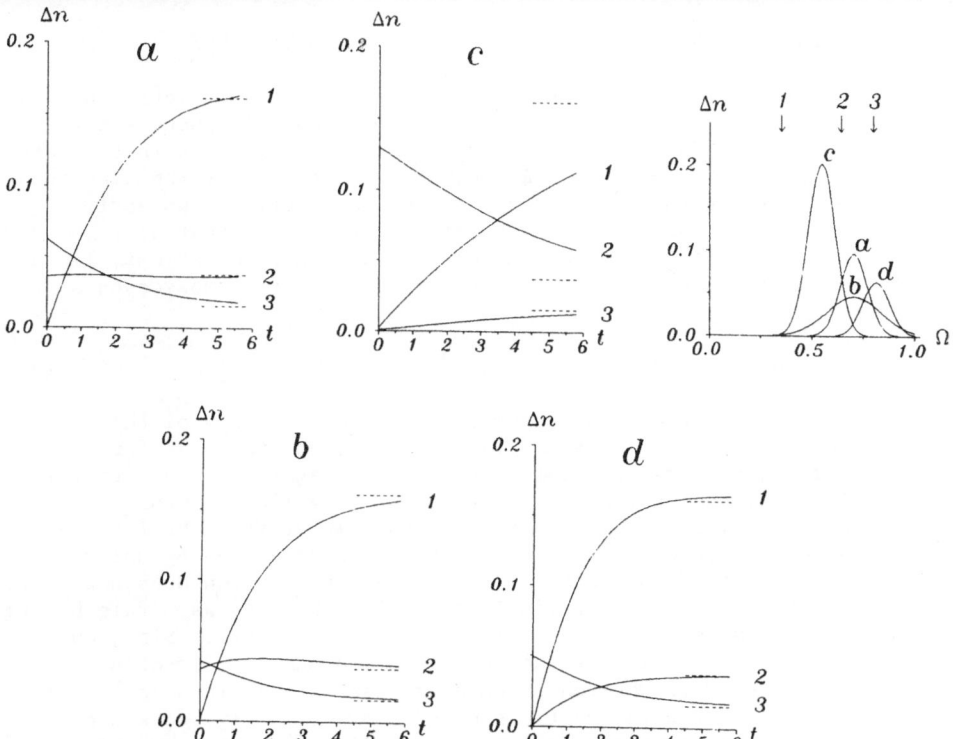

Fig. 3. The calculated phonon occupation numbers for different frequencies Ω as functions of time: 1-14 cm^{-1}, 2-25 cm^{-1}, 3-32 cm^{-1} resulting from different initial distributions $n_o(\Omega)$, depicted in the upper right corner. The dashes indicate the final $\Delta n(\Omega)$ after the equilibrium temperature T_f = 10.5K is established.

ANALYZING EXPERIMENTAL RESULTS BY MEANS OF THE MODEL

To fit the experimental curves with the calculated ones we changed simultaneously the parameters C in the expression of the type of [3], conserving the ratio between $\ell{\leftrightarrow}t{+}t$ and $\ell \leftrightarrow \ell{-}t$ processes by using $C_{\ell{\leftrightarrow}t{+}t} = C_{\ell{\leftrightarrow}\ell{+}t}$. Such a change is conveniently represented as a change of the Debye phonon lifetime $\tau_D = \tau_{decay}(\Omega_D)$. The best fit was achieved for τ_D = 1 ns.

The second adjustable parameter was the spectrum of the acoustic phonons, initially generated by optical ones, i.e., the $n_o(\Omega)$ distribution. We changed its maximum position Ω_{max} as well as its halfwidth $\Delta\Omega$. The calculations show the relaxation pattern to depend essentially on the form of $n_o(\Omega)$. The best agreement was achieved with Ω_{max} = 0.6 Ω_D and $\Delta\Omega \leq \Omega_D/4$ (see Fig. 3). A value of Ω_{max} = 28 cm^{-1} is consistent with the decay of the lowest optical phonon into two acoustic phonons with approximately half the frequency being the main source of the initial phonon distribution[5], Ω_{min}^{opt} 49 cm^{-1} in anthracene.

The possibilities of the model described are not restricted to those demonstrated above. In a number of experiments acoustic phonons are generated in semiconductors at low temperatures by light, as in the case of anthracene dealt with above. A theoretical analysis of the kinetic equations describing the behavior of the phonon system in the early stages of its evolution was given in Refs. 6, 7. For a spatially homogeneous excitation it was shown that the system of equations, linearized in the small quantity $n(\Omega,t)$, corresponding to (2), possesses a scaling solution of the form

$$n(\Omega,t) = A\,\Omega^{-3}t^{1/5}f_{sc}(\eta), \quad \eta = \Omega t^{1/5}, \quad \Omega \ll 1. \tag{4}$$

Here A is a constant that depends on the total energy of the phonons, and $f_{sc}(\eta)$ is a certain universal function, the same for different imparted energies. Later an order of magnitude estimate for the time interval over which the solution (4) is valid was made in Ref. 3, the deviation of the true occupation numbers from the form (4) being regarded as connected with the typical $n(\Omega,t)$ achieving unity. Meanwhile, the fact of the existence of a scaling solution does not mean, generally speaking, that a certain initial distribution, as a rule having nothing to do with the form (4), will acquire such a form. Since our model enables us to obtain the phonon occupation numbers directly as functions of time, with the coalescence processes taken into account, it can be used to answer both questions posed above. Namely, does a certain initial distribution in fact approach the scaling form (4) and, if so, what is the time interval over which the true solution preserves scaling.

To answer these questions in the framework of our model we calculated from the obtained functions $n(\Omega,t)$ the following function $f(\eta)$:

$$f(\eta) = \frac{f^*(\eta)}{\int_0^\infty f^*(\eta)d\eta}, \quad f^*(\eta) = \begin{cases} n(\Omega,t)\Omega^3/t^{1/5}, & \eta = \Omega\cdot t^{1/5} \\ 0, & \eta > t^{1/5} \end{cases}. \tag{5}$$

For the system linearized in $n(\Omega,t)$ it coincides with $f_{sc}(\eta)$ from (4). Our normalization corresponds to the energy conservation law in the form

$E = A\int_0^\infty f(\eta)d\eta = $ const. and enables us to eliminate the constant A. The $f(n)$ calculated from the linearized system ceases to change at $t \approx 10$. This result led us to the conclusion that the linearized solutions approach a scaling form at $t \geq 10$, and the obtained $f(\eta)$ for $t \geq 10$ is assumed to be $f_{sc}(\eta)$. Now we conclude that the solution of the nonlinear system (2) approaches the scaling form if the corresponding $f(\eta)$ coincides with $f_{sc}(\eta)$. The quantitative measure of deviation is determined as follows

$$S(t) = \int_0^\infty |f(\eta)-f_{sc}(\eta)|d\eta. \tag{6}$$

The real solution is estimated as haveing a scaling form if it yield $S(t) \ll 1$.

Figure 4 shows $S(t)$ for the above described solution (see Fig. 2), resulting from an initial δ-function-like distribution with different excitation energies, i.e., different final temperatures T_f. One can see that in the time interval $1 \leq t \leq 10$ the true solution $n(\bar{\Omega},t)$ in fact approaches a scaling form, practically independently of the energy of the

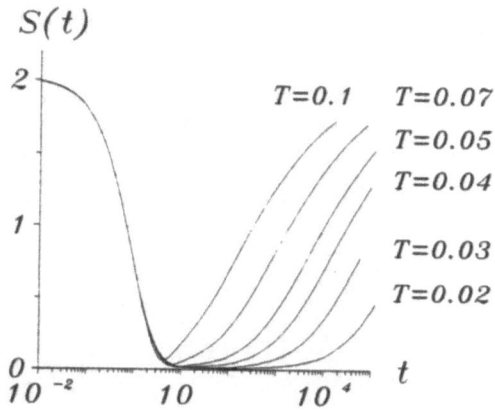

Fig. 4. The time dependence of S(t) function for several final temperature T_f values.

system. However, in qualitative agreement with [3], the time interval in which the solution has the scaling form depends strongly on the energy. From Figure 4 it can be estimated as proportional to T^{-5}.

PROPAGATION OF NONEQUILIBRIUM PHONONS

We also investigated the propagation of high frequency phonons, generated in anthracene crystals (d = 10-50 μm) by short laser pulses. In the first experiments[8], performed in the plane geometry by means of a luminescence phonon detector, a propagation regime, attributed to hydrodynamic phonon flow, was found to exist in addition to quasi-ballistic propagation. We assumed the existence of such a regime at low temperatures is due to peculiarities of high frequency phonon propagation accompanied by their frequency down-conversion, resulting in their occupation numbers becoming unity, $n(\Omega,t) \simeq 1$.

In more recent experiments, performed by means of a phonon spectro-meter, we studied the spatial evolution of the nonequilibrium phonon energy distribution in the course of their propagation along the crystal surface[9]. In general, these experiments confirmed typical features of high frequency phonon propagation with frequency down-conversion, reveal-ed in the "generation" model. In addition they revealed a significant change in the nature of the "spreading" of the nonequilibrium phonon packet, which had a velocity of the order of $1 \cdot 10^5$ cm/s immediately after the pumping pulse, slowing down markedly at $t_D \simeq 40$ ns (see the inset in Fig. 5). We believe the observed change of the phonon propagation regime is connected with the increase of the occupation numbers in the high frequency tail of the phonon distribution due to coalescence of phonons of lower frequencies. Consequently, this intensifies the Umklapp scat-tering processes and, according to our estimates, the mean free path for the principal phonon group decreases to approximately 1 μm. If we assume this principal group to have typical frequencies $\Omega \simeq 10$ cm$^{-1} \simeq \Omega_D/4$, then in terms of the model described the time for Debye phonons to form the mode temperature $T_{\Omega_D} \simeq T_f$ appears to be $\simeq 10$ ns (see Fig. 6). This value is reasonably consistent with the experimental data. It

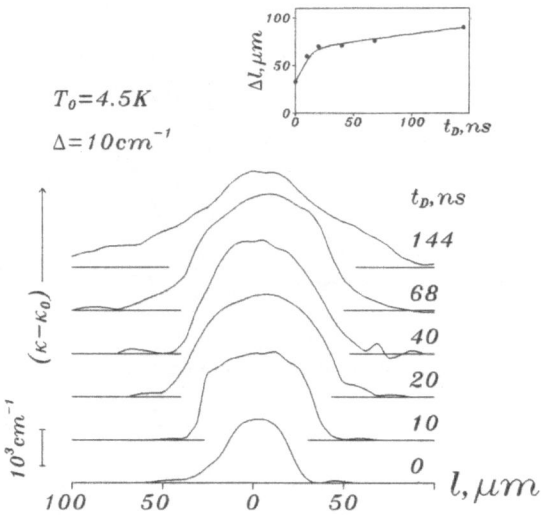

Fig 5. Spatial distribution of the light absorption coefficient k-k$_o$
($\Delta=\omega_T$-ω) at different t$_D$; $\Delta\ell$ - halfwidth of the phonon spatial
distribution.

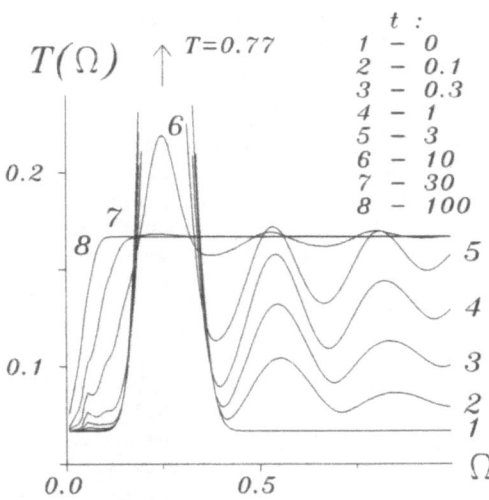

Fig. 6. Mode temperature as function of frequency at several times
t for initial distribution with maximum at Ω = 0.25 (calculation).

should be mentioned that the observed transport of energy from the excited region of the crystal into cold regions resembles the mechanism of nonlocal heat transfer proposed by Levinson in Ref. 10.

REFERENCES

1. U. Happek, K. F. Renk, Y. Ayant, R. Buisson, Phonon Scattering in Condensed Matter V., Proc. 5th Int. Conf., Urbana, Illinois, June 1986, ed. by A. C. Anderson and J. P. Wolfe, Berlin: Springer-Verlag, 1986, p. 347; Europhys. Lett., 1987, 3, 1001.
2. R. Baumgartner, M. Engelhardt, K. F. Renk, Phys. Rev. Lett., 1981, 47, 1403.
3. D. V. Kazakovtsev, Y. B. Levinson, Zh. Eksp. Teor. Fiz., 1985, 88, 2228, [Sov. Phys. JETP, 1985, 61, 1318]; see also Physics of Phonons, ed. by T. Paszkiewicz, Berlin: Springer-Verlag, 1987, p. 276 (Lecture Notes in Physics, 285).
4. A. A. Maksimov, I. I. Tartakovskii, Zh. Tekh. Fiz., Pis'ma, 1986, 12, 112; [Sov. Tech. Phys. Lett., 1986, 12, 46].
5. B. K. Rhee, W. E. Bron, Phys. Rev. B. 34 1986, 7107.
6. D. V. Kazakovtsev, Y. B. Levinson, Zh. Eksp. Teor. Fiz., Pis'ma, 1978, 27, 194 [Sov. Phys. JETP Lett., 1978, 27, 181].
7. D. V. Kazakovtsev, Y. B. Levinson, Phys. Stat. Sol. (b), 1979, 96, 117.
8. V. L. Broude, N. A. Vidmont, D. V. Kazakovtsev, et al., Zh. Eksp. Teor. Fiz. 1978, 74, 314 [Sov. Phys. JETP, 1978, 47, 161].
9. A. A. Maksimov, I. I. Tartakovskii, Zh. Eksp. Teor. Fiz., Pis'ma, 1985, 42, 458 [Sov. Phys. JETP Lett., 1985, 42, 568].
10. Y. B. Levinson, Zh. Eksp. Teor. Fiz., 1980, 79, 1394 [Sov. Phys. JETP, 1980, 52, 704].

EVOLUTION IN REAL TIME AND SPACE OF SHORT POLARITON PULSES IN CRYSTALS

F. Vallée, G. Gale and C. Flytzanis

Laboratoire d'Optique Quantique du C.N.R.S.
Ecole Polytechnique
91128 Palaiseau Cedex, France

ABSTRACT

We demonstrate that short polariton pulses in real time and space can be directly studied by a nonlocal time resolved nonlinear technique. One can directly determine the propagation characteristics and decay of coherence of both bare and dressed polariton pulses in crystals. The technique in particular allows one to differentiate the impact of phonon anharmonicity from that of the dielectric disorder on the polariton coherence.

1. INTRODUCTION

We address here the problem of the spatiotemporal propagation and decay of very short polariton pulses in crystals. Polaritons are electric dipole allowed propagating modes of light and matter motion characterized by a dispersion relation that connects the frequency ω and the wave vector $\underline{k}$ from which one can extract their group velocity

$$v_g = \partial\omega/\partial k \tag{1}$$

This vastly varies with frequency (or wave vector), by several orders of magnitude as the polariton character changes from matter like excitation to photon like along its dispersion curve and is related to the electromagnetic signal propagation in a material medium close to its dipole allowed collective transition [1]. In this respect the polariton propagation characteristics in real time and space and the time and space scales over which one stores energy and coherence in a polariton pulse are of both technological and fundamental relevance.

Although these modes in essence arise from collective motion of two level systems on a periodic lattice it is quite clear that their energy and coherence decay cannot be addressed with the techniques that have been devised to study these processes in an assembly of localized two level entities. Here we have for instance in mind the time or frequency resolved Coherent Anti-Stokes Raman Scattering (CARS) technique that is being extensively used to study vibrational relaxation in liquids or other local excite-and-probe techniques ; in these cases the energy and coherence relaxation occur within the immediate environment of the excited point and the excitation and probe stages must coincide. In the case of polariton where the group velocity is essentially that of the light the relaxation process within the immediate environment can take place concurrently with the escape of the excitation from the excited zone and its resonant transfer to other identical entities. Non local techniques where excitation and probe are separated in space and in time are needed to disentangle these propagation effects from the intrinsic relaxation processes and the retardation effects must be taken into consideration.

At presently there are two techniques that allow one to address this problem : the electro-optic Cerenkov effect [2] and the nonlocal time resolved CARS [3] ; in addition the transient optical grating technique can be extended to study relaxation and diffusion of delocalized excitations [4]. The electrooptic Cerenkov technique exploits the large frequency width of a femtosecond pulse in the visible to generate, through the optical rectification (or inverse electro-optic) effect in a non centrosymmetric crystal, an extremely fast far infrared electromagnetic transient, with essentially the same frequency width as that of the initial femtosecond visible pulse, and excite the infrared active polariton in the same crystal. Because of the vastly different group velocities in the visible and far infrared this produces a Cerenkov like cone which is exploited to study the polariton in real space by probing the concomittant induced birefringence with a second femtosecond pulse parallelly displaced with respect to the first. The selectivity of this technique is limited and the separation of temporal and spatial features is not straightforward. More importantly the method is restricted to low-frequency polaritons with an upper frequency limit given by the infrared femtosecond pulse bandwidth which cannot presently exceed 200 - 300 cm^{-1}.

We have proposed and demonstrated a different technique, the non local time resolved CARS technique, which does not suffer from these limitations and allows one to directly follow the evolution in real space and time of a short picosecond polariton pulse in a non centrosymmetric crystal. Below we summarize some applications and extensions of this technique.

2. NON LOCAL TIME RESOLVED CARS

In essence this technique consists in separating in space the coherent excitation and the anti-Stokes probe stages in a time resolved CARS experiment and in appropriately adjusting the time delay with the space separation by taking into consideration the group velocity and the phase matching condition [3]. The principle of the experiment is outlined in Fig.1. Coherent excitation of the polariton at $t = 0$ and position $X = 0$ is provided by two time-coincident optical picosecond pulses of wave vectors k_L and k_S and frequencies ω_L and ω_S, respectively. The polariton wave vector is $\underline{k}_\pi = \underline{k}_L - \underline{k}_S$ and, hence, its propagation direction is imposed by adjusting the phase matching angle between the two beams ; resonant excitation at $\omega_\pi(k_\pi)$, the frequency of the polariton of wave vector k_π, is obtained by frequency tuning ω_L such that $\omega_L - \omega_S = \omega_\pi$. Coupling with the light fields in the coherent excitation stage is described by the phenomenological energy density

$$W = - d_E \, E_L \, E_S^* \, E_\pi^* - d_Q \, E_L \, E_S^* \, Q_\pi^* + c.c$$

where E_L and E_S are the electric fields at ω_L (laser) and ω_S (Stokes), respectively, E_π and Q_π represent the electric field and transverse optical mechanical vibration associated with the polariton ; d_E and d_Q are the coupling parameters which vary only slowly with polariton frequency [5]. From symmetry requirements, it is clear that this coupling can only occur in a noncentrosymmetric crystal. This coupling produces a picosecond duration polariton wave packet at $t = 0$ and $X = 0$ which then propagates in the crystal in a direction determined by the polariton wave vector $\underline{k}_\pi = \underline{k}_L - \underline{k}_S$ and at a speed given by the

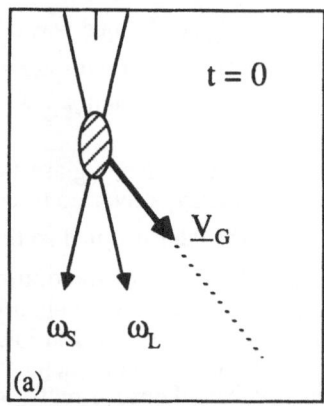
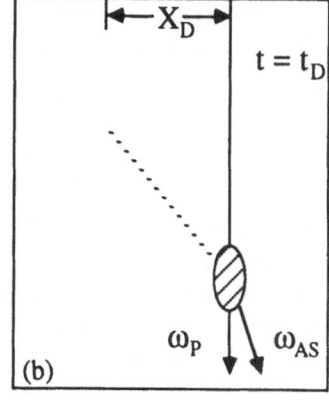

Fig.1. Principle of the time and space resolved CARS technique : (a) coherent
polariton excitation, (b) coherent polariton detection

polariton group velocity $v_g(k_\pi)$. The temporal and spatial evolution of the propagating
excitation is followed by phase matched coherent anti-Stokes scattering at $\omega_a = \omega_p + \omega_\pi$
of a third pulse ω_p displaced (see fig.1) with respect to the excitation in time (by t_D) and
space (by X_D). The measured dependence of the spatial displacement X_D, where the
signal is maximum, on time delay t_D is directly related to the energy propagation
characteristics of the polariton wave packet while its relaxation is measured by the
temporal (and spatial) evolution of the signal intensity maximum. Because the frequencies
ω_L, ω_S and ω_p can be chosen independently albeit using a single initial laser source one
can actually study polariton over a most wide range of frequencies ; actually there is no
intrinsic limitation.

As stated this technique can only be used in non centrosymmetric crystals however
if the Raman effect, exploited in the coherent excitation and probe stages is replaced by
the hyper-Raman effect polaritons in centrosymmetric crystals like NaCl can also be
studied ; this essentially amounts in replacing the virtual transition with one ω_L photon by
a virtual transition with two ω_L photons and tune ω_L (or ω_S) so that $2\omega_L - \omega_S = \omega_\pi$ and
similarly for the ω_p photon virtual transition in the probe stage so that one follows the
propagation with the phase matched coherent anti-Stokes hyperscattering at $\omega_a = 2\omega_p + \omega_\pi$.

Polaritons in their simplest form arise from the mutual coherent interaction
between a single polar material mode and an electromagnetic mode. In addition to these
bare polaritons there are also more complex hybrid excitations involving polaritons with
other modes which we will designate by dressed polaritons.

3. BARE POLARITON PROPAGATION AND DAMPING

a) Polariton propagation

An example of the use of the technique to directly follow the propagation of a
polariton and determine its characteristics is that of the NH_4Cl crystal, a cubic crystal
obtained from the NaCl one by replacing Na^+ with NH_4^+ which is an intrinsically
noncentrosymmetric molecular entity. At low temperature the ammonium ions are all
ordered the same way but at higher temperatures their orientation becomes disordered and

the crystal undergoes an order disorder phase transition. The technique was used to study the polariton related to the v_4 vibrational mode of the NH_4^+ radical with $v_{4T} = 1400$ cm^{-1} and $v_{4L} = 1418$ cm^{-1} for the transverse and longitudinal modes respectively [3].

The experimental system is driven by a mode-locked Nd^{3+}/glass laser system producing a single 5-ps pulse at 1.054 μm which is frequency converted to generate two fixed frequencies (ω_p and ω_S) and a variable one (ω_L) ; the later is tuned so that $\omega_L - \omega_S = \omega_\pi$ the polariton frequency. The slope of the linear dependence of the optimum spatial displacement X_D on probe delay t_D precisely give the polariton group velocity. This was found to vary dramatically with polariton frequency (Fig.2), from $\approx c/2$ at low frequency to $\approx c/50$ at high frequency, which reflects the change in polariton character from photon like to phonon-like as we sweep the dispersion relation. These directly determined experimental values were found to perfectly agree with the ones calculated from the dispersion law which has been extensively studied by Raman scattering. This is the first direct determination of a polariton group velocity.

b) Polariton damping

If the spatial or temporal spreading (due to velocity dispersion) of the "tracked" polariton packet is small the intensity decrease of its peak coherent anti-stokes signal with time delay and space displacement gives directly the loss of coherence (T_2) of the polariton packet. Except for extreme cases, for instance when the polariton interacts with the crystal surface [6], the time decay of the polariton coherence was found to be exponential throughout its dispersion curve and for a wide range of temperatures up and close to the order disorder transition temperature. The value of T_2 however exhibits a strong resonant behavior with frequency and a characteristically critical behavior with the temperature which we wish to discuss below.

The damping of the non propagating longitudinal component of the v_4-mode at 1418cm^{-1} was also determined and was found to possess a similar behavior.

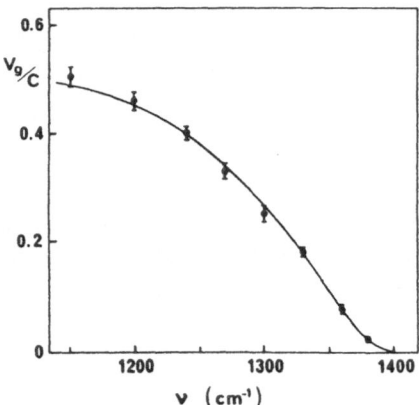

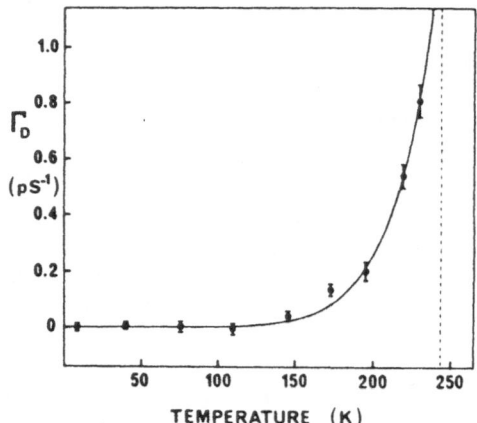

Fig.2. Measured polariton group velocity in NH_4Cl normalized to the speed of light in vacuum as a function of polariton frequency. The full line is calculated by differentiating the polariton dispersion curve.

Fig.3. Temperature dependence of the disorder induced part Γ_D of the coherent polariton decay rate at 1360 cm^{-1} in NH_4Cl. The full line is a fit using $\Gamma_D = \gamma_D(1-L^2)$

c) Anharmonicity vs Disorder

The polariton is an admixture of an electromagnetic (photon) and a material (phonon) part and its damping Γ_π is certainly related to the way these two components are affected by the different lattice interactions and imperfections. Because this admixture arises through a coherent interaction of an electromagnetic and a material mode any damping mechanism will indiscriminately affect both parts ; however under certain assumptions one may differentiate their origin.

In an ordered crystal the damping of polaritons is mainly due to the finite lifetime of their mechanical component Q_π, which is anharmonically coupled to many-phonon bands. This mechanism is the only one that has been considered in the literature [7]. In ammonium chloride at low temperature this mechanism is the most important one and in particular explains the frequency variation of the polariton damping rate which can be traced to a many-phonon band degenerate with the polariton. This process gives only a weak temperature-dependence of the relaxation rate Γ_A, via Bose-Einstein occupation numbers, as experimentally observed between 7 and 130 K.

In a crystal with disorder an additional damping mechanism is introduced which can directly affect the electromagnetic part of the polariton as well as its mechanical part. The effect on the electromagnetic part is similar to the one produced on the light scattered of a medium with random dielectric constant : light suffers an attenuation in the forward propagation mode and a loss of coherence as a consequence to its scattering into other modes. This damping depends on the amplitude of the fluctuations of the random dielectric constant.

To a first approximation we may assume that the two damping processes, the one due to temporal fluctuations and the other due to spatial fluctuations are statistically uncorrelated and hence we may write for the total loss of coherence rate for the polartion

$$\Gamma_\pi = \Gamma_A + \Gamma_D$$

where Γ_A and Γ_D are the contributions of the temporal (anharmonic) and spatial disorder scattering respectively.

In the case of NH_4Cl such a spatial disorder can be gradually introduced for instance by increasing the temperature and approaching the order-disorder transition temperature $T_c \approx 243$ K. The effect of the disorder can be globally measured with the order parameter L which can be deduced from thermodynamic data. As can be seen in Fig. 3, the disorder induced part of the damping rate, Γ_D, can be described by the simplest law for a disorder related process : $\Gamma_D = \gamma_D(1-L^2)$. This gives an excellent fit to the experimental data and supports the assumption that dielectric disorder makes a strong contribution to the loss of coherence.

4. DRESSED POLARITONS

a) Polariton Fermi resonance

With this technique it is now possible to investigate some new aspects of the interaction of polaritons with other collective excitations in a crystal. A typical example is the polariton Fermi resonance : this is produced whenever a polariton branch crosses a many-phonon band, most often a two-phonon band, of same symmetry as the bare polariton. This situation is frequently encountered in crystals because both polaritons and two-phonon states span large regions in frequency and wave vector space. For a strong anharmonic coupling, the resulting new excitation can be described in terms of a dressed polariton with substantially different characteristics compared to the bare polariton.[8]

The main consequences of polariton Fermi resonance are a partial localization, or slowing down, of the polariton which results from the opening of a new gap in the polariton dispersion curve and a strong, frequency dependent damping of the resulting dressed polariton. These manifestations can be theoretically analyzed by use of the Green's function method, and, in particular, it can be shown that the line shape of the dressed polariton is still lorentzian on the edges of the two-phonon band (where the density of states is relatively low) albeit strongly broadened by the opening of a new direct relaxation channel into the two-phonon continuum [9].

This lorentzian broadening, directly proportional to the two phonon density of states can be directly evidenced and measured in the time and space resolved CARS experiments by the exponential behavior in time of the polariton-induced anti-Stokes signal. The investigation was performed in ammonium chloride (NH$_4$Cl) in the frequency region where the polariton is in strong interaction with the polar $2\nu_4$ two-phonon band which extends roughly from 2800 to 2900 cm^{-1} [9]. As expected we have observed an exponential decrease of the coherent signal over several orders of magnitude (up to seven) which demonstrates the lorentzian character of the polariton broadening. The dramatic frequency dependence of the polariton dephasing rate for polariton frequencies close to the edges of the $2\nu_4$ band is depicted in Fig.4. The rapid increase of the damping rate inside the $2\nu_4$ band is a clear indication that polariton desintegration into two isoenergetic free phonons constitues a very efficient polariton relaxation channel. Outside the $2\nu_4$ band indirect phonon assisted processes are operative for the loss of coherence and damping of the polariton by up or down conversion to the $2\nu_4$ continuum. Taking into account these direct and indirect processes the complete frequency dependence of the measured damping rate can be precisely mapped out. Additional support for the correctness of the chosen mechanisms is provided by the temperature dependence of the damping rate in the range 10-120 K ; at higher temperatures the gradual appearance of disorder leads to a strong increase of the polariton dephasing rate very similar to the one discussed previously.

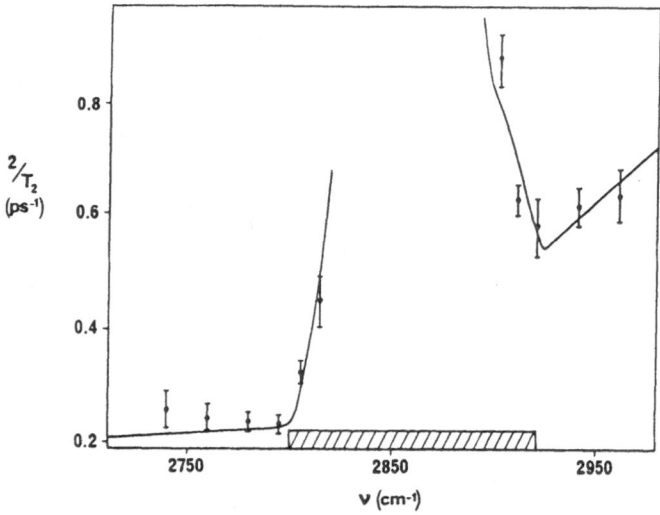

Fig.4. Measured polariton dephasing rate in NH$_4$Cl as a function of polariton frequency under conditions of Fermi resonance with the $2\nu_4$ band (shaded region). The full line is theoretical

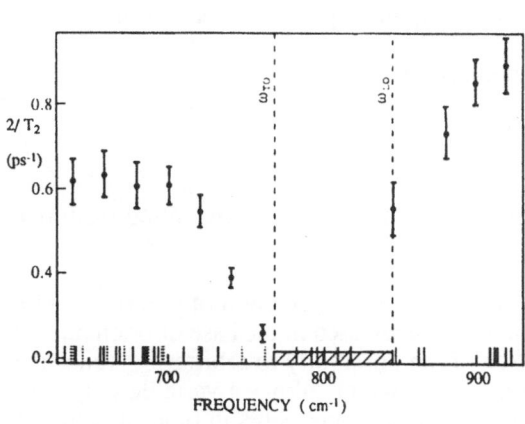
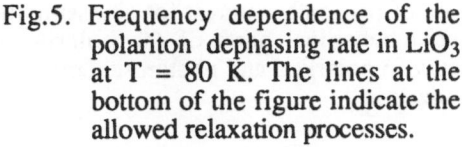

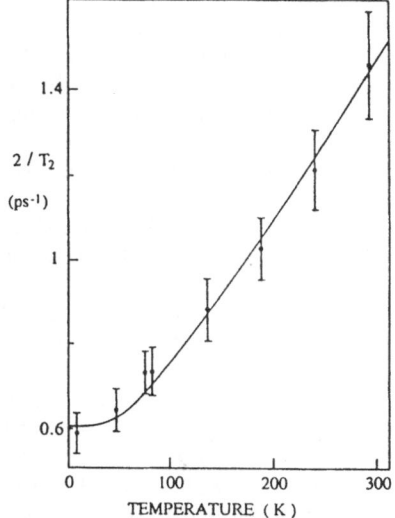

Fig.5. Frequency dependence of the polariton dephasing rate in LiO_3 at T = 80 K. The lines at the bottom of the figure indicate the allowed relaxation processes.

Fig.6. Temperature dependence of the 880 cm^{-1} polariton dephasing rate (full line is calculated)

b) Polaritons in uniaxial crystals.

In cubic crystals, the wave-vector restrictions associated with Raman techniques limit the accessible polariton region to apart of its lower branch. This limitation can be circumvented in uniaxial crystals, where the birefringence allows the entire polariton dispersion curve to be observed. We have exploited this fact and extended our time and space resolved CARS technique in uniaxial crystals, like $LiIO_3$, where we have investigated the upper and lower ordinary E_1 polariton, on both sides of the highest frequency Reststrahlen band ($\omega_{TO} = 768$ cm^{-1}, $\omega_{LO} = 843$ cm^{-1}).

Here too our group velocity measurements are in good agreement with the indirect value obtained from (1). The frequency dependence of the measured polariton dephasing rate is depicted in Fig.5 and shows a minimum in the vicinity of the Reststrahlen band.

This can be easily interpreted by letting the ω_{TO} - ω_{LO} frequency region play a role similar to the two-phonon band for the indirect mechanism in the case of polariton Fermi resonance (see above), acting as a reservoir of accessible states for phonon assisted up- or down- conversion of the polariton.

As for the polariton Fermi resonance in NH_4Cl, the temperature dependence of the damping rate is imposed by this interpretation of polariton relaxation, which then can be tested by temperature dependent measurements. This is indeed the case as shown in Fig.6 where the theoretical curve was obtained without fitting any parameters.

5. POLARITON OPTICS. CONCLUSIONS

This technique opens up new possibilities for the investigation of fundamental problems associated with pulses of collective excitations in crystals regarding their propagation characteristics, dephasing and energy loss processes. In this respect the problem of the polariton pulses is of central interest since it is connected with electromagnetic signal propagation close to a resonance.

The space and time resolved CARS technique allows one to track a polariton pulse at any "point" inside the crystal and analyse its phase and energy content and assess the impact of temporal and spatial disorder and, in particular, the effect of boundaries. The later is of particular importance for understanding polariton optics inside a crystal :

reflection and transmission by plane boundaries
polariton Fabry-Perot cavities
polariton total reflection and tunnelling
polariton nonlinear propagation

We have thus observed reflected polaritons and polaritons transmitted from one crystal to another separated by air.

A problem of particular fundamental interest is the interaction of polaritons with random spatial disorder. This problem has only been addressed in the case of orientational disorder in NH_4Cl where advantage was taken of the possibility to arbitrarily "tune" the crystal disorder by changing the crystal temperature, which allows a variable degree of disorder to be probed in the same sample. However, measurements in other disordered systems, such as isotopically disordered ones, are of particular interest for assessing the mutual coherence of the electromagnetic and mechanical parts of the polariton ; furthermore since there is a close connection between polariton and photon scattering by disorder one can expect to observe polariton localization.

As stated above the intrinsic limitation of the Raman technique to non centrosymmetric crystals can be circumvented by the use of the hyper-Raman configuration for coherent excitation and probe in crystals like NaCl. Here the decrease in efficiency of the hyper-Raman processes can be counterbalanced by using the very high peak power delivered by picosecond lasers and the much higher damage threshold in shortening the light pulses in such ionic crystals. Furthermore, one could also address the problem of propagation and relaxation of vibrational polaritons in highly disordered media such as glasses or liquids [10].

REFERENCES

1. R. Loudon, J.Phys.A3, 233 (1970)
2. K.P. Cheung and D.H. Auston, Phys.Rev.Lett. 55, 2152 (1985)
3. F. Vallée, G.M. Gale and C. Flytzanis, Phys.Rev.Lett. 57, 1867 (1986)
4. M.D. Fayer, Am.Rev.Phys.Chem.33, 63 (1982)
5. C.H. Henry and C.G.B. Garett, Phys.Rev.171, 1058 (1968)
6. G.M. Gale, F. Vallée and C. Flytzanis, in, "Dynamics of Molecular Crystals", ed. J. Lascombe (Elsevier Science, 1987) p.81
7. S. Ushioda, J.D. McMullen and J.J. Delaney, Phys.Rev.B8, 4634 (1973)
8. V.M. Agranovich and I.I. Lalov, Sov.Phys.Usp. 28, 484 (1985)
9. F. Vallée, G.M. Gale and C. Flytzanis, Phys.Rev.Lett. 61, 2102 (1988)
10. V.N. Denisov, B.N. Mavrin and V.B. Podobedov, Phys.Rep. 151, 1 (1987)

QUASIELASTIC ELECTRONIC LIGHT SCATTERING IN SEMICONDUCTORS AT LOW CONCENTRATIONS OF CURRENT CARRIERS

B. H. Bairamov, I. P. Ipatova, V. V. Toporov, V. A. Voitenko

A. F. Ioffe Physico-Technical Institute, Academy of Sciences of the USSR, Leningrad 194021, USSR

INTRODUCTION

The Energy and momentum conservation laws in an elementary process of quasielastic electronic light scattering can be expressed in the form[1]

$$E_{\vec{p}+h\vec{k}} - E_{\vec{p}} = h\omega , \tag{1}$$

where $E_{\vec{p}}$ is the energy of an electron with quasimomentum $\vec{p}$, while $h_{\vec{k}}$ and $h\omega$ are the momentum and energy transferred in the light scattering process. The majority of the electronic light scattering experiments have been carried out on heavily doped semiconductors with high current carrier concentrations $n \gtrsim 10^{16} cm^{-3}$, when the condition for strong screening holds

$$kr_s \ll 1 , \tag{2}$$

where r_s is the electron screening radius. Light scattering is possible in this case only due to special mechanisms providing the neutrality condition. Quasielastic light scattering spectra have a Lorentzian shape which indicates the diffusive motion of carriers in heavily doped materials[2].

In the case of low electron concentrations $n \lesssim 10^{15} cm^{-3}$, the opposite inequality

$$kr_s \gg 1 \tag{3}$$

holds, which means that there is no screening. In this case light scattering occurs due to fluctuations of the electron density.

Early experiments by Mooradian[1] dealt with GaAs crystals with low carrier concentrations $n \sim 3 \times 10^{15} cm^{-3}$. It was shown that the light scattering spectra have a Gaussian shape. Such spectra are reminiscent of the quaselastic light scattering spectra from classical collisionless plasmas where the Gaussian shape of the light scattering cross section results from the Maxwellian distribution of free electrons in velocities[3]. It was shown by Abramson et al.[4] that under the condition of the experiment in Ref. 1 there is a small deviation of the light scattering cross section from a Gaussian form.

The low concentration samples used in Ref. 1 were obtained by compensation. There is a high concentration of donors and acceptors in them. The electron collisions with impurities lead to a low electron mobility, which indicates that the charge carriers are not free carriers.

In order to clarify the situation we have studied quasielastic electron light scattering in n-InP samples with very low carrier concentrations $n \leq 10^{15}$ cm^{-3}. It is shown that the spectra are Gaussians with a halfwidth which increases almost linearly with the impurity concentration N. Theoretical consideration shows that this dependence is the result of glancing (small-angle) collisions of the carriers with the large-scale impurity potential created by charged impurities in doped and compensated samples.

EXPERIMENT

Quasielastic light scattering spectra were excited by a YAG:Nd^{3+} cw laser using the λ = 1064 nm line and registered by a double grating spectrometer equipped with a cooled photomultiplier in the photon-counting regime. The spectrometer scan was linear in wavelength, and the spectral resolution was 4 cm^{-1}.

Samples of n-type InP with electron concentration from 1×10^8 cm^{-3} (semi-insulating sample) up to 1×10^{16} cm^{-3} were studied at room temperature, T = 300 K. All samples were single crystal, x-ray oriented, rectangular parallelepipeds with typical sizes of about 0.5 x 2 x 2 mm^3. The electron concentration and mobility were measured by resistivity and Hall measurements. Raman spectra were taken with the laser beam incident along the <112> direction of the crystal, and the scattered light was detected at 90° along the <111> direction. The $\vec{e}_I \| \vec{e}_s$ polarization has been used to find light scattering from charge density fluctuations, where $\vec{e}_I$, $\vec{e}_s$ are the polarization vectors of the incident and scattered light, respectively. Typical continuum unshifted bands within -200 cm^{-1} with finite intensity at zero frequency from inelastic light scattering are shown in Fig. 1 for three samples with concentrations $n = 5 \times 10^{15}$ cm^{-3}, 7×20^{14} cm^{-3}, 5×10^{14} cm^{-3}. It is clearly seen that the absolute integral intensity decreases along with the electron concentration. The spectrum with $n = 1 \times 10^8$ cm^{-3} consists of a well-shaped band with maxima at 29 and 37 cm^{-1}. It is the two phonon difference band caused by short wavelength optical phonons corresponding to the boundary of the Brillouin zone. By subtracting the two-phonon spectrum from the observed quasielastic electronic spectrum one obtains the spectra shown in Fig. 1 by the solid lines. These spectra are Gaussians centered near the exciting laser frequency. The solid curves can be fitted over the entire frequency range by Gaussians contours which are shown in Fig. 1 by open circles. The small experimental uncertainty at frequencies below 15 cm^{-1} is due to the presence of the exciting laser light. The widths of the Gaussian contours are dependent on the concentration of impurities N. This dependence is given in Table 1.

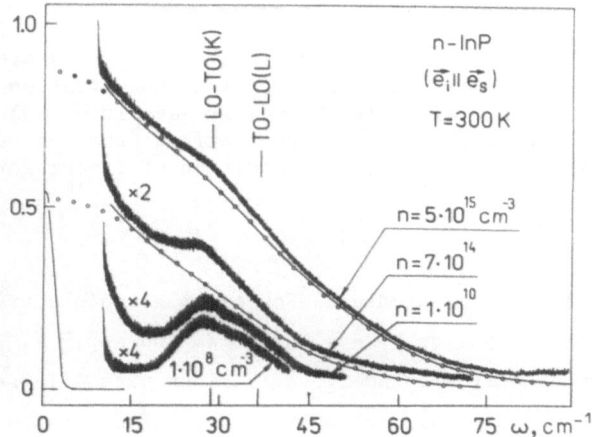

Fig. 1. Quasielastic electronic Stokes light scattering spectra in n-InP
with n = $1 \cdot 10^8$ cm^{-3}, $7 \cdot 10^{14}$ cm^{03}, $5 \cdot 10^{15}$ cm^{-3}. Solid lines show the
quasielastic spectra obtained by subtraction of the two phonon
contribution. Calculated spectra are represented by open circles.
Spectral resolution equals 4 cm^{-1}, T = 300 K.

The increase of the electron concentration up to 10^{16} cm^{-3} in n-type
InP leads to strong screening. Light scattering from the charge density
fluctuations is screened out. It was shown by Wolf[5] that the unscreen-
ed mechanism of light scattering is connected with electron energy fluctuations in nonparabolic materials. The quasi-free motion of the
electrons is replaced at these concentrations by the collision controlled
motion. The condition of frequent collisions $q\ell \ll 1$ holds, where ℓ is
the electron mean free path[2]. The shape of the light scattering
spectrum changes dramatically. Gaussians with the width $\Gamma = kv_T$ narrow
with an increase of the concentration into Lorentizians with the width

$$\Gamma \; k^2 D \; = \; k\ell k v_T \ll k v_T \; .$$

Collisional narrowing of this type has been observed in samples of n-InP
with 10^{15} cm^{-3} < n < 10^{17} cm^{-3}. Similar narrowing is well known in the
physics of classical atomic plasmas. It is called Dicke-narrowing[6].

Table 1. Dependence of the width of the Gaussian lineshape on the
impurity concentration

N(cm^{-3})	Γ(cm^{-1})
5×10^{15}	42.3
7×10^{14}	34.3
5×10^{14}	33.8

THEORY

Samples of n-InP with low concentrations of electrons are obtained by compensation. This means that there is high concentration of charged impurities in the crystal. It follows from the condition (3) that there is no screening of the long wavelength excitations produced by the exciting light. At the same time, the condition of linear screening of charged impurities,

$$nr_s^3 \gg 1 \ , \tag{4}$$

holds for the samples of InP studied. For example, an electron concentration $n \approx 10^{14}$ cm^{-3} leads to $r_s = 4 \times 10^{-5}$ cm and $nr_s^3 \approx 7$. Due to the condition (4) current carriers undergo collisions with a large-scale impurity fluctuation potential $\gamma(\vec{r})$ only. The characteristic size of γ is the order of the screening radius r_s [7].

$$\gamma(r_s) \approx \frac{e^2}{\epsilon r_s} \left(\tilde{N} r_s^3\right)^{1/2} \ , \tag{5}$$

where $e^2/\epsilon r_s$ is the potential of an isolated impurity centre, and $\tilde{N}$ is the total number of impurities within the fluctuation area. Since the momentum of a carrier $\vec{p}$ is much larger than the momentum transfer hk, it follows from (3) that

$$\frac{p}{h} r_s \gg 1 \ . \tag{6}$$

In n-InP at room temperature $(1/h) pr_s \approx 20$. The inequality (6) means that the large-scale impurity fluctuation potential scatters electrons through small angles $\theta \approx \frac{h}{pr_s} \ll 1$. It allows the reduction of the problem of carrier motion in the random field of impurities to the problem of the quasi-free motion of carriers in the large-scale, quasiclassical impurity potential.

It is known (see Sections 78-80 of Ref. 8) that at low concentrations of electrons the main mechanism for light scattering is connected with fluctuations of the electron charge density $e\delta n$, where δn is the fluctuation of the electron number density. The corresponding light scattering cross section has the form [8]

$$\frac{d^2\Sigma}{d\omega d\Omega} = \frac{1}{2\pi} \left(\frac{e^2}{m^* c^2}\right)^2 (\vec{e}_I \cdot \vec{e}_S)^2 \langle|\delta n_{\vec{k}\omega}|^2\rangle. \tag{7}$$

Here the angular brackets denote both an average over the random positions of the impurities and a statistical average, and $\delta n_{\vec{k}\omega}$ is the Fourier-transform of the electron density fluctuation

$$\delta n_{\vec{k}}(t) = \sum_\ell e^{i\vec{k}\cdot\vec{r}_\ell(t)} \ , \tag{8}$$

where $\vec{r}_\ell(t)$ is the position of the ℓth electron. In carrying out the averaging in (7) one should take into account that due to the inequalities (3) and (6) there is no correlation in the motions of different electrons. This means that the contributions to the light scattering cross section from different electrons do not interfere. Neglecting in (7) terms nondiagonal with respect to the index of the electron one can obtain the light scattering cross section in the form

$$\frac{d^2\Sigma}{d\omega d\Omega} = n\,\frac{1}{\pi}\left(\frac{e^2}{m^*c^2}\right)^2\left(\vec{e}_I\cdot\vec{e}_s\right)^2\ \text{Re}\int_0^\infty d\tau\ e^{-i\omega\tau}\ <e^{i(\vec{k}\Delta\vec{r}(\tau))}>\ ,\tag{9}$$

where $\Delta\vec{r} = \vec{r}(t+\tau)-\vec{r}(t)$ is the displacement of the classical electron during the time τ. In the presence of the quasiclassical random impurity potential the carriers are subjected to a random force $\vec{F}$. The uniformly accelerated motion of an electron is described by the displacement

$$\Delta\vec{r}(\tau) = \vec{v}\tau + \frac{\vec{F}}{2m}\,\tau^2.\tag{10}$$

The random distribution of impurities in the crystal leads to a random renormalization of the chemical potential of the electrons, ς. Due to the condition of the linear screening (4) the fluctuations of $\vec{F}$ and ς are small. They are described by the Gaussian distribution[9]

$$\Phi(\varsigma,\vec{F}) = \frac{1}{2\pi^2}\,\frac{1}{\gamma\rho_o^{3/2}}\,e^{-\frac{1}{2}\left(\frac{\varsigma^2}{\gamma}+\frac{F^2}{3\rho_o}\right)}\ ,\tag{11}$$

where γ and ρ_o are the mean square fluctuations of ς and $\vec{F}$. On taking the average over random quantities and over a Maxwellian velocity distribution of the carriers one obtains

$$<e^{i\vec{k}\Delta\vec{r}(\tau)}> = \int d\varsigma\int d^3F\int\frac{2d^3p}{(2\pi h)^3}\ f_o(p)\Phi(\varsigma,\vec{F})e^{i(\vec{k}\cdot(\vec{v}\tau+\frac{\vec{F}}{2m}\tau^2))}\tag{12}$$

where $f_o = \frac{n}{(2\pi m^*T)^{3/2}}\exp\left[(\varsigma-E_{\vec{p}})/T\right]$ is the Maxwellian distribution.

Substitution of (11) in (12) leads after averaging to the light scattering cross section in the form

$$\frac{d^2\Sigma}{d\omega d\Omega} = n\left(\frac{e^2}{m^*c^2}\right)^2\left(\vec{e}_I\cdot\vec{e}_s\right)^2\int\frac{dt}{2\pi}\cos\omega t\ e^{-\frac{(kv_T t)^2}{4}-\frac{\rho_o k^2 t^4}{24 m^{*2}}}\ ,\tag{13}$$

where $v_t = (2T/m^*)^{1/2}$ is the thermal velocity of the carriers.

In taking the average over the Maxwellian distribution the contributions of the kinetic and potential energies are separated in the quasiclassical approximation. An average over the fluctuations of the chemical potential ς leads to a renormalization factor of the order of unity. The average over fluctuations of the random force leads to the second exponential factor with ρ_o in the integrand of (13). If we neglect force fluctuations then the calculation of the integral in (13) results in a Gaussian contour with a width that is well known in the physics of light scattering from collisionless classical plasmas[3],

$\Gamma = kv_T$. When fluctuations of the random force $\vec{F}$ are taken into account the shape of the light scattering spectrum differ only slightly from Gaussian, but its width appears to be a linear function of the impurity

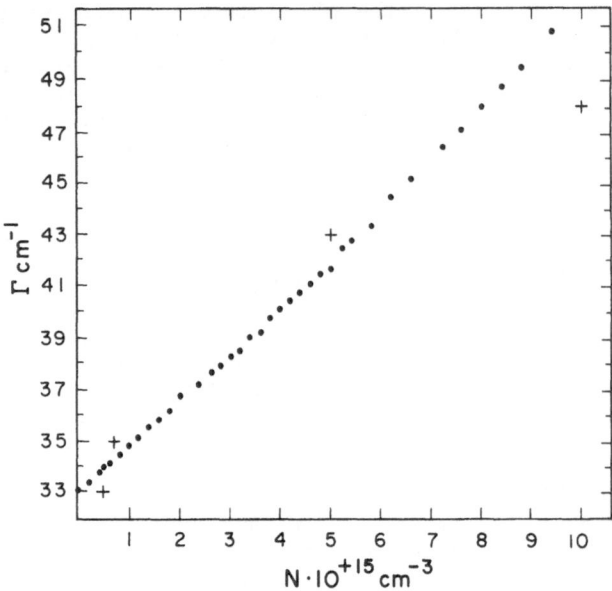

Fig. 2. Calculated dependence of the Gaussian width on the concentration of impurities. Experimental data are shown by crosses.

concentration N. The relative contribution of the two exponentials in the integrand of (13) is defined by the parameter

$$a = \frac{\rho_o k^2}{24m^2(kv_T)^4} \approx \frac{4\pi N \int r^2 dr (\nabla \varsigma)^2}{24(Tk)^2} . \tag{14}$$

This factor decreases with an increase of the frequency of the incident light.

When the electron concentration decreases, condition (4) fails first. It defines the low concentration boundary of the theory. If the concentration of electrons increases up to 10^{16} cm^{-3} conditions (2) and (6) are violated They restrict the theory from the side of high concentrations.

The integral in (13) has been found numerically for 40 values of the parameter a given by (14). The parameter kv_T was chosen from a comparison of the calculated spectrum with the spectrum of collisionless plasmas with the lowest concentration studied, $n = 5 \times 10^{14}$ cm^{-3}. The value obtained $kv_T = 33.8$, leads to an effective mass of the carriers which participate in the quasielastic light scattering of $m^* = 0.1\ m_o$, where m_o is the mass of a free electron. This value of m^* is larger than the value $m^* = 0.07\ m_o$ obtained from standard measurements for the bottom of the conductivion band at $T = 0$ K. This fact reflects the considerable nonparabolicity of the electron energy bands in InP. The width of the corresponding Gaussians is shown in Fig. 2 as a function of the impurity concentration N. The dependence is almost linear. Experimental values are shown by crosses.

REFERENCES

1. Mooradian A., in Light Scattering Spectra of Solids, ed. G. B. Wright, (Springer-Verlag, New York, 1960), pp. 285-295.
2. Bairamov, B. H., Voitenko, V. A., Ipatova, I. P., Subashiev, A. V., Toporov, V. V., Jahne, E., Sov. Phys. Solid State $\underline{28}$, 754 (1986).
3. Du Bois, D. F., Gilinski, U., Phys. Rev. $\underline{133}$, A1308 (1964).
4. Abramson, D. A., Tsen, K. T., Bray, R., Phys. Rev., B$\underline{26}$, 6571 (1982).
5. Wolff, P. A., Phys. Rev. $\underline{171}$, 436 (1968)
6. Dicke, R. H., Phys. Rev. $\underline{89}$ 472 (1953).
7. Shklovskii, B. I., Efros, A. L., Electronic Properties of Doped Semiconductors (Springer, Heidelberg, 1984).
8. Landau, L. D., and Lifshitz, E. M. The Classical Theory of Fields (Pergamon, Oxford, 1962).
9. Ginzburg, S. L., Sov. Phys. - JETP $\underline{63}$, 2264 (1972).

CONDENSED MATTER SCIENCE WITH FAR INFRA RED FREE ELECTRON LASERS

Vincent Jaccarino

Department of Physics, University of California, Santa Barbara, CA, 93106 U.S.A.

ABSTRACT

   The Free Electron Laser (FEL) offers the opportunity to study nonlinear and nonequilibrium phenomena with a coherent, high power, broadly tunable source of radiation. Following a discussion of the different kinds of FELs presently available for scientific applications, a review will be given of the experiments that have been performed and are being contemplated using the FIR FEL at the University of California at Santa Barbara.

NONEQUILIBRIUM TERAHERTZ RANGE ACOUSTIC PHONONS AND LUMINESCENCE OF EXCITONS IN SEMICONDUCTORS

A. V. Akimov, A. A. Kaplyanskii, and E. S. Moskalenko

A. F. Ioffe Physico-Technical Institute
Academy of Sciences of the USSR
194021 Leningrad, USSR

INTRODUCTION

Acoustic phonons in the terahertz range, with energies from a few meV to tens of meV interact effectively with various electronic states (electrons, holes, excitons, local states), and thus participate in many electronic processes in semiconductors.

In the single event of exciton-phonon collision the kinetic energy E and momentum k of the exciton change due to the absorption (emission) of an acoustic phonon. The total energy and momentum are conserved in this process. An elementary analysis of the interaction of two quasi-particles, an exciton with a dispersion law $E = h^2k^2/2M_{exc}$ (M_{exc} is the translational mass of the exciton), and an acoustic phonon with a linear dispersion law $\omega = s|\vec{q}|$ (ω is the phonon frequency, s is the speed of sound, $\vec{q}$ is the wave vector of the phonon), shows that only the phonons with frequencies lower than

$$h\omega^{max} = 2M_{exc}s^2 + 2(2M_{exc}s^2E)^{\frac{1}{2}} \tag{1}$$

are active in the exciton-phonon interaction[1].

Usually in experiments on exciton luminescence in semiconductors the excitons are excited optically in crystals having the fixed temperature T. In this case the phonon system is characterized by the equilibrium occupation numbers

$$\overline{n(\omega)} = \left(e^{h\omega/kT}-1\right)^{-1} . \tag{2}$$

If the lifetime of the excitons is large enough, the processes of inelastic exciton-phonon scattering result in the establishment of equilibrium between the exciton and phonon gases, the kinetic energy distribution of excitons in the band being described by the Maxwell-Boltzmann (MB) law,

$$N(E) \sim \sqrt{E} \cdot e^{-E/kT} , \tag{3}$$

with a temperature equal to the lattice temperature T. The energy distribution of excitons is reflected in the spectra of phonon assisted exciton luminescence[2].

The present paper summarizes the results of recent studies of the exciton luminescence in semiconductors in situations when the phonon system is far from equilibrium[3-6]. This situation may be easily realized in samples, immersed in liquid helium (LH), by the injection of heat pulses from a surface film heater or by strong optical pumping of the sample.

A very large effect of nonequilibrium phonons on exciton luminescence was observed, which permitted the use of the exciton luminescence as an effective optical detector of nonequilibrium phonons, and to obtain by means of this detector important information on various properties of high frequency acoustic phonons in semiconductors.

2. THE EFFECT OF HEAT PHONON PULSES ON EXCITON LUMINESCENCE IN SILICON

Silicon serves as a model crystal in the physics of nonequilibrium phonons in solids due to the possibility of having pure and perfect single crystals with a large mean free path of subterahertz phonons at low temperatures. Many experiments in Si with ballistic phonon beams generated and detected mostly by superconducting tunnel junction techniques have been carried out (see, for example, Ref. 7). The first studies of nonequilibrium phonons in Si by means of optical techniques were reported in Refs. 3 and 4.

2.1. Experimental Technique

In typical experiments (Fig. 1) an ultrapure Si single crystal ($\rho \sim 10\Omega\cdot$cm) immersed in LH (T = 1.8K) is illuminated by a cw-Ar-laser beam, which produces e-h pairs near the surface, which bind into free excitons (FE). The majority of the FE annihilate near the surface due to strong surface recombination. Other FE diffuse into the sample forming an excitonic cloud (EC), whose thickness is $\Gamma \sim \sqrt{D\tau_o} \sim 1$ μm, where D = 100 cm^2/s [8] is the FE diffusion constant, and τ_o = 1 μs [9] is the FE bulk lifetime. The latter is determined mainly by the capture of FE by residual impurities, $\tau_o \sim W_{FB}^{-1}$, which results in the formation of bound excitons (BE), where W_{FB} is the FE $\rightarrow$ BE capture probability.

A thin metal film "h" on the opposite face of the sample is heated by short (200 ns) current pulses, phonon heat pulses being injected into the sample. The effect of these nonequilibrium phonon pulses on the EC is studied by measuring the phonon-induced temporal changes $\Delta J(t)$ in the intensity of the lines of the EC luminescence spectrum. The spectral emission of FE (LO, TO assisted transitions 1.099, 1097 eV [10]) and of BE (LO assisted transitions 1.093 eV [10]) was measured.

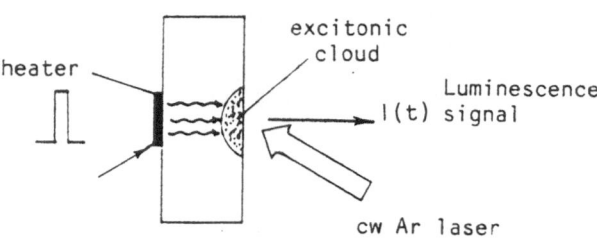

heater

excitonic cloud

Luminescence I(t) signal

cw Ar laser

Fig. 1.   Scheme of experiments.

2.2. Observation of Phonon-Induced Drag of Excitons

It was established that phonon induced relative changes of the FE luminescence intensity are of several tens of percent and drastically depend on the boundary conditions on the surface with EC, namely on the nature of the media in contact with this surface (superfluid LH, He gas, vacuum) and on the perfection of the Si surface.

Figure 2 shows the phonon-induced luminescence pulses $\Delta I^F(t)$ by FE and $\Delta I^B(t)$ for BE for the case of standard ("ordinary") Si surfaces with the thin (70 Å) oxide film usually formed by oxidation of surfaces in the open air atmosphere. In the case of a surface-vacuum boundary a phonon-induced enhancement of FE-luminescence is observed (Fig. 2a). This "positive" pulse has a sharp leading edge, coinciding in time with the time of ballistic flight of phonons from the heater (h) to the detector (EC). In the case of a surface-LH boundary the sign of the pulse $\Delta I^F(t)$ is negative, indicating the phonon-induced quenching of FE-luminescence (Fig. 2b). It can be seen (Fig. 2b) that the amplitude of the negative pulses increases with the increase of heat pulses power. Above some injected energy threshold, when the LH around the sample boils and a helium bubble is formed, the sign of $\Delta I^F(t)$ is reversed and becomes positive (Fig. 2b).

Generally the negative sign of $\Delta I^F(t)$ is observed for all "ordinary" oxidized Si surfaces in contact with LH, the relative signal I/I_o being independent of the specific type of surface preparation (etching, polishing), which strongly affects the absolute value of the stationary FE-luminescence signal I_o.

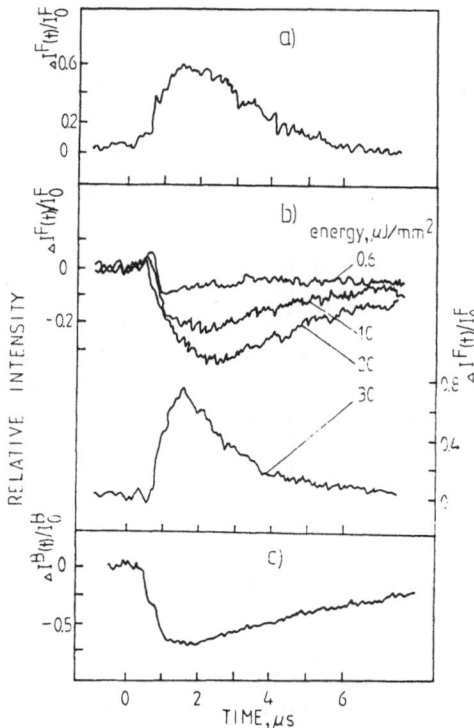

Fig. 2. Phonon induced luminescence pulses for a Si sample with an ordinary surface in contact with vacuum (a), liquid helium (b,c); a,b - FE, c - BE.

While discussing the observed effect of phonon pulses on EC

Finally, the sign of the phonon induced pulses of BE luminescence $\Delta I^B(t)$ observed in all experiments is negative, irrespective of the above boundary conditions near the Si surface with EC (Fig. 2c). luminescence we have to consider the interaction of acoustic phonons coming from the heater "h" both with FE (i) and BE (ii).

(i) The inelastic FE-phonon scattering processes result in a redistribution of energy and momenta between the phonon and exciton gases. The energy transfer and FE heating by a phonon pulse does not change the total number of FE in EC and thus must not change the intensity of FE luminescence. If, however, the distribution of momenta in the phonon pulse remains anisotropic in the EC region, the transfer of momenta from phonons to FE may lead to a drag of FE towards the surface, where they effectively recombine, the FE luminescence being quenched. In Si only the low frequency acoustic phonons with $h\omega < 1.8$ meV (0.4 THz) are active in FE drag, because of the energy and momentum conservation conditions (1). Quantitatively, the phonon flux induced FE drag may be characterized by the exciton drift velocity $V \ s \cdot \overline{\Delta n}/n$, where s is the sound velocity, $\overline{n}$ is the mean occupation number for low frequency phonons (< 0.4 THz), and $\Delta n = \overline{n}^+ - \overline{n}^-$ describes the anisotropic distribution of phonon occupation numbers in a flux with phonon momentum projections directed normally towards the surface ($\overline{n}^+$) and back ($\overline{n}^-$) (11).

(ii) Dissociation of bound excitons BE → FE which may be induced by "high-frequency" acoustic phonons with frequencies $h\Omega > \Delta E$, where $\Delta E = 3.8$ meV (0.9 THz) is the binding energy of BE in Si, results obviously in an enhancement of FE-luminescence and in the quenching of BE luminescence. The efficiency of BE → FE processes may be characterized by the value of the real bulk lifetime τ of FE in EC in the presence of high-frequency phonons. This time appears to be increased in comparison with the previously considered bulk lifetime τ_o in the absence of phonons due to repetitive BE → FE processes of liberation of captured (FE → BE) excitons, $\tau = \tau_o(1+W_{BF} \cdot W_o^{-1})$. Here $W_o^{-1} \approx 1 \ \mu s$ (9) is the reciprocal probability of Auger annihilation of BE, while $W_{BF} \sim \overline{N}$ is the probability of phonon-induced dissociation of BE → FE, with $\overline{N}$ the occupation number of active phonons with $\Omega > 0.9$ THz.

Thus, there are two phonon frequency depending contributions of the opposite sign to the total phonon induced FE-luminescence signal $\Delta I^F(t)$. In the "quasistationary" approximation (drift velocity v = const, bulk FE lifetime τ = const) the ratio of the intensities of FE-luminescence in the presence (I) and absence (I_o) of phonons is (3)

$$\frac{I}{I_o} \approx \left(\frac{\tau}{\tau_o}\right)^{1/2} \cdot \left(\beta + (\beta^2+1)^{1/2}\right)^{-1}, \tag{4}$$

the coefficient $\beta = \frac{1}{2} v \sqrt{\tau}/d$ characterizes the relative contribution of the processes of FE drag (v) and of dissociation BE → FE (τ). Depending on the relative value of τ and v both the increase ($I/I_o > 1$) and quenching ($I/I_o < 1$) of the intensity of FE-luminescence is possible.

The relative role of the FE drag and BE → FE dissociation processes strongly depends on the experimental conditions, which determine the transmission and reflection of acoustic phonons at the crystal surface with EC. Indeed, if the reflection of acoustic phonons from the crystal boundary is high enough, the back phonon flux compensates the incident one and the net directed phonon flux in the EC region becomes negligible. Hence, no phonon flux induced FE drag towards the surface occurs in this case and the resulting signal ΔI^F of FE-emission must be positive since it is due exclusively to the BE → FE dissociation induced by phonons with $\Omega > 0.9$ THz. This explains the positive sign of ΔI^F observed for Si crystal-vacuum (Fig. 2a) and crystal-He gas (Fig. 2b) boundaries, where 100% reflection of phonons takes place. The negative sign of the signals ΔI^F observed for all oxidized Si surfaces in contact with LH (Fig. 2b) can be attributed to the predominant role of luminescence quenching due to FE drag accompanied by a strong surface recombination of FE. The FE drag occurs here because of the known strong phonon transmission through such boundaries (Kapitza anomaly)[7] resulting in the emergence of an uncompensated phonon flux towards the crystal surface. Finally, the universal negative sign of BE-luminescence pulses $\Delta I^B(t)$ (Fig. 2c) is explained by phonon-induced BE → FE dissociation processes which occur in all situations.

Thus for the first time the phonon flux induced by drag of FE in Si crystals was observed. Earlier, the drag of FE by phonons was observed in Ge at 2K [12] and in CdS at 77 K [11].

2.3. Reflection of Phonons From Crystal-Liquid Helium Boundary and Kapitza Anomaly.

As was mentioned in Section 2.2, the sign of the phonon induced FE-luminescence signal is sensitive to the behavior of phonons on the crystal surface with EC. It allows the use of the luminescence technique described above for studying the nature of the anomalously low heat resistance of the solid-liquid helium boundary which is not consistent with the predictions of the acoustic mismatch theory (Kapitza anomaly). As was shown earlier in the experiments on the reflection of nonequilibrium phonons from surfaces performed for some crystals by the tunnel junction technique[7] the anomalously high phonon conductance through the crystal-LH boundary is due to imperfections of different kind of real crystal surfaces. Indeed, the atomically pure surfaces of alkali fluorides in LH give practically 100% reflectance of phonons (290 GHz) which indicates the absence of the Kapitza phonon conductance for such surfaces[13]. It seemed to be interesting to study phonon reflection from "ideal" as-cleaved Si surfaces in LH.

Figure 3a shows phonon induced FE-luminescence pulses $\Delta I^F(t)$ measured for an atomically pure (111)-surface of Si, which was prepared by cleavage of the sample in situ in LH. A positive $\Delta I^F(t)$ pulse is observed with a steep leading edge corresponding to the ballistic arrival of phonons in the EC region. Figure 3b shows the $\Delta I^F(t)$ pulse measured under the same conditions but after 20 hours oxidation of this cleaved (111)-surface in the open air at room temperature. It is seen that oxidation of a freshly cleaved surface results in the inversion of sign of the main part of the signal $\Delta I^F(t)$, which becomes negative. These results (Fig. 3) directly indicate the existence of strong phonon reflection from a freshly cleaved atomically pure Si surface in contact with LH, the phonon reflectance decreasing significantly with the oxidation of this surface. These results correlate with those obtained for alkali fluorides[13] and correspond fully to the above mentioned conclusions about the close relation between the anomalous Kapitza phonon conductance and irregularities of the structure of real surfaces. It is interesting to note that strong phonon reflection from high quality Si surface prepared by laser annealing was observed in Ref. 14.

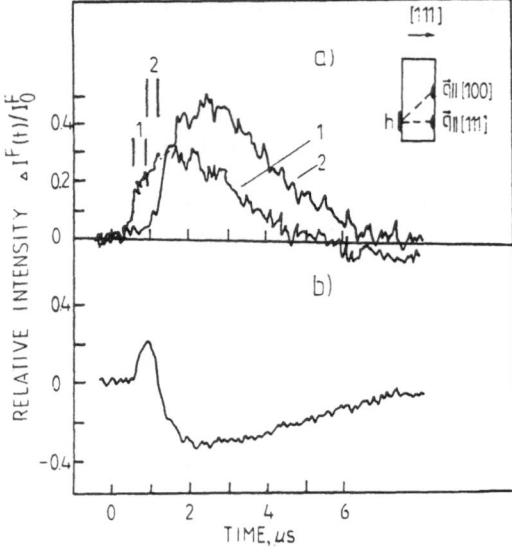

Fig. 3. Phonon induced FE luminescence pulses for a freshly cleaved
(a) and oxidized (b) Si surface (111) in contact with LH. The h-EC
direction ∥ [111] (a-curve, 1,b) and ∥ [100] (a-curve 2). The
arrival times for LA and TA balistic phonon are indicated by vertical
lines.

2.4. Isotopic Scattering of Terahertz Phonons

The propagation of acoustic phonons with frequencies as high as 1
THz and greater was studied in Si. The fact was used that under the
conditions of compensation of phonon flux, when the FE drag is small, the
positive $\Delta I^F(t)$ signal is due to bound exciton dissociation BE → FE
induced by high frequency phonons with $\Omega > 0.9$ THz (3.8 meV).

Figure 3a shows FE-luminescence pulses $\Delta I^F(t)$ observed in the sample
with a freshly cleaved (111) surface for two directions of phonon
propagation from the heater "h" to EC - along [111] and along [001]. In
the latter case "h" and EC on opposite faces of sample are shifted and
not situated on the same normal to the faces. The leading edges of the
$\Delta I^F(t)$ pulses correspond to the shortest times of phonon arrival from
"h" to EC. This indicates ballistic phonon propagation, as well as the
observation of focussing effect for TA-phonons propagating along [001];
indeed, the pulse with $\vec{q}\|[001]$ is stronger than with $\vec{q}\|[111]$, although
the distance "h-EC" is twice as long in the former case (Fig. 3a, inset).

In addition to the ballistic component in propagation, prominent
bulk scattering was observed for high frequency phonons with
$\Omega > 0.9$ THz. This scattering reveals itself in the geometry of the
experiment, in which both the heater (h) and detector (EC) are on the
same face of the sample, with a deep cut between them (Fig. 4, inset),
preventing the direct ballistic flight of phonons from "h" to EC. In
such a geometry the luminescence pulses from EC are induced by phonons
which are scattered in the bulk or reflected from the opposite face of
the sample. The observed sign of ΔI^F is negative in the case of full
reflection at the crystal-vacuum boundary of the low frequency phonons
participating in FE drag (Fig. 4b). But when effective escape of such
phonons in LH occurs, a positive pulse is observed (Fig. 4a), which is
induced by high frequency phonons, which possess strong ($\sim\Omega^4$) bulk
Rayleigh scattering.

42

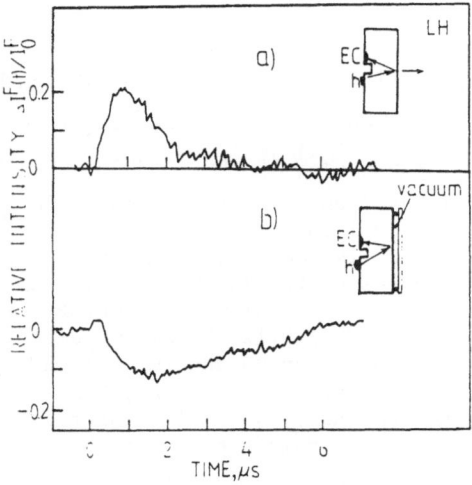

Fig. 4. Phonon induced FE luminescence pulses for Si in back
geometry of experiment.

The first observation of a prominent ballistic component for
$\Omega = 0.9$ THz phonons, as well the observation of strong bulk scattering of
phonons with $\Omega > 0.9$ THz, show that the mean free path l for THz phonons
in pure silicon is of the order of several mm. This value coincides well
with the theoretical estimate[15] $l = 4$ mm for the Rayleigh scattering
of 1 THz phonons from Si isotopes in natural material (4.7% ^{29}Si,
3.05% ^{30}Si).

3. PHONON HOT SPOT AND EXCITON LUMINESCENCE OF CUPROUS OXIDE CRYSTALS

In the case of optical excitation of semiconductors the major part
of the absorbed energy is transformed into phonons which are emitted into
the lattice in the processes of energy relaxation and recombination of
photoexcited carriers and excitons. The intense interband excitation of
the crystal at $T = 2$ K, when light penetration is small (1 μm), results
in the formation of a so-called phonon "hot spot" (HS), which is a small
local region near the surface strongly overheated relative to the cold
bulk. Such HS may also be formed in the crystal under the injection of
sufficiently powerful heat pulses from the metal film. A large
$(10^{-6} - 10^{-5}$ s) time of cooling, which depends on the power injected, is
characteristic for the HS. It is due to processes of inelastic phonon-
phonon scattering and of elastic isotope scattering ($\sim\omega^4$) which prevent
the ballistic escape of phonons, especially of high-frequency ones, from
the HS. The majority of papers relates to studies of low frequency
phonons emitted by HS into the cold bulk.

The question arises whether the frequency distribution of phonons in
a HS is an equilibrium one and can be described by a definite temperature. It is possible in principle to get the answer by measuring the
energy distribution of excitons interacting with the phonons in a HS
region.

The studies[5,6] were performed on 1s-ortoexcitons in cuprous oxide
crystals. The spectral shape I(E) of the luminescence band in the 614 nm
region, which is due to the radiative annihilation of 1s-excitons (Γ_5^+)
with the emission of a 109 cm^{-1} optical phonon (Γ_3^-), was studied. As was
shown earlier[16], the shape of I(E) directly reflects the kinetic energy
distribution f(E) of excitons in the 1s-band,

$$I(E) \sim \rho(E) \cdot f(E) , \qquad\qquad (5)$$

where $\rho(E) \sim \sqrt{E}$ is the density of states in the exciton band. The value of E is measured as the distance from the spectral frequency ν_o^1-109 cm^{-1}, where $\nu_o^1 = 16400$ cm^{-1} is the frequency of the pure electronic quadrupole 1s-exciton transition. The lifetimes of the 1s-ortoexcitons in Cu$_2$O are equal to $10^{-9} - 10^{-8}$ s [17].

The Cu$_2$O single crystal is immersed in pumped LH (T$_o$ = 1.7 K) - see Fig. 5, inset. Strong pulses of the 2nd harmonic of a YAG:Nd laser (λ = 530 nm, pulse duration 200 ns, maximal pulse power P = 100 W/mm^2, repetition rate 5 kHz) pump a 0.1x0.1x0.01 mm^3 volume of the sample, exciting simultaneously both phonon HS and luminescence of 1s-excitons. by using a weak probe cw-Ar-laser beam (514 nm, 0.2 W/mm^2) which excites the luminescence (but not the HS!), we could measure the luminescence spectrum I(E) with various time delays Δt after the action of the main YAG-pulse.

Figure 5 shows typical line shapes I(E) in the luminescence spectrum measured during the time of YAG-pulse action (Δt = 0) at low (P = 0.4 W/mm^2, curve 1) and high (P = 100 W/mm^2, curve 2) levels of pumping density. A broadening of the I(E) lineshape with increase of P is seen, which indicates the heating of the exciton gas. Figure 6 shows exciton luminescence spectra measured with different time delays after YAG-pulse action. It is seen that the lineshape I(E) remains broadened during long times, $\sim$ 10 μs, after the YAG pulse action. These times strongly exceed the lifetimes of ortoexcitons ($\tau_o \approx 10^{-9}$ s) and of all other possible electronic states in Cu$_2$O, excited by the YAG-laser. Hence the broadening of FE under strong optical excitation may be due only to the heating of the phonon system in the excited volume (HS!), which may remain over microsecond time scales. The heating of the exciton gas reflected in the broadening of I(E) is produced obviously by the interactions of the 1s-excitons with acoustic phonons of the HS.

It is important to stress, that the shape of I(E) broadened by strong optical pumping cannot be approximated by a single Maxwell-Boltzmann (MB) distribution [3] with any fixed temperature (see Fig. 5, dotted line). Hence the experimental exciton energy distribution function f(E) is not a Boltzmann exponent ($e^{-E/kT}$). In Fig. 5, f(E) is

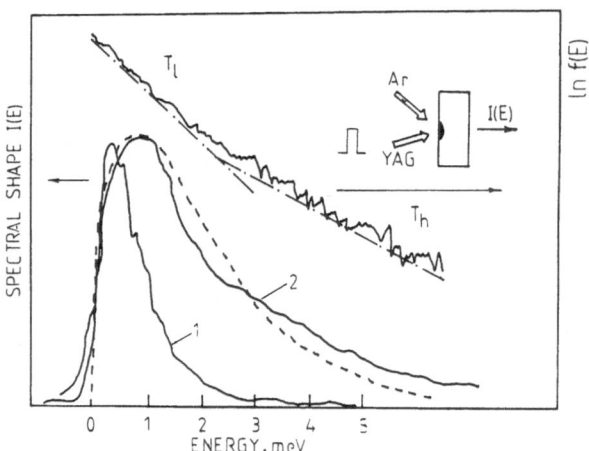

Fig. 5. Experimental line shapes I(E) and $\ln(I(E)/\sqrt{E})$ in the exciton luminescence spectra of Cu$_2$O (see text). Theoretical distributions: MB spectrum for T = 16 K (dashed line) and $\ln$ f(E) for T = 14 K, T = 22 K (dashed-dotted).

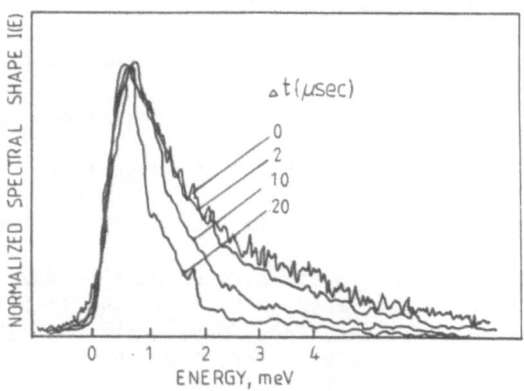

Fig. 6. Exciton luminescence spectra of Cu_2O measured with time
delay after intense pulse excitation.

approximated by two Boltzmann distributions with different temperatures
for the low energy (T_ℓ ≈ 14 K) and high energy (T_h ≈ 22 K) parts, the
value of T_ℓ being significantly smaller than T_h. The crossover energy
between the two distributions with T_ℓ and T_h is nearly 2 meV.

The observed absence of an equilibrium MB distribution in an exciton
gas interacting with a HS phonon gas means that the frequency distribution of the phonons in an HS is also nonequilibrium (non-Planckian). The
fact that the temperature for the "low energy" part of the exciton distribution I(E) is lower than for the "high energy" part ($T_\ell < T_h$) shows
that in a phonon gas interacting with excitons the effective temperature
for low frequency phonons is lower than for high frequency ones. Indeed,
the expression (1) for Cu_2O parameters (m_{exc} = 3m_o, s_{LS} = 4.8 × 10^5 cm/s)
shows that excitons with energy E ≈ 2 meV interact inelastically with low
frequency (ω < 3.5 meV) phonons, whereas excitons with E ≈ 6 meV interact
predominantly with high frequency phonons (ω ≈ 6 meV). Of course, the
description of the phonon spectrum by means of two temperatures
($T_\ell < T_h$) is very crude and serves only as a semiquantitative illustration of the nonequilibrium, non-Planckian nature of the phonon
spectrum in an HS, which has a deficit of low frequency phonons. This
deficit originates from the frequency dependence of anharmonic and defect
phonon scattering processes, which permits low frequency phonons to
escape the HS region more easily.

In conclusion, it should be noted that the results obtained on Si
and Cu_2O crystals demonstrate the possibilities of a luminescence
technique, based on the exciton-phonon interaction, in studies of
properties of terahertz range acoustic phonons in semiconductors. It
should be noted that at the present time the optical luminescence
technique is widely used in studies of nonequilibrium phonons in
insulating crystals (see, for example Ref. 18), the dominant technique in
phonon physics of semiconductors being an electric technique, based on
superconducting junction devices.

REFERENCES

1. M. D. Sturge, in "Excitons" eds. E. I. Rashba and M. D. Sturge,
   (North-Holland, Amsterdam, 1982), p. 1.

2. E. F. Gross, S. A. Permogorov, and B. S. Razbirin, Sov. Phys. Uspekhi, 14, (1971), 104.
3. A. V. Akimov, A. A. Kaplyanskii, E. S. Moskalenko, and R. A. Titov, JETP 94 (11) (1988).
4. A. V. Akimov, A. A. Kaplyanskii, and E. S. Moskalenko, J. Luminescence (1989), Proceedings of DPC-89.
5. A. V. Akimov, A. A. Kaplyanskii, and E. S. Moskalenko, Sov. Phys. Solid State 29, (1987), 288.
6. A. V. Akimov and A. A. Kaplyanskii in "Phonon Physics" eds. J. Kollar et al., 1985, p. 449.
7. D. Marx, W. Eisenmenger, Z. Physik B-Cond. Matt., 1982, 48, 277.
8. M. A. Tamor and J. P. Wolfe, Phys. Rev. Lett. 44, (1980) 1703.
9. R. B. Hammond and R. N. Silver, Appl. Phys. Lett. 36 (1980) 68.
10. P. J. Dean, J. P. Haynes, and W. Flood, Phys. Rev. 161, (1967) 711.
11. N. N. Zinovyev, I. P. Ivanov, V. I. Kozub, and I. D. Yaroshetskii, JETP, 84 (1983) 1761.
12. B. Etienne, M. Voos, and C. Benoit a la Guillaume In Proc. 14th Int. Conf. on Physics of Semiconductors, Edinburgh, 1978, p. 387.
13. J. Weber, W. Sandman, W. Dietshe, H. Kinder, Phys. Rev. Lett. 40, (1978), 1469.
14. H. C. Basso, W. Dietsche, H. Kinder, and P. Leiderer in "Phonon Scattering in Condensed Matter" eds. W. Eisenmenger, K. Lassmann, and S. Dottinger, Springer-Verlag, 1984, p. 212.
15. D. V. Kazakovtsev and I. B. Levinson, phys. stat. sol. (b) 136 (1986) 425.
16. F. I. Kreingold and B. S. Kulinkin, Optika i Spektroskopiya 33 (1972) 706; A. Compaan and H. Z. Cummins, Phys. Rev. B6 (1972) 4753.
17. J. W. Weiner, N. Caswell, and P. Y. Yu, Sol St. Comm. 46 (1983) 105.
18. W. E. Bron in "Nonequilibrium Phonons in Nonmetallic Crystals," ed. by W. Eisenmenger and A. A. Kaplyanskii. (North-Holland, Amsterdam, 1986), pp. 227-275.

NONLINEAR OPTICAL STUDIES OF MOLECULAR ADSORBATES

Y. R. Shen

Department of Physics, University of California
Materials and Chemical Sciences Division, Lawrence Berkeley Lab.
Berkeley, California 94720

In recent years, the possibility of employing laser interaction with surfaces as a surface diagnostic tool has attracted a great deal of attention. Optical second harmonic generation (SHG) and sum-frequency generation (SFG) have been proven to be most effective and versatile for surface and interfacial studies.[1] By symmetry, these processes are forbidden in media with centrosymmetry, and can therefore be highly surface-specific. They have the advantages of being capable of high spatial, temporal, and spectral resolutions, suitable for *in-situ*, remote sensing of samples in hostile environment, and applicable to all interfaces accessible to light. Indeed, they have been applied with great success to a large variety of surface and interfacial problems[1]: probing adsorption and desorption of molecules from surfaces, measuring average molecular orientation of adsorbates, monitoring surface symmetry and surface phase transitions, conducting surface microscopy and spectroscopy, and many others. Here, we shall describe a few experiments recently carried out in our laboratory using these techniques.

We first discuss the use of SHG to study surface diffusion.[2] This is a subject of great importance in modern surface science, as surface diffusion often plays a major role in limiting a surface reaction.[3] Research in this area is, however, still rather limited because of the lack of convenient tools.[4] Most exciting techniques are either applicable only to metal surfaces or vulnerable to perturbation during measurements. The SHG technique has a number of important advantages over the common methods and provides many new opportunities, as we shall see.

Our technique is based on the following designed steps. First, a monolayer grating of adsorbed molecules is created on the sample surface by laser desorption using two-beam interference.[5] Second, diffraction of surface SHG by the monolayer grating is used to probe the grating.[5,6] Third, the smearing-up of the monolayer grating by surface diffusion is monitored by the decay of the diffracted SH signal.[2] The results analyzed by the one-dimensional diffusion equation then yield the diffusion constant for the surface diffusion.

Our demonstrating experiment used CO on Ni(111).[2] The sample was in an ultrahigh vacuum chamber with a base pressure of 0.9×10^{-10} torr. The Ni surface was initially dosed with a saturated CO coverage of $\theta = \theta_s = 0.5$. The optical arrangement for the creation of a CO monolayer grating by laser desorption and the detection of the grating by SH diffraction is shown in Fig. 1. It involved a single-mode Q-switched Nd:YAG laser at 1.06 μm. For the creation of a CO monolayer grating, the 1.06-μm beam was split into two and recombined at incident angles $\phi = \pm 1.50$ on the CO-covered Ni surface over an area of 2-3 mm in diameter. Desorption of CO by the spatially modulated beam intensity as a result of interference produced a monolayer grating

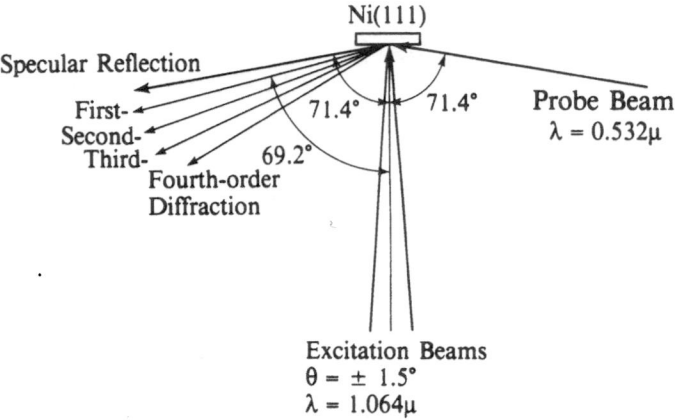

Fig. 1. Experimental arrangement for creation of a CO monolayer grating on Ni(111) by the excitation beams and detection of the grating by second harmonic diffraction of the probe beam.

with a spacing $2a = \lambda/2 \sin\phi \approx 20\mu m$. The grating profile was determined by the energy distribution in the two beams.[5] This can be predicted if the CO desorption versus the laser pulse energy impinging on the surface is known. In our experiment, the latter relation was obtained by using SHG to probe quantitatively the remains of CO on the surface (calibrated against thermal desorption spectroscopy) after laser desorption with different energies.[7] The monolayer grating could be probed by diffraction of surface SHG[5,6] (Fig. 1). For this, we used the frequency-doubled beam at 0.53 μm for the laser. It was incident on the sample at 71.4° and covered the entire desorption area. The nth-order diffraction of SHG with n = 1-4 appeared at $\Delta\theta$ = 2.22°, 4.24°, 6.10°, and 7.84°, respectively, off the specularly reflected direction. With the surface coverage given by

$$\theta(x) = \theta_0 + \sum_{n=1}^{\infty} 2\theta_n \cos(n\pi x/a) \tag{1}$$

it is easily seen that the nth-order SH diffraction signal S_n is directly proportional to θ_n^2, or more rigorously,

$$S_n \propto \int \theta_n^2 I_p^2 dxdy \tag{2}$$

where I_p is the probe beam intensity and $\hat{x}$ - $\hat{y}$ is the surface plane. Therefore, knowing θ_n from laser desorption, S_n can be calculated and directly compared with experiment. For our experiment, this is presented in Fig. 2.[5] The good agreement between measured and calculated results show clearly that we not only can have quantitative control over the creation of a monolayer grating by laser desorption but also can quantitatively characterize the monolayer grating using SHG.

For our surface diffusion studies,[2] a CO monolayer grating was first created on Ni(111) at a sufficiently low temperature (< 200K). The sample temperature was then raised to initiate surface diffusion of CO across the grating. This one-dimensional diffusion is expected to follow the Fick's law

$$\partial\theta/\partial t = (\partial/\partial x)(D \partial\theta/\partial x). \tag{3}$$

If the diffusion constant D is assumed to be independent of θ, then the solution of Eq. (3) yields

$$\theta_n(t) = \theta_n(0)\exp(- n^2\pi^2 Dt/a^2). \tag{4}$$

We then have

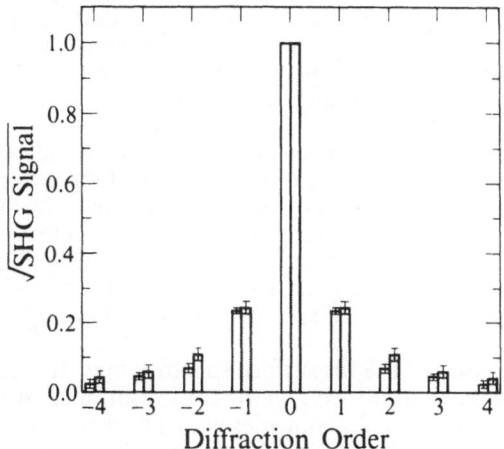

Fig. 2.   Measured (unshaded) and calculated (shaded) amplitudes of various orders of second harmonic diffraction from a monolayer grating of CO on Ni(111).

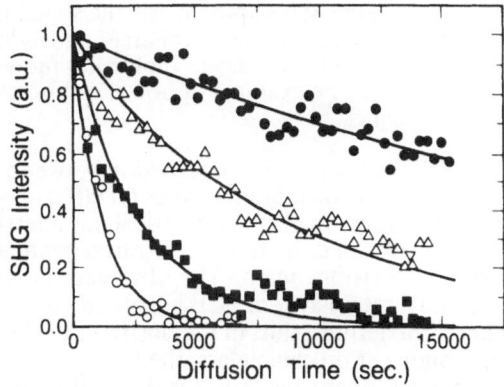

Fig. 3.   Normalized first-order diffracted second harmonic signal from CO monolayer gratings on Ni(111) as functions of time at various sample temperatures. Solid circles: T = 219K; open triangles: T = 247K; solid squares: T = 261K; open circles: T = 273K. The solid curves are least square fits using single exponential functions.

$$S_n(t) = S_n(0)\exp(-2n^2\pi^2 Dt/a^2) . \tag{5}$$

Thus, from the measured $S_n(t)$ and the known value of a, we can obtain D. In our experiment, we monitored the first-order SH diffraction signal $S_1(t)$. The results are shown in Fig. 3 for four different sample temperatures 219, 247, 261, and 273K. The diffusion constants deduced from the data are plotted against the temperature in Fig. 4. They can be fit by the Arrhenius expression $D(T) = D_0\exp(-E_{diff}/kT)$, with $D_0 = 1.2 \times 10^{-5}$ cm^2/sec and $E_{diff} = 6.9$ KCal/mole. In the present case, the adsorbed CO molecules occupy the bridge sites. The above value of E_{diff}, together with the adsorption energies of CO at bridge and top sites, suggests that the pathway of CO diffusion between bridge sites on Ni(111) is via hopping over a top site.

The present technique measures macroscopic diffusion of adsorbed molecules on a surface. It has a few very attractive features. The analysis of the experimental results is simple. Anisotropy of surface diffusion can be readily measured by orienting the grating in different directions. A large dynamic range of kinetic parameters can be studied by properly adjusting the grating spacing. Finally, the technique is applicable to a large variety of surfaces and interfaces, including those of semiconductors and insulators. Moreover, it can be applied to studies of surface diffusion of energy, momentum or other types of excitations.

We now discuss the use of SFG for surface spectroscopy. It is clear that with a tunable laser, SHG can serve as a surface spectroscopic tool since resonant enhancement of SHG is expected when ω or 2ω coincides with a transition frequency. This has been demonstrated in a number of cases.[8,9] As an example, we show in Fig. 5 the SH spectrum of a retinal monolayer on water.[9] The 335-nm peak is due to a resonant one-photon transition; in membranes, it is shifted to the green and is responsible for the green sensitivity of the eye. The 360-nm peak arises from a two-photon resonance, which has never been observed in linear absorption measurements.

The SHG surface spectroscopy is, however, limited to probing electronic transitions in the visible because the available photodetectors in the infrared generally are not sensitive enough to detect a surface monolayer. Thus, infrared-visible SFG which is an extension of SHG, must be used in order to probe surface vibrational or electronic transitions in the infrared.[10] In the process, a tunable infrared laser at ω_1 excites the resonant transition and a visible laser beam at ω_2 up-converts the signal to a sum-frequency output at $\omega_1 + \omega_2$ in the visible. The technique has been employed to obtain, for example, vibrational spectra of CH stretch modes of molecular monolayers adsorbed at various interfaces.[11] Here, we show one example of how it can be used to probe interaction between coadsorbed molecules.[12]

The experiment was on detecting possible interaction between cyano-biphenyl liquid crystal molecules (8CB) and surfactant molecules $CH_3(CH_2)_{17}(Me)_2N^+(CH_2)_3$ $Si(OMe)_3Cl^-$ (DMOAP) coadsorbed on glass. An optical parametric amplifier pumped by a picosecond Nd:YAG laser provided the tunable infrared beam while frequency-doubling of the laser provided the visible beam. The observed spectra of a full monolayer of 8CB on clean glass and a monolayer of 8CB adsorbed on DMOAP-coated glass are depicted in Fig. 6. There is a definite shift in the positions of the C-H peaks in the two spectra. For the 8CB monolayer on clean glass, the C-H stretching frequencies are nearly the same as those of the gas phase, whereas for the 8CB monolayer on DMOAP-coated glass, they appear close to those of the condensed phase. The red shift of the peaks from one to the other is a clear indication of existing interaction between DMOAP and 8CB.

SFG surface spectroscopy has great potential for many exciting research opportunities. Infrared vibrational or electronic transitions of buried interfaces or bare surfaces can now be studied. With time-resolved SFG using picosecond or femtosecond laser pulses, surface reactions can be probed *in-situ* with the possibility of identifying intermediate species. Selective surface dynamics can also be investigated. For example, it is possible to study surface diffusion of selected molecules in the presence of other molecular species on the surface.

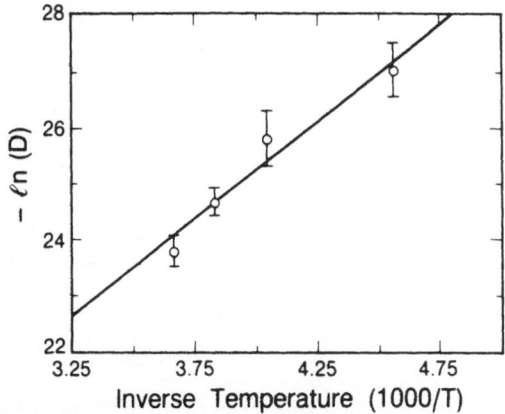

Fig. 4.   The Arrhenius plot (open circles) of the diffusion constant obtained from the results of Fig. 3. The solid line is a least square fit to the data.

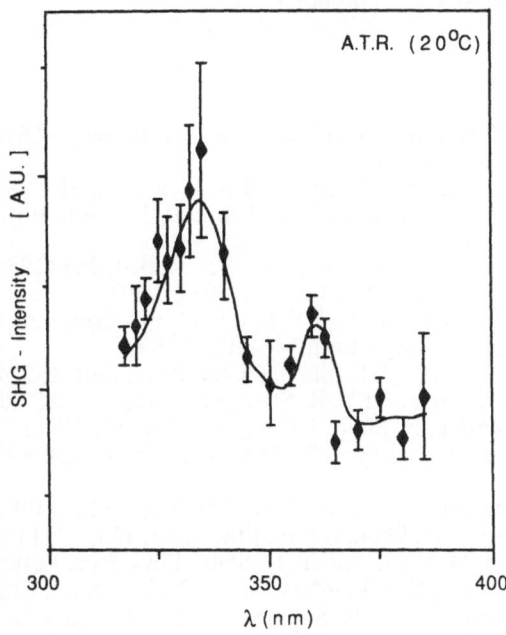

Fig. 5.   Second harmonic spectrum of an all-trans retinal monolayer at an air/water interface.

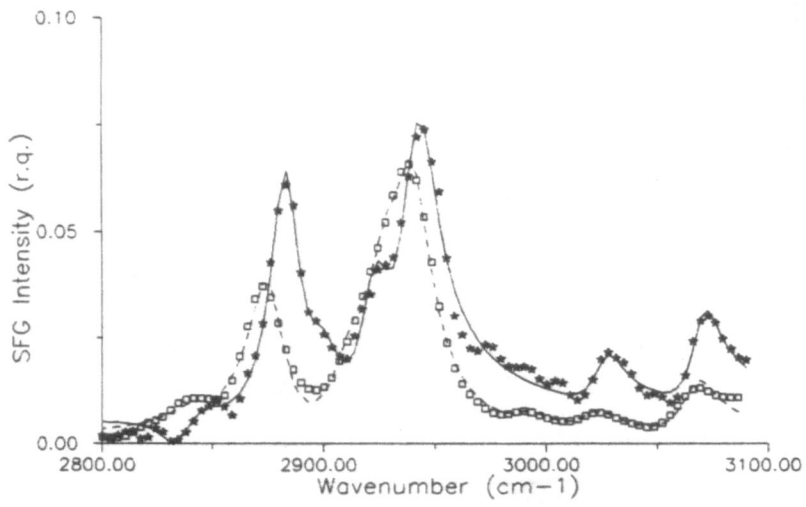

Fig. 6.   Sum-frequency vibrational spectrum of an 8CB monolayer on clean glass (solid curve) is compared with that of an 8CB monolayer deposited on DMOAP-coated glass (dashed curve).

ACKNOWLEDGEMENTS

This work was supported by the Director, Office of Energy Research, Office of Basic Energy Sciences, Materials Sciences Division, of the U.S. Department of Energy under Contract No. DE-ACO3-76SF00098.

REFERENCES

1. See, for example, Y. R. Shen, Ann. Rev. Mat. Sci. **16**, 69 (1986); Nature **337**, 519 (1989), and references therein.
2. X. D. Zhu, Th. Rasing, and Y. R. Shen, Phys. Rev. Lett. **61**, 2883 (1988).
3. E. G. Seebauer, A. C. f. Kong, and L. D. Schmidt, J. Chem. Phys. **88**, 6597 (1988), and references therein.
4. A. G. Naumovets and Yu. S. Vedula, Surf. Sci. Rep. **4**, 365 (1985).
5. X. D. Zhu and Y. R. Shen, Optics Lett. **14**, 503 (1989).
6. G. A. Reider, M. Huemer, and A. J. Schmidt, Optics Commun. **68**, 149 (1988); T. Suzuki and T. F. Heinz, Optics Lett. **14**, 1201 (1989).
7. X. D. Zhu, Th. Rasing, and Y. R. Shen, Chem. Phys. Lett. **155**, 459 (1989).
8. T. F. Heinz, H. W. K. tom, and Y. R. Shen, Phys. Rev. A **28**, 1883 (1983); G. Berkovic, Th. Rasing, and Y. R. Shen, J. Opt. Soc. Am. **B4**, 945 (1987).
9. Th. Rasing, J. Y. Huang, A. Lewis, T. Stehlin, and Y. R. Shen, Phys. Rev. A **40**, 1684 (1989).
10. X. D. Zhu, H. Suhr, and Y. R. Shen, Phys. Rev. B **35**, 3047 (1987); J. H. Hunt, P. Guyot-Sionnest, and Y. R. Shen, Chem. Phys. Lett. **133**, 189 (1987).
11. P. Guyot-Sionnest, J. H. Hunt, and Y. R. Shen, Phys. Rev. Lett. **59**, 14 (1987); A. L. Harris, C. E. D. Chidsey, N. J. Levinos, and D. N. Loiacono, Chem. Phys. Lett. **141**, 350 (1987); P. Guyot-Sionnest, R. Superfine, J. H. Hunt, and Y. R. Shen, Chem. Phys. Lett. **144**, 1 (1988); R. Superfine, P. Guyot-Sionnest, J. H. Hunt, C. T. Kao, and Y. R. Shen, Surf. Sci. **200**, L445 (1988).
12. R. Superfine, J. Y. Huang, and Y. R. Shen (to be published).

ENHANCEMENT OF EXCITON TRANSITION PROBABILITIES IN ULTRATHIN FILMS OF CADMIUM TELLURIDE

V. S. Bagaev, D. V. Kazantsev, A. G. Poiarkov, and A. A. Serov

P. N. Lebedev Physical Institute, Academy of Sciences of the USSR, 117333, Leninskii Prospekt 53, Moscow, USSR

INTRODUCTION

A great deal of interest is now focused on properties of thin semiconductor layers. It is caused by the success of film growth techniques, which are now capable of preparing films with thicknesses in the nanometer range. The two main techniques are molecular beam epitaxy (MBE) and metal-organic compound vapour deposition (MOCVD). At low film thickness, carrier parameters such as gap width, effective mass, and exciton binding energy differ from those of a bulk material. With the change of gap width, caused generally by the quantum size effect [1,2], the carrier effective masses begin to depend on the film thickness, thus changing the spectra of Wannier-Mott excitons. In addition, when the film thickness becomes smaller than the excitonic Bohr radius (a_B), the exciton wave function, which is initially spherical in bulk isotropic crystals, suffers deformation, caused by the confinement of motion along the z-direction, and changes completely to the 2D case at zero thickness. Theoretical studies predict that the wavefunction deformation due to the decrease of thickness results in an enhancement of exciton optical transition probabilities, i.e., in other words, in the enhancement of the oscillator strength [1,3]. The dependence of the positions of the excitonic peaks on film thickness was well studied both theoretically and experimentally [1-7]. As for the transition probabilities, there are the following difficulties: the poor quality of the samples makes quantitative measurements of luminescence impossible; small thickness causes difficulties in absorption measurements (because the variations om absorption caused by exciton features in a Cadmium Telluride film whose thickness is 2 nm are of the order of 0.3-0.7% magnitude). Thus, the experimental study of the probability of optical transitions involving exciton creation, and the comparison of these date with theoretical predictions, are certainly of interest.

EXPERIMENT

To investigate exciton features in the transmittance spectra of thin Cadmium Telluride films, we used a specially designed twin-beam spectro-photometer. The light of a band lamp passes through a grating mono-chromator (600 grooves/mm, 500 mm focusing mirrors, 0.03-0.6 mm slot), and is focused with a spherical collecting mirror on a chopper, which cuts it down to two separate beams, opening alternatively with a

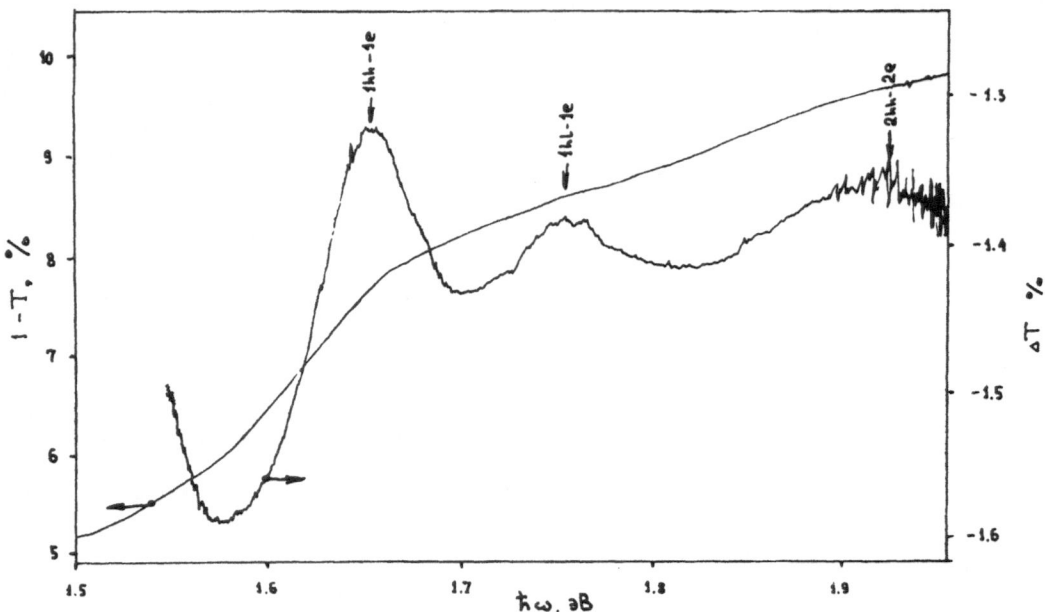

Fig. 1. Transmittance spectrum of a 9.8 nm thick CdTe sample (left scale) at T=2.1 K. Right scale - difference between the spectra at helium and at room temperatures. Deep left is caused by the temperature change of gap width. The three peaks are exciton absorption features at liquid helium temperature.

frequency of about 125 Hz. Then the beams are focused on a sample placed in a helium optical cryostat. The sample covers half the substrate, so that one beam passes through the sample-covered substrate, while the other one passes through the bare substrate. Both beams are then focused on a PMT cathode, and the photocurrent is processed with corresponding electronics. Thus the sensitivity achieved in transmittance measurements was as high as 0.015-0.02% at about 100% transmittance in our experiments. The Cadmium Telluride films were grown by laser evaporation in vacuum onto KBr(100) substrates (giving a small mismatch in lattice constants) [8,9]. Then the substrates were dissolved in water and the films were mounted on a Magnesium Fluoride substrate. MgF_2 was chosen because it matches CdTe in the value of its thermal expansion coefficient [10]. The transmittance spectrum of a 9.8 nm thick sample at 2.1K is plotted in Fig. 1. The difference spectrum obtained from the subtraction of two spectra measured at liquid helium and room temperatures has several peaks of about 0.3-2%. We assume them to be of excitonic nature, because the dependence of the peak position on film thickness is in agreement with the results of a theory that takes into account the quantum size effect and the change of the Coulomb interaction caused by a decrease in the thickness for excitons [6]. Taking into account more subtle factors such as polaronic effects and Coulomb image interaction forces between the electron and hole [11] further improved the agreement with experimental results. For film thicknesses greater than 30 nm these peaks merge into a single peak, whose position corresponds to the position of the exciton peak in bulk crystals [12]. At lower film thicknesses, the peak splits into 3-4 peaks, each shifted toward shorter wavelengths with positions well predicted by theory [6,11]. The peaks are polarization-identified by using oblique incidence of light [5].

RESULTS

It is well known that the integral of absorption curve within the line of a certain transition corresponds to the probability of transition per unit time. The absorption probability depends on the wavefunction for the relative motion of the electron and hole as $|\phi(0)|^2$, and is maximal for excitons of small radius. Thus the peaks corresponding to the creation of the 1st quantum size subband of the heavy hole exciton (transition 1hh-13) have the largest magnitude. The measured 1hh-1e transition probability vs. film thickness is plotted in Fig. 2. The experimental value is normalized to the film thickness to get the probability per unit volume, and by its value for 30 nm film in which size effects are considered to be weak. The film thickness was measured ellipsometrically on Magnesium Fluoride substrates and additionally calculated from energy distances between the exciton peaks for the first and the second quantum size subbands of the heavy-hole exciton (1hh-13 and 2hh-23) [6].

DISCUSSIONS

It is interesting that the increase of transition probability takes place at thicknesses of the order of the Bohr radius. In our opinion, this means that here the exciton wave function, initially spherical in bulk Cadmium Telluride, changes its shape. In the case of dipole allowed transitions the calculation of the transition probability is given by the square of the modulus of the overlap integral

$$K=\left|\frac{1}{NN}\int d^3r_e d^3r_h\ \psi(r_e,r_h)\delta(r_e-r_h)\right|^2 , \qquad (1)$$

where $\psi(r_e,r_h)$ is the wave function of electron and hole, and NN is the normalizing factor, whose exact form depends on analytic expression for the wave function. Looking at this formula, we can say that the lower the exciton size (i.e., a parameter like the Bohr radius), the higher is the transition probability. The increase of the exciton transition probability per unit volume is discussed in many works, in which analytical [2,4] as well as numerical [3] calculations were carried out.

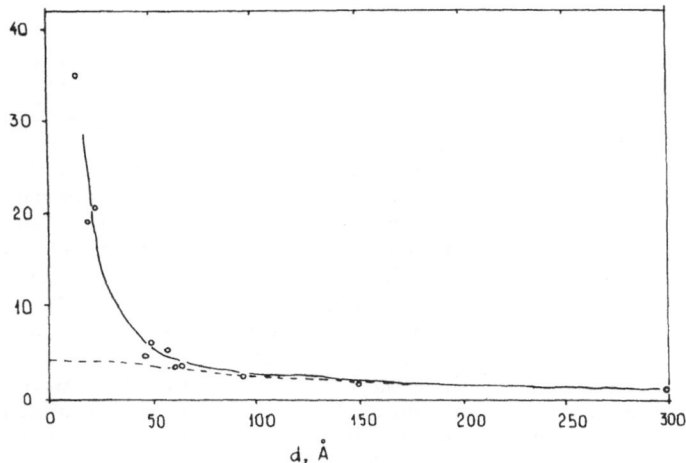

Fig. 2. Normalized area under the absorption peak of the 1hh-1e exciton transition as a function of thickness (circles), results of calculations based on the overlap integral (solid curve), and results of electrodynamical calculations (dashed).

Generally we can point out these effects: 1) increase of the oscillator strength caused by the exciton deformation at 3D-2D wave function transfer for thickness comparable to the free exciton a_B in bulk [2,3]; 2) enhancement of the Coulomb interaction in the exciton (the high ϵ of surrounding bulk CdTe is replaced by the low one of MgF$_2$ substrate, [6]) resulting in decrease of a_B; 3) the mass renormalization (caused by quantum forbidden gap broadening due to the size effect) least also to the decrease of a_B. Unfortunately, the thickness is about a_B for free excitons (in bulk CdTe) we have no analytical form of the wave function to get exact theory based on these assumptions. On the other hand, the following simple theory with analytical form of wave function [4] (with mass renormalization by [13,14] gives a good fit to experiment. By approximating the exciton wave function as

$$\psi(r_e, r_h) = \Phi(R)\phi(r) , \qquad (2)$$

where r and R are the relative and center of mass coordinates, respectively, one can rewrite (1) as

$$K = \left| \frac{1}{NN} \int d^3R \, \Phi(R) \int d^3r \phi(r)\delta(0) \right|^2 = \left| \frac{1}{NN} \phi(0) \int d^3R \, \Phi(R) \right|^2 , \qquad (3)$$

and remove the value of the wave function for the relative motion of the electron and hole at zero separation outside integral sign.

By using the wave function of the exciton presented in [4] and carrier masses calculated according to [13], one can plot the dependence of the squared modulus of the overlap integral on film thickness. After normalizing to its value for ahe thick film (for a 30 nm thick film in our case) we plotted it also in Fig. 2. The following assumptions were made here: the film thickness is much smaller than a_B; the quantum size energy shift exceeds the exciton binding energy; the a_B quantum size energy shift exceeds the exciton binding energy; the a_B for a heavy hole is much smaller than the thickness. In fact, all these values are comparable in our case. Still one can see a good fit in Fig. 2. The variational calculation with a wave function for the relative motion of the electron and hole of the form

$$\phi(r) = e^{\kappa r} \qquad (4)$$

reported in [3] leads to a similar result.

The increase of the observed peak intensity may be explained partially as follows. The high index of refraction of the film (n=3.5 for Cadmium Telluride) and the low one of the surrounding materials (n=1.5 for Magnesium Fluoride and n=1.0 for liquid helium) cause the film to be a Fabry-Perot etalon. A conventional electrodynamical calculation of the modulation of the transmitted light caused by the absorbance variation for such a Fabry-Perot system (i.e. film) predicts an increase of the modulation at resonant frequencies, even at zero thickness $(2nd/\lambda) \rightarrow 0$. The dashed line in Fig. 2 shows the film-on-substrate transmittance signal change when the absorbance constant α ($I = I_0 \exp(-\alpha d)$) changes from 0 to 20000 cm^{-1}. The curve was calculated for incident light whose wavelength was 700 nm, and then was normalized to a 30 nm thickness curve for the sake of simplicity. To plot this curve the approach described in [15] was used. The main idea was to match 5 optical fields at the boundaries of the sample. The 5 fields are: the incident wave, the reflected one, two waves in the film (i.e., forward and backward), and the transmitted wave in the substrate.

CONCLUSIONS

When the thickness of a film of Cadmium Telluride on a Magnesium Fluoride substrate decreases down to the exciton Bohr radius and less, the optical exciton transition probability calculated per one electron increases. This can be explained by the deformation of the exciton wavefunction by the potential influence of the well walls. An interesting fact is that electrodynamical calculations of light transmittance through a weakly absorbing film with a high refraction coefficient on a transparent substrate with usual refraction coefficient predicts a small [2-5 times] enhancement of the role of the film's absorbtion coefficient in transmittance when the film thickness vanishes.

ACKNOWLEDGEMENTS

We thank N. N. Salaschenko, F. V. Garin, A. V. Kochemasov, and L. V. Paramonov for growing the samples.

REFERENCES

1. L. V. Keldysh, JETP Lett. $\underline{30}$, (4) (1979), pp. 244-248.
2. L. V. Keldysh, JETP Lett. $\underline{29}$ (11) (1979), pp. 716-719.
3. M. Matsuura, T. Kamizato, 2nd Int. Conf. (Yamada Conference) on modulated semiconductor structures (Kyoto, 1985), pp. 103-110.
4. A. L. Efros, FTP $\underline{6}$ (1984), p. 1382.
5. N. A. Babaev, V. S. Bagaev, F. V. Garin, A. V. Kochemasov, L. V. Paramonov, A. G. Poyarkov, N. N. Salaschenko, V. B. Stopachinskii, JETP Lett., $\underline{40}$ (5) (1984), pp. 190-193.
6. N. A. Babaev, V. S. Bagaev, A. G. Poyarkov, N. N. Salaschenko, V. B. Stopachinskii, 17th Int. Conf. on Phys. of Semicond. (Pergamon Press, 1985), pp. 371-373.
7. S. I. Beril, E. P. Pokatilov, I. S. Cheban - FTT $\underline{25}$ (9) (1983), pp. 2561-2565.
8. S. V. Gaponov, N. N. Salaschenko, Electronnaya Promyshlennost' $\underline{1}$ (49) (1978), pp. 11-20 (in Russian).
9. S. V. Gaponov, Vestnik AN USSR, $\underline{12}$ (1984), pp. 3-10.
10. N. A. Babaev, V. S. Bagaev, B. D. Kolylovskii, A. G. Poyarkov, N. N. Salaschenko, V. B. Stopachinskii - FTT $\underline{26}$, (12) (1984), pp. 3611-3617.
11. E. P. Pokatilov, S. I. Beril, V. M. Fomin, V. V. Kalinovskii - Kishinew (Moldavia) State Univ., Moldavian NIINTI, 1988.
12. Z. C. Feng, et al., Appl Phys. Lett. $\underline{47}$ (1985) p. 24.
13. B. L. Gelmont, FTP $\underline{7}$ (10) (1975), pp. 1912-1919.
14. M. I. D'yakonov, A. V. Khateskii, JETP $\underline{82}$ (5) (1982), pp. 1584-1590.
15. L. D. Landau, E. M. Lifshits, Electrodynamics of Continuous Media (in Russian), (Nauka, Moscow, 1982), pp. 393-414.

STUDIES OF SEMICONDUCTOR SURFACES AND INTERFACES BY THREE-WAVE MIXING

SPECTROSCOPY

T.F. Heinz, F.J. Himpsel, M.M.T. Loy, and E. Palange*

IBM Research Division, T.J. Watson Research Center
Yorktown Heights, New York   10598

E. Burstein

Department of Physics, University of Pennsylvania
Philadelphia, PA   19104

ABSTRACT

The surface-sensitive nonlinear optical processes of second-harmonic generation (SHG) and sum-frequency generation (SFG) are powerful tools for investigating the structural and electronic properties of semiconductor surfaces and interfaces. Information about the symmetry and structural properties of the surface or interface can be obtained from an analysis of the polarization dependence of the nonlinear optical response, as has been illustrated for clean Si(111) surfaces prepared with different surface reconstructions. The approach has been applied to real-time, in-situ studies of thin film growth and annealing, from which barriers for adatom motion can be inferred. Information concerning the electronic properties of surfaces and interfaces can be obtained from spectroscopic measurements of resonant three-wave mixing (SHG or SFG). The method has been utilized, for example, to probe electronic transitions at the insulator/semiconductor interface formed by CaF_2/Si(111). These measurements led to a determination of the "interface" band gap, i.e. the minimum energy separation between filled and empty interface states. The interface band gap is found to differ strongly from that of either of the bulk materials, highlighting our ability to probe the distinctive properties of this quasi-two-dimensional region.

*Permanent Address: University of Rome

TIME-RESOLVED RESONANT REFLECTION OF LIGHT

J. Aaviksoo[1], J. Kuhl[2], and I. Reimand[1]

[1] Institute of Physics, Estonian Academy of Sciences
Riia 142, 202400 Tartu, Estonia, USSR

[2] Max-Planck Institut für Festkörperforschung
Heisenbergstrasse 1, 7000 Stuttgart 80, WEST GERMANY

1. INTRODUCTION

Light reflected from a dielectric interface shows distinct spectral features near material resonances. Proceeding from the Fresnel formula this spectrum can be related to the dielectric function ϵ and has been widely used to study the corresponding material parameters. The frequency dependence of the reflection coefficent implies that transients should be observed in reflection if light pulses are incident on the interface. In the general case, we can consider a light pulse $E_I(t)$ incident on a semi-infinite dielectric medium, characterized by a dielectric function $\epsilon(\omega,k)$, under angle α, and split into a reflected pulse $E_R(t)$ and a transmitted pulse $E_T(t,z)$ (Fig. 1). The problem of transient reflection is in relating the temporal properties of the reflected pulse to the material parameters of the medium and lacks an analytic solution in the general case. Let us note that these transients appear in a linear reflection process and are treated apart from nonlinear cases where ϵ itself is modified by a strong (additional) incident field. We prefer to call the latter case transient reflectivity and to use the term transient reflection to denote linear response.

Transient reflection was first analyzed by Elert[1] almost 60 years ago. He considered a truncated nonresonant monochromatic wave incident on a semi-infinite local medium and showed that transients arise at the leading edge of the reflected field. The characteristic time-scale of the transients was comparable to the period of the light wave. Later, Birman et al.[2,3] analyzed transient reflection from nonlocal media near

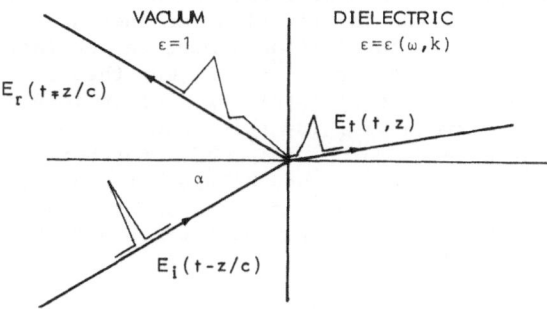

Fig. 1. The transient reflection scheme.

exciton-polariton resonances. It was shown that transients are associated with the leading and trailing edges of the truncated incident field and have a characteristic duration of ~ 0.1 ps under certain circumstances. Spatial dispersion was shown to enhance the transients, and the manifestation of additional boundary conditions (ABC) was treated. In a recent paper[4] we have considered analytical incident pulses to check the experimental observability of transients in transmission and reflection. A numerical analysis predicted the appearance of extended (over tens of picoseconds) pulse tails in the case of Gaussian incident pulses reflected from a GaAs crystal surface near the exciton resonance.

In the present paper we give a more detailed theoretical analysis of the transient reflection problem, and present the corresponding experimental results for GaAs[5], InP[6], and CdSe. We have focused our attention on the underlying physical mechanisms and tried to relate material parameters to the observed transients.

2. THEORY

The linear reflection problem can be treated by the Fourier analysis method -- the reflected light pulse $E_R(t,z=0)$ is the Fourier transform of the product of the incident pulse spectrum $E_I(\omega)$ and the reflection coefficient $r(\omega)$

$$E_R(t,z=0) = \int E_I(\omega) \cdot r(\omega) \cdot e^{-i\omega t} d\omega. \tag{1}$$

The reflection coefficient $r(\omega)$ can be determined proceeding from the dielectric function $\epsilon(\omega,k)$, which for a classical nonlocal medium is

$$\epsilon(\omega,k) = \epsilon_0 + \frac{4\pi\alpha\omega_0^2}{\omega_0^2 - \omega^2 - i\omega\Gamma + (h\omega_0/m)k^2} \tag{2}$$

In the above formula ϵ_0 is the background dielectric constant due to higher lying resonances, ω_0 is the transverse exciton frequency, $4\pi\alpha$ is the polarizability of the resonance, m is the exciton effective mass, and Γ is a phenomenological damping constant. It can be shown[7], that Eq. (2) can also be obtained as a special case of the quantum theory. In the following we proceed entirely from these two formulas.

The reflected pulse can also be calculated as a convolution of the incident pulse with the response function of the reflecting surface r(t). The latter is the Fourier transform of the reflection amplitude r(ω), and is equal to the reflected pulse if the incident pulse is infinitely short $(E_I(t) \sim \delta(t))$.

In order to calculate the reflection coefficient for a nonlocal medium one needs to solve the ABC problem, which we do, following Pekar[8], by assuming that the polarization vanishes at the crystal surface. Additionally, dead layer effects must be considered in the case of excitons in semiconductor crystals[9]. In our theoretical analysis we have avoided this problem by setting the angle of incidence α equal to the Brewster angle $\alpha = \arctan\sqrt{\epsilon_0}$. By this means the dead layer does not contribute to reflection and can be neglected. Furthermore, reflection due to higher lying resonances vanishes altogether and only the resonant contribution to the reflection amplitude is preserved. This makes the interpretation of the results more transparent. Under these conditions the reflection coefficient is

$$r(\omega) = \frac{1 - n^*(\omega)}{1 + n^*(\omega)}, \tag{3}$$

where $n^*(\omega)$ is the effective index of refraction, which can be calculated from Eq. (2) with the help of the formulas given in Ref. 10.

To get an idea of the reflection kinetics we have calculated the reflected pulse shapes ($|E_R(t)|^2$, see Fig. 2) by using the parameters of a GaAs crystal ($\epsilon_o = 12.6$, $\omega_o = 1.2221.5$ cm^{-1}, and $4\pi\alpha = 10^{-3}$ [11]). The incident Gaussian pulse has a FWHM of 1.2 ps and its spectrum covers the resonant reflection region, i.e. it is short enough to introduce no distortions to the transient response under study. It is clearly seen that the reflected pulses have significant tails extending over tens of picoseconds depending on the damping and the effective mass of the excitons. Three other observations are of importance. First, increasing the damping parameter Γ also increases the decay of the reflected pulse and leads to an almost exponential decay for large Γ. Second, for negligible damping increasing the spatial dispersion (decreasing the effective mass m^*) also shortens the reflected pulse. And third, even for no damping and spatial dispersion the reflected pulse decays (in an oscillatory manner).

a. Radiative decay. If $m^* = \infty$ and $\Gamma = 0$ the surface polarization decays only due to the coherent reemission of the excitation in the form of reflected and refracted pulses. This process is characterized by the polarizability $4\pi\alpha$ of the resonance. It can be shown analytically that the first moment of the reflection curve is inversely proportional to the LT-splitting of the resonance

$$\tau = \Delta_{LT}^{-1} = (2\pi\alpha\omega_o/\epsilon_o)^{-1} .\qquad (4)$$

From the numerical analysis we got the following equality $\tau(ps) = 10.12\cdot\Delta_{LT}^{-1}(cm^{-1})$. For GaAs this yields $\tau = 127$ ps in accordance with the curves in Fig. 2.

b Dephasing. This process is characterized by the phenomenological damping constant Γ. If we put $m^* = \infty$ and let $\Delta_{LT}/\Gamma \to 0$ we get for the Brewster angle of incidence an analytical solution[4]

$$r(\omega) = \frac{n_o - n}{n_o + n} = \frac{\pi\alpha\omega_o}{\epsilon_o}\cdot\frac{1}{\omega_o - \omega - i\Gamma/2} ,\qquad (5)$$

i.e. in this limiting case the reflection response is an exponential with a decay constant $\Gamma/2$ corresponding to a decay time $\tau = 1/\Gamma$ of the reflected pulse.

c. Spatial dispersion. If we put $\Gamma = 0$ and let $\alpha \to 0$ the reflected pulse exhibits a non-exponentially decaying tail with the first moment τ proportional to the effective mass of the excitons. The numerical modeling yields the following proportionality factor (for GaAs parameters) $\tau(ps) = 32.5\cdot m^*$ (free electron masses). The observed decay in this case is clearly related to exciton propagation -- the calculated time τ is crudely equal to λ/π divided by the group velocity of the exciton normal to the crystal surface

$$\tau = \lambda/\pi v_g = \frac{m^* c^2 (1+\epsilon_o)^{1/2}}{h\omega_o^2 \epsilon_o} .\qquad (6)$$

So, transient reflection can be related to certain decay mechanisms of the surface polarization in these limiting cases. We have modelled the decay also for more complicated intermediate situations, where only one of the decay mechanisms can be neglected. The results are depicted in Figs. 3, 4, and 5, where the first moment of the reflection response τ is plotted as a function of $\tau(\Delta_{LT},\Gamma)$ for $m^* = \infty$, $\tau(m^*,\Delta_{LT})$ for $\Gamma = 0$, and $\tau(m^*,\Gamma)$ for $\Delta_{LT} \to 0$. Together with the direct numerical data we have

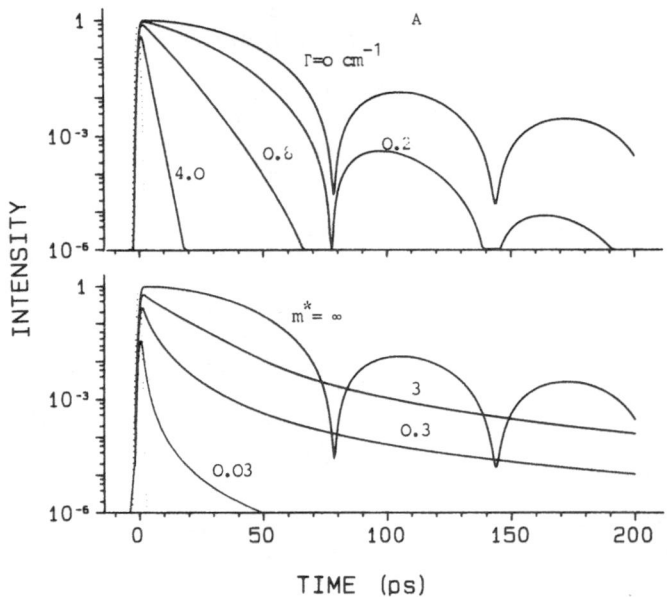

Fig. 2. Transient reflection from a GaAs crystal under Brewster angle. A: Dependenence on $m^* = \infty$. B: Dependence on exciton effective mass m^* for $\Gamma = 0$. The incident pulse is depicted at zero delay by dashed curve.

plotted the τ values calculated from the following phenomenological formula

$$\tau^{-1} = (\Delta_{LT}/10.12) + \Gamma + (0.031/m^*), \qquad (7)$$

which relies on the assumption that the three decay mechanisms are independent. We can see that the deviation of the direct numerical data from the estimates according to (7) is less than 20% for Figs. 3 and 4, indicating that the corresponding decay channels are almost independent of each other. In Fig. 5 the deviation is larger ($\approx$50%) in the intermediate region of parameters, where both damping and spatial dispersion contribute to the decay. This may be understood by noticing that increasing Γ also modifies the dispersion curve of polaritons and influences (enhances) polarization propagation thereby. The increase of group velocity with an increase of damping has been observed experimentally in Ref. 12. Based on our numerical anaysis we draw the following conclusions. Transient reflection manifests the polarization decay process in the reflecting surface layer. Three decay channels can be distinguished --exponential decay due to polarization damping, characterized by the damping parameter Γ, oscillatory decay due to reemission of light, characterized by the value of the LT-splitting of the resonance, and non-exponential ("convex") decay due to polarization propagation, characterized by the effective mass of the excitons. These three decay channels are almost independent and can be summarized by the phenomenological formula (7).

These results can be compared to the results presented in Refs. 2,3,12. The general predictions of the two approaches coincide -- transients appear as decaying tails of abrupt features of the incident pulse. For common semiconductors these tails may extend over several picoseconds making their experimental observation feasible, in contrast to the first treatment of transient nonresonant reflection[1]. However, two principal differences are also evident. An enhancement of the

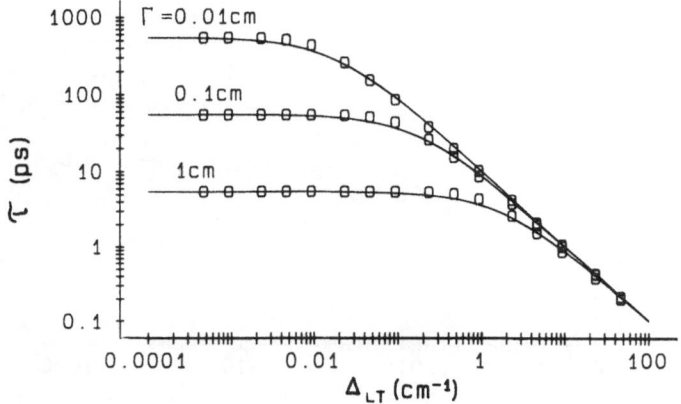

Fig. 3. Decay constants of transient reflection dependent on the LT-splitting and damping parameter Γ $(m^* = \infty)$.

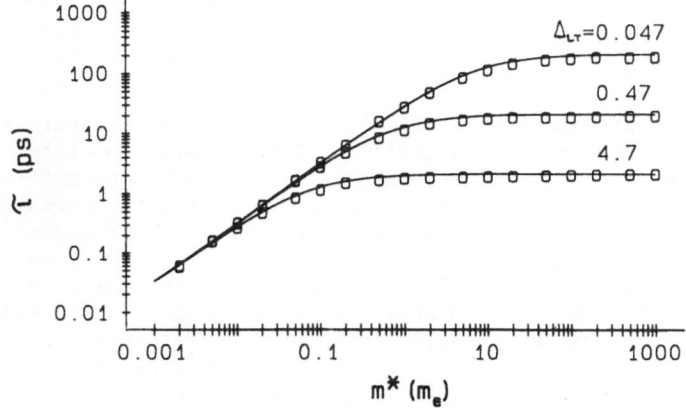

Fig. 4. Decay constants of transient reflection dependent on the effective mass of excitons and LT-splitting ($\Gamma = 0$).

transients is claimed to take place if spatial dispersion is included[2], while we observe a distinct reduction of the transients due to the intro-duction of a new decay mechanism. Second, a crossover from an exponential to a slow inverse power law is predicted at longer times in Ref. 2. Our analysis, on the contrary, predicts that non-exponential (inverse power law?) decay can be observed at early times if spatial dispersion is the main cause of the surface polarization decay. However, at long enough times exponential decay due to dephasing must still prevail.

Further theoretical analysis as well as experiments should elucidate the reasons for these discrepancies.

3. EXPERIMENTS

3.1. Experimental

The experimental set-up was based on synchronously pumped dye lasers. The reflected pulses were analysed by measuring their cross-correlation functions with the incident pulses using a synchronous detection technique. The laser beam was focused onto the samples to a $\approx$ 0.1mm spot by an f = 250 mm lens. All the measurements were carried out in an immersion cryostat below the λ-point of liquid helium (T $\approx$ 2K).

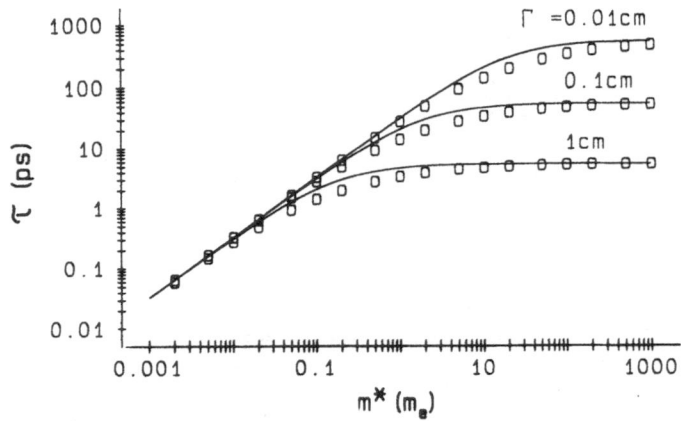

Fig. 5. Decay constants of transient reflection dependent on the effective mass of excitons and damping ($\Delta_{LT} \to 0$).

3.2. GaAs[5]

The expitaxial GaAs samples (R560) showed a characteristic reflection dip near the exciton frequency[11]. Time resolution of the reflected pulse (see Fig. 6), measured at normal incidence and low excitation densities reveals a long non-exponential tail, which starts at about a 1% intensity level and can be detected over 30 ps beyond the pulse maximum. The long term decay time of the tail was 5.4 ps as depicted by the solid line. This general shape of the reflected pulse profile can be divided into a nonresonant contribution, which follows the exciting pulse, and resonant reflection, which is responsible for the delayed tail. We have tried to model the observed decay theoretically by making use of the two sets of material parameters which were used to fit the steady-state reflection spectrum[11], and the results are also given in Fig. 6. Strong overestimation (three times) of the long term intensity of the pulse tail is evident, i.e. the fast initial decay of the resonant contribution is not reproduced by both models. Comparing the two models we see that the model with no spatial dispersion predicts a "concave" nature of the exponential decay, which is not observed experimentally and cannot be altered by varying the parameters of the model. The inclusion of spatial dispersion predicts a "convex" shape of the nonexponential tail and we therefore prefer it as the more appropriate one to describe the observed transients. The fit of the experimental curve can be considerably improved by slightly increasing the damping parameter. We note that the fast initial decay not reproduced by any of the models, may result from additional damping processes near the crystal surface, which has been suggested earlier[14]. The strong dependence of the transients on incident intensity is evidently caused by increased damping due to the induced exciton-exciton collisions. Let us add that the long term decay time of 5.4 ps is reasonably close to the estimate of 3.8 ps from (7).

3.3. InP [6]

The expitaxial InP samples revealed a characteristic excitonic reflection dip at 11441.7 cm^{-1} in accordance with earlier reflection measurements[15]. The results of time resolved reflection measurements are presented in Fig. 7. The measured CCF show clearly that the reflected pulses are delayed and have tails which extend several picoseconds beyond the exciting pulses. The other apparent feature is the dependence of the pulse shape and delay on the incident light intensity. We have modelled the experimental curves numerically (inserting the following

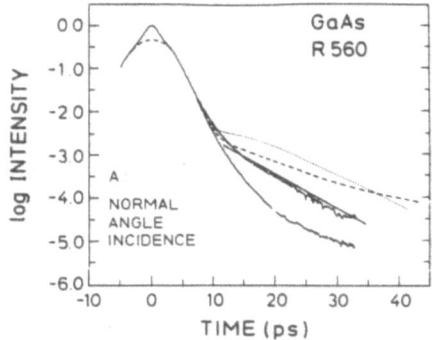

Fig. 6. Experimental cross-correlation functions of the reflected pulses with incident pulses (solid lines, weak excitation - upper curve, strong excitation - lower curve). Dotted line - theoretical fit neglecting spatial dispersion. Dashed line - theoretical fit including spatial dispersion. The slope of 5.4 ps is given by the straight line.

parameters for InP ϵ_o = 12.1, ω_o = 11441 cm^{-1}, and $4\pi\alpha$ = 2.3 · 10^{-3}[15]. The damping constant Γ and the effective mass of the excitons m* were varied in the curve fitting procedure. The fitting procedure yielded two important results: 1) no reasonable fit could be obtained neglecting spatial dispersion, whereas the inclusion of spatial dispersion provided a good fit over three orders of magnitude of the reflection decay, and 2) at higher excitation intensities the decay curves could be modelled by increasing the dephasing rate, however, the fit was clearly worse than in the low intensity limit. The best fitting parameters were Γ = 1.5 cm^{-1} and m* = 0.2 m$_o$.

3.4. CdSe

The transient reflection of vapor grown thin platelets of CdSe was studied near the lowest exciton resonance at ω_o = 14729 cm^{-1}. The crystals exhibited well pronounced steady-state reflection and polariton emission spectra. The corresponding time resolved normal angle incidence reflection curve is presented in Fig. 8. The incident laser pulse had an autocorrelation FWHM of 650 fs. The reflected pulse includes an ultra-fast response at zero delay, which is caused by the nonresident back-ground reflection, and a long non-exponential tail, beginning at about the 30% level, which is related to the resonant excitonic polarization. We can estimate the first moment of the resonant component of the reflected pulse making use of (7) and inserting the following material parameters for CdSe (ϵ_o = 8, Δ_{LT} = 1.3...0.5 meV and Γ = 0.05 meV[16]). We get $\tau \simeq$ 1 ps, which is close to the estimate τ = 1.2 ps given in Fig. 8. We may notice that in this case the radiative process contributes significantly to the observed decay, and spatial dispersion seems to be responsible for the strongly nonexponential nature of the decay. A more detailed numerical fit and experiments at the Brewster angle of incidence are needed to yield unambiguous material parameters from this transient reflection measurement.

4. CONCLUSIONS

We have analyzed transient resonant reflection of light pulses and related it to the decay of resonant polarization induced by the incident pulse in the surface layer. Three almost independent decay mechanisms have been distinguished: 1) radiative decay, 2) polarization damping, and

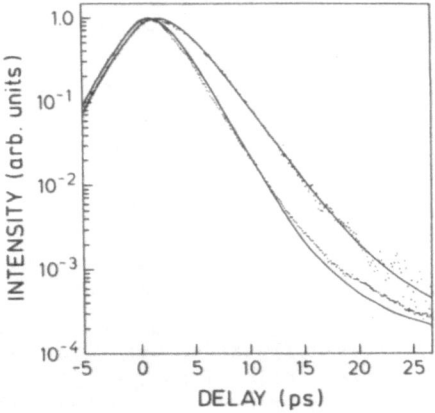

Fig. 7. Logarithmic plot of experimental cross-correlation functions of the reflected pulses (dotted) and theoretical fits to the data (continuous line) at two excitation intensities (incident laser power 30 μW - upper, and 1000-μW lower curves) for InP under Brewster angle incidence.

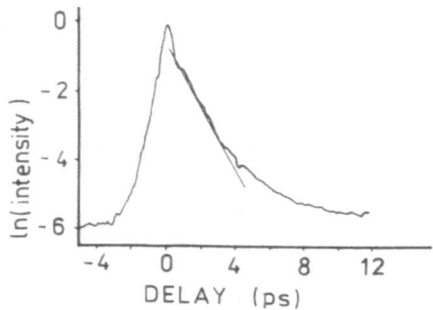

Fig. 8. Logarithmic plot of the experimental cross-correlation function of the reflected pulse from a CdSe crystal atr normal incidence. The slope of 1.2 ps is indicated by the straight line.

3) propagation of the polarization into the crystal. We have reported on the first experimental observations of the transient reflection effect in GaAs, InP, and CdSe. The experimental results can be described by numerical models as well as by general predictions of the theoretical analysis. Material parameters could be determined from the fit of the data in the case of InP.

ACKNOWLEDGMENTS

Thanks are due to E. Bauser, H. Scholz, and P. Lavallard who kindly supplied the samples. One of us (J.A.) thanks the A. von Humboldt Stiftung for the Fellowship to carry out the experiments.

REFERENCES

1. D. Elert, A.. Phys. Ser. 5, $\underline{7}$, 65 (1930).
2. G. P. Agrawal, J. L. Birman, D. N. Pattanayak, A. Puri, Phys. Rev. B$\underline{25}$, 2715 (1982).
3. S. V. Branis, K. Arya, J. L. Birman, Phys. Rev. B$\underline{39}$, 8371 (1989).
4. J. Aaviksoo, J. Lippmaa, J. Kuhl, J. Opt. Soc. Am. B$\underline{5}$, 1631 (1988).
5. J. Aaviksoo, J. Kuhl, IEEE J. Quant. Electr. $\underline{QE-25}$, (1989).
6. J. Aaviksoo, J. Kuhl, I. Reimand, Solid State Commun. $\underline{71}$, (1989).
7. W. C. Tait, Phys. Rev. B$\underline{5}$, 648 (1972).
8. S. I. Pekar, Sov. Phys. JETP $\underline{6}$, 785 (1958).
9. J. J. Hopfield, D. G. Thomas, Phys. Rev. $\underline{132}$, 563 (1963).
10. S. A. Permogorov, A. V. Selkin, V. V. Travnikov, Sov. Phys. Solid State $\underline{15}$, 1215 (1973).
11. D. D. Sell, S. E. Stokowski, R. Dingle, J. V. DiLorenzo, Phys. Rev. B$\underline{7}$, 4568 (1973).
12. J. Aaviksoo, J. Lippmaa, A. Freiberg, A. Anijalg, Izv. AN SSSR. Ser Fiz. $\underline{48}$, 550 (1984).
13. S. V. Branis, J. L. Birman, in: Laser Optics of Condensed Matter (Plenum Press, New York, 1988) p. 303.
14. L. Schultheis, J. Lagois, Phys. Rev. B$\underline{29}$, 6784 (1982).
15. F. Evangelisti, J. U. Fishbach, A. Frova, Phys. Rev. B$\underline{9}$, 1516 (1974).
16. T. Itoh, P. Lavallard, J. Reydellet, C. Benoit a la Guillaume, Solid State Commun. $\underline{37}$, 925 (1981).

FEMTOSECOND PHOTOEMISSION STUDIES OF IMAGE POTENTIAL AND ELECTRON DYNAMICS IN METALS

R. W. Schoenlein[*], J. G. Fujimoto, G. L. Eesley[†], and W. Capehart[†]

Department of Electrical Engineering and Computer Science and Research Laboratory of Electronics, Massachusetts Institute of Technology, Cambridge, MA 02139

[†]Physics Department, General Motors Research, Warren, MI 48049

ABSTRACT

The combination of femtosecond laser generation techniques with photoemission electron spectroscopy can permit the measurement of surface state and electron dynamics on a femtosecond time scale. We describe the application of these techniques to perform the first time resolved measurements of image potential state dynamics in metals. Image potential states occur when an electron outside a metal surface is bound in a Coulombic potential produced by its image charge in the metal. These states form a Rydberg-like energy series and resemble a two dimensional electron gas. The lifetimes of the $n = 1$ and $n = 2$ states in Ag(100) are 15-35 fs and ~ 200 fs, respectively. In Ag(111) the lifetimes of the $n = 1$ and $n = 2$ states are less than ~ 20 fs. Experimental results are in qualitative agreement with theoretical models of image potential electron dynamics.

SUMMARY

The combination of femtosecond optical measurement techniques with surface science diagnostics such as photoemission spectroscopy can provide new approaches for the measurement of surface state and electron dynamics on the femtosecond timescale. In this paper, we describe the application of femtosecond time resolved photoemission spectroscopy to perform the first time-resolved measurements of image potential states in metals. Using pump and probe photoemission spectroscopy, the dynamic evolution of the excited electron states can be measured directly. Experimental measurements of the $n = 1$ and $n = 2$ image potential states on Ag(100) and Ag(111) are in qualitative agreement with theoretical predictions of image potential dyanmics. Extensions of femtosecond optical measurement techniques to other surface science diagnostics hold the promise of performing a wide range of studies on a femtosecond time scale.

Image potential states occur in a wide variety of systems[1-3]. The existence of these states was first postulated by Shockley over 50 years ago[4]. In simple physical terms, the image potential state can be viewed as a state in which an electron localized

---

[*]Current address: Lawrence Berkeley Laboratory, Berkeley, CA 94720.

outside the surface of a metal is bound by a Coulombic potential to its image charge in the bulk. The image potential states resemble a two-dimensional electron gas and are and are somewhat analogous to quantum well states in semiconductors which exhibit one-dimensional confinement.

The physical properties and dynamics of image potential states can be described using a simple single particle picture. The potential at the crystal surface consists of a potential barrier which approaches the vacuum level outside the crystal and becomes a period potential in the bulk. An electron can be confined in the potential barrier outside of the crystal if it sees an effective repulsive barrier at the crystal surface. This can arise if the electron has a momentum such that it is Bragg reflected from the crystal lattice. Alternately, this requires a gap of bulk state for the given $k_\perp$ direction and energy.

Theoretical calculations of the image potential state may be performed using a single particle wavefunction and scattering phase shift model[2]. Scattering resonances then give rise to two classes of bound states. In a standard crystal-induced surface state, the electron wave function is localized inside the crystal and depends on the bulk crystal properties. In contrast, the image potential state is a barrier-induced state with the electron wave function localized predominantly outside of the crystal. The electron energy levels form a Rydberg-like series in the Coulombic potential with binding energies $E_n \sim -1/n^2$. Since the image potential states are localized in only one dimension, they can also have transverse momentum with $k_\parallel \neq 0$.

The existence of image potential states has been experimentally confirmed in a variety of metals using inverse photoemission measurements as well as two-photon photoemission spectroscopy[5-11]. The $n = 1$ and $n = 2$ Rydberg states have been measured on single crystal surfaces of Ag, Cu, and Ni. The binding energies and dispersion of the image states have also been studied. Typical binding energies for the $n = 1$ image potential state are in the range of several hundred meV. It is interesting to note that since the binding energy of electrons in the image potential state is measured with respect to the vacuum, these states have relatively high energies of a few eV above the Fermi level.

Since the electron wave function is localized outside of the bulk crystal, the lifetime of the image potential state is expected to be relatively long compared to states with comparable energy in the bulk. A simple prediction of the lifetime of image potential states may be obtained by considering the wavefunction overlap with the bulk. Using a hydrogenic wavefunction approximation, higher order states in the Rydberg series exhibit longer lifetimes because wavefunctions are localized progressively further from the crystal surface. This model predicts lifetime which scales as n^3 where n is the order of the Rydberg state. More detailed calculations, using self-energy and accounting for many body effects, can also be performed[12-14]. If the image potential state energy and momentum occurs in a gap of bulk states, the image potential decays through inelastic, electron-hole, scattering processes into the bulk. In contrast, if the image potential state is resonant with bulk states, relaxation can also occur through electron-electron, elastic scattering processes with bulk states. These calculations yield excited state lifetimes on the order of several tens of femtoseconds. Measurements of image potential lifetime from the energy broadening linewidth of the photoemission spectra have been difficult because of limited energy resolution, as well as possible inhomogenous broadening mechanisms.

The development of femtosecond pump and probe photoemission spectroscopy techniques now provides a direct approach for measuring image potential dynamics. Figure 1 shows a schematic of the measurement technique. In order to perform measurements on a femtosecond time scale, it is necessary to employ optical pump and probe techniques. In this case, an ultraviolet femtosecond pulse is used to populate the image potential state via transitions from the bulk. The image potential state then evolves in time and after a given time delay a visible femtosecond probe pulse is used to photoemit the electron from the image potential state. The resulting photoemitted electrons are analyzed using a standard photoemission spectrometer. By varying the time delay between the ultraviolet pump pulse and visible probe pulse, the transient photoemission spectrum can be measured and the evolution and dynamics of the image potential state determined. This technique permits a measurement resolution limited only by the duration of the pump and probe optical pulses, and the transient behavior of the photoemission spectrum can be measured on the femtosecond time scale.

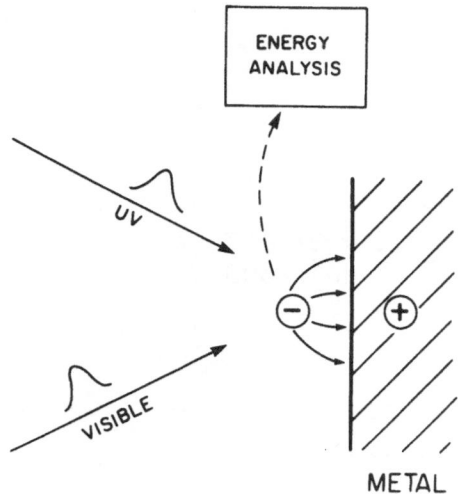

Fig. 1. Schematic of femtosecond photoemission spectroscopy using pump probe technique. After a femtosecond UV pump pulse populates the image potential state, a delayed visible pulse probes the time evolution by photoemission from this state.

The laser system used for these studies consisted of a colliding pulse modelocked (CPM) ring dye laser[15] and a high repetition rate copper vapor laser (CVL) pumped amplifier[16]. This system produced pulse durations of 50 fs at 620 nm (2.0 eV) with pulse energies of a few microjoules at a repetition rate of 8 kHz. Ultraviolet pump pulses were generated by focusing the amplified femtosecond pulses into a thin, 100 μm, KDP crystal. In order to compensate for dispersive pulse broadening effects in the optics which can be very severe in the ultraviolet, a pair of quartz prisms were incorporated in the ultraviolet pump beam delay line. The pulse duration of the UV pulses at 4.0 eV was ~ 50 fs. A portion of the amplified laser output at 2.0 eV was directed through a variable delay line to function as the probe pulse. The time delay between the pump and probe pulses was adjusted using a computer controlled 0.1 μm resolution stepping motor stage. Both beams were then coupled into an ultrahigh vacuum (UHV) chamber which contained the metal sample.

The ultrahigh vacuum chamber was maintained at 10^{-10} Torr. The samples were cut and mechanically polished with a surface orientation of $\pm 2°$. Prior to photoemission measurements, the samples were sputter ion cleaned and annealed. Surface qualtity was verified using Auger spectroscopy and low energy electron diffraction (LEED). The pump and probe femtosecond pulses were directed onto the metal specimen in a colinear geometry using a concave focusing mirror. The photoemitted electrons were analyzed with a double pass cylindrical mirror analyzer (CMA) which had an energy resolution of ~ 180 meV. Although measurements were not performed with true angle resolved photoemission, the acceptance angle of the photoemission spectrometer was ~ 10°. The output of the CMA was detected using an electron multiplier and all measurements were performed in the electron counting limit to avoid space charge effects.

The first measurements of image potential dyanmics were performed on the Ag(100) surface because this surface provides a simple model system with image potential energies which corresponded well with our available laser excitation energies[17-19]. Measurements were performed on photoelectrons emitted normal to the surface with $k_\parallel = 0$. A schematic of the Ag(100) bandstructure is shown in Fig. 2. The energy of the vacuum level is 4.43 eV, and the binding energies of the $n = 1$ and $n = 2$ image potential states are 0.53 and 0.16 eV, respectively. For the Ag(100) surface orientation, the image potential energies occur in gap of bulk states, and therefore the lifetime of the image potential states is expected to be relatively long.

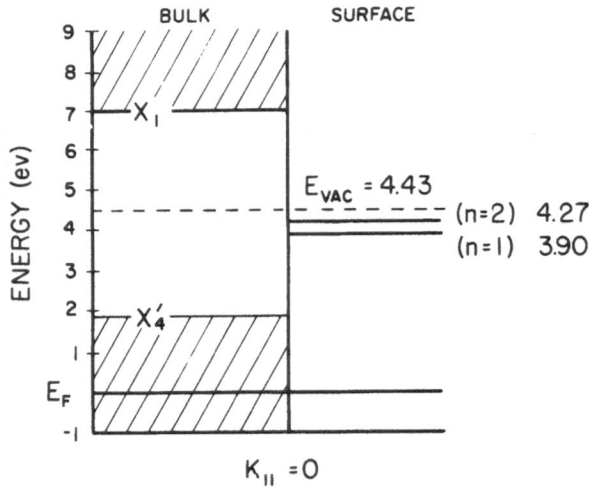

Fig. 2. Schematic of energy level structure of Ag(100) surface. The energies of the $n = 1$ and $n = 2$ image states are 3.90 and 4.27 eV, respectively. An ultraviolet pump pulse populates the state while a visible probe pulse photoemitts electrons to the vacuum. Note that the vacuum level and image potential states lie in a gap of bulk states.

Time-resolved measurements of the image potential state were performed using the pump and probe photoemission technique. For Ag(100) the UV second harmonic of the laser system at 4.0 eV has sufficient energy to populate the $n = 1$ image state at 3.90 eV through excitation from bulk states below the Fermi surface. The populated image potential state then can subsequently be probed by photoemitted to the vacuum, using a 2.0 eV visible probe pulse. In the photoemission spectrum, the image potential will then be observed at 2.0 eV minus the $n = 1$ image potential binding energy of 0.53 eV or ~ 1.5 eV. Thus the image potential state can be investigated using the two step, two photon UV-visible process where the UV photon populates the state and the visible photon probes the state.

It is important to note that there are also other multiphoton photoemission processe which can arise from the ultraviolet or visible pulses. For the UV pulse alone, a 2-photon UV photoemission process is possible in which the image potential state is both populated and photoemitted with UV pulses. In this case, the image potential state will be observed at 4.0 eV minus the $n = 1$ binding energy or ~ 3.5 eV. It is also possible to observe a 3-photon photoemission process arising from visible pulses alone. The image potential state is populated by 2-photon excitation from the bulk and subsequently photoemitted via a third visible photon to produce a photoemission signal at ~ 1.5

eV. These 2-photon ultraviolet and 3-photon visible processes will scale as the square of the ultraviolet intensity and the cube of the visible intensity, respectively.

By appropriately choosing the pump and probe intensities, it is possible to obtain a nearly background-free photoemission signal, where the dominant contributions arise from the pump and probe process. The advantage of using ultraviolet and visible pump and probe wavelengths is that the symmetry of excitation and probing is broken, and therefore, the transient photoemission spectrum can be measured by delaying the pump and probe.

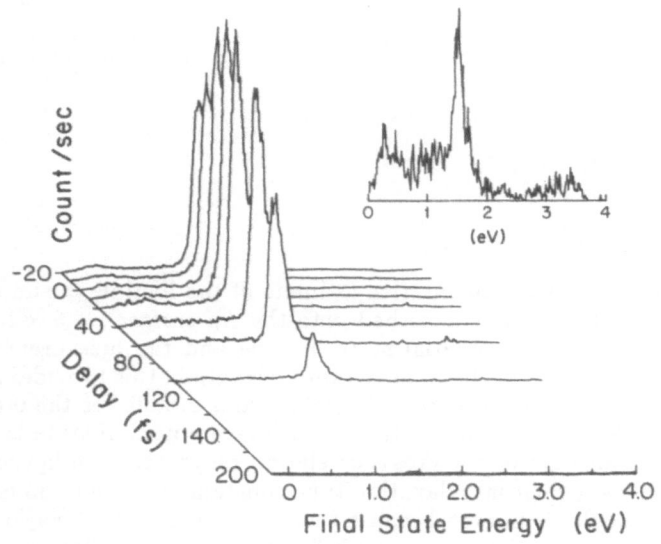

Fig. 3. Femtosecond photoelectron spectra showing the evolution of the $n = 1$ image potential state in Ag(100) taken using a 4.0 eV pump and 2.0 eV probe. The inset, maginfied 100x, shows the background photoemission spectrum.

Figure 3 shows the transient photoemission spectrum of the $n = 1$ image potential state on Ag(100). Measurements were performed using UV and visible fluences of $\sim$ 60 $\mu J/cm^2$ and 0.7 mJ/cm^2, respectively. The count rate was several hundred counts per second. Both UV and visible beams were P-polarized in order to satisfy the selection rules for populating and photoemitting from the image potential state. The inset shows a background trace obtained by setting the probe to arrive $\sim$ 200 fs after the pump and shows the two-photon UV and three-photon visible photoemission processes. The background is $\sim$ 100x below the pump probe photoemission signal. The peak of the image potential state is observed in the pump probe photoemission spectrum at $\sim$ 1.5 eV.

The dynamics of the image potential state can be observed by examining different photoemission spectra obtained at different time delays of the probe relative to the pump. In order to measure the dynamics of the $n = 1$ potential state directly, the photoemission spectrometer can be tuned to the peak of the $n = 1$ image potential state energy, and the pump and probe delay scanned. Figure 4 shows a measurement of the resulting $n = 1$ image state potential dynamics. This measurement indicates an image potential state lifetime which is on the order of the laser pulse duration.

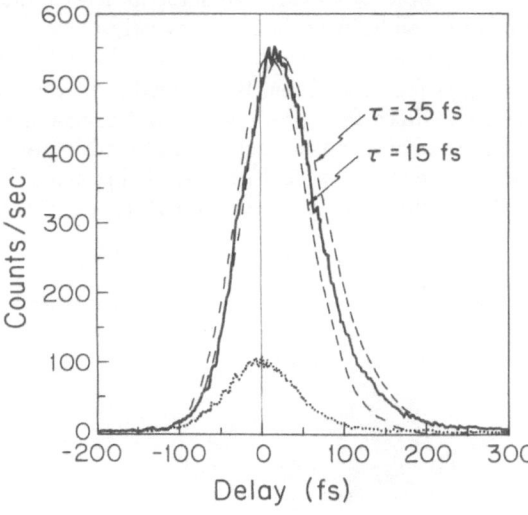

Fig. 4. Lifetime of the $n = 1$ image potential state in Ag(100). Solid lines show fits using single exponential time constants of 15 and 35 fs.

In order to obtain a more quantative estimate of the lifetime, a simple fitting proceedure was used. Oxygen was adsorbed onto the Ag surface ($\sim 3 \times 10^{-4}$ Torr s) in order to quench the image potential state lifetime and the measurement performed under identical conditions. In the limit of rapid relaxation, this provides an estimate of the ultraviolet and visible cross-correlation width and establishes the 0 delay between pump and probe. The decay time of the $n = 1$ image potential state is observed as a peak shift and broadening of the cross-correlation response. The figure shows calculated fits assuming exponential relaxation time constants of 15 and 35 fs, respectively. The accuracy of the fits is limited because the image potential relaxation time is close to our experimental resolution but nevertheless the relaxation time can be estimated to be in the range 15-35 fs. We believe that this is the first time resolved measurement of the image potential state in a metal[20].

In order to test the theoretical models of the image potential state relaxation time, it is interesting to examine the higher energy states in the Rydberg series. In Ag(100) this requires the generation of shorter wavelength ultraviolet pulses in order to populate these states by transitions from the bulk states. Tunable femtosecond pulses can be generated from our amplified femtosecond laser system using continuum generation and frequency doubling techniques[21]. Amplified pulses are focused into a thin jet of ethylene glycol to generate a broadband femtosecond continuum, using self phase modulation. A portion of the this continuum is selected and then frequency doubled using a 300 μm thick crystal of KDP. In order to populate the $n = 2$ image potential state in Ag(100), wavelengths near 570 nm are selected from the continuum to generate an ultraviolet wavelength of 285 nm or 4.35 eV. This is used as the pump beam. A pair of quartz prisms is required to compensate for dispersion due to optical elements[22], and pulse durations as short as approximately 90 femtoseconds can be obtained in the UV. These high energy photons will populate both the $n = 2$ and $n = 1$ image potential states. The image potential states can be probed by photoemission using a visible femtosecond pulse at 2.0 eV.

Transient photoemission spectra of the $n = 2$ and $n = 1$ image potential states are shown in Fig. 5. Measurements were performed using UV and visible fluences of 30 μJ/cm^2 and 0.6 mJ/cm^2, respectively. The different spectra correspond to different time delays of the probe after the ultraviolet pump pulse. The peaks at ~ 1.5 eV and 1.8 eV correspond to the $n = 1$ and $n = 2$ image potential states. Note that the spectra at longer time delays show a decrease in the ratio of $n = 1$ to $n = 2$ photoemission signal. This indicates that the $n = 1$ image potential state relaxes more rapidly than the $n = 2$ state. There is also evidence of a narrowing and shifting of the $n = 1$ line to lower energy. For these photon energies, electrons can be excited into the $n = 1$ state with nonzero transverse momentum. These higher energy electrons relax more rapidly than states with $k_\parallel = 0$.

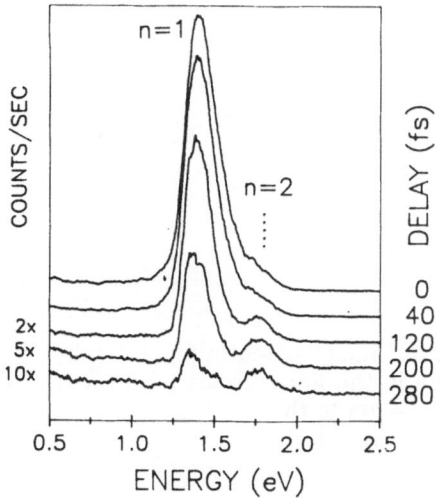

Fig. 5. Femtosecond photoelectron spectra of the $n = 1$ and $n = 2$ states in Ag(100). Spectra taken using 4.35 eV pump and 2.0 eV probe.

In order to measure the image potential state lifetimes directly, the photoemission spectrometer can be tuned to the peak of the $n = 2$ or $n = 1$ signal, and the time delay between the pump and probe pulse can be varied. The resulting measurements of $n = 2$ and $n = 1$ image potential dynamics are shown in Fig. 6. The resulting measurements show that the $n = 2$ state is significantly longer-lived than the $n = 1$ state. The cross-correlation between the visible and ultraviolet pulses was estimated by repeating the measurment and oxygen-dosing the sample or alternately, by roughening the surface using sputter ion cleaning. This decreases the lifetime of the image potential states and thereby provides an estimate of a cross-correlation and zero delay between the pump and probe.

The lifetime of the $n = 2$ state can be estimated by fitting with a simple exponential time constant. In this case, the fitting is somewhat more complicated because the linewidths of the $n = 2$ and $n = 1$ photoemission spectra overlap. Tuning the photoemission spectrometer to the $n = 2$ peak detects a portion of the $n = 1$ signal and thus, it is necessary to subtract off a component of the $n = 1$ dynamic response to obtain a fit to the $n = 2$ experimental data. In our case, an amplitude of 0.8 was chosen for the $n = 1$ response. Exponential time constants of 160, 180 and 200 fs and their fit to the $n = 2$ response are shown in the figure. The best fit is obtained by using an $n = 2$ image potential lifetime of 180 fs.

These results are consistent with previous theoretical calculations which predict an increase in the liftime of the higher order image potential states. Using a simple hydrogenic model predicts lifetimes which increase as n^3. Higher order states have an electron wave function localized further from the bulk and therefore overlap with bulk states which produce scattering processes is reduced. Recent calculations including self-energy and many-body effects predict n^3 lifetime of scaling for higher order states where $n \geq 5$. The model is most accurate for higher order states; however, for low order states the model predicts lifetime scaling of less than n^3 with the ratio of lifetimes $\tau_{n=2}/\tau_{n=1} \sim 4$. Our measured ratio of lifetimes is in the range of 4 to 8 and thus is in relatively good agreement with theoretical predictions[23].

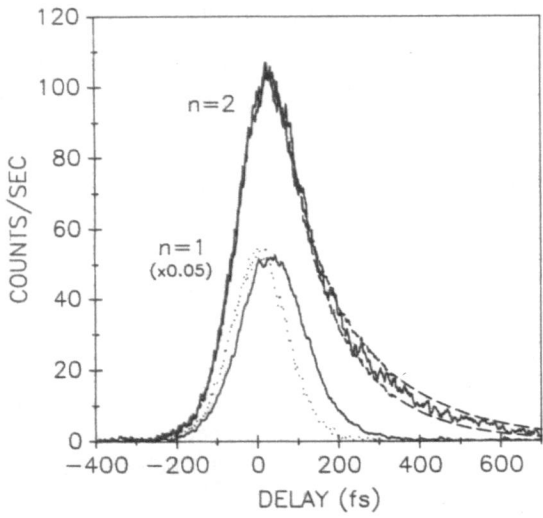

Fig. 6. Lifetime of the $n = 1$ and $n = 2$ image potential states in Ag(100). The dotted line is the normalized cross correlation. The dashed lines are fits to the $n = 2$ state using a 0.8 contribution from the $n = 1$ state and seponential time constants of 160 and 200 fs.

The Ag(100) surface is characterized by the existence of a gap of bulk states in the $k_\parallel = 0$ direction. The effect of this gap in bulk states on the image potential dyanamics can be investigated by studying other surfaces. For the Ag(111) surface, the image potential states lie near the top of the gap of bulk states. In addition, the allowed transitions are more complicated because a populated surface state is present.

Figure 7 shows a schematic diagram of the bandstructure for Ag(111) [19,24]. The work function for Ag(111) is 4.49 eV. The gap of bulk states extends from 0.31 eV below the Fermi level to 3.85 eV above the Fermi level. In contrast to Ag(100), where the vacuum is centered in a gap of bulk states, the vacuum for Ag(111) coincides with a continuum of bulk states. For Ag(111) the $n = 1$ image potential state occurs at 3.72 eV and is just below the top of the bulk state energy gap. The $n = 2$ state at 4.26 eV lies above the gap and is resonant with bulk states.

In addition to differences in the bulk state energies, the Ag(111) has surface has an occupied surface state at 0.12 eV below the Fermi level. Transitions from this surface state can significantly enhance the two photon image potential signal since the image potential state can be populated by resonant excitation from this state rather than by transitions from the bulk.

Transient photoemission spectra were measured using a 4.4 eV ultraviolet pump pulse with a 2.0 eV probe pulse. Figure 8 shows the time resolved photoemission spectra of the $n = 1$ and $n = 2$ image potential states on Ag(111). Measurements were performed with ultraviolet and visible fluences of 25 $\mu J/cm^2$ and 1.2 mJ/cm^2 respectively. Note that the relative peak heights of the $n = 1$ and $n = 2$ states remain relatively constant as the probe pulse is delayed relative to the pump.

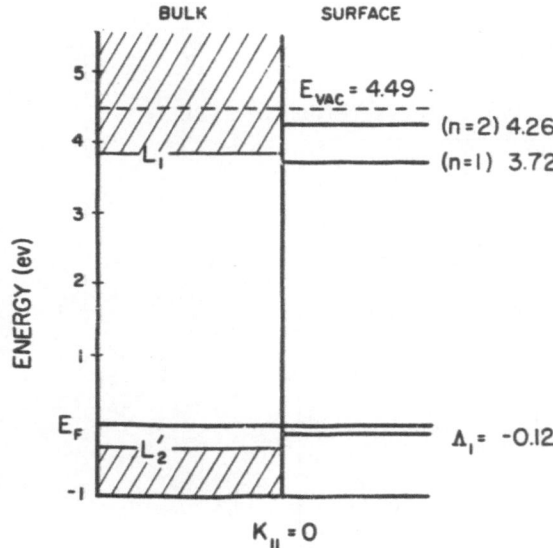

Ag (III)

Fig. 7. Schematic of energy level structure of Ag(111) surface. The energies of the $n = 1$ and $n = 2$ image states are 3.72 and 4.26 eV, respectively. The $n = 2$ state is resonant with bulk states. A surface state is present at 0.12 eV below the Fermi level.

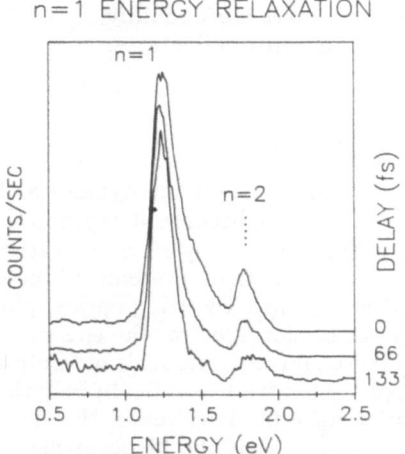

n=1 ENERGY RELAXATION

Fig. 8. Femtosecond photoelectron spectra of the $n = 1$ and $n = 2$ states in Ag(111). Spectra taken using a 4.41 eV pump and 2.0 eV probe.

Measurements of the image potential lifetimes can be performed for the $n = 1$ and $n = 2$ states by tuning the photoemission spectrometer to 1.3 eV and 1.8 eV. Figure 9 shows the corresponding measurement of $n = 1$ and $n = 2$ image potential state relaxation. In this case, note that the $n = 2$ state relaxes more rapidly than the $n = 1$. The associated lifetimes of both states are comparable to our experimental resolution and estimate to be less than ~ 20 fs.

The observed behavior is in qualitative agreement with theoretical predictions based on the many body self energy model. For the $n = 1$ state, the relaxation is dominated by inelastic, electron-hole, scattering which is determined by the penetration of the wavefunction into the bulk. For the $n = 2$ state, where the energy lies above the gap of surface, the relaxation is dominated by elastic, electron-electron, scattering which scatters the electron from the image state into resonant bulk states.

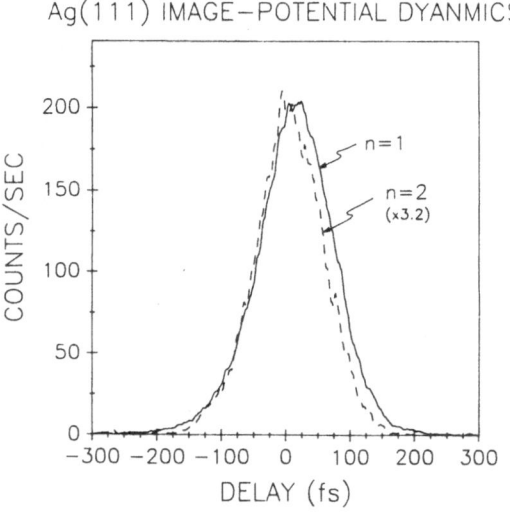

Fig. 9. Lifetime of the $n = 1$ and $n = 2$ image potential states in Ag(111). The $n = 2$ state relaxes more rapidly than the $n = 1$ because it is resonant with bulk states. Lifetimes are estimated to be less than 20 fs.

In summary, we have performed the first measurements of the dynamics of image potential states. The image potential state is of interest because it represents a two-dimensional electron gas system and in fact exhibits relatively slow relaxation times compared to other states at comparable energies in metallic systems. Studies were performed using femtosecond photoemission spectroscopy. By using optical pump and probe techniques and measuring transient photoemission spectra, the energy distribution of the excited electronic states can be measured on the time scale of femtoseconds. Results of measurements are in qualitative agreement with theoretical predictions. For the case of Ag(100), the $n = 1$ image potential state lifetime was in the range of 15 to 35 fs with the $n = 2$ lifetime 180 fs. This confirms that the higher order states in the image potential Rydberg series exhibit longer lifetimes in agreement with single particle and many body theoretical models. In contrast, for measurements performed on Ag(111), where the image potential states have energies approaching resonance with bulk states, relaxation times are extremely rapid. The $n = 2$ image potential state is resonant with bulk states and relaxes more rapidly than that of $n = 1$ suggesting that its dynamics are dominated by elastic scattering with resonant bulk states. The image potential dyanmics in Ag(111) cannot be described by the single particle wavefunction model but is commensurate with many body self energy models.

More detailed studies using angle resolved photoemission should permit the investigation of energy dispersion and intraband relaxation effects in the image potential states. Further studies using time-resolved photoemission spectroscopy can provide an opportunity to measure a wide range of electronic relaxation processes in both bulk as well as surface states. Extensions of these techniques which combine optical measurements with other surface science diagnostics hold the promise of performing a wide range of time-resolved surface studies with femtosecond resolution.

Acknowledgements

This research was supported in part by the Air Force Office of Scientific Research under Contract F49620-88-C-0089, the Joint Services Electronics Program under Contract DAAL03-89-C-0001, the AT&T Foundation, and a grant from IBM. RWS gratefully acknowledges support from the Newport Corporation. JGF gratefully acknowledges support from the National Science Foundation Presidential Young Investigator Award 8552701-ECS.

References

1. M. W. Cole and M. H. Cohen, "Image-potential induced surface bands in insulators," Phys. Rev. Lett. 23:1238 (1969).

2. P. M. Echenique and J. B. Pendry, "The existence and detection of Rydberg states at surfaces," J. Phys. C. 11:2065 (1978).

3. W. Steinmann, "Spectroscopy of image-potential states by two-photon photoemission," Appl. Phys. A 49:365 (1989).

4. W. Shockley, "On the surface states associated with a periodic potential, Phys. Rev. 56:317 (1939).

5. P. D. Johnson and N. V. Smith, "Image-potential states and energy-loss satellites in inverse photoemission spectra," Phys. Rev. B 27:2527 (1983).

6. D. Straub and F. J. Himpsel, "Identification of image-potential surface states on metals," Phys. Rev. Lett. 52:1922 (1984).

7. N. Garcia, B. Reihl, K. H. Frank, and A. R. Williams, "Image states: Binding energies, effective masses, and surface corrugation," Phys. Rev. Lett. 54:591 (1985).

8. D. Straub and F. J. Himpsel, "Spectroscopy of image-potential states with inverse photoemission, Phys. Rev. B 33:2256 (1986).

9. K. Giesen, F. Hage, F. J. Himpsel, H. J. Riess, and W. Steinmann, "Two-photon photoemission via image-potential states," Phys. Rev. Lett. 55:300 (1985).

10. K. Giesen, F. Hage, F. J. Himpsel, H. J. Riess, and W. Steinmann, "Hydrogenic image-potential states: A critical examination, Phys. Rev. B 33:5241 (1986).

11. K. Giesen, F. Hage, F. J. Himpsel, H. J. Riess, and W. Steinmann, "Binding energy of image-potential states: Dependence on crystal structure and material," Phys. Rev. B 35:971 (1987).

12. P. M. Echenique, F. Flores, and F. Sols, "Lifetime of image surface states," Phys. Rev. Lett. 55:2348 (1985).

13. P. de Andrés, P. M. Echenique, and F. Flores, "Lifetime in a two-dimensional image-potential-induced electron band," Phys. Rev. B 35:4529 (1987).

14. P. de Andrés, P. M. Echenique, and F. Flores, "Calculation of the lifetimes for intermediate Rydberg states," Phys. Rev. B 39:10356 (1989-I).

15. J. A. Valdmanis, R. L. Fork, and J. P. Gordon, "Generation of optical pulses as short as 27 femtoseconds directly from a laser balancing self-phase modulation, group velocity dispersion, saturable absorption and saturable gain", Opt. Lett. 10:131 (1985).

16. W. H. Knox, M. C. Downer, R. L. Fork, and C. V. Shank, "Amplified femtosecond optical pulses and continuum generation at 5-kHz repetition rate," Opt. Lett. 9:552 (1984).

17. N. E. Christensen, "The band structure of silver and optical interband transitions," Phys. Stat. Sol. B 54:551 (1972).

18. B. Reihl, K. H. Frank, and R. R. Schlittler, "Image-potential and intrinsic surface states on Ag(100)," Phys. Rev. B 30:7328 (1984).

19. A. Goldmann, V. Dose, and G. Borstel, "Empty electronic states at the (100), (110), and (111) surfaces of nickel, copper, and silver," Phys. Rev. B 32:1971 (1985).

20. R. W. Schoenlein, J. G. Fujimoto, G. L. Eesley, and T. W. Capehart, "Femtosecond studies of image-potential dynamics in metals," Phys. Rev. Lett. 61:2596 (1988).

21. R.L. Fork, C.V. Shank, C. Hirlimann, and R. Yen, "Femtosecond white-light continuum pulses," Opt. Lett. 8:1 (1983).

22. R. L. Fork, O. E. Martinez, and J. P. Gordon, "Negative dispersion using pairs of prisms," Opt. Lett. 9:150 (1984).

23. R. W. Schoenlein, J. G. Fujimoto, G. L. Eesley, and T. W. Capehart, "Femtosecond dynamics of the $n = 2$ image-potential state on Ag(100)," Phys. Reb. B, rapid communications (in press).

24. S. L. Hulbert, P. D. Johnson, N. G. Stoffel, and N. V. Smith, "Unoccupied bulk and surface states on Ag(111) studied by inverse photoemission," Phys. Rev. B 32:3451 (1985).

High-Intensity, Ultrashort Pulse Laser Heated Solids

R.W. Falcone, M.M. Murnane and H.C. Kapteyn

Department of Physics
University of California at Berkeley
Berkeley, CA 94720

Intense pulses of laser light with a duration of about 100 femtoseconds can be focused onto the surface of a solid to produce high energy electrons and highly ionized material just inside the surface.[1,2,3] Immediately following the laser pulse the electrons will cool and rapidly recombine due to both thermal conduction to the bulk of the solid and inelastic collisions. The density of the solid will also rapidly decrease due to expansion of the high temperature material.

This process was recently demonstrated using a colliding-pulse-mode-locked laser which was amplified to an energy of 3.5 mJ at a pulse length of about 160 fsec.[1,2] Laser pulses were focused onto a solid target at power densities in excess of 10^{16} W cm^{-2}. Peak electron temperatures of approximately 350 eV were produced in a silicon target and soft x-ray emission was observed. Reflectivity measurements confirmed that laser energy was coupled into a near-solid density plasma. Experimental reflectivity values are in agreement with the predictions of a collisionally damped, free electron model of the dielectric constant of the solid, assuming an ionization fraction given by equilibrium calculations. An x-ray streak camera was used to measure the duration of the x-ray emission.[4] The temporal response of the streak camera is 2 picoseconds; assuming a reasonable instrumental resolution, the x ray pulse duration (thus the lifetime of the high temperature, solid density plasma) is on order of 1 picosecond or less.

Figure 1 shows the soft x-ray emission as a function of time. Accurate calibration of the time axis was obtained using fiducials obtained from a portion of the laser pulse which was then focused onto the x-ray streak camera photocathode.

Figure 2 shows the reflectivity of a short pulse and a long pulse laser as a function of energy fluence on a flat silicon target . The high reflectivity of the short pulse (a) indicates that the reflection is coming from a sharp interface between the vacuum and a high density, high temperature plasma. Longer laser pulses exhibit lower reflection from the lower density plasma blow-off formed in front of the solid.

Figure 3 is the resulting spectrum from 2 nm to 12 nm from the short pulse laser heated silicon. Both line and continuum emissions are evident.

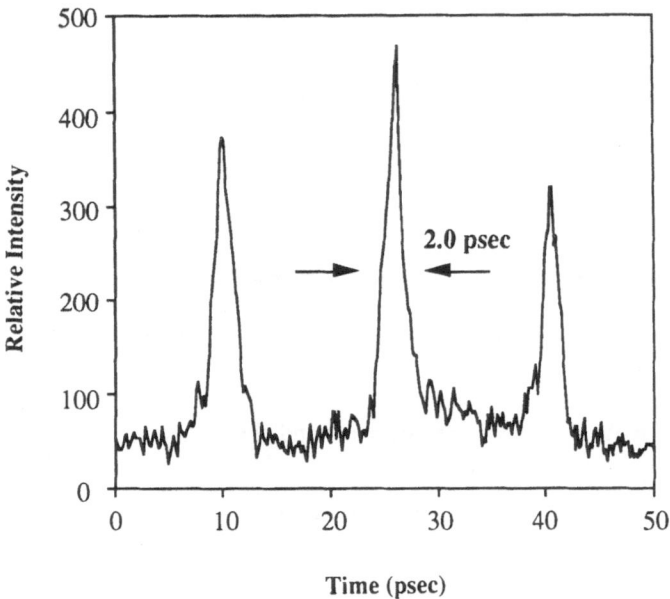

Fig. 1 X-ray streak camera data for emission from short pulse laser heated silicon. The x-ray pulse is in the center, bracketed by two timing fiducials obtained from the laser pulse.

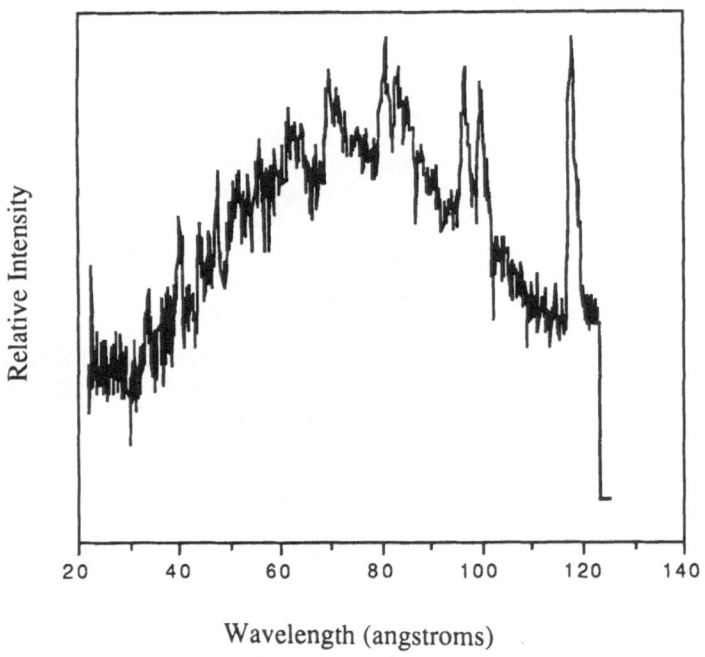

Fig. 2 Emission spectrum of silicon heated by the 160 fsec laser pulse at high intensity.

The use of more intense laser pulses (to produce more and shorter wavelength x-rays) will be limited by the reduced coupling of the laser into the solid as the electrons are made hotter and more numerous by increased heating. The collisionality of the electrons decreases at higher energies and the plasma frequency increases at higher density, leading to both higher reflectivity and reduced absorption. A solution to this dilemma appears to be the use of structured surfaces to lower the effective dielectric constant at the surface. For example, if the surface has grooves with characteristic dimensions less than a wavelength of the laser light, and the polarization of the laser is orthogonal to the grooves, the effective

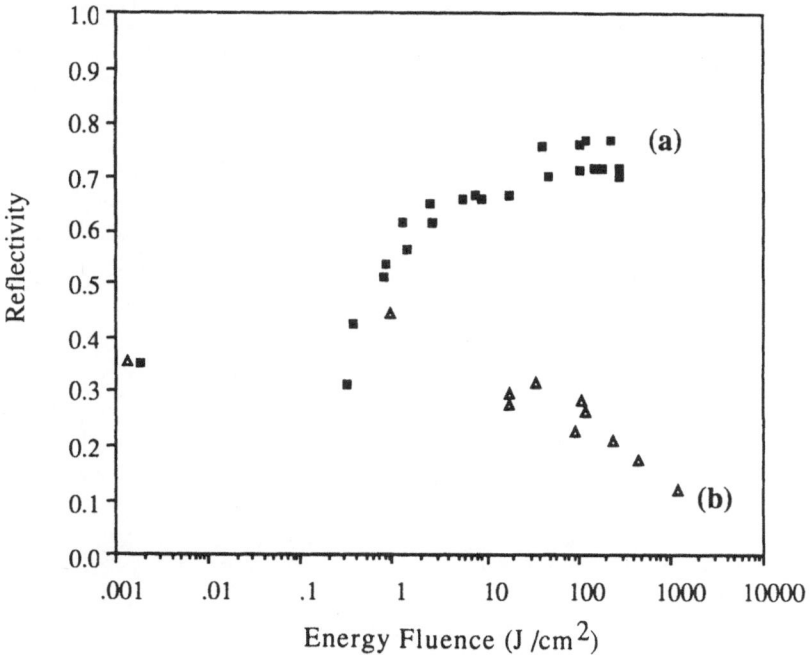

Fig. 3 Reflectivity of a silicon target as a function of energy fluence. (a) Laser pulse with a duration of 160 fsec. (b) Laser pulse with a duration of 7 nsec.

dielectric constant and the Fresnel reflectivity of the surface are reduced and the laser fields penetrate into the waveguide type structure of the grooves. New absorption mechanisms may now be invoked since the electric field is now orthogonal to the interior surface of the grooved target.

This work was supported by the U.S. Air Force Office of Scientific Research, the National Science Foundation, and through a collaboration with Lawrence Livermore National Laboratory under the auspices of the U.S. Department of Energy under contract #W-7405-ENG-48. M.M. Murnane acknowledges support from a University of California President's Postdoctoral Fellowship.

References

1. M.M. Murnane, H.C. Kapteyn, R.W. Falcone, "High-Density Plasmas Produced by Ultrafast Laser Pulses," Phys. Rev. Lett. **62**, 155 (1989).

2. M.M. Murnane, H.C. Kapteyn, R.W. Falcone, "Generation and Application of Ultrafast X-Ray Sources," IEEE Jour. Quant. Elect. **25**, 2417 (1989).

3. H.M. Milchberg, R.R. Freeman, S.C. Davey and R.M. More, "Resistivity of a Simple Metal from Room Temperature to 10^6 K," Phys. Rev. Lett. **61**, 2364 (1988).

4. M.M. Murnane, H.C. Kapteyn, R.W. Falcone, "X-Ray Streak Camera with 2 Picosecond Response," (to be published in Applied Physics Letters).

CONTROL OF TRANSVERSAL INTERACTIONS IN NONLINEAR OPTICS: NEW SPATIO-TEMPORAL EFFECTS IN NONLINEAR WAVE DYNAMICS

S. A. Akhmanov, M. A. Vorontsov, A. V. Larichev

Department of Physics, Moscow State University, 119899 Moscow
USSR

ABSTRACT

Control of the topology and the scale of transverse interactions leads to a new class of spatio-temporal light beam instabilities in media with cubic nonlinearity.

In this paper we present the results of theoretical and experimental investigations of instabilities and spatio temporal nonlinear wave structures which are driven by large scale coherent transverse interactions.

In Refs. 1 and 2 we have suggested and implemented some arrangements with so-called two-dimensional (2D) feedback, in which the characteristic scale of transverse interactions $L_\perp \sim d$, where d is the beam width. Depending on the parameter values of the nonlinear medium and the initial boundary conditions, a complete nonlinear wave dynamics hierarchy including steady-state dissipative structures, rotating waves (optical reverberators), spiral waves, and optical turbulence can be observed. With optical reverberators we observed the new effect of light field nonlinear dynamics - the hysteresis of rotating nonlinear structures (see also Ref. 3). The theory, based on the nonlinear parabolic equation with a shifted spatial argument, explains the experimental data quantitatively. Some very new experimental results on spatially nonlinear structures, locked by the external light, are also presented. Optical turbulence scenarios in systems with 2D feedback are discussed in comparison with 1D dynamical optical chaos. Among the applications of the considered phenomena the generation of classical spatial squeezed states and optical modeling of neural networks will be mentioned (Ref. 4).

REFERENCES

1. S. A. Akhmanov, M. A. Vorontsov, D. V. Pruidze, and V. I. Shmalgausen, preprint No. 33, Department of Physics, Moscow State University, (1986).
2. S. A. Akhmanov, M. A. Vorontsov, and V. Yu. Ivanov, JETP Letters, 47, 707 (1988).
3. S. A. Akhmanov, M. A. Vorontsov, and A. V. Larichev, Proceedings of the VI Rochester Conference on Quantum Optics, Rochester, NY, (1989).
4. S. A. Akhmanov, M. A. Vorontsov, A. S. Chirkin, et al., new physical principles of optical signal processing, (Moscow, Nauka 1989); English translation: Cambridge University Press, Cambridge, 1990.

SYNCHRONIZATION OF ATOMIC QUANTUM TRANSITIONS BY LIGHT PULSES

V. P. Chebotayev, V. A. Ulybin, V. M. Klementiev, O. I. Pyltzyn,
and V. F. Zakhariash

Institute of Thermophysics
Siberian Branch of the USSR Academy of Sciences
Novosibirsk-90, SU-630090
USSR

A new spectroscopic method based on using pairs of pulses of radiation with a linewidth which is broader than the measured quantum-transition frequency interval but with a stable interpulse time has been considered in [1]. The duration of the interaction between a particle and the single pulse is shorter than the oscillation period of the quantum transition. Hence, the result of the atomic interaction with two short pulses is defined by the phase of free oscillations of a dipole moment at time of arrival of the pulse. In other words, the particle makes the transition from one energy level into the other synchronously with the atomic oscillations. The synchronized quantum transitions are very accurately determined in time and may be applied to precision direct measurements of time and frequency.

1. Qualitative Analysis

The phenomenon of synchronization of quantum transitions is not so sensitive to the nature of a pulse perturbation. Hence, we shall explain it by a simple example of an atomic interaction with two DC pulses of an electric field. We write the last one in the form

$$E(t) = E\, g(t) + E'\, g(t-T), \tag{1}$$

where $g(t)$ is the pulse shape, and T is the interpulse time. The field-induced dipole moment of a two-level atom may be written in the form

$$d(t) = a_{21} d_{21}^{*} d_{21} \exp(-i\omega_{21}t) + c.c., \tag{2}$$

where a_{21}, d_{21} and ω_{21} are the probability amplitude, the dipole moment and the frequency of the transition $|1> \to |2>$, respectively (the atom is considered to be in the state $|1>$ at $t = -\infty$). Both $d(t)$ and d_{21} are assumed to be projections onto the field direction. According to [2] we have

$$a_{21} = (i/h) d_{21} \int_{-\infty}^{t} dt'\, E(t') \exp(i\omega_{21}t'). \tag{3}$$

If the duration τ of the atomic interaction with a single pulse is much shorter than the period of atomic oscillations $2\pi/\omega_{21}$, then the function $g(t)$ in (1) and (3) may be replaced by $\tau\delta(t)$ where $\delta(t)$ is Dirac's delta-function. This means that we probe the atomic dipole moment at times t = 0 and t = T. For t > T we find from (1)-(3) the dipole moment as the superposition of dipole moments excited at t = 0 and t = T

$$d(t) = id_{21}^* \tau \Omega_{21} \exp(-i w_{21} t)$$
$$\times \left[1 + (E'/E) \exp(i w_{21} T) \right] + c.c., \tag{4}$$

where $\Omega_{21} = E d_{12}/h$, and the probability of the transition $|1> \to |2>$

$$|a_{21}|^2 = (\tau |\Omega_{21}|)^2 \left[1 + (E'/E)^2 + 2(E'/E) \cos w_{21} T \right]. \tag{5}$$

As we see from (4) and (5), both $d(t)$ and $|a_{21}|^2$ have maxima at times T that are multiples of the period $2\pi/w_{21}$. Thus, when an atom interacts with two short pulses the quantum transition $|1> \to |2>$ may be synchronized with its natural oscillations. The effect is also shown to be very distinctive with equal pulse amplitudes (E' = E).

Instead of DC pulses of electric or magnetic fields, AC pulses may be used as well. The case of AC pulses is of interest for optical transitions. It is clear, that a carrier frequency of AC pulses may be of any value satisfying the inequality $w \gg \tau^{-1}, w_{21}$. There are only two pulse parameters of principal importance, i.e. the duration of a perturbation pulse should be shorter than an atomic-oscillation period, and the interpulse time should be stable and a multiple of the period.

The aim of the present work is to consider the phenomenon of synchronization of quantum transitions by light pulses. At the present time, the advanced methods of ultrashort laser pulse generation provide the possibility of obtaining light pulses of 10 - 100 fs duration [3]. This allows us to carry out the experiments for synchronization of IR and FIR quantum transitions and the direct measurements of time with an accuracy of the order of $10^{12} - 10^{13}$. As the light frequency $w \gg w_{21}$, the excitation of the quantum transition is realized by the two-photon Raman process. We start the consideration with the analysis of the Raman interaction between an atom and a single light pulse of any intensity. This analysis is of independent interest.

2. Interaction Between an Atom and an Ultrashort Light Pulse

Let $|1>$ and $|2>$ be the ground and metastable atomic states which obey the selection rules for a two-photon transition, see Fig. 1. We write the light pulse in the form

$$E(t) = E g(t) \exp(-i w t) + c.c.,$$

where 2E is the amplitude of the electromagnetic wave pulse. The frequency w is nonresonant to the intermediate transitions $|1> \to |\alpha>$ and $|2> \to |\alpha>$. The duration of the pulse is the form

$$\tau = \int_{-\infty}^{\infty} dt \, |g(t)|^2,$$

and obeys $w^{-1} \ll \tau \lesssim w_{21}^{-1}$. We assume that the light pulse is of a symmetric shape $g(-t) = g(t)$.

The equations for the density matrix elements which describe the stimulated Raman scattering in the field $E(t)$ were reduced as in [4] to the equations for a two-level atom in an effective nonoscillating field

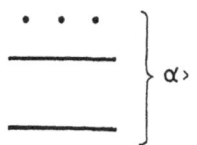

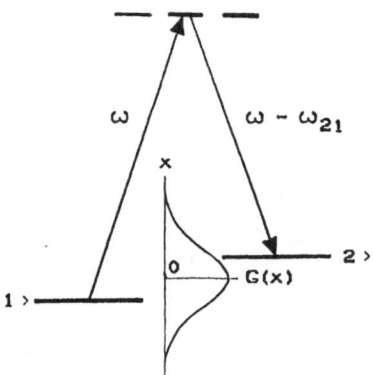

Fig. 1. Two-photon Raman interaction between an atom and a single ultrashort light pulse. G(x) is the Fourier transform of the pulse shape $|g(t)|^2$, x is a frequency.

$$\left\{ d/dt + i\left[\omega_{21} - |g(t)|^2\Delta \right] + \Gamma \right\}\rho_{21} = (i/2)\Omega|g(t)|^2(\rho_{11} - \rho_{22}),$$

$$(d/dt + 2\Gamma)\rho_{22} = (-i/2)\Omega|g(t)|^2\rho_{21} + c.c.,$$ (6)

$$\rho_{11} + \rho_{22} = 1,$$

where 2Γ is the spontaneous decay rate of the upper state $|2\rangle$, $\Delta = E^2(D_{22} - D_{11})$ is the difference of the optical Stark shifts of the levels $|2\rangle$ and $|1\rangle$, $\Omega = 2E^2D_{21}$ is the effective (two-photon) Rabi frequency, $D_{ik} = D_{ik}(\omega) + D_{ik}(-\omega)$, and $D_{ik}(\omega) = \sum_\alpha d_{i\alpha}d_{\alpha k}h^{-2}(\omega_{\alpha 1} - \omega)^{-1}$ is two-photon matrix element.

The following calculations will be carried out with $\Delta = 0$ in (6) because the Stark light shift Δ is small in comparison with the effective Rabi frequency Ω (for close atomic levels $|2\rangle$ and $|1\rangle$, $D_{22} - D_{11} \ll D_2$).

Before the interaction at time $t = 0$ with the light pulse an atom was in the ground state $|1\rangle$. Hence, using the initial condition $\rho_{11}(-\infty) = 1$ we find the coherence and the upper-level population probability.

$$\rho_{21}(t) = (i/2)\exp\left[-(\Gamma + i\omega_{21})t\right]\sin 2\theta,$$ (7)

$$\rho_{22}(t) = \exp(-2\Gamma t)\sin^2\theta,$$ (8)

where $t \gg \tau$,

$$\theta = (\Omega/2)\, G(\omega_{21}),$$

$$G(x) = \int_{-\infty}^{\infty} dt\, |g(t)|^2\exp(ixt).$$

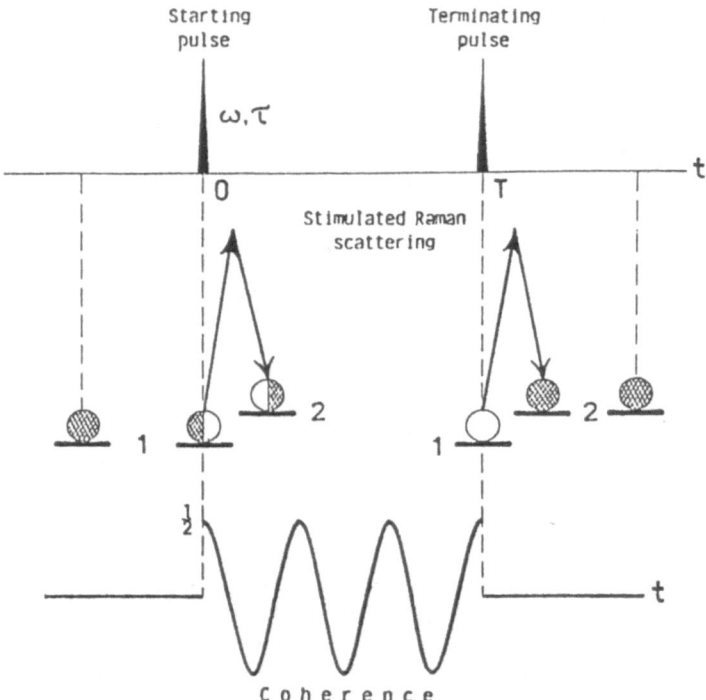

Fig. 2.   Interaction between an atom and two ultrashort light pulses
($\theta = \pi/4$).

The parameter θ determines the power of the pulse perturbation: the case
of $\theta \ll 1$ ($\Omega \ll \omega_{21} \lesssim \tau^{-1}$) corresponds to a weak perturbation and the
case $\theta \simeq 1$ ($\Omega \simeq \tau^{-1} \gtrsim \omega_{21}$) to a strong one. The function $G(\omega_{21})$ is the
Fourier transform of the light-pulse shape at the frequency $\tilde{\omega}_{21}$. It
determines the atom-field interaction efficiency.

In the case of the Gaussian pulse $|g(t)|^2 = \pi^{-1/2} \exp(-t^2/\tau^2)$ we have
$G(\omega_{21}) = \tau \exp\left[-(\omega_{21}\tau/2)^2\right]$. The atom-field interaction is shown to be
efficient if $\tau \lesssim 2/\omega_{21}$. Increasing the pulse duration τ leads to the
vanishing of the spectral components at $\omega \pm \omega_{21}$ in a light pulse, so the
two-photon interaction becomes impossible.

For the rectangular light-pulse shape we have $G(\omega_{21}) = \sin(\omega_{21}\tau/2)/(\omega_{21}/2)$. The efficiency is high for $\tau = \pi(2n-1)/\omega_{21}$, with n
an integer here and elsewhere in this paper. The result of the
interaction of the atom with a light pulse of duration $\tau = \pi/\omega_{21}$ (n = 1)
is the same as in the case of a long pulse (n $\gg$ 1). It means that only
sharp fronts of the light pulse may give rise to the two-photon Raman
process. The upper limit of τ will be defined by an accuracy, which is
required for the time localization of the atomic transition $|1\rangle \to |2\rangle$.

In this analysis we neglected the one-photon excitation of the
intermediate levels $|\alpha\rangle$, see Fig. 1. This process reduces the number of
atoms involved in the two-photon process. For $\omega_{\alpha 1} - \omega \gg \tau^{-1}$ the α-level
population is of the form

$$\rho_{\alpha\alpha} = (\Omega_{\alpha 1}^2/2) \, \gamma_{\alpha 1}\tau(\omega_{\alpha 1} - \omega)^{-2},$$

where $\Omega_{\alpha i}$ and $\gamma_{\alpha i}$ (i = 1, 2) are the one-photon Rabi frequency and the spontaneous decay rate of the transition $|\alpha\rangle \rightarrow |i\rangle$, respectively. The population ρ_{22} induced by the two-photon process in a weak field is $\rho_{22} \approx \theta^2$, see (8). So, the inequality $\rho_{22} \gg \sum_\alpha \rho_{\alpha\alpha}$, which corresponds to suppressing the one-photon process, is reduced to

$$\Omega_{\alpha 2}^2 \gg \tau\gamma_{\alpha 1}G^2(\omega_{21}) \sim \tau\gamma_{\alpha 1}\omega_{21}^2,$$

where α is referred to the intermediate level, which may be used to estimate the two-photon matrix element D_{21}.

To conclude this discussion we note the following. First, the effect of the Stark light shift which we neglected above may be shown to be distinct in the strong field alone and leads to a decrease in the dynamical variation of the coherence amplitude $|\rho_{21}|$ and the probability ρ_{22} in proportion to the ratios Ω/Ω_0 and $(\Omega/\Omega_0)^2$, respectively, with $\Omega_0 = (\Omega^2 + \Delta^2)^{1/2}$. Second, equations (7) and (8) may be applied to the interaction between an atom and an ultrashort $(\tau \leq \omega_{21}^{-1})$ DC pulse of an electric field as well (for forbidden transitions). To use them one should equate the frequency ω in the two-photon matrix elements $D_{ik}(\omega)$ to zero.

3. Interaction Between an Atom and Two Light Pulses

From among the physical quantities which describe a light field only the field power $\sim E^2|g(t)|^2$ appears in (6). So, the interaction between an atom and a pair of the time-separated pulses will depend on the time delay T but not on the difference of their optical phases.

We have considered the excitation of an atomic natural oscillation with the frequency ω_{21} by one ultrashort light pulse, see (7). Another light pulse delayed by the time T (terminating pulse) will interact with the atom, when the phase of the oscillations equals $\omega_{21}T$. It is the phase that determines the result of the atomic interaction with the pair of ultrashort pulses.

Using (7) and (8) at the time t = T as initial conditions for the interaction between the atom and the second light pulse we find from (6) the coherence and the upper-level population probability in the forms

$$\rho_{21}(t) = (i/2)\sin2\theta \exp\left[-(\Gamma + i\omega_{21})(t - T)\right]$$
$$\times \left\{1 + \exp\left[-(\Gamma + i\omega_{21})T\right] - 2\sin^2\theta \exp(-\Gamma T)\right.$$
$$\left. \times \left[\exp(-\Gamma T) + \cos \omega_{21}T\right]\right\}, \tag{9}$$

$$\rho_{22}(t) = \sin^2\theta \exp\left[-2\Gamma(t - T)\right]\left[1 + \exp(-2\Gamma T)\cos2\theta\right.$$
$$\left. + (1 + \cos2\theta)\exp(-\Gamma T)\cos \omega_{21}T\right], \tag{10}$$

where t - T $\gg \tau$. We see that the variation of the density matrix during the small time τ of the terminating pulse duration is synchronous with the atomic oscillations at the frequency ω_{21}. This is considered as the synchronized stimulated quantum transition. There is a peculiarity of the transitions induced by the weak and strong perturbations, hence we consider them separately.

For a field which is too weak to saturate the two-photon transition $|1\rangle \rightarrow |2\rangle$ ($\theta \ll 1$), we see from (7) and (8) that the starting pulse

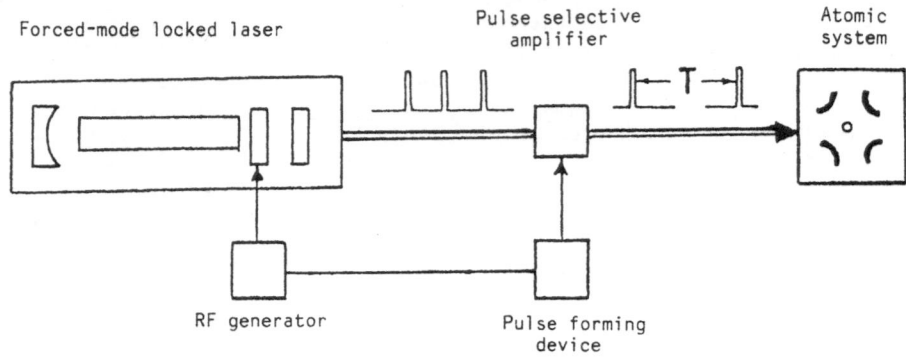

Fig. 3. A possible spectrometer scheme to observe the synchronized quantum transitions.

induces the atomic coherence ρ_{21} ~ $E^2 \tau$ and probability $\rho_{22} = |\rho_{22}|^2$. If the second pulse delayed in time arrives synchronously with the atomic oscillations $(T = 2\pi n/\omega_{21})$, then the amplitude of the previous one is doubled (for $\Gamma T \ll 1$) and the upper level population probability is quadrupled, see (9) and (10), due to the interference of the probability amplitudes. If a delay time is equal to a half-integer multiple of the atomic periods $(T = \pi(2n-1)/\omega_{21})$, the natural oscillations are annihilated and the atom returns into the ground state. Equations (9) and (10) for a weak field are similar to (4), (5).

For the strong light field $(\Omega \simeq \tau^{-1} \gtrsim \omega_{21})$ we emphasize the case of $\pi/4$-pulse $(\theta = \pi/4)$ which yields the maximum amplitude for the variations of the synchronized density matrix. From (7) - (8) we see that the starting pulse induces the probability $\rho_{22} = 1/2$ and atomic oscillations with the amplitude $|\rho_2| = 1/2$. The terminating pulse arriving synchronously with the atomic oscillations $(T = 2\pi n/\omega_{21})$ doubles the upper-level population probability and returns the atom into the ground state if $T = \pi(2n-1)/\omega_{21}$, with the natural oscillations being annihilated in both cases, see Fig. 2.

4. Atomic Ensemble: Rarefied Gas

To find the upper-level population for an atomic ensemble we must take into account the atomic motion and average the population probability $\langle \rho_{22} \rangle$ over coordinates and velocities. Atomic motion is accounted for by replacing the value T on the right-hand side of (10) by the interpulse time $T_a = T + c^{-1}[z(T_a) - z_0]$ in the rest frame of the moving atom, where z_0 and $z(T_a)$ are the atomic coordinates along the light wave propagation direction at the times of the starting and terminating ultrashort pulses, respectively. The ensemble-averaging of the upper-level population probability $\langle \rho_{22} \rangle$ is reduced to averaging the oscillating factor $\cos \omega_{21} T_a$.

For a free atom $z(T_a) - z_0 \approx v_z T$, hence after the averaging over velocities v_z with the equilibrium distribution function we find

$$\langle \cos \omega_{21} T_a \rangle = \cos \omega_{21} T \exp\left[-(T\delta/2)^2\right] , \qquad (11)$$

where $\delta = \omega_{21} v_0/c$ is the Doppler shift at the Raman frequency, with v_0 being the thermal velocity. We see that for a rarefied gas the observation of the quantum transitions synchronized with the atomic oscillations at the frequency ω_{21} is limited by the time interval $T_1 \lesssim \delta^{-1}$ which yields an accuracy of time measurements of order $(\tau\delta)^{-1} \simeq c/v_0$.

The Doppler effect influence may be decreased by elastic collisions in a sufficiently dense gas [5]. In this case the diffusive distance which a particle covers during the interpulse time should be of the order of the wave-length associated with the atomic transition $|1> \rightarrow |2>$.

Note that in an atomic ensemble a spin-echo-like effect may be observed, i.e., the coherence $\rho_{21}(t)$ is seen to contain a term which is proportional to $\exp\left[-i\omega_{21}(t - 2\tau)\right]\sin2\theta\sin^2\theta$, see (9).

5. Ensemble of Trapped Atoms

During the interval between the starting and terminating pulses, an atom trapped in a finite volume may be displaced not more than the macroscopic oscillation amplitude v_{zmax}/ω_z, where ω_z is the frequency of atomic oscillations in a trap. Ensemble-averaging the factor $\cos \omega_{21}T_a$ we see that the parameter $\mu = \delta/\omega_z$ arises instead of $T\delta$ in (11). Furthermore, the Doppler effect does not limit the delay time T if the latter is a multiple of the macroscopic oscillation period. In this case the terminating pulse finds an atom at the same point as the starting one.

Let an atomic ensemble be trapped in a harmonic potential, with the principal axis of trap symmetry coinciding with the light direction Z. The equation of motion of a single atom is of the form

$$z(t) = z_0\cos \omega_z t + (v_{z0}/\omega_z)\sin \omega_z t.$$

Averaging over the initial coordinates z_0 and velocities v_{z0} we find

$$<\cos \omega_{21}T_1> = \cos \omega_{21}T \exp\left[-\mu^2\sin^2(\omega_z T/2)\right].$$

This confirms our statements above.

6. Laser Spectrometer Scheme

In the method under consideration the light source has a very large linewidth ($\Delta\omega \simeq 1/\tau$), and the value of its carrier frequency is unimportant. Only the stability of the pulse delay time is significant. There are at least two possibilities to realize the spectrometer. In the first case, an optical delay line may be used to form the terminating pulse. The delay time $T = L/c$ (where L is the length of the delay line) may be changed continuously with high accuracy. Unfortunately, the absolute accuracy of the delay time measurement is limited here by the accuracy of the length measurement. If the transition frequency is known, then according to the new definition of the meter, the delay time T and the length L may be directly measured. The second possibility is given by the laser spectrometer scheme shown in Fig. 3. This spectrometer allows one to measure the absolute values of both the delay time T and the transition frequency ω_{21}. The spectrometer is based on the use of ultrashort pulses generated by the forced-mode locked laser. The pulse repetition frequency is determined by the frequency ν of the RF generator, which controls the intracavity amplitude modulator. The optical pulse amplifier locked to the RF generator allows one to form the time-separated pulses. Their delay time will be a multiple of the laser interpulse time, i.e., $T = n\nu^{-1}$. Usually $\nu \sim 10^8$ Hz. Time tuning may be

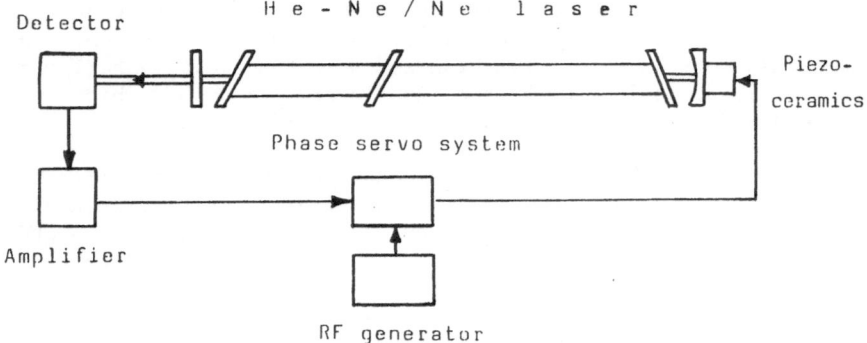

Fig. 4. Experimental setup for stabilizing the interpulse time.

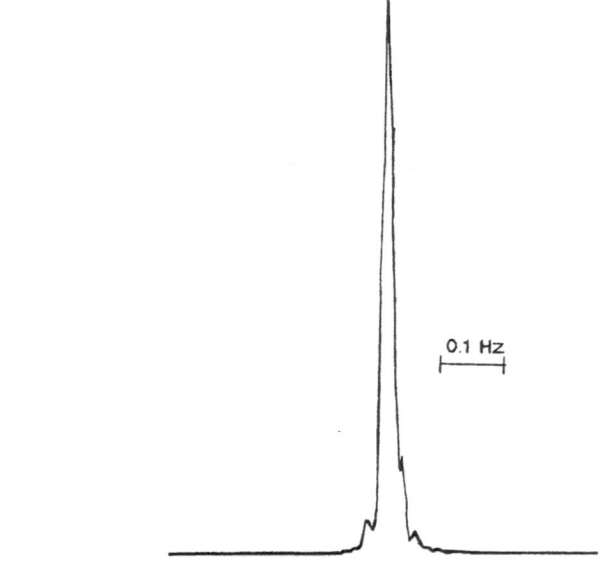

0.1 Hz

Fig. 5. Spectral signal of intermode beats for the stabilized, self-mode-locked laser.

realized by tuning the frequency ν. Such a system may serve as a standard for time and frequency simultaneously. As the frequency ω_{21} is stable, it is possible to stabilize the delay time T and consequently the Rf generator frequency ν.

We have considered another possibility to produce short light pulses with stabilized interpulse time. The possibility is based on the phase locking of a pulse repetition frequency by an external RF generator with a very high frequency stability. The experimental setup is shown in Fig. 4. We used the self-mode-locked He-Ne/Ne laser of 4.8 m length. The laser radiation was detected and the signal at intermode beat frequency used in a phase servo system to produce an error signal by comparing it with a signal from an external frequency synthesizer locked to the Rb frequency standard. The tuning of the laser frequency was realized by a piezo-ceramics.

The spectrum width of beats at 28.5 MHz for the He-NE/Ne laser in a free-running regime was approximately 50 kHz, and in a self-mode-locking regime we observed a 30 Hz width with the total number of modes being 43-45. Using the phase servo system we achieved the stabilization of the intermode beat frequency of the self-mode-locked He-Ne/Ne laser and consequently stabilized the interpulse time. The spectrum width turned out to be equal to the apparatus function width (0.02 Hz) of a spectrum analyzer, see Fig. 5. The intermode beat frequency was measured to be 28463806.90 Hz, that correspond to the interpulse time $T = 3.513233502 \times 10^{-8}$ s.

Conclusion

We have shown that for a single atom or an atomic ensemble the dynamics of the natural Raman oscillations may be investigated by the synchronization of atomic quantum transitions with these oscillations. This synchronization may be realized, for example, by time-separated ultrashort light pulses. The proposed effect may be used in super-high precision frequency measurements, developing new foundations for the standards and magnetometers, the measurement of atomic spectroscopy constants, the selective excitation of atoms and molecules, etc. The possibility of using this phenomenon to develop high-speed atomic systems is also of interest.

References

1. V. P. Chebotayev, Pis'ma JETF, _49_, 429 (1989).
2. L. D. Landau, E. M. Lifshitz; _Quantum Mechanics_, (Nauka, Moscow, 1974), p. 177.
3. S. A. Akhmanov, V. A. Vysloukh, A. S. Chirkin, _Optics of Femtosecond Laser Pulses_. (Nauka, Moscow, 1988); W. Kaiser (ed.), _Ultrashort Laser Pulses and Applications_, Topics Appl. Phys. _60_ (Springer, Berlin, Heidelberg, 1988).
4. E. V. Baklanov, B. Ya. Dubetsky, Kvantovaya Elektronika, _5_, 99 (1978).
5. R. H. Dicke: Phys. Rev. _89_, 472 (1953).

CHAOS IN NONLINEAR OPTICS

Robert W. Boyd and Alexander L. Gaeta

Institute of Optics
University of Rochester
Rochester, N.Y. 14627

There has recently been great interest in the field of deterministic chaos. By deterministic chaos, one refers to the situation in which the output of a physical system fluctuates erratically in time, even though the system is governed by deterministic equations. It is known theoretically that chaotic behavior can occur only in nonlinear systems. We have been particularly interested in studying chaos in nonlinear optical systems.[1] Our motivation for studying chaos in nonlinear optical systems has been two-fold: One is that studies of deterministic chaos can provide some insights into the origin of uncertainty in physics, and optical systems provide good systems in which to perform exacting studies of such effects. The other reason is that chaos can lead to limitations in the performance of practical optical devices. To illustrate this latter point, we consider some hypothetical optical device which was intended to provide a steady output but which instead produces the highly erratic output illustrated in Fig. 1. In order to stabilize the output of such a device, it is necessary to know whether the fluctuations which appear in the output are the result of random noise or of deterministic chaos, because the proper procedure for stabilizing the output would be different in the two cases. If the fluctuations are due to random noise, one might stabilize the output by shielding the system from its environment, whereas if these fluctuations are due to deterministic chaos one would need to decrease the magnitude of the nonlinearity in order to reduce the fluctuations.

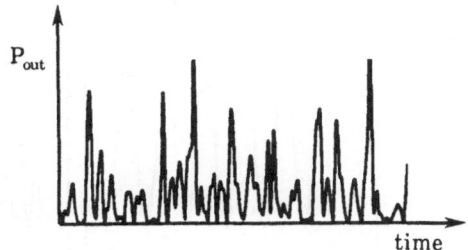

Fig. 1. Fluctuations in the output of an optical system.

There are several different methods for characterizing the instabilities that can occur in a nonlinear optical system. *Pure gain instabilities*, also known as *convective instabilities*, lead to the exponential growth of a perturbation in space. Examples of gain instabilities include stimulated Raman and stimulated Brillouin scattering. These instabilities are to be contrasted with *dynamical,* or *absolute, instabilities*, that is, to instabilities that lead to fluctuations in time. Examples of dynamical instabilities include laser instabilities (such as those studied by Arecchi[2] and by Haken[3]) and instabilities in bistable optical devices (such as the Ikeda[4] instability). These instabilities result from feedback that is is provided by the external cavity that is part of the laser or bistable optical device. In contrast, our own research has stressed the study of *intrinsic nonlinear optical instabilities*, that is, of instabilities that do not require external feedback for their development. Examples include instabilities due to the mutual interaction of laser beams, either in a counterpropagaing geometry or in the four-wave mixing geometry of optical phase conjugation.

There has been considerable recent work on the stability characteristics of two counterpropagating laser beams, as illustrated schematically in Fig. 2. Even though this interaction is conceptually very simple, it can lead to extremely complicated dynamical behavior, including periodic and chaotic fluctuations in the output intensities. Interest in this interaction stems both from the fact that chaotic behavior can occur in such a simple system and from the fact that counterpropagating laser beams are present in many types of optical devices (such as lasers, phase conjugate mirrors, and bistable optical devices), and any instabilities that develop in these laser beams could lead to a degradation in the performance of the device.

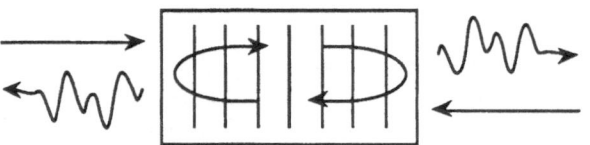

Fig. 2. Interaction of counterpropagating laser beams.

Figure 3 shows the sort of behavior that is typically observed as the input intensities of the two interacting beams are gradually increased. For low input intensities, the system is stable and the output intensities remain constant in time. If the input intensities are increased somewhat, the system makes a transition into a different state in which the output intensities fluctuate periodically in time. At still higher intensities, the system makes a transition into a chaotic state in which the output intensities fluctuate erratically. The behavior shown in Fig. 3 is typical in the sense that the progression shown is known to occur for a large variety of mechanisms leading to nonlinear coupling. In particular, chaotic instabilities are known to occur for the case of a sluggish Kerr medium,[5] for a tensor Kerr medium (not necessarily sluggish),[6] for coupling due to the nonlinear response of an atomic vapor,[7] and for coupling due to the Brillouin scattering mechanism.[8]

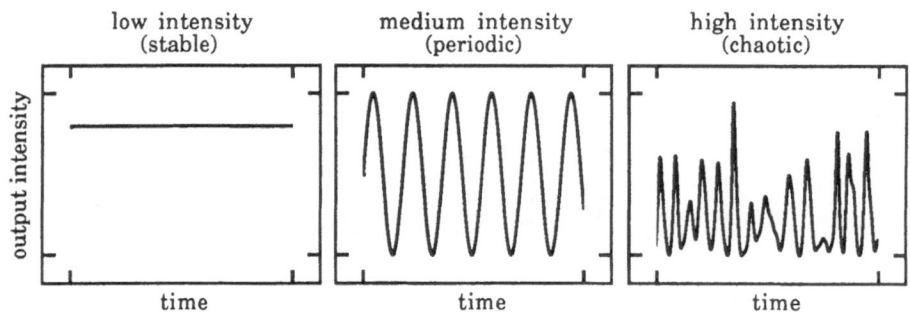

Fig. 3. Typical dynamical behavior of counterpropagating laser beams.

One of the first theoretical studies of the stability of counterpropagating laser beams was that of Silberberg and Bar-Joseph.[5] These authors treated the interaction in the scalar approximation for the case of a sluggish Kerr medium, that is, for a medium in which the nonlinear contribution to the refractive index n_{NL} obeys a Debye relaxation equation of the form

$$\tau \dot{n}_{NL} + n_{NL} = n_2 I \tag{1}$$

where n_2 denotes the nonlinear refractive index, τ denotes the response time of the nonlinearity, and I is the total intensity. These authors found that for sufficiently intense input waves the intensities of the transmitted waves could become unstable and that for very large input intensities these fluctuations could become chaotic. They also found that the threshold for instability increased without bound as the response time τ approached zero. Thus, instabilities are predicted within the scalar approximation only for a system with a memory. Silberberg and Bar-Joseph also presented a heuristic model for understanding the origin of the instability. They showed that the instability was a form of parametric oscillation due to a gain-feedback mechanism. Feedback is present due to distributed feedback: The two counterpropagating waves interfere to form a sinusoidal intensity distribution which leads to a modulation of the refractive index of the medium due to its nonlinear response. The forward going wave can scatter off of this grating into the backward direction and the backward going wave can scatter off of this grating into the forward direction to provide feedback, as illustrated schematically in Fig. 2. Gain is present in the sense that sidebands to the laser frequency detuned by an amount $\delta\omega \approx 1/\tau$ experience amplification for the same reason that there is gain in stimulated Rayleigh-wing scattering.

Our own work has dealt with the polarization characteristics of counterpropagating light beams. We assume that the medium is isotropic and that the nonlinear response is described by the tensor relation

$$\chi_{ij} = (A - \tfrac{1}{2}B)(\mathbf{E} \cdot \mathbf{E}^*)\delta_{ij} + \tfrac{1}{2}B(E_i^* E_j + E_i E_j^*) . \tag{2}$$

Note that we assume that the medium responds instantaneously. We then derive coupled amplitude equations for the x and y polarization components of the forward- and backward-going waves. We have found that the solution to these equations predicts bistability and instability in the polarizations of the transmitted fields.[6] Note that there is no need to assume that the medium has a memory, because polarization provides an additional degree of freedom not present in the scalar analysis of Silberberg and Bar-Joseph.

Figure 4 shows our theoretical prediction of polarization bistability. We assume that the input waves are both linearly polarized in the x direction. We assume that the forward-going input wave is turned on at time $t = 10$ and is subsequently left on. We also assume that the backward-going wave is turned on at time $t = 0$, is turned off at time $t = 30$, and is turned on again at time $t = 50$. We solve the coupled amplitude equations numerically for this

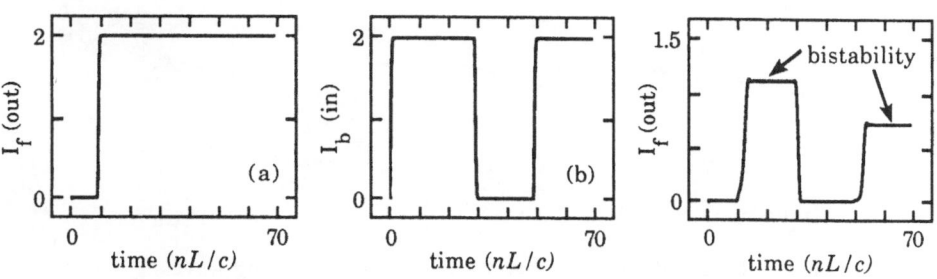

Fig. 4. Theoretical prediction of polarization bistability.

choice of boundary conditions and find that the y component of the forward-going output wave has the form shown in the last panel in the figure. We see that, for identical input conditions, the output wave can have two different values, depending upon which of the two input waves was turned on first. Thus, the system is predicted to display hysteretic bistability.

We have also conducted theoretical studies of the stability characteristics of the polarizations of counterpropagating light waves. We find that instabilities can occur when the input intensities are larger than those for which bistability is predicted. We also find that the threshold for the occurrence of these instabilities is several times lower than that of the scalar instability. Figure 5 shows the behavior as predicted by a numerical integration of the complete set of coupled amplitude equations. We assume that the input waves have linear and parallel polarizations. We see that periodic oscillations occur at low intensities and that chaotic behavior occurs at higher intensities.

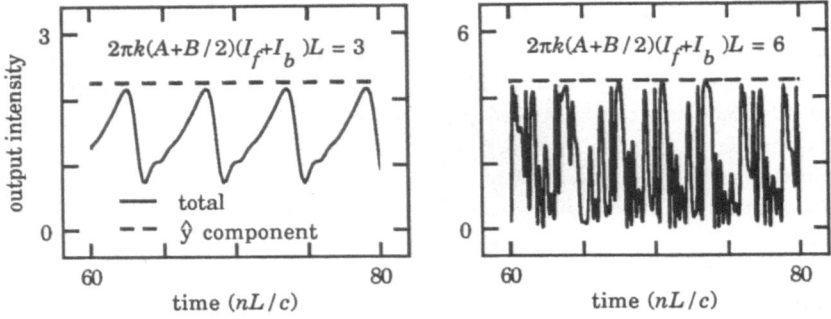

Fig. 5. Prediction of periodic and chaotic instabilities of the polarizations of counterpropagating light waves, for the case $I_f = 3I_b$ and $B/A = 6$.

We have studied these effects experimentally using the nonlinear response of an atomic sodium vapor. Two laser beams with linear and parallel polarizations interact in a 5-cm-long cell containing $\sim 10^{13}$ atoms per cubic centimeter. We use a polarizing beamsplitter and a photodetector to detect the presence of any light in the orthogonal polarization component. We find that bistability and instability do occur, and that their presence or absence depends sensitively upon experimental parameters such as atomic number density, input intensities, and laser frequency.

Our experimental results on polarization bistability[9] are shown in Fig. 6. With the

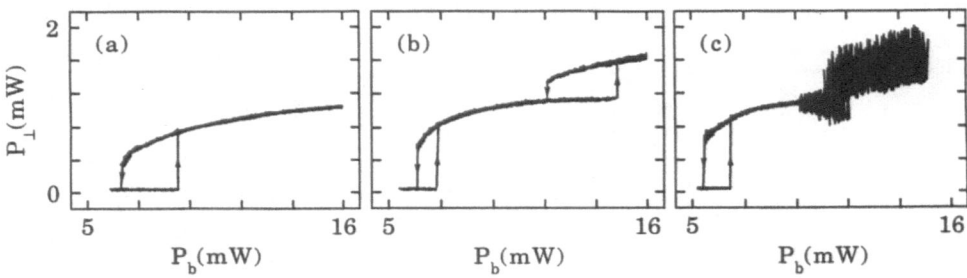

Fig. 6. Polarization bistability with counterpropagating waves.

intensity of the forward-going wave held fixed at 74 mW, the intensity of the backward-going wave is ramped slowly up to 16 mW and then is slowly ramped back down. The intensity of the transmitted forward-going wave associated with the polarization component orthogonal to that of the input waves is plotted on the vertical axis. For an atomic number density of 1.6×10^{13} cm^{-3}, a single hysteresis loop is observed. At a somewhat larger number density of 1.9×10^{13} cm^{-3}, two loops are formed. At our largest number density of 2.6×10^{13} cm^{-3}, we observe bistability at low intensities and instability at high intensities. Note that it is possible to avoid the unstable regime either through use of low number densities or low intensities.

We have also studied the nature of the instabilities that occur in the unstable regime shown in the last panel of Fig. 6. In Fig. 7, we show the temporal evolution of the power in the orthogonal polarization component on the left and the corresponding phase space trajectory on

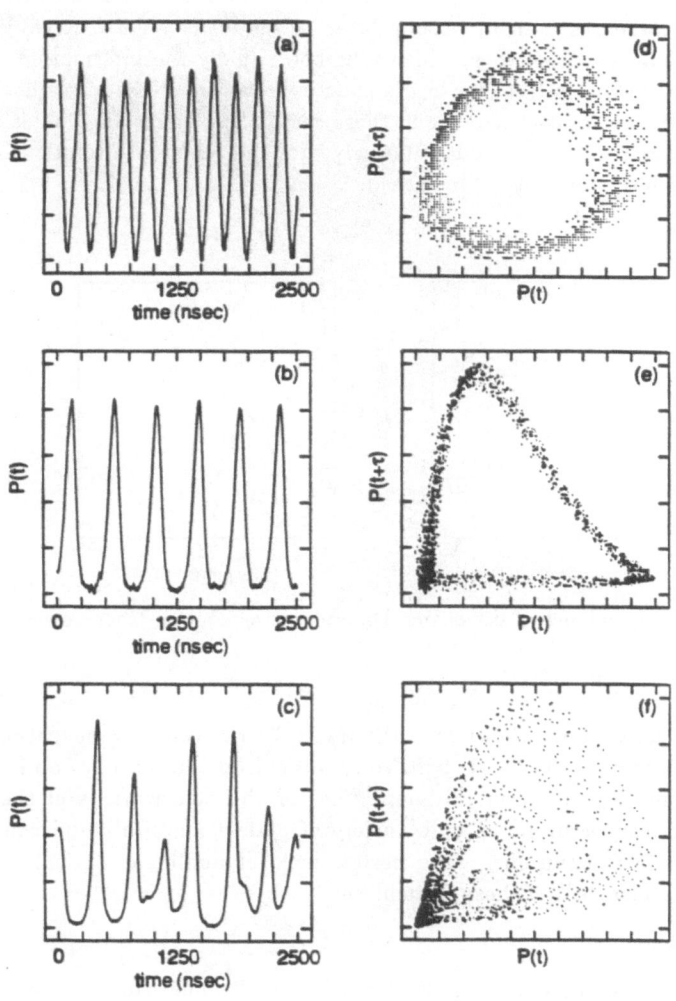

Fig. 7. Polarization instabilities with counterpropagating waves.

the right.[9] The phase space trajectory is obtained by plotting the power measured at time t against the power measured at time $t + \tau$. Three different values of the intensity of the backward-going wave are studied, with the intensity of the forward-going wave held fixed at 160 mW. For the lowest power shown, the output oscillates sinusoidally at approximately 4 MHz, leading to a circular phase space trajectory. For a slightly higher value of the input power (26 mW), a period-doubling bifurcation has occurred, leading to self-pulsing with twice the former period, and the phase space trajectory is more complicated than in the first case. For the highest input intensity shown, the output fluctuates chaotically, leading to what appears to be a strange attractor in phase space.

In order to establish that the the fluctuations shown for $P_b = 29$ mW in Fig. 7 are in fact chaotic and that the complicated phase space trajectory is in fact a strange attractor, we have made use of the method of Grassberger and Procaccia[10] to estimate the fractal dimensions and metric entropies of the system. In Fig. 8 we plot, as functions of the laser power of the backward-going beam, the the correlation dimension D_2 and the order-2 Reyni entropy K_2. The correlation dimension D_2 is an estimate of the Hausdorf or fractal dimension. Below the threshold for instability, the fractal dimension is zero, implying that the system occupies a single point in phase space. For the case of periodic orbits in phase space, the dimension is 1, and for higher laser powers the system becomes chaotic and the fractal dimension increases with increasing laser power. The order-2 Reyni entropy is an estimator of the Kolmogorov entropy, which tells how rapidly the information regarding the initial state of the system is lost due to the chaotic behavior. Nonzero values of K_2 hence demonstrate that the evolution is chaotic in these cases. The increasing values of D_2 and K_2 with increasing laser power shows quantitatively that the temporal evolution becomes more complicated as the laser intensity is increased.

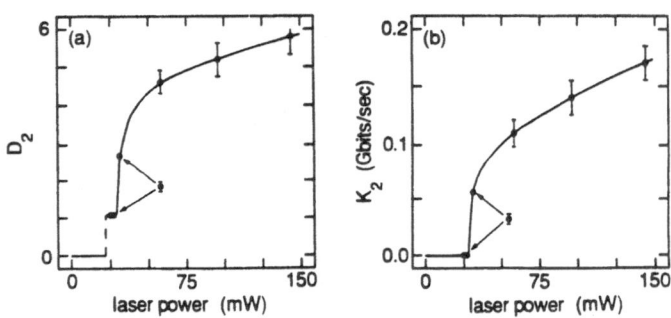

Fig. 8. Dependence of dimension D_2 and entropy K_2 on laser ower.

In conclusion, we have shown that the interaction of counterpropagating laser beams can lead to complex dynamical behavior, including bistability and instabilities. Instabilities can develop either in the amplitudes or the polarizations of the light waves. These instabilities can occur for a wide variety of optical materials and can lead to both periodic and chaotic fluctuations. The performance of nonlinear optical devices can be degraded by the development of these instabilities.

Acknowledgement

This work was supported by the National Science Foundation, the U S Army Research Office University Research Initiative, and the New York State Center for Advanced Optical Technology.

References

1. See for example, *Optical Instabilities*, edited by R. W. Boyd, M. G. Raymer, and L. M. Narducci (Cambridge University Press, Cambridge, 1986).
2. F. T. Arecchi, R. Meucci, G. P. Puccioni, and J. R. Tredicce, Phys. Rev. Lett. **49**, 1217 (1982).
3. H. Haken, Phys. Lett. **53A**, 77 (1975).
4. K. Ikeda, H. Daido, O. Akimoto, Phys. Rev. Lett. **45**, 709 (1980).
5. Y. Silberberg and I. Bar-Joseph, Phys. Rev. Lett. **48** 1541 (1982).
6. A. L. Gaeta, R. W. Boyd, J. R. Ackerhalt, and P. W. Milonni, Phys. Rev. Lett. **58**, 2432 (1987).
7. D. J. Gauthier, M. S. Malcuit, and R. W. Boyd, Phys. Rev. Lett. **61**, 1827 (1988).
8. P. Narum, A. L. Gaeta, M. D. Skeldon, and R. W. Boyd, J. Opt. Soc. Am. B **5**, 623 (1988).
9. D. J. Gauthier, M. S. Malcuit, A. L. Gaeta, and R. W. Boyd, Phys. Rev. Lett. **64**, 1721 (1990).
10. P. Grassberger and I. Procaccia, Phys. Rev. Lett. **50**, 346 (1983).

SELF-ORGANIZATION AND SPATIO-TEMPORAL CHAOS IN PHASE-LOCKED SEMICONDUCTOR
LASER ARRAYS

Herbert G. Winful

Department of Electrical Engineering and Computer Science, University
of Michigan, Ann Arbor, MI 48109-2112, U.S.A.

ABSTRACT

    Phase-locked semiconductor laser arrays are of great technological
interest as sources of high output power in a well-collimated beam. In
this talk we discuss the dynamic stability of these arrays, the mechanism
of synchronization, the formation of spatial patterns, and the onset of
spatio-temporal chaos. It is shown that laser arrays can be accurately
modeled by a set of coupled van der Pol type oscillators. The nonlinearity responsible for both synchronization and instability is the carrier
density dependent refractive index. Some experimental results will be
presented.

COMPETITIVE AND COOPERATIVE DYNAMICS IN OPTICAL NEURAL NETWORKS

Dana Z. Anderson

Department of Physics and Joint Institute for Laboratory Astrophysics
University of Colorado, Boulder, CO 80309-0440, U.S.A.

ABSTRACT

Real-time holographic materials are in many ways ideally suited to neural network implementation. In an associative memory, for example, information may be stored in a distributed manner so that, from a partial or distorted input, an entire pattern can be recalled. When several distinct items of information are stored together in a distributed memory, how do the various information chunks become separated again when something is to be recalled? In particular, if an input resembles more than one stored pattern, how does a system decide on a single output? Such questions lead us to consider decision making in neural networks as a dynamical process that involves a competitive and cooperative inter-action among the stored items in the memory. The mathematics describing this mode interaction is familiar to several areas of physics. In particular, the competitive cooperative process is aptly described by general Lotka-Volterra equations. Photorefractive two-beam coupling can be employed to establish a wide array of dynamical mode interactions. I will discuss the photorefractive flip-flop as the prototype example of a competitive system. The mode interaction of the two-state flip-flop can be extended to a multimode system to establish what is known in neural network parlance as "winner take all" dynamics. Cooperation can also be implemented with photorefractive two-beam coupling. We demonstrate the application of cooperative dynamics to "playback" a time sequence of modes in an optical ring resonator.

TRANSITIONS BETWEEN ORDERED AND DISORDERED SOLID-MELT PATTERNS FORMED ON SILICON BY CONTINUOUS LASER BEAMS: COMPETITION BETWEEN ELECTRODYNAMICS AND THERMODYNAMICS

K. Dworschak, J.E. Sipe and H.M. van Driel

Department of Physics, University of Toronto
and Ontario Laser and Lightwave Research Centre
Toronto, Ontario, Canada   M5S 1A7

INTRODUCTION

The melting of a semiconductor is usually accompanied by a discontinuous change in linear optical properties. If then a light beam is used to melt a semiconductor the accompanying change in optical properties is a highly nonlinear function of intensity. For example, for silicon at 10µm, the reflectivity of the (dielectric) solid is approximately 0.3 up to the melting temperature ($T_m = 1685K$). However, uniformly molten (metallic) silicon has a reflectivity which is approximately 0.9. As silicon is heated from just below the melting point to just above it with a small increase in laser intensity, the dramatic change in reflectivity will prevent a uniform molten or solid state from existing since neither state is consistent with the amount of absorbed power. The nonlinear changes in reflectivity accompanying the melting transition therefore leads to a lateral or transverse instability, in which a uniform incident beam will cause a transversely inhomogeneous state to form on the surface.[1-3]

A second type of transverse instability is well known in pulsed laser-solid interactions (in which the apparent translational symmetry of a surface can be broken by surface roughness, material inhomogeneity, etc.). A linearly polarized incident beam can therefore give rise to a coherent surface wave which interferes with the incident beam to produce periodic energy deposition and (at the threshold of melting or vaporization) permanent periodic structures.[3-7] At normal incidence the spacing of the grating is equal to the incident wavelength and its orientation (given by its wavevector) is parallel to the incident field. More complex grating structures occur for different polarizations and angles of incidence but all these effects can be simply understood in terms of the electrodynamic response of the surface alone. That is to say, in most situations of interest involving the interaction of nanosecond or picosecond pulses with solids, heat flow is not important on less than a 10 nsec time scale and the patterns merely reflect the optical response of the material.

When continuous laser beams are used to induce melting there are then two possible sources of transverse instabilities: the one associated with the solid-melt transition discussed in the first paragraph above (which can occur independent of the coherence properties of the incident beam) and the one associated with the coherent interaction between incident and surface scattered beams discussed in the previous paragraph. In both cases heat flow in the sample has the possibility of influencing the pattern of any spatially inhomogeneous power deposition and helps to define the steady-state melt/solid morphology. For incident $\lambda = 10.6\mu m$ laser light on silicon we have previously shown[3] how the occurrence and stability of metal-dielectric grating structures with spacings of λ, 2λ, etc. can be explained on the basis of a physical (as opposed to geometrical) optics analysis which takes into account that the size and spacing of the molten elements is comparable to the incident wavelength. Here we wish to offer an explanation as to why it is also possible to form a disordered state consisting of isolated lamellae. In earlier experiments we noted that these tend to form for larger laser spot sizes and lower intensities. We will show here that these lamellae can be considered as "particles" which form a liquid-like state. As the laser spot size is reduced these lamellae form the grating structure in which an apparent phase transition occurs to the ordered state. This locking phenomena can be viewed as a laser-induced freezing of the disordered state.

EXPERIMENTAL DETAILS

Fig. 1 shows a schematic diagram of the experimental arrangement we have used to produce and observe the melt/solid patterns for silicon-on-sapphire. The laser is a cw, 14 W CO_2 linearly polarized laser operating at 10.6 µm in the TEM_{00} mode. To control the polarization of the beam which reaches the silicon sample we use CdS quarter and half-wave plates. The spot size on the sample controlled with a 10 cm focal length lens. The samples are 2.5×2.5 mm^2 pieces of 2 µm thick polycrystalline Si on a 1 mm thick Al_2O_3 substrate. To observe the solid/melt patterns a microscope with a 10x objective is used to resolve the visible component of the black-body radiation emitted by the hot silicon. The image is recorded by film or videotape. To reach the melting temperature we proceeded in two stages. Once the silicon film is brought to ~ 800 C by resistive heating, thermally generated electrons and holes can absorb sufficient radiation at 10.6 µm to raise the temperature to the melting point. In various studies we have used film thicknesses in the range 0.5-3 µm with qualitatively the same types of patterns forming.

In a second set of experiments we have used continuous radiation from a $\lambda = 1.06$ µm laser with maximum power of 20 W. An appropriate set of lenses, waveplates, etc. was used to prepare the beam before it was incident on the sample.

RESULTS

We begin first by discussing the typical patterns generated by $\lambda = 10.6$ µm light. Extensive experiments have been carried out as function of angle of incidence and polarization, but here we will consider only the normal incidence experiments. Fig. 2a shows the melt pattern generated with a laser spot size of 500 µm and an intensity of 2.7 kW/cm^2. Alternating strips of solid and liquid appear, with a spacing $\Lambda = 10.6$ µm $\pm$ 0.1 µm, and oriented with the wavevector of the grating parallel to the polarization of the laser beam. Approximately 40% of the surface is molten. Unlike the gratings which form in pulsed laser-semiconductor interactions, those formed by cw beams have strips which are virtually parallel, illustrating the fact that feedback in the light scattering process has allowed the strongest component of the scattered light field distribution to dominate the pattern formation. For a higher laser intensity (3.2 kW/cm^2) the simple grating structure is replaced with a grating which has twice the period[3] and with

75% of the surface molten. Of course, as the intensity increases one would expect the amount of molten area to increase. That we have observed, however, is that the increase does not occur smoothly but rather occurs in discrete jumps, with the simple grating replaced by a double-period grating.[3] Within a given grating structure, an increase in laser intensity simply results in a redistribution of the deposited power and the temperature field is changed without any additional melting. In this sense one has a very stable morphological "phase", which is somewhat insensitive to the spot

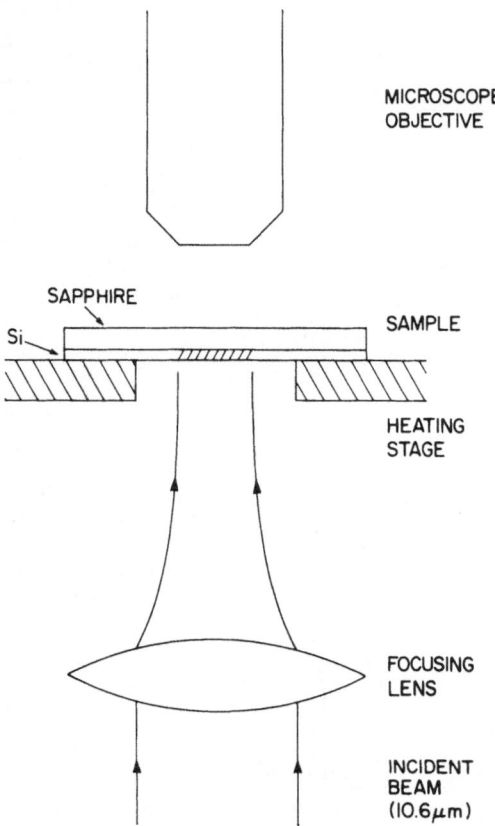

MICROSCOPE
OBJECTIVE

SAPPHIRE

Si

SAMPLE

HEATING
STAGE

FOCUSING
LENS

INCIDENT
BEAM
(10.6μm)

Fig. 1. Diagram of the experimental setup used to observe pattern formation in silicon-on-sapphire by a 10.6 μm, cw laser beam.

size and laser intensity. Fig. 2a shows direct evidence of this with the width of molten and solid strips constant across a spot illuminated with a Gaussian intensity profile.

For spot sizes larger than approximately 500 μm, the melt/solid morphology changes into an irregular pattern, with liquid "dots" or "strings" in a solid background. At higher intensities this pattern is replaced with solid dots in a molten background. In all we have observed at least five different ordered or disordered "morphological phases" which are stable

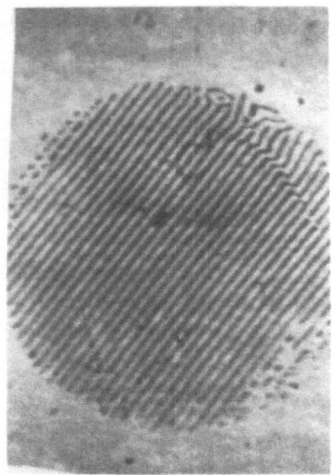

Figure 2A. Solid-melt pattern
formed at a power density of
2.7 kW/cm^2 with a spot size
of 500 µm; grating spacing
10.6 ± 0.1 µm; dark regions
are molten.

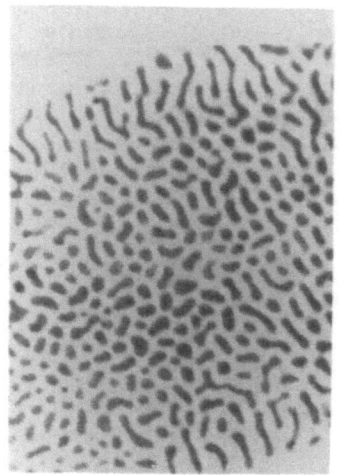

Figure 2b. Solid-melt pattern
formed with a normal incidence
beam at a power density of 1.7
kW/cm^2 and a spot size of 700
µm.

within a range of laser spot sizes and intensities. These phases are summarized in Fig. 3. The dynamics of these patterns have also been studied
albeit qualitatively. Near the "phase boundaries" it is possible, by rapidly scanning the beam or rotating the polarization, to initiate "defects" in
the otherwise stable structures. It is also possible to change disordered
structures into ordered ones.

For the ordered patterns only one parameter, the spatial periodicity,
is required to describe the structure. As noted above this spatial periodicity is λ or some multiple there of. A Fourier transform of the periodic
structures therefore gives a series of dots spaced by $2\pi/\lambda$ for the simple
grating, π/λ for the "doubled grating", etc. In an effort to determine the
spatial characteristics associated with the disordered structures, we have
numerically carried out a Fourier transform of the patterns shown in Figure
2b. The result is shown in Fig. 4. It reveals a characteristic diffuse
circle reminiscent of a Debye-Scherrer x-ray diffraction pattern of an
amorphous or "liquid-like" state. The radius of the circle, which is characteristic of the average separation between the particles of the system,
is approximately $2\pi(14\mu m)^{-1}$

In experiments carried out with a 1.06 µm light source we were only
able to produce disordered structures. For no combination of spot size or
laser intensity were we able to observe ordered patterns. Rather, we were
only able to observe disordered structures with Fourier transforms similar
to that indicated in Fig. 3. Indeed, even the same characteristic spacing
as observed indicating that the wavelength of light is at best playing a
secondary role in determining the morphology. Analysis of the Fourier
transforms for the difference disordered patterns formed by λ = 10.6 or
1.06 µm light reveals that the average spacing between the molten regions
increases with laser spot size but further work must be done to reveal the
functional dependence.

DISCUSSION

To be able to explain <u>ab initio</u> all the details of the ordered and disordered morphological phases is complicated by the complex electrodynamic effects that occur. The problem of treating the interaction of light with a microscopically inhomogeneous metal-dielectric is difficult enough. The disordered melt phases make the calculation of scattered light fields and energy deposition intractable while heat flow considerations further

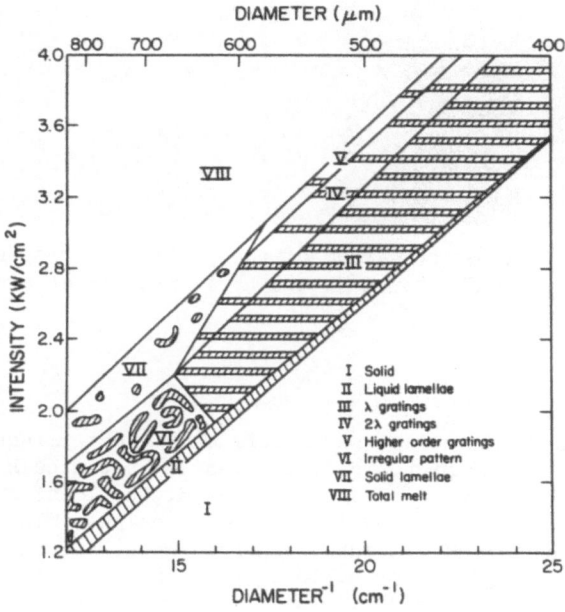

Fig. 3. "Phase Diagram" illustrating the spectrum of solid-melt morphologies formed by a normally incident beam.

complicate the problem. At the present time we have set ourselves the modest goals of determining, for the disordered or ordered structures, the reasons for their existence and why an apparent phase transition can be induced between the structures.

To address the electrodynamic aspects of the problem the theoretical approach we have used is based on the simple model of an inhomogeneous metal/dielectric thin film. We assume that induced polarizations in the liquid are confined to the top and bottom layers for light at normal incidence. A uniform polarization is assumed to exist in the bulk of the solid regions. These polarization densities are used with an electrodynamic Green's function for the thin film to calculate fields everywhere.

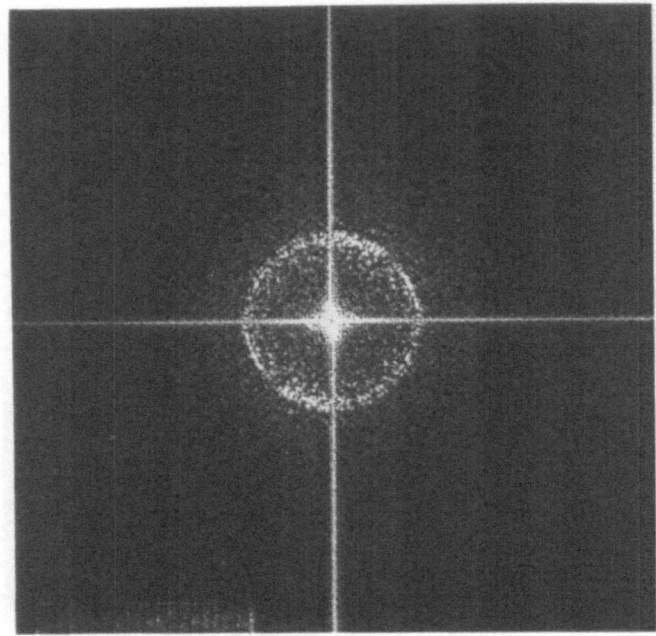

Fig. 4. Fourier transform of the pattern corresponding to Fig. 2b; vertical and horizontal lines are
artifacts.

The three (polarization density) parameters can be determined by the conditions that electric and magnetic fields vanish in the bulk of the liquid.
Once the polarization parameters are calculated, the local power deposition
can be calculated in the melt and in the solid regions. This of course
depends on the structure of the inhomogeneous phase, and only becomes equivalent to that of bulk liquid or solid in the limit that the molten or
solid areas have an extent which is large compared to the wavelength of
light. To emphasize the importance of the microscopic structure in determining local power deposition we have found it convenient to express our
results in terms of "structural absorption coefficients" for the liquid and
solid regions. These give the ratio of the actual power deposition in an
infinite bulk region of the same composition.

(i) Ordered Structures

In spite of the simplicity of the overall approach, we have found
that many of the salient experimental features of the ordered structures
can be explained in a self-consistent manner. We begin by assuming the
existence of a melt/solid grating on a thin silicon film with the orientation of the grating perpendicular to the polarization of a normally incident, linearly polarized 10.6 μm beam. The structural absorption coefficient for this morphology is shown in Fig. 5 as a function of grating
spacing assuming different fractions (f_m) of the surface to be molten.
When approximately 40% of the surface is molten one finds that the difference between power deposition in liquid and solid regions is highest when
the grating spacing matches the incident light wavelength. Note as well
that the structural absorption coefficient for the liquid (solid) at this
spacing is well above (below) that for a grating with $\Lambda \to \infty$, corresponding
to alternating infinite regions of solid and liquid. This illustrates that

for the microscopic grating with $\Lambda = \lambda$, the absorption in the liquid region is enhanced over what a bulk liquid would give, while that of the solid is reduced. This is exactly the opposite to what one would expect from a geometric optics picture in which the liquid has the higher reflectivity. The actual structural absorption coefficients can be understood in terms of the large scattered fields which arise from a metal-dielectric grating. When these interfere with the incident field there is destructive interference in the solid but constructive interference in the liquid. In a sense the liquid shields the solid from the incoming radiation; we have referred to this effect as "interference shielding".[3]

If we now assume that 80% of the surface is molten (Fig. 5) the structural absorption coefficient of the solid is greater than that of the

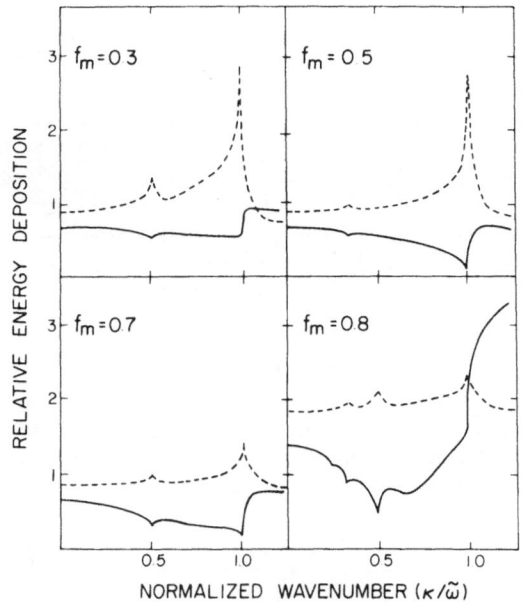

Fig. 5. Structural absorption coefficient as a function of normalized grating wavenumber for different fractions of the surface molten. ($\kappa/\tilde{\omega} \equiv \lambda/\Omega$); dashed lines (melt), solid lines (solid).

liquid for $\Lambda = \lambda$, indicating that the simple grating structure is no longer stable. However, for $\Lambda = 2\lambda$, the difference between solid and melt structural absorption coefficients is at a maximum. This is consistent with our observation of the double period grating when approximately 75% of the surface is molten. It is worth reminding the reader at this point that no attempt has been made to predict, as a function of the amount of surface

that is molten, which particular structure will form. Without the benefit
of a variational principle or some other principle to predict the most
stable structure all we are attempting to do is show that our calculated
results are consistent with what is observed.

(ii) Disordered Structures

From the analysis of the Fourier transforms of the disordered structures
we note that the molten dots can be considered as forming a liquid-like phase
with an average separation that scales with the spot size and which has a
value of 10-20 µm. It is particularly noteworthy that this is close to
but not equal to the incident 10.6 µm wavelength. On the other hand similar separations are observed when incident 1.06 µm radiation is used. This
scale length is therefore undoubtedly set by mechanisms unrelated to physical optics. Indeed, it is set by some combination of the geometrical optical (reflectivity of solid and melt) and thermodynamic parameters. Although
a more complete argument awaits detailed calculations we can here qualitatively rationalize the spot size dependence and the approximate scale length.
For simplicity, since they obviously can't play a key role, we will ignore
the structural absorption coefficient effects.

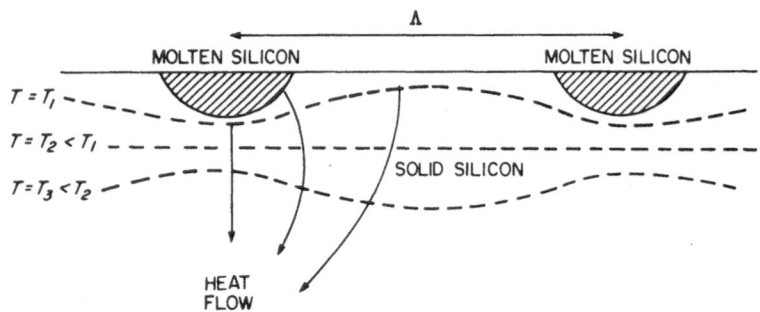

Fig. 6. Schematic diagram of local solid-melt cross-section associated with the disordered phase. Dashed lines of contours
of constant temperature (T).

We note from Fig. 3 that the intensity required to reach the threshold
of melting varies inversely as the spot size (r). This is not unexpected
since the power deposition must of course equal the heat flow out of the
spot in steady state. But if the heat flow is primarily lateral in the
thin film, then the heat flow or power deposition is proportional to the
parameter and hence the diameter of the spot. The relationship between
power (in Watts) and intensity (Watt/cm^2) then allows one to conclude that
the local intensity will scale inversely with spot diameter at the melting
threshold. This, however, is a global argument and locally the description
of a disordered structure is as illustrated in Fig. 6 which shows molten
dots formed in a solid silicon background. The absorption depth of the
liquid is hundreds of Angstroms whereas that of the solid silicon is a
few microns. On the other hand the absorbance of the liquid is 7 times
less than that of the solid. Therefore less power is deposited (but it is
deposited closer to the surface) in the liquid than in the solid. One expects then, that behind the molten region the temperature will fall off more
quickly than it does in the solid region giving rise to contours of constant

temperature as shown in Fig. 6. This will then lead to heat flow from the solid region to beneath the liquid region and an effective cooling of the solid region. This effect helps to stabilize the disordered phase. At the same time one expects that near (but above) the threshold for melting the temperature of the solid near the surface is slightly less (T_{m-}) than the melting temperature and that of the liquid is slightly higher (T_{m+}). Locally, the deposited intensity will result in heat flow which occurs through a temperature gradient. One expects that

$$I \propto r^{-1} = \kappa \nabla T \propto \frac{T_{m+} - T_{m-}}{\Lambda}$$

where κ is the thermal conductivity. It is therefore not unreasonable that $\Lambda \propto r$ as observed experimentally. The actual length scale is governed by the magnitude of the temperature gradient which will be approximately several optical absorption depths in the solid and hence be in the range of 10 μm as observed.

One can then see that as the spot size decreases the separation between dots does as well to the point where it matches the wavelength of the 10.6μm laser. At that point the physical optics and structural absorption effects take over and induce a locking into the ordered state. A phase transition or laser-induced freezing is observed. The fact that one never observes such an effect for 1.06 μm light is consistent with the fact that this wavelength is far from the thermodynamic length scale which governs the separation of the molten dots.

CONCLUSIONS

Much quantitative work remains to be done before we can fully understand the cw induced structures, in particular the stability of the disordered or "amorphous" structures, the nature of the transitions between amorphous and ordered structures. Nonetheless we feel we have identified, at least qualitatively, the salient features of the problem and the basic reasons for the existence of the ordered and disordered phases. An additional important fundamental question is whether the symmetry breaking we see here is of the nature of the "true" Broken Symmetry[9] observed in dissipative systems.

ACKNOWLEDGEMENTS

We gratefully acknowledge support for this work from the Natural Sciences and Engineering Research Council of Canada and the Ontario Laser and Lightwave Research Centre.

References

1. M.A. Bosch and R.A. Lemons, Phys. Rev. Lett. 47, 1151 (1981).

2. W.G. Hawkins and D.K. Biegelsen, Appl. Phys. Lett. 42, 358 (1983).

3. J.S. Preston, J.E. Sipe and H.M. van Driel. Phys. Rev. Lett. 58, 69 (1987); J.S. Preston, H.M. van Driel, and J.E. Sipe, Phys Rev. B, 40, 3942 (1989), J.S. Preston, J.E. Sipe, H.M. van Driel, and J. Luscombe, Phys. Rev. B, 40, 3931 (1989).

4. H.M. van Driel, J.E. Sipe and J.F. Young, Phys. Rev. Lett. 49, 1955 (1982).

5. J.F. Young, J.E. Sipe and H.M. van Driel, Phys. Rev. B, 30 2001 (1984).

6. P.M. Fauchet and A.E. Siegman, I.E.E.E. J. Quan, Elec. 22, 1389 (1986).

7. S.A. Ahkmanov, V.I. Emel'yanov, N.E. Koroteev and V.N. Seminogov, Usp. Fiz. Nauk. 147, 675 (1985)[Sov. Phys. Usp. 28, 1989 (1985)].

8. K. Dworschak, J.E. Sipe and H.M. van Driel, J. Opt. Cos. Am. B (to be published) (1990).

9. P. Anderson in Order and Fluctuations in Equilibrium and Non-equilibrium Statistical Mechanics, Nicolis, Dewel, Turner Eds. (Wiley, New York, 1981) P.289.

LIGHT SCATTERING IN OXIDE SUPERCONDUCTORS

P. E. Sulewski, P. A. Fleury, and K. B. Lyons

AT&T Bell Laboratories
Murray Hill, NJ 07974

INTRODUCTION

As is well known, the unique structural feature connecting the various high T_c materials is the presence of copper oxide planes.[1] By varying the out-of-plane constituents these planes may be doped into the metallic and eventually superconducting state. In the insulating phase, the Cu 2+ sites have a spin of 1/2 and order antiferromagnetically in the ground state, as determined, for example, by neutron scattering.[2] Since La_2CuO_4 exhibits the simplest crystal structure of the high T_c materials, much effort has been focused on this and isostructural systems, in the hope of avoiding irrelevant complications. The schematic phase diagram, shown in Fig. 1, however, demonstrates that even the nominally simpler La_2CuO_4 system exhibits a rich variety of behaviors.

While, in principle, light scattering can probe phonons, charge and spin excitations, and, in the superconducting state, the energy gap, this paper focuses on the contributions of light scattering to the current understanding of the spin degrees of freedom in these materials. The majority of the discussion will concentrate on the undoped antiferromagnetic (AFM) materials, since here the samples are of high quality and there is a clear theoretical understanding of the spectral features. With the insights gained in studying the undoped crystals, the discussion will proceed into the doped region of the phase diagram. While intense effort has been directed toward these materials in the last three years, the work presented here represents a rare meeting of theory and experiment, resulting in a new understanding of the two dimensional (2D) spin-1/2 quantum AFM, in which the dynamics are dominated by quantum fluctuations.[3] Properly accounting for fluctuations leads to quantitative agreement with the magnon-pair spectra. From this agreement, a value of ~ 1000 cm^{-1} (~ 1500 K) is found for the exchange parameter, J, within the planes for the various cuprates. This large energy sets the scale for the 2D spin correlations. The much smaller interplane coupling leads to three dimensional (3D) ordering at a Néel temperature of 260 K. Although no long range order exists above the Néel temperature, the strong AFM correlations within the planes persist to temperatures of order J. The presence of spin fluctuations with such a large characteristic energy scale provides a tantalizing potential mechanism for superconductivity.

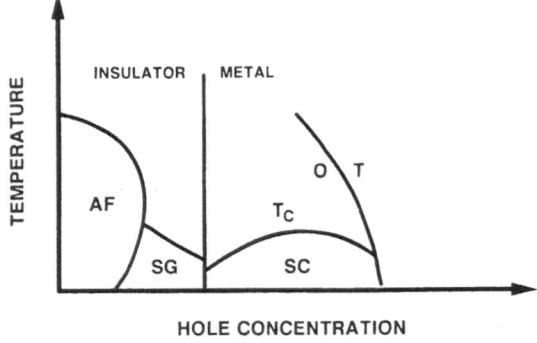

FIGURE 1. Schematic phase diagram for the La_2CuO_4 system.

UNDOPED AFM MATERIALS

The interaction of the spins on the Cu sites is well-described by the nearest neighbor Heisenberg Hamiltonian:

$$H = J\sum_{(ij)}\vec{S}_i \cdot \vec{S}_j. \tag{1}$$

In the classical picture, the AFM ground state is the Néel state, and the excitations above this ground state are spin-waves. The spin fluctuation dynamics are probed by inelastic light scattering, occurring through magnon-pair creation.[4] Light from an argon laser is incident on a single crystal sample, and the inelastically scattered light is collected and frequency analyzed with a spectrometer. The components of the Raman tensor are identified by using various combinations of incident and scattered polarizations and crystal orientations. In order to conserve momentum, since the photons have essentially zero momentum on the scale of the Brillouin zone, the two magnons must have equal and oppositely directed q-vectors. The value of q, though, in contrast to a one-magnon process, can lie anywhere in the zone. This two-magnon process thus probes the entire dispersion curve. To first order, the resulting spectrum reflects the joint density of states (DOS) of the dispersion curve. Since the DOS is greatest at the zone boundary (ZB), high-q excitations make the dominant contribution to the scattering cross section, and the spectrum peaks at twice the ZB value of zSJ, where S is the spin and z is the coordination number. In practice, magnon-magnon interactions, readily calculable within spin-wave theory,[5] lead to a shifting and broadening of this peak.

This simple picture works remarkably well for K_2NiF_4,[6,7] which is isostructural to La_2CuO_4. The two materials differ in that the Ni sites have spin-1, while the Cu sites have spin-1/2. The solid curve in Fig. 2 is the experimental data[6] for K_2NiF_4, while the dashed curve represents the joint DOS for the magnons. The filled circles are the spectrum calculated by Parkinson,[7] using spin-wave theory and the effective scattering Hamiltonian:

$$H_R = \sum_{(ij)}(\vec{E}_{inc} \cdot \vec{\sigma}_{ij})(\vec{E}_{sc} \cdot \vec{\sigma}_{ij})\vec{S}_i \cdot \vec{S}_j, \tag{2}$$

where $\vec{E}_{inc}$ and $\vec{E}_{sc}$ are electric field vectors for the incident and scattered photons, and $\vec{\sigma}_{ij}$ is a unit vector connecting spin sites i and j. For the spin-1/2 system, the spin-pair spectrum calculated from spin-wave theory appears quite similar to the spin-1 case, with the peak shifted to slightly lower energy.

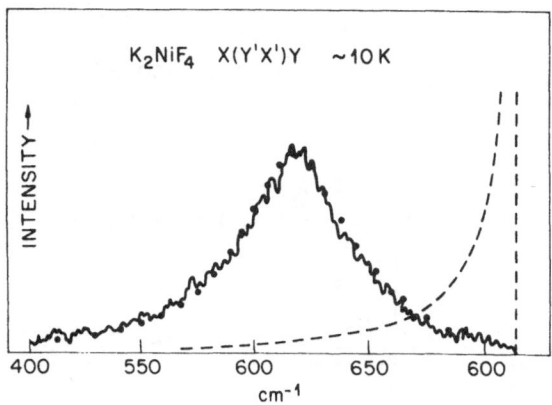

FIGURE 2. Magnon-pair mode in K_2NiF_4. Dashed curve neglects magnon-magnon interactions, while dotted curve includes these effects.

In calculating the scattering, the B_{1g} component is weighted most heavily at the zone boundary points. Since this weighting emphasizes the short wavelength magnons, the scattering can be described to first order by a local picture, in which the photon scatters by flipping a pair of nearest neighbor spins. Since only six bonds are broken, the peak appears at $3J$. Also, from this local picture, the spectra are primarily sensitive to the short range order. When the spins are well separated, 8 bonds are broken, which leads to a high energy cutoff at $4J$. The A_{1g} component of the scattering Hamiltonian (Eq. (2)) commutes with the Heisenberg Hamiltonian, and so no A_{1g} scattering is expected.

Early spectra,[8] taken with a photomultiplier and a conventional scanning spectrometer, exhibited a B_{1g} mode with a width three times that predicted by spin-wave theory. Poor signal-to-noise and questions of sample integrity and fluorescence contributions, which pose a problem when dealing with such broad features, led to a decision to exploit the advantages of multichannel detection by employing a charge-coupled-device (CCD) array detector. Another advantage of the CCD is its high quantum efficiency in the infrared, resulting in spectra covering energy shifts of up to 1 eV. Since over the range of 1 eV the spectrometer response varies substantially, the system is carefully calibrated in order to obtain accurate lineshapes, essential for comparison with theory. Fig. 3 (a) is an example of a spectrum on La_2CuO_4 taken with the CCD. Although of much higher quality than the initial spectra, there is still strong disagreement with the width calculated from spin-wave theory. Also, a great deal of spectral intensity exists beyond the classical cutoff of $4J$. In addition to the expected B_{1g} scattering, well-defined modes of A_{1g} and B_{2g} symmetry, which are classically forbidden, are observed,[9] as seen in Fig. 3 (b).

The disagreement between the spin-wave theory and the spectra reveals that spin-wave theory begins with an incorrect assumption concerning the ground state. In fact, the true ground state is not the Néel state, but rather includes strong quantum fluctuations. The ground state wave function may be written as a sum of the Néel state, the first fluctuation contribution, which has a single pair of spins flipped, and so on. In principle, knowing the ground state and all of the excited states, the spectrum may be calculated exactly for $T = 0$:

$$I(\omega) = \sum_i \delta(\omega-(E_i-E_o)) \, |\langle\psi_i| H_R |\psi_0\rangle|^2. \qquad (3)$$

123

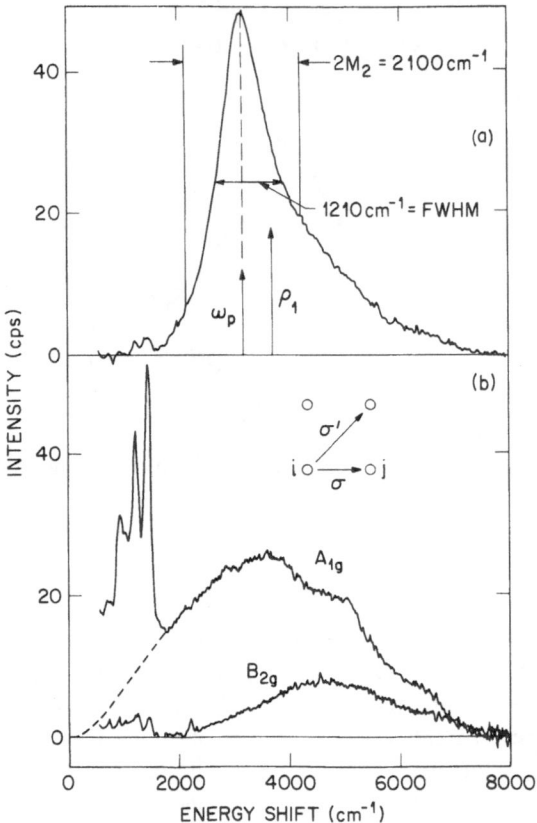

FIGURE 3. Raman spectra for La_2CuO_4 excited with 4880 Å light. (a) B_{1g} ($x'y'$) spectrum. (b) A_{1g} ($y'y'-yx$) and B_{2g} (yx) spectra.

Since, in practice, this calculation is extremely difficult, the spectral moments,

$$\rho_n = \int \omega^n I(\omega)d\omega/I_T, \tag{4}$$

with

$$I_T = \int I(\omega)d\omega, \tag{5}$$

are calculated instead. The moments require calculating matrix elements involving only the ground state wave function and commutators of the interaction and Heisenberg Hamiltonians.[3] For example, the first moment is given by

$$\rho_1 = \langle \psi_0 \mid H_R[H,H_R] \mid \psi_0 \rangle. \tag{6}$$

While this approach works extremely well, it is only feasible for the first few moments, and a complete calculation of the lineshape could reveal important nuances of the system.

In order to calculate these moments, Singh, et al.,[3] used an Ising series expansion technique. Beginning with the Ising model, the Heisenberg limit is approached by adding the xy piece perturbatively:

$$H = J_z \sum_{(ij)} S_i^z S_j^z + J_{xy} \sum_{(ij)} (S_i^x S_j^x + S_i^y S_j^y). \tag{7}$$

The expansion parameter, x, is the ratio of the xy piece to the Ising piece ($x \equiv J_{xy}/J_z$). Matrix elements involving the spin operators are then calculated by developing series in x, using clusters of a size appropriate for the order in x. For example, the first moment, given as a series in x, is:

$$\frac{\rho_1}{J} = 3 + 0x + \frac{1}{2}x^2 + 0x^3 + 0.0286x^4 + 0.0372x^5 + \cdots \tag{8}$$

Taking the limit $x \to 1$, the moment appropriate to the Heisenberg case is obtained. The divergences which might occur in taking such a limit are due to the long wavelength Goldstone modes.[10] Since the two-magnon scattering process is primarily a short wavelength probe, these divergences do not effect the convergence of the series.

The description of the lineshape may be equivalently recast in terms of cumulants,

$$(M_n)^n = \int (\omega - \rho_1)^n I(\omega) d\omega / I_T, \tag{9}$$

which have a more direct interpretation in the spectral lineshape. The first cumulant is a measure of the position of the mode, the second cumulant measures the width, and the third measures the skewness or asymmetry of the mode. From the series expansion calculations, the first three cumulants are found to be: $M_1 \equiv \rho_1 = (3.6 \pm 0.1)J$, $M_2 = (0.8 \pm 0.1)J$, and $M_3 = (1.0 \pm 0.2)J$. In ratios of these cumulants, the exchange parameter cancels, thus comparing ratios of the experimental and theoretical cumulants provides a parameter-free test of the role of quantum fluctuations in the 2D quantum AFM. Table I presents the ratios of M_1 to M_2 for La_2CuO_4 at three laser excitation wavelengths. The excellent agreement demonstrates that the width of the peak is dominated by quantum fluctuations. The agreement is especially good in comparison to the ratio calculated from spin-wave theory, which is a factor of three smaller. The value of $J = 1030 \pm 50$ cm^{-1} extracted from this comparison is in excellent agreement with the most recent high energy neutron scattering results of Aeppli, et al.[11] In addition to describing the width of the B_{1g} mode, the agreement of the third cumulant indicates that all of the important scattering has been observed. If significant scattering occurred beyond 1 eV, the third cumulants would be quite different.

Thus, by considering nearest neighbor (NN) spin flips, a good description of the B_{1g} scattering is obtained. Considering scattering of photons by diagonal next neighbor (DNN) spin flips (see inset to Fig. 3 (b)) provides an explanation for the A_{1g} and B_{2g} features. In the Néel state, the diagonal spins are aligned, and flipping a pair of diagonal spins changes the total spin by 2. Since $\vec{S}_i \cdot \vec{S}_j$ is a spin zero operator, this process is forbidden. The strong quantum fluctuations, however, imply that there is a large probability that the diagonal spin will be antialigned, thus allowing the DNN process. In particular, the first fluctuation component gives rise to two classes of excitations. Starting with the flipped pair, the spin operators can flip an adjacent spin. Since only six bonds are broken, the peak is expected at $3J$. The other excitation class flips a diagonal spin further away, resulting in two isolated flipped spins, eight bonds broken, and a peak at $4J$. The first process exhibits pure A_{1g} symmetry, while the second contains both A_{1g} and B_{2g} components. Returning to the data in Fig. 3 (b), this heuristic argument explains why the B_{2g} feature peaks higher in frequency than the A_{1g} feature, and why the A_{1g} feature has a broader lineshape. When the complete series expansion calculation is performed, the calculated moments, using the value of J extracted from the B_{1g} feature, are in good agreement with those obtained from the spectra, as shown in Table I.

Since the coupling of the light to the magnons arises from the modulation of J by the electric fields, the matrix elements (M) might be expected to scale with the ground state J values:

$$M \sim \frac{\delta^2 J}{\delta E^2} \sim J. \tag{10}$$

Table I. Ratios of cumulants for A_{1g}, B_{1g}, and B_{2g} scattering in La_2CuO_4.

Sample	λ_L (Å)	B_{1g} M_2/M_1	A_{1g} M_1/J	A_{1g} M_2/M_1	B_{2g} M_1/J	B_{2g} M_2/M_1
La_2CuO_4	5145	0.23	3.65	0.38	4.12	0.22
	4880	0.26	3.62	0.39	4.65	0.25
	4579	0.28	4.19	0.39	5.05	0.23
Theory		0.23	3.5	0.34	3.9	0.28

In this case, the large spectral weight observed for the DNN scattering would appear anomalous, since the ground state diagonal exchange is much smaller than the NN exchange. Such a conclusion would hold true as long as the virtual excitations are within the manifold of the Heisenberg system. However, since the incident photons have energies much greater than J, consideration of the microscopic basis for H_R reveals that other eigenstates are important, and the matrix elements do not scale with the ground state J values.

Microscopically, the interaction of the photons and magnons described by the effective Hamiltonian in Eq. (2) can be derived from an excited state exchange model. For the case of simple direct exchange,[4] in which the photon scatters inelastically leading to an exchange of the spins at site i and j, the Raman matrix element has terms of the form:

$$M_{ij} = \sum_{\mu\nu} \frac{<\phi_{i\downarrow}\phi_{j\uparrow} \mid e\vec{E}_{sc}\vec{r}_j \mid \phi_{i\downarrow}\phi_{\nu\uparrow}>V_{ij}^{\mu\nu}<\phi_{\mu\uparrow}\phi_{j\downarrow} \mid e\vec{E}_{in}\vec{r}_i \mid \phi_{i\uparrow}\phi_{j\downarrow}>}{(\varepsilon_\nu-\omega_{in})(\varepsilon_\mu-\omega_{in})}, \tag{11}$$

where

$$V_{ij}^{\mu\nu} \equiv <\phi_{i\downarrow}\phi_{\nu\uparrow} \mid V_{ex} \mid \phi_{\mu\uparrow}\phi_{j\downarrow}>, \tag{12}$$

with

$$V_{ex} = \frac{e^2}{r_{ij}}, \tag{13}$$

and where the energy ε_ν includes the energy of the flipped spin on site i. Conceivably, the photon induced virtual excitations which enter into Eq. (11) could couple nearly equally well to both NN and DNN Cu sites. For example, in the high photon energy limit, where the intermediate states are plane waves, the Raman spin-pair scattering amplitude involves contributions from spin-pairs of all separations.

One set of intermediate states which would couple to NN and DNN copper sites is the set of La levels. To test this idea, the spin-pair scattering was measured[12] for Nd_2CuO_4 and Sm_2CuO_4 crystals, which have a slightly different out-of-plane structure, and have the La replaced by Nd or Sm. The spectra for the three symmetries are remarkably similar among the three crystals. If the La site were central to the excited state exchange process, dramatic changes would occur in the resonance behavior, since from the band structure[13] the levels of Sm and Nd are known to differ by 2 eV. Instead, the resonance profiles, the lineshapes, and even the intensities are quite similar, indicating that the scattering reflects only the excited state manifold of the CuO_2 planes. Also, if the A_{1g}

and B_{2g} were due to charge transfer excitations, the variations in crystal structure might change the positions of the features for the different crystals. The observed similarity gives additional confidence that these features are indeed magnetic excitations.

Finally, turning to the insulating $Ba_2YCu_3O_6$ material, Fig. 4 (a) shows very similar spectra to those in Fig. 3, even though the crystal structure is quite different from that of La_2CuO_4. The similarities among these spectra indicate that the light scattering probes the dynamics of the spin fluctuations within the planes, and that the out-of-plane structure has little influence on the spectra.

While much has been learned from the Ising series expansion, since the calculation begins with an effective Hamiltonian, it provides no prediction of the relative size of the NN and DNN components. Also, the calculation contains no information on the resonance behavior. Perhaps most importantly, the technique is restricted to the undoped systems. To calculate these properties, a more sophisticated model is needed, such as cluster calculations,[14,15] which can in principle calculate the resonance profiles, the size of the DNN to NN, and deal with doped systems. Since the spectra are dominated by the electronic structure within the CuO_2 planes, these spectra represent a well-defined challenge to theorists, and provide an excellent test for models of the CuO_2 planes.

EFFECTS OF DOPING

With the understanding of the 2D spin-1/2 quantum AFM gained in studying the undoped system, light scattering from spin fluctuations in doped materials may provide insight into the nature of the coupling of spin and charge excitations. Since we have focused on the undoped La_2CuO_4 system, a study of $La_{2-x}Sr_xCuO_{4+\delta}$ would be the logical extension into the metallic and superconducting regimes. However, although single crystals of $La_{2-x}Sr_xCuO_{4+\delta}$ with $0 < x < 0.2$ have been prepared, they do not exhibit bulk superconductivity, even for $x=0.15$, which gives $T_c = 33$ K in ceramic samples. Thus, at this point, further materials research is needed to clarify and control the relative roles of Sr doping (x) and oxygen stoichiometry (δ) in establishing superconductivity in La_2CuO_4.

In the $Ba_2YCu_3O_{6+\delta}$ system, on the other hand, the oxygen content (δ) in single crystals is under better control, and thus the spectra are more definitive. For the $Ba_2YCu_3O_{6+\delta}$ system, varying the oxygen content introduces holes. The initial observation of spin-pair scattering in $Ba_2YCu_3O_{6+\delta}$ led to the identification of an exchange parameter $J \sim 1000$ cm^{-1} for this system as well.[16] The series of spectra in Fig. 4 indicates the changes which occur in the spin fluctuation scattering as the oxygen content is increased. In the insulating phase, Fig. 4 (a), the well-defined two-magnon peak is observed in the B_{1g} symmetry. Introducing holes leads to a 60 K superconductor when $\delta = 0.6$ (Fig. 4 (b) [spectra shown in (b) were taken at 92 K]). The B_{1g} broadens and shifts to somewhat lower frequency, while the A_{1g} feature broadens and the peak shifts to zero frequency (Fig. 4 (c)). Further doping into $Ba_2YCu_3O_7$ leads to a 90 K superconductor, and to a complete broadening of the B_{1g} feature, which now peaks at zero frequency. The scattering has not vanished however, as seen by comparing the B_{1g} component to the B_{2g} component. The most obvious interpretation of these spectra is that while holes act as mobile defects which modify the dynamics of the spin fluctuations, the highly energetic spin fluctuations certainly persist in the doped materials.

Similar effects are seen in the La_2CuO_4 system, shown in Fig. 5. As carriers are doped into the material, the long range order is destroyed. For low doping the crystals are still in the insulating phase, and very little broadening is observed (Fig. 5 (b)). Further doping, into the metallic phase leads to a dramatic broadening of all features (Fig. 5 (c)).

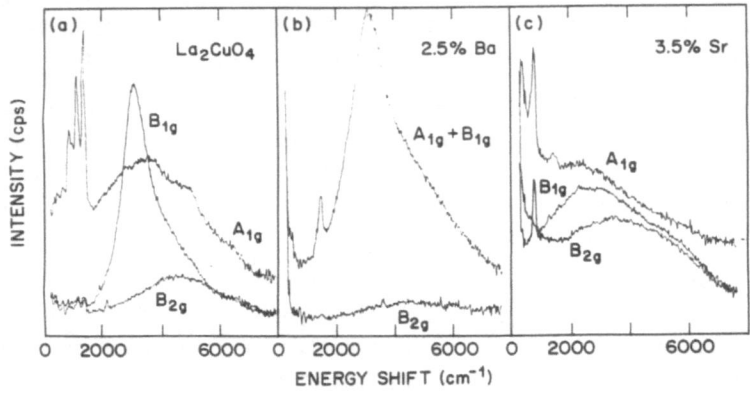

FIGURE 4. Raman spectra of $Ba_2YCu_3O_{6+\delta}$ for various oxygen dopings.

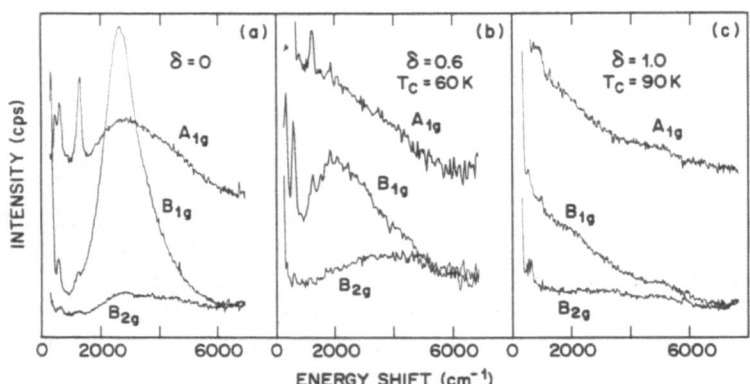

FIGURE 5. Raman spectra of doped and undoped La_2CuO_4.

As indicated, however, further materials research is needed and in progress in order to fabricate high quality single crystals exhibiting bulk superconductivity.

Holes can also be introduced into La_2CuO_4 by a high pressure oxygen anneal.[17] Spectra on these oxygen doped crystals also exhibit dramatically broadened features. While these oxygenated samples are superconducting at 40 K, the Meissner fraction is very small (< 2%), and the samples are probably multiphase. Indeed, neutron powder diffraction of oxygenated $La_2CuO_{4+\delta}$ samples reveals the presence of two different phases.[18] Further materials research on these systems is necessary in order to develop well-characterized single phase samples.

Doping Pr_2CuO_4 with Ce reputedly leads to an electron-doped superconductor with $T_c \sim 25$ K.[19] At this time, the electronic nature of these materials is still a point of controversy. Nominally, though, the effect of electron doping can be studied in the Pr_2CuO_4 system. At low Ce doping, very little broadening is observed in the spectra of Fig. 6 (a). At metallic concentrations of electrons, however, the B_{1g} mode in Fig. 6 (b) has broadened, but in a way somewhat different from the hole doped case. Instead of being uniformly broadened, the peak develops a flat top, but the edges of the feature are preserved. This asymmetry between the effect of hole and electron doping on spin fluctuations may be related to the asymmetry observed in the destruction of the AFM state with doping.[20] These materials are known to be multiphase,[21] however, and these results should be viewed as very preliminary, but nonetheless intriguing.

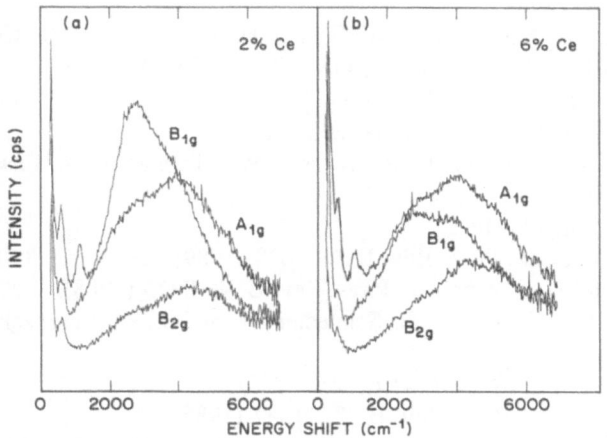

FIGURE 6. Spin fluctuation scattering in $(Pr,Ce)_2CuO_4$ for two dopings

CONCLUSION

Through inelastic light scattering from spin fluctuations a quantitative understanding of the spin dynamics in the 2D quantum AFM has been achieved. The cumulants of the observed spectral features are in quantitative agreement with the calculations which properly account for quantum fluctuations. From this agreement, values of $J \sim 1000$ cm^{-1} are determined for various AFM compounds possessing CuO_2 planes. In addition, the quantum fluctuations allow a DNN scattering process which leads to scattering in the A_{1g} and B_{2g} channels.

As carriers are doped into the materials, the spectral weight persists in the measured spectra, even into the superconducting phase. All of the spin fluctuation features broaden, weaken, and shift with doping. There appears to be a qualitative difference between the effects of localized and itinerant carriers. Some differences are also observed between doping with holes and doping with electrons. Clearly, there is a need for understanding the 2D quantum AFM with mobile carriers, and many theorists have begun to take up the challenge.

ACKNOWLEDGMENTS

We have benefited from discussions with M. S. Hybertsen, P. B. Littlewood, L. F. Mattheiss, A. Millis, K. M. Rabe, M. Schluter, R. R. P. Singh, and C. M. Varma. Samples have been provided by S-W. Cheong, A. S. Cooper, G. P. Espinosa, Z. Fisk, D. Rytz, L. F. Schneemeyer, and J. V. Waszczak. We thank J. Schirber for the high pressure oxygen anneal of the La_2CuO_4 crystals. We also thank H. L. Carter for technical assistance.

REFERENCES

1. S. A. Sunshine, *et al.*, Phys. Rev. B **38**, 893 (1988).
2. D. Vaknin, *et al.*, Phys. Rev. Lett. **58**, 2802 (1987).

3. R. R. P. Singh, P. A. Fleury, K. B. Lyons, and P. E. Sulewski, Phys. Rev. Lett. **62**, 2736 (1989).

4. P. A. Fleury and R. Loudon, Phys. Rev. **166**, 514 (1968).

5. R. J. Elliott and M. F. Thorpe, J. Phys. C **2**, 1630 (1969).

6. P. A. Fleury and H. J. Guggenheim, Phys. Rev. Lett. **24**, 1346 (1970).

7. J. B. Parkinson, J. Phys. C **2**, 2012 (1969).

8. K. B. Lyons, P. A. Fleury, J. P. Remeika, A. S. Cooper, and T. J. Negran, Phys. Rev. B **37**, 2353 (1988).

9. K. B. Lyons, P. E. Sulewski, P. A. Fleury, H. L. Carter, A. S. Cooper, G. P. Espinosa, Z. Fisk, and S-W. Cheong, Phys. Rev. B **39**, 9693 (1989).

10. R. R. P. Singh, Phys. Rev. B **39**, 9760 (1989).

11. G. Aeppli, *et al.*, Phys. Rev. Lett. **62**, 2052 (1989).

12. P. E. Sulewski, P. A. Fleury, K. B. Lyons, S-W. Cheong, and Z. Fisk, Phys. Rev. B **41**, 225 (1990).

13. L. F. Mattheiss, unpublished.

14. M. S. Hybertsen, *et al.*, Phys. Rev. B **39**, 9028 (1989).

15. E. B. Stechel and D. R. Jennison, Phys. Rev. B **38**, 8873 (1988).

16. K. B. Lyons, P. A. Fleury, L. F. Schneemeyer, and J. V. Waszczak, Phys. Rev. Lett. **60**, 732 (1988).

17. J. E. Schirber, *et al.*, Physica C **152**, 121 (1988).

18. J. D. Jorgensen, *et al.*, Phys. Rev. B **38**, 11337 (1988).

19. H. Takagi, S. Uchida, and Y. Tokura, Phys. Rev. Lett. **62**, 1197 (1989).

20. S. Uchida, *et al.*, Physica C **162-164**, 1677 (1989).

21. A.C.W.P. James and D. W. Murphy, in *Chemistry of Superconducting Materials*, T. A. Vanderah, ed., (Noyes, 1990).

RAMAN SCATTERING IN HIGH-T_c SUPERCONDUCTORS $YBa_2Cu_3O_x$ WITH DIFFERENT

OXYGEN CONTENTS

V. N. Denisov, A. F. Goncharov, B. N. Mavrin, and V. B.
Podobedov

Institute of Spectroscopy, USSR Academy of Sciences, Troitsk
Moscow Region 142092 USSR

Physical properties of the high-T_c superconductors $YBa_2Cu_3O_x$
significantly depend on the oxygen content in this compound. Raman
spectroscopy is an effective method to study the influence of the oxygen
content on lattice dynamics and, hence, on the physical properties of
$YBa_2Cu_3O_x$. It should be noted that a careful investigation of lattice
dynamics remains actual for understanding superconductivity in high-T_c
superconductors. The results of earlier Raman studies of high-T_c superconductors stage are summarized in several reviews[1,2]. In addition to
the investigations of the atomic vibrations of the lattice, Raman
spectroscopy of $YBa_2Cu_3O_x$ has been useful for a study of the energy gap,
two-magnon excitations, vibrational anisotropy, and anomalous changes in
the vibrational spectra at the superconducting transition. In this paper
we shall consider some correlations between Raman spectra and superconducting properties of $YBa_2Cu_3O_x$ single crystals with different oxygen
content ($6 \leq x \leq 7$).

We studied Raman scattering from freshly cleaved lateral planes of
$YBa_2Cu_3O_x$ single crystals at grazing incidence of the exciting radiation.
In order to reduce the laser heating the samples were pressed into indium
(they made contact with bulk metal along all sides except those being
studied), and the average laser power did not exceed 20 mW (the
excitation power density was ≤ 0.02 mW/μm)^2. According to our estimates
in this case the sample heating was less than 1 K per 3-5 mW at continuous laser excitation. Raman spectra were detected by a multichannel,
triple-stage polychromator with a resolution of about 4-5 cm^{-1}.[9] This
spectrometer enabled us to study Raman spectra of superconductors from
20-25 cm^{-1}.

The Raman fundamental bands of $YBa_2Cu_3O_x$ single crystals are
observed in the frequency region below 600 cm^{-1} (Fig. 1). Well-defined
spectra are obtained only in the zz and xx(yy) scattering geometries
where one should expect only totally symmetric A_g-modes for the orthorhombic phase ($x \geq 0.5$) and the $A_{1g}+B_{1g}$ modes for the tetragonal phase
($x < 0.5$)[4]. The spectra of these phases differ in some shift of the
corresponding bands and also in the appearance of new bands in the
spectra of the orthorhombic crystal, for example, at 230 cm^{-1}. Raman
spectra of the tetragonal phase transform gradually into spectra of the
orthorhombic phase with increasing x. Note that the total number of

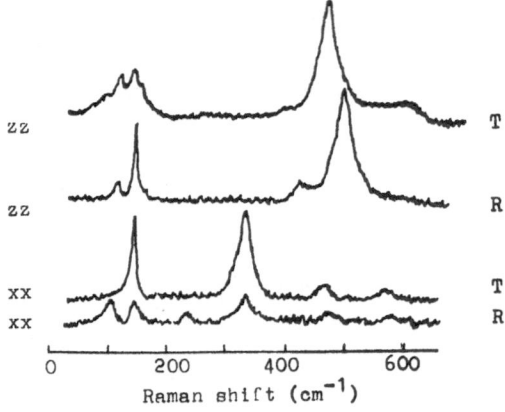

Fig. 1. Polarized Raman spectra of tetragonal (T) and orthorhombic (R) $YBa_2Cu_3O_x$ single crystals.

bands in these scattering geometries exceeds the expected number. If the additional band at 590 cm^{-1} is due to lattice disorder, since its intensity depends on a crystal quality, then an appearance of other additional bands can be completely explained by a doubling unit cell,[4] which is confirmed by many electron diffraction[5] studies.

The Raman spectra of $YBa_2Cu_3O_x$ are accompanied by a background due to electronic Raman scattering[6]. At temperatures below T_c a significant decrease of the background intensity in the frequency region below than 300 cm^{-1} is found which was interpreted as due to the appearance of the energy gap for electrons near the Fermi surface in the superconducting state of $YBa_2Cu_3O_x$. We have found that in the xx(yy) scattering geometries this electronic Raman background gradually decreases with decreasing x (Fig. 2), and its intensity correlates approximately with the carrier density in $YBa_2Cu_3O_x$ with different oxygen contents.

As is seen from Fig. 2, the intensity of the 339 cm^{-1} band that is attributed to out-of-plane vibrations of 02 and 03 oxygens in the CuO_2 planes (the B_{1g}-mode in the tetragonal phase) gradually decreases as x increases. This vibration is active only in the xx(yy) scattering geometries. The decrease of the band intensity may be related directly with the growth of the carrier concentration (of holes on 02 and 03 oxygens) in the crystal. Actually, it is known[7] that the appearance of a hole on oxygen gives rise to a contraction of the oxygen p-orbitals and, hence, may lead to a decrease of the overlap integral S_σ of the pdσ type (the overlap of the $d_{x^2-y^2}$-orbital of Cu2 with the $p_{x,y}$-orbitals of 02 and 03 in the CuO_2 plane). Since the 339 cm^{-1} band intensity in the xx(yy) scattering geometries should be proportional to $(S_\sigma)^4$ [8], then a decrease of S_σ with the hole doping will cause a weakening of the Raman intensity with increasing x.

Now the assignment of the main Raman bands in the spectra of $YBa_2Cu_3O_x$ to specific atomic vibrations is well known[1,4]. The most intense bands are attributed to vibrations with atomic displacements along the Oz-axis. With increasing x the most pronounced changes in the Raman spectra are inherent to band frequencies near 140 and 500 cm^{-1} that are assigned to out-of-phase vibrations (along the Oz-axis) of two Cu2 atoms (ν_2-mode) located in neighboring CuO_2 planes, and of two 04 bridging oxygens (ν_1-mode) connecting the CuO_2 planes with the CuO_3

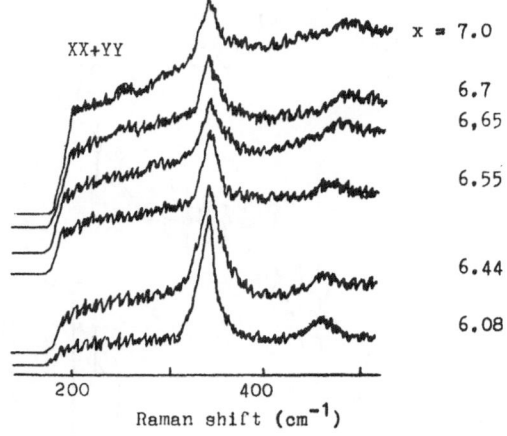

XX+YY

x = 7.0

6.7
6,65

6.55

6.44

6.08

Raman shift (cm^{-1})

Fig. 2. Raman spectra of $YBa_2Cu_3O_x$ with different oxygen content x. A step on the left of the spectra shows the background level.

chains, respectively. If the ν_2-mode shifts toward higher frequencies by about 10 cm^{-1} with increasing x, then the ν_1-mode shifts by about 30 cm^{-1} (1,2,4,9-11)

In $YBa_2Cu_3O_x$ crystals the lattice parameter c and the temperature T_c the of superconducting transition also depend on the oxygen content. We have measured the dependences of the ν_1-mode frequencies, temperatures T_c, and lattice parameters c on oxygen content for a great set of $YBa_2Cu_3O_x$ single crystals prepared by various methods (Fig. 3). It should be emphasized that the frequencies and the lattice parameters represented in Fig. 3 were measured at room temperature. It is seen from Fig. 3 that the $\nu_1(x)$ and $T_c(x)$ dependences are very similar, and one can show[16] that for $x \geq 6.5$ the ratio of the frequency shift $\Delta\nu$ ($\Delta\nu = \nu_1(x)-\nu_1$ (x=6.0)) to T_c is approximately constant. One can assume that the striking similarity of these dependences is due to the same reason responsible for their trends.

It should be noted that the behavior of $\nu_1(x)$ and $T_c(x)$ depends on the sample preparation method. For example, the plateau in these dependences was absent at x = 6.5-6.75 for single crystals prepared by annealing and quenching in helium or oxygen without subsequent low-temperature heat treatment. However, in all cases the dependence of the ν_1-mode frequency on temperature T_c, in contrast to $\nu_1(x)$, was smooth and independent of the sample preparation method[13].

Since the existing lattice-dynamical calculations are not satisfactory for analyzing the frequency shift with gradually increasing oxygen content in $YBa_2Cu_3O_x$ we have used a linear-chain model to show that changes of interatomic distances with increasing x cannot cause the observed increase in the ν_1-mode frequency. We have considered a chain of atoms located on the same Oz-axis: ...Cu2-O4-Cu1-O4-Cu2...[13]. It seems that this model gives a reasonable description of the nonpolar ν_1 and ν_2 modes. Actually, from solution of the equations of motion [13] one can obtain two vibrations with preferential displacements of O4 and Cu2 atoms, respectively, in accordance with the expected forms of

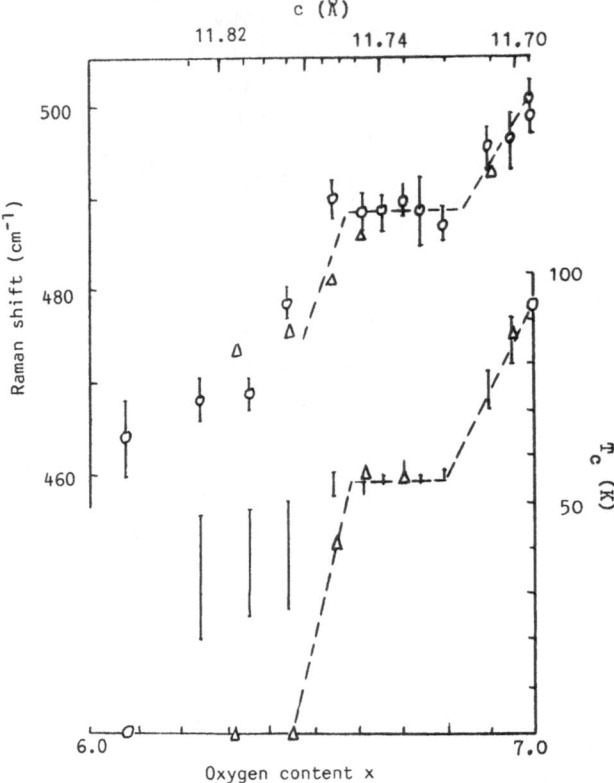

Fig. 3. Dependence of the ν_1-mode frequency, temperature T_c and the lattice parameter c on oxygen content in $YBa_2Cu_3O_x$ single crystals.

the ν_1 and ν_2 modes, and the ratio of the calculated frequencies is close to its experimental value. This model also describes, in agreement with experiment, a frequency shift of the ν_1-mode due to isotopic substitution of O4 and also due to changing the Cu_1O4 bond lengths in $LnBa_2Cu_3O_x$ when the Y atom is replaced by other lanthanides.

Within the framework of this linear-chain model one can obtain the result that the frequency of the ν_1-mode should decrease slightly (by about 1%) as x increases from 6 to 7 in $YBa_2Cu_3O_x$, while it increases by about 30 cm^{-1} in the Raman experiment (Fig. 3). Hence, a change of the Cu1-O4 and O4-Cu2 bond lengths with increasing x cannot account for the increase in the ν_1-mode frequency. Therefore the appearance of additional interactions in $YBa_2Cu_3O_x$ with inreasing oxygen content must be assumed.

One can arrive at such a conclusion also from an analysis of the mode Grüneisen parameter γ for the ν_1-mode for x close to 7 [13]. From the data for the dependence of the frequency of the ν_1-mode on pressure[14] and on the linear compressibility β_c along the Oz-axis[15] one can find the linear mode Grüneisen parameter for the ν_1-mode:

$$\gamma_p = \frac{1}{\beta_c} \frac{d\ln\omega}{dp} = 2.2. \tag{1}$$

On the other hand, since

$$\beta_c = - \frac{d\ell nc}{dp} ,$$

Eq. (1) may be represented as follows:

$$\gamma_c = - \frac{d\ell n\omega}{dc} . \tag{2}$$

For $YBa_2Cu_3O_x$ the $\omega(c)$ dependence may be measured without applying pressure as the lattice parameter c changes with inreasing x (Fig. 3). One can find from Fig. 3 that for the ν_1-mode $\gamma_c = 7 \pm 2$ for x > 6.75, which is significantly larger than γ_p. From the inequality $\gamma_c > \gamma_p$ it follows that the ν_1-mode shifts with increasing x not only due to a change of interatomic distances but also due to the appearance of some additional mechanism of the frequency shift of the ν_1-mode.

As was suggested in Ref. 16, an additional frequency shift of the vibrational bands in the spectra of superconducting cuprates is possible due to the interaction between the vibrations and the effective charge carriers. This shift is proportional to the carrier density n(x):

$$\Delta\nu \propto n(x). \tag{3}$$

In accordance with our measurements of $\nu_1(x)$ and $T_c(x)$ for x > 6.5 (Fig. 3) we have

$$\Delta\nu \propto T_c. \tag{4}$$

If this frequency shift of the ν_1-mode is completely due to the interaction of this mode with the effective charge carriers we obtain from Eqs. (3) and (4):

$$T_c \propto n(x). \tag{5}$$

Note that it is difficult to explain Eq. (5) within the framework of the Cooper pairing of the carriers[16], for which T_c is determined by the density of the electronic states at the Fermi level. At the same time, this relationship was also found from muon-spin rotation experiments on high-T_c superconductors[17].

The assumption that the frequency shift of the ν_1-mode is due to its interaction with the carriers is consistent with the conclusion[12] about the strong electron-phonon coupling for this mode. We consider that the strong electron-phonon interaction is the main reason for the unusual dependence of the Raman spectra of $YBa_2Cu_3O_x$ for x < 6.5 on the exciting power density[3,4]. We have observed a low-frequency shift, broadening, and weakening of the ν_1-mode intensity, and also the appearance of new bands in the 550-600 cm^{-1} region with increasing power density. These changes may be due to an interaction of phonons with photoinduced carriers that are generated during the absorption of the exciting radiation in semiconducting $YBa_2Cu_3O_x$.

Now we should like to draw attention to a change in the intensity of the I_b background that is present in Raman spectra (Fig. 2). In spite of difficulties in measuring I_b accurately, one can find from Fig. 2 and 3 that for x > 6.5 this intensity is also approximately proportional to T_c, which also gives rise to Eq. (5), since one should expect that $I_b \propto n(x)$.

In conclusion, we have found and studied some changes in the Raman spectra of $YBa_2Cu_3O_x$ as x increases from 6 to 7. With increasing oxygen content the background, due to Raman scattering by effective charge carriers gradually increases, the intensity of some bands decreases due to increase in the concentration of holes on oxygen atoms, and the frequencies of some bands may shift because of the interaction of phonons with the effective charge carriers. It is shown that the temperature T_c is most likely proportional to the carrier density in $YBa_2Cu_3O_x$.

REFERENCES

1.   R. Feile, Physica C 159, 1 (1989).
2.   P. Choudhury, Indian J. Phys. A 63, 221 (1989).
3.   A. F. Goncharov, V. N. Denisov, B. N. Mavrin, and V. B. Podobedov, Zh. Eksp. Teor. Fiz. 94, 321 (1988).
4.   V. N. Denisov, B. N. Mavrin, V. B. Podobedov, I. V. Alexandrov, A. B. Bykov, A. F. Goncharov, and O. K. Melnikov, Phys. Lett. A. 130, 411 (1988).
5.   Y. Kubo and H. Igarashi, Phys. Rev. B39, 725 (1989).
6.   F. Slakey, S. L. Cooper, M. V. Klein, J.P. Rice, and D. M. Ginsberg, Phys. Rev. B39, 2781 (1989).
7.   J. E. Hirsch and S. Tang, Sol. St. Comm. 69, 987 (1989).
8.   E. I. Rashba and E. Ya. Sherman, Pisma Zh. Eksp. Teor. Fiz., 47, 404 (1988).
9.   R. M. Macfarlane, H. J. Rosen, E. M. Engler, R. D. Jakowitz, and V. Y. Lee, Phys. Rev. B38, 284 (1988).
10. I. V. Alexandrov, A. B. Bykov, A. F. Goncharov, V. N. Denisov, B. N. Mavrin, O. K. Melnikov, and V. B. Podobedov, Pisma Zh. Eksp. Teor. Fiz. 47, 184 (1988).
11. A. F. Goncharov, V. N. Denisov, I. P. Zibrov, B. N. Mavrin, V. B. Podobedov, A. Ya. Shapiro, and S. M. Stishov, Pisma Zh. Eksp. Teor. Fiz. 48, 453 (1988).
12. T. Jarlborg, Sol. St. Comm. 71, 669 (1989).
13. V. N. Denisov, B. N. Mavrin, V. B. Podobedov, and A. F. Goncharov, Phys. Rep., to be published.
14. V. D. Kulakovski, O. V. Misochko, V. B. Timofeev, M.I. Eremets, E. S. Itskevich, and V. V. Stuzhkin, Pisma Zh. Eksp. Teor. Fiz. 47, 536 (1988).
15. I. V. Alexandrov, A. F. Goncharov, and S. M. Stishov, Pisma Zh. Eksp. Teor. Fiz. 47, 357 (1988).
16. V. M. Agranovich, V. N. Denisov, V. E. Kravtsov, B. N. Mavrin, and V. B. Podobedov, Phys. Lett. A134, 186 (1988).
17. H. Keller, IBM J. Res. Develop. 33, 314 (1989).

RAMAN SCATTERING FROM HIGH TEMPERATURE SUPERCONDUCTORS*

M. V. Klein

Department of Physics, Materials Research Laboratory,
and Science and Technology Center for Superconductivity
University of Illinois at Urbana-Champaign
104 S. Goodwin Avenue, Urbana IL 61801, USA

ABSTRACT

Raman scattering has been an unusually fruitful method of studying high temperature superconductors. Two-magnon excitations in antiferromagnetic samples, the broad electronic continuum in superconducting samples, and vibrational modes and their sensitivity to doping and to temperature have all been extensively investigated. The properties of the continuum are discussed, including its coupling to phonons, its universality, and its redistribution upon cooling well below T_c. Possible origins of the continuum, its possible non-Fermi liquid behavior, and its connection with the superconducting gap will be considered.

INTRODUCTION

It was predicted in 1961 that electronic Raman scattering from a superconductor should reveal the energy gap 2Δ.[1,2] The original calculations assumed that the superconducting coherence length ξ_0 was greater than the optical penetration depth δ, but the first experimental observations were on systems such as 2H-NbSe$_2$[3] and Nb$_3$Sn[4] or V$_3$Si[5] for which $\xi_0 << \delta$. In that case BCS theory says that for <u>intra</u>band scattering there should be a sharp, asymmetric peak in the Raman spectrum at 2Δ [6,7]. If the Fermi surface consists of intersecting sheets, <u>inter</u>band Raman scattering should also be observed, and the normal state should show a linear rise extending to energies of order of the band width. In the superconducting state at low temperatures, the rise is completely suppressed for energy shifts below 2Δ. There is an overshoot at 2Δ, followed by a gradual approach to the normal state response at higher energy shifts. A superposition of intraband and interband scattering gives a good quantitative fit to the Raman data from Nb$_3$Sn.[4] A plausible argument for interband scattering can be based on calculations of band structure.

Electronic Raman continua have been observed at room temperature in YBa$_2$Cu$_3$O$_{7-\delta}$ (YBCO).[8] They are remarkably strong in the sense that the integrated scattering is enormous (See below.). Some results are shown in Fig. 1. These samples were multiply twinned, so that, e.g., (xx) really denotes an average of (xx) and (yy) spectra. One sees phonon features in the (xx) spectra at 115, 150, 330, 420, and 505 cm^{-1}, assignable to the five expected modes of A$_g$ symmetry in the orthorhombic YBCO structure.[9]

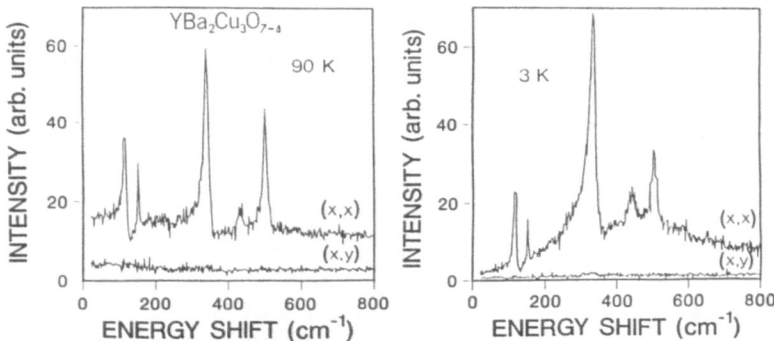

Fig. 1. Spectra [(xx) = orthorhombic A$_g$, and (xy) = orthorhombic B$_{1g}$] for single crystal YBa$_2$Cu$_3$O$_{7-\delta}$ at 90K (left) and at 3K (right). [Cooper et al.[8]]

Continuum for Temperatures above T$_c$

The continuum starts at zero energy (shift) and continues to 4000 cm^{-1}, in published spectra[10,11,12] It is reported to extend at least to 8000 cm^{-1}.[13] It has the wrong temperature dependence and is too structureless to be a two-phonon continuum. We conclude that the continuum must be electronic in origin and will discuss some possible origins for it below. Results from an untwinned single crystal of YBCO show that the strength of the continuum is greater when both photon polarizations are along the y (or b) direction than along the x (or a) direction.[14]

The fluctuation-dissipation gives for the Raman spectral density

$$S(\omega) = R''(\omega) \, (1 - e^{-\omega/T})^{-1} , \qquad (1)$$

where R''(ω) is the imaginary part of the response function for Raman scattering. For intraband electronic Raman scattering from a clean metal, Fermi liquid theory says that R'' should be proportional to energy ω from ω=0, peaking near qv$_f$, where v$_f$ is the Fermi velocity and where q is the wave-vector transferred. The latter is "smeared" due to the finite penetration depth $\delta \approx$1000Å, and this leads to an estimate of "q"v$_f \approx$20 cm^{-1}.

Within a conventional Fermi liquid picture the only other electronic source for the continuum would be <u>interband</u> scattering. This would require the Fermi surface to be multi-sheeted. If the sheets intersect in a line, $q \approx 0$ inter-sheet scattering would be described by an R" that would be proportional to ω at small ω. If the sheets touch in a line, R" would depend on a higher power (cubic) of ω. In the case of cubic or hexagonal symmetry, it is relatively easy to imagine a Fermi surface with intersecting sheets. For the orthorhombic structure of YBCO, that is not possible. Monien and Zawadowski have recently discussed a model electronic structure for YBCO where one branch of the Fermi surface, derived from the Cu-O planes is essentially tangent to another branch, derived from the Cu-O chains and the bridging oxygen atoms.[15] If this occurs at the zone boundary in the b*-direction, inversion symmetry will prevent the bands from hybridizing.

Continuum for Temperatures $T \ll T_c$

<u>Gap-like features.</u> As the temperature is lowered below T_c, the continuum changes shape, giving a broad peak in the 300-500 cm^{-1} region. (See Fig. 1.) It is tempting to associate the position of this peak with the value 2Δ of the gap. Figure 2 shows how the continuum looks using Raman selection rules appropriate for the Cu-O planes, namely tetragonal.[16] The (y'y') geometry gives tetragonal A_{1g} symmetry, and the (y'x') geometry gives tetragonal B_{1g} symmetry. The 330 cm^{-1} phonon has almost pure tetragonal B_{1g} symmetry, and the 115 cm^{-1} phonon has pure tetragonal A_{1g} symmetry. Note the symmetry-dependence of the continuum. For $YBa_2Cu_3O_{7-\delta}$ the position of the peak is higher (≈ 550 cm^{-1}) in B_{1g} tetragonal symmetry than in A_{1g} symmetry (≈ 350 cm^{-1}). This behavior might be interpreted as evidence for gap anisotropy.

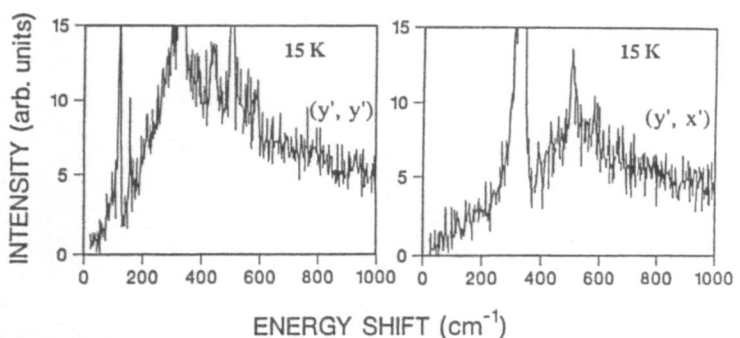

Fig. 2. Spectra [(y'y') = tetragonal A_g, and (y'x') = tetragonal B_{1g}] for single crystal $YBa_2Cu_3O_{7-\delta}$ at 15K.[16]

<u>Linear rise at small ω.</u> The low frequency portion of the Raman response at low temperature shows a linear rise that is unexpected for conventional BCS-type superconductors having a narrow distribution of gaps. A nodal structure (lines or points of zeros) of the gap such as discussed for the cases of anisotropic pairing would reasonably be expected to give a frequency dependence stronger

than linear. This suggests that some form of gaplessness is present in YBa$_2$Cu$_3$O$_{7-\delta}$ and, as we shall see, in Bi$_2$Sr$_2$CaCu$_2$O$_{8+\delta}$. For the former case, Monien and Zawadowski suggest that if the gap function $\Delta(\mathbf{k})$ has the appropriate p-wave or d-wave symmetry, it will vanish at the zone boundary, essentially where the interband Raman transitions are taking place.[15] This picture could explain the persistence of low frequency scattering at low temperatures.

<u>Antiresonances with Phonons</u>

In YBa$_2$Cu$_3$O$_{7-\delta}$ the phonons at 330 cm^{-1} and 112 cm^{-1} show temperature-dependent changes in position and line-shape.[17,18,19] This is shown in expanded detail in Fig. 3. The solid lines represent fits to a Fano line shape, wherein R" is given by a Lorentzian multiplied by an anti-resonance factor: The phonon is coupled to the continuum, thereby acquiring a width and suffering a shift in frequency. Light can scatter from the continuum through two channels, one direct, and the other indirect, via the phonon. At the antiresonance frequency (here greater than the renormalized phonon frequency) the amplitudes for the two channels cancel.

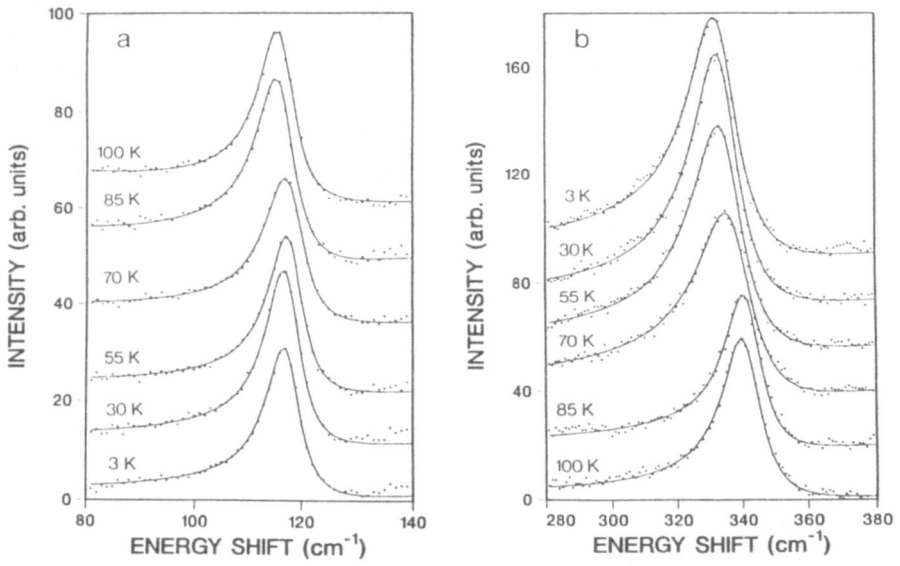

Fig. 3. Temperature dependences of the A$_{1g}$-tetragonal 115 cm^{-1} phonon (left) and B$_{1g}$-tetragonal 330 cm^{-1} phonon. Lines give Fano lineshapes.[19]

The 115 cm^{-1} phonon, which has pure A$_{1g}$-tetragonal symmetry, couples with the A$_{1g}$ part of the continuum, the major strength of which is well above 115 cm^{-1}; thus the renormalization by the coupling to the continuum will soften this mode. As the temperature is reduced through T$_c$ to nearly zero, both the broadening and the softening are seen to reduce. The former effect is a sign that the continuum at 115 cm^{-1} is weakening, and the latter effect is a sign that the effective frequency of the continuum is shifting upwards. The 330 cm^{-1} phonon both softens and broadens as the temperature is lowered. Broadening implies that the continuum near 330 cm^{-1} is becoming stronger; softening may imply that the effective frequency of the continuum at some average value of $2\Delta(T)$ is shifting upwards. We see indirectly via these phonon effects and directly from temperature-dependent spectra [Cooper et al., Ref. 8] that the continuum is redistributing when the temperature is lowered through T_c. We attribute this change to the formation of the superconducting gap, namely an increase in R" in the frequency range of the most probable value of 2Δ, and a reduction in R" at lower frequencies. Theoretical discussions of the effect of superconductivity on the phonon self energy have been given by Klein and Dierker using BCS theory[6] and by Zeyher and Zwicknagl using strong-coupling theory.[20]

UNIVERSALITY OF CONTINUUM

One naturally wants to know how "universal" is this continuum--its extension to very high frequencies, its redistribution due the formation of a gap, and its gapless, low-frequency behavior.

LSCO. A continuum has been observed at room temperature in superconducting $(La_{1-x}Sr_x)_2CuO_4$ by Sugai.[21] It extends to beyond 4000 cm^{-1}, and has a shape very similar to that in YBCO. No gaplike features have been identified in the 100 cm^{-1} range at low temperatures. They are probably masked by disorder-induced one-phonon scattering.

BSCCO. The continuum has also been seen in $Bi_2Sr_2CaCu_2O_{8+\delta}$ (BSCCO).[22,23] The sample used for Ref. 23 was freshly cleaved just before mounting in the cryostat for the measurements. BSCCO and YBCO give semiquantitatively similar spectra--broad "2Δ" peaks, the higher-frequency one having B_{1g} symmetry, with scattering all the way to (nearly) zero energy.

Sugai compares the room temperature continua of superconducting YBCO, LSCO, and BSCCO from 0 to 4000cm^{-1}.[21] They are remarkably similar.

BKB. Sugai also shows a nearly flat continuum from 0-4000 cm^{-1} in the non-cuprate superconductor $Ba_xK_{1-x}BiO_3$.[21] Unlike the other cases, the parent phase, $BaBiO_3$ is not an antiferromagnet but a charge-density-wave insulator.

TCBCO. A flat continuum for $T>T_c$ has recently been reported in $Tl_2Ca_2Ba_2Cu_3O_{10}$ by Krantz et al.[24] At low temperatures they observe the familiar linear rise to a rather weak, broad peak near 450 cm^{-1}. Results have recently been reported on $Tl_2CaBa_2CuO_8$ by Maksimov et al.[25] They show a weaker continuum and weaker gap-like structures than in the Y- and Bi-based superconductors. The Chernogolovka group has also observed continua in $Tl_2Ba_2CuO_{6-x}$, $Tl_2CaBa_2Cu_2O_{8+x}$, and $TlCaBa_2Cu_2O_{7-x}$.[26]

Oxygen-deficcient YBCO. In recent work Slakey et al. have reported a weak continuum in a single crystal of $YBa_2Cu_3O_{6.5}$ having $T_c = 60K$.[27] This work has continued, and some new results will be reported here. At room temperature the continuum is flat; at 150K it is still flat, as can be seen from the top spectrum of Fig. 4. The B_{1g}-tetragonal spectra shown in the figure then evolve with further cooling into a peak near 550 cm^{-1} and a shoulder at 200 cm^{-1}. Note that there are definite signs of the peak at 100K, well above T_c. A comparison of smoothed data with the phonon features removed is given in Fig. 5 between this sample and a fully oxygenated sample having $T_c=90K$. The curves have been arbitrarily matched at 800 cm^{-1}. Note that in A_{1g}-tetragonal symmetry the gap-like feature has slightly hardened and considerably broadened in the 60K sample with respect to the 90K sample.

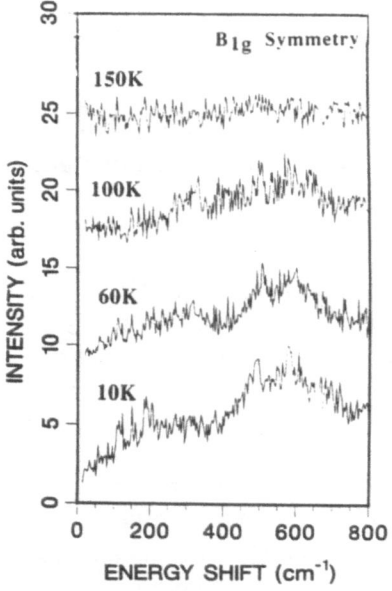

Fig. 4. Raman spectra of B_{1g}-tetragonal symmetry from $YBa_2Cu_3O_{6.5}$ having T_c= 60K.

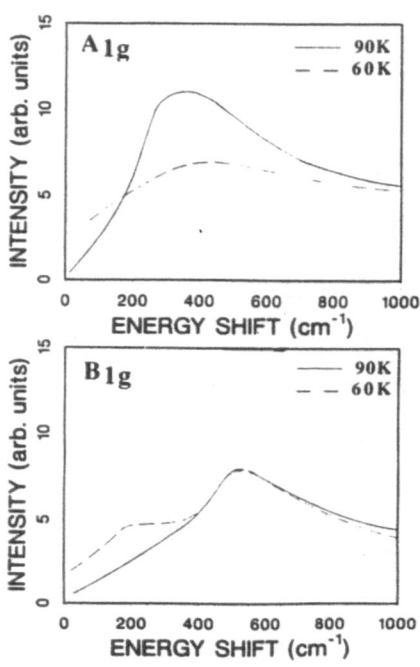

Fig. 5. Comparison of smoothed Raman spectra at 10 K from $YBa_2Cu_3O_{7-\delta}$ samples having T_c= 60K (dashed lines) and T_c=90K (solid lines).

DISCUSSION

Possible extrinsic origin for linear rise. Since $S(\omega) \propto \omega$ at small ω,T for a normal metal, one might attribute the linear rise to a metallic, non-superconducting near-surface layer. Results from a superconducting epitaxial YBCO thin film about twice as thick as the optical penetration depth show the same general behavior observed with YBCO single crystals.[28] The outer surface of the laser-ablated film is quite different from that of the single crystals. Freshly-cleaved BSCCO also shows the same behavior; it should not have such a normal-state "dead layer." One can probably rule out such an explanation for the continuum and its changes with temperature. The possibility that the result $[S(\omega) \propto \omega$ at small ω,T] in YBCO results from inhomogeneities or impurities is highly unlikely. The antiresonance of the 115 cm^{-1} phonon in that material is almost complete. This indicates that the excitations that make up the continuum must couple to the phonon displacements in practically every unit cell of the crystal. This argument does not apply to BSCCO or TCBCO, where antiresonances are absent or much less pronounced.

Flat normal state continuum. In the normal state of these materials the inelastic transport scattering rate is very strong, of order T,[29] whereas elastic scattering is much weaker.[30] We have suggested that inelastic scattering from fluctuations in the highly correlated electronic liquid is responsible for the normal-state continuum.[28] The nearly constant ω-dependence of $S(\omega)$ combined with Eq. (1) leads to

$$R''(\omega) \approx S_0 \, \omega/T \, , \text{ for } \omega << T, \tag{2a}$$
$$R''(\omega) \approx S_0 \, , \qquad \text{for } \omega >> T, \tag{2b}$$

where S_0 is a constant. If we assume the light scattering from fluctuations to be relaxational in nature, and if the scattering rate is Γ, then

$$R''(\omega) \propto \omega\Gamma/(\omega^2 + \Gamma^2) \, . \tag{3}$$

Equations (2a,b) and (3) will be obtained if Γ is a function of ω and T of the form

$$\Gamma(\omega,T) \approx Max(\omega,T) \, . \tag{4}$$

Equation (4) is consistent with the resistivity measurements.[29,30] A similar conclusion has been drawn from far-infrared conductivity.[31] Such behavior implies that the metallic state is a marginal Fermi liquid, as pointed out by Varma et al.,[32] who make a bold assumption that the Raman, infrared and a host of other anomalies can be universally explained in terms of an unusual, nearly q-independent charge and spin polarizability, obeying Eqs. (2a,b).

Ruvalds and Virosztek have argued that the Raman continuum can be explained as due to a nested nearly 2d Fermi surface.[33] Their result relies on assumptions about behavior of integrals used to obtain Green's functions and self energies under assumed nesting conditions. These assumptions need to be independently verified.

Finally, one would like to know the connection, if any, between the flat continuum discussed here for superconducting materials and the broadly-peaked two-magnon spectra seen in the antiferromagnetic insulating state.[34] The observation of the continuum in superconducting BKB implies that at least in that material with its nonmagnetic parent state the continuum is nonmagnetic.

Persistence of gap-like features above T_c = 60K in YBCO. The persistence of the 550 cm^{-1} peak seen in Fig. 5 reminds one of an edge or knee seen in the infrared reflectivity at 435 cm^{-1},[35] which has been seen above T_c in oxygen-reduced YBCO. Are features such as these evidence of pairing in the normal state prior to formation of a superfluid state?

An intriguing possibility is that the 200 cm^{-1} shoulder seen in Figs. 4 or 5 is a true measure of the gap. More Raman results are needed from a wider variety of oxygen-deficient samples in order to track these features with T_c.

Acknowledgement

This work was supported by the NSF under grants DMR 87-15103, DMR 86-12860, and DMR 87-14555 and received generic support through the Science and Technology Center for Superconductivity under NSF DMR 8809854.

References

* Collaborative work with F. Slakey, S. L. Cooper, J. P. Rice, E. D. Bukowski, and D. M. Ginsberg.

1. A. A. Abrikosov and L. A. Fal'kovskii, Zh. Eksp. Theor. Fiz. **40**, 262 (1961). [Transl.: Sov. Phys. JETP **13**, 179 (1961)]; A. A. Abrikosov and V. M. Genkin, Zh. Eksp. Teor. Fiz. **65**, 842 (1973). [Transl.: Sov. Phys. JETP **38**, 417 (1974)].
2. S. Y. Tong and A. A. Maradudin, Mat. Res. Bull. **4**, 563 (1969).
3. R. Sooryakumar and M. V. Klein, Phys. Rev. Letters **45**, 660 (1980); R. Sooryakumar and M. V. Klein, Phys. Rev. B **23**, 3213, (1981).
4. S. B. Dierker, M. V. Klein, G. W. Webb, and Z. Fisk, Phys. Rev. Letters **50**, 853 (1983).
5. R. Hackl, R. Kaiser, and S. Schicktanz, J. Phys. C **16**, 1729 (1983).
6. M. V. Klein and S. B. Dierker, Phys. Rev. B **29**, 4976 (1984).
7. A. A. Abrikosov and L. A. Fal'kovsky, Physica C **156**, 1 (1988).
8. A. V. Bazhenov, et al., Pis'ma Zh. Eksp. Teor. Fiz. **46**, 35 (1987) [JETP Lett. **46**, 32 (1987)]; K. B. Lyons et al., Phys. Rev. B **36**, 5592 (1987); S. L. Cooper et al., Phys. Rev. B **37**, 5920 (1988); R. Häckl et al., Phys. Rev. B **38**, 7133 (1988).
9. For a review of Raman scattering in high temperature superconductors see: C. Thomsen and M. Cardona in Physical Properties of High Temperature Superconductors I, edited by D. M. Ginsberg (World Scientific, Singapore, 1989), pp. 409-507.
10. I. Bozovic, D. Kirillov, A. Kapitulnik, K. Char, M. R. Hahn, M. R. Beasley, T. H. Geballe, Y. H. Kim, and A. J. Heeger, Phys. Rev. Letters **59**, 2219 (1987).
11. S. Sugai, Solid State Commun. **72**, 1193 (1989).
12. D. M. Krol, M. Stavola, L. F. Schneemeyer, J. V. Waszczak, H. O'Bryan, and S. A. Sunshine, Phys. Rev. B **38**, 11346 (1988).
13. Private communications from I. Bozovic and S. Sugai.
14. F. Slakey, S. L. Cooper, M. V. Klein, J. P. Rice and D. M. Ginsberg, Phys. Rev. B **39**, 2781 (1989).
15. H. Monien and A. Zawadowski, Phys. Rev. Letters **63**, 911 (1989).
16. S. L. Cooper, F. Slakey, M. V. Klein, J. P. Rice, E. D. Bukowski, and D. M. Ginsberg, J. Opt. Soc. Am. B **6**, 436 (1989).
17. R. M. Macfarlane, H. J. Rosen, and H. Seki, Solid State Commun. **63**, 831 (1987).
18. C. Thomsen, M. Cardona, B. Gegenheimer, R. Liu, and A. Simon, Phys. Rev. B **37**, 9860 (1988).
19. S. L. Cooper, F. Slakey, M. V. Klein, J. P:. Rice, E. D. Bukowski, and D. M. Ginsberg, Phys. Rev. B **38**, 11934 (1988).
20. R. Zeyher and G. Zwicknagl, Solid State Commun. **66**, 617 (1988); submitted to Z. Physik.
21. S. Sugai, Solid State Commun. **72**, 1193 (1989).
22. A. Yamanaka, T. Kimura, F. Minami, K. Inoue, and S. Takekawa, Jpn. J. Appl. Phys. **27**, L1902 (1988).
23. F. Slakey, M. V. Klein, S. L. Cooper, E. D. Bukowski, J. P. Rice, and D. M. Ginsberg, IEEE J. Quantum Elect., **25**, 2394 (1989); F. Slakey, M. V. Klein, E. D. Bukowski, and D. M. Ginsberg, Phys. Rev. B, to be published.
24. M. C. Krantz, H. J. Rosen, J. Y. T. Wei, and D. E. Morris, Phys. Rev. B **40**, 2635 (1989).
25. A. A. Maksimov, I. I. Tartakovskii, and V. B. Timofeev, Physica C **162-164**, 1243 (1989); A. A. Maksimov, I. I. Tartakovskii, L. A. Fal'kovskii, and V. B. Timofeev, preprint.
26. V. B. Timofeev, A. A. Maksimov, O. V. Misochko, and I. I. Tartakovskii, Physica C **162-164**, 1409 (1989).
27. F. Slakey, M. V. Klein, J. P. Rice, and D. M. Ginsberg, Physica C **162-164**, 1095 (1989).
28. M. V. Klein, S. L. Cooper, F. Slakey, J. P. Rice, E. D. Bukowski, and D. M. Ginsberg, in Strong Correlation and Superconductivity edited by H. Fukuyama, S. Maekawa, and A. P. Malozemoff, (Springer-Verlag, Berlin, Heidelberg, 1989) pp. 226-235.
29. P. A. Lee and N. Read, Phys. Rev. Letters **58**, 2691 (1987).
30. Resistivity measurements are reviewed by P. B. Allen, Z. Fisk, and A. Migliori in Ref. 9, Ch. 5.
31. R. T. Collins, Z. Schlesinger, F. Holtzberg, P. Chaudari, and C. Feild, Phys. Rev. B **39**, 6571 (1989).
32. C. M. Varma, P. B. Littlewood, S. Schmitt-Rink, E. Abrahams, and A. E. Ruckenstein Phys. Rev. Letters **63**, 1996 (1989).
33. J. Ruvalds and A. Virosztek, preprint.
34. See article by P. Sulewski, this volume.
35. S. L. Cooper, G. A. Thomas, J. Orenstein, D. H. Rapkine, M. Capizzi, T. Timusk, A. J. Millis, L. F. Schneemeyer, and J. V. Waszczak, Phys. Rev. B **40**, 11358 (1989).

DECAY OF EXCITON GRATINGS IN ANTHRACENE: ANISOTROPY OF LOWEST EXCITON

BANDS AND COEXISTENCE OF LONGPATH AND SHORTPATH WAVEGUIDE MODES

V. M. Agranovich and T. A. Leskova

Institute of Spectroscopy
USSR Academy of Sciences
Troitsk, Moscow Region
142902, USSR

INTRODUCTION

An anthracene crystal is one of convenient model systems that has
been used already for many decades for systematic investigations of
electronic excitation energy transfer by Frenkel excitons (small-radius
excitons). Laser spectroscopy methods have significantly enriched this
field of investigations. In particular, within the last few years the
methods of transient gratings (TG) has aroused special interest.

This method is known to be used widely for the study of various
kinetic parameters of condensed media (liquids, semiconductors, etc.,
see, e.g. [1]). For the study of exciton transport in molecular crystals
TG experiments were proposed in [2-5] and afterwards were performed on
anthracene thin films in [6]. The results obtained attracted the
attention of theoreticians (see, e.g. [7-14] and references there), and
experiments on exciton transport in anthracene were continued (see
[15-17]). Of special interest here are, first, investigations of the
exciton transport at low temperatures (T = 1-4 K) when the anisotropy of
the exciton bands must be manifested in full degree.

Note that the anisotropy of exciton bands in anthracene is mainly
caused by a long-range resonance dipole-dipole interaction. In cubic
crystals of this kind the interaction leads to the so-called longitudinal-transverse splitting of exciton bands. In the case of an
anthracene crystal (space group C_{2h}^5) this interaction is responsible for
the appearance of a nonanalytical dependence of the exciton energy on the
wave vector $\mathbf{k}$ for small $\mathbf{k}$. The lowest electronic transition in an
anthracene molecule, from which the lowest exciton bands in an anthracene
crystal originate, corresponds to a rather high oscillator strength
(F = 0.24). Therefore, contributions to the exciton energy nonanalytic
in $\mathbf{k}$ turn out to be quite substantial, up to $\Delta = 400$ cm^{-1}, and at
sufficiently low temperatures they make the motion of excitons
practically quasi-two-dimensional [18,19]. However, for the same reason
(the high oscillator strength) a discussion of the mobility of electronic
excitations in anthracene at low temperatures with the use of Coulomb
states (Coulomb excitons) ceases to be justified. The inclusion of the
retarded part of the interaction is known to lead to the renormalization
of the exciton spectrum into a spectrum of polaritons which is rather
anisotropic in an anthracene crystal in the region of the lowest energies

we are interested in. In addition, this spectrum differs appreciably from that of Coulomb excitons (see below).

Nevertheless, in a series of numerous theoretical works (see, e.g. [7-14]) neither the anisotropy of the exciton bands nor paritonic effects were taken into account in the analysis of mechanisms of the decay of exciton gratings in anthracene at low temperatures. Besides, the majority of them were carried out within the nearest neighbour approximation and for one-dimensional models (for their inapplicability in the discussion of the decay kinetics of exciton gratings in anthracene see, also, [14]).

Note that in [6] the energy of coherent laser photons generating the exciton grating was approximately equal to $\hbar\omega = E_b + \hbar\Omega$, where $\hbar\Omega$ is the energy of the vibrational quantum $\Omega = 1400$ cm^{-1}. The lifetime of vibronic states of such energy is two or three orders smaller than the decay time of the exciton grating τ_f observed in [6] (according to [20] $\tau_f < 1$ps; it is clear that τ_f characterizes the time of decay of the vibrational quantum $\hbar\Omega$ into lattice phonons). Therefore, when analyzing the experiments [6] the state which results from the decay of the vibronic states can be considered as the initial state. Since many phonons can participate in the above-mentioned decay processes and, besides, as the width of the optical phonon band is much smaller than that of the lowest exciton band in anthracene, there are no physical reasons restricting the wave vector of the exciton after the decay of the vibrational quantum. However, as we consider the region of low temperatures T << ΔE, where ΔE is the exciton band width (T ≈ 1K, ΔE ≈ 200K), excitons also undergo rapid relaxation into the region of low energies. However, in the region of low energies excitons turn into polaritons, and after rapid relaxation (i.e. for t < 10^{-11}s) polaritons must accumulate in the "bottle neck" region. States of the same type must, apparently, appear after rapid relaxation in the case when the frequency of the laser beams is confined to the exciton resonance (as was the case in [15,16]). In this frequency region the mean free path of photons is short, $\ell \approx 10^{-6}$ cm, so that the uncertainty of the wave vector of excited polaritons turns out to be quite large (much larger than ω_b, $\frac{\omega_b}{\beta} \approx 1.5 10^5$ cm^{-1}).

The polariton concentration at each point of the crystal at the initial moment (i.e. for t $\gtrsim 10^{-11}$s) must be proportional to the total energy absorbed at the same point. As this energy is proportional to $|\vec{E}(\vec{r},t)|^2$ ($\vec{E}(\vec{r},t)$ is the electric field of the incident pulses, $|\vec{E}(\vec{r},t)|^2 \propto 1 + \cos(2\pi x/L)$), an initial inhomogeneous (periodic) distribution of the concentration of polaritons is formed which is afterwards smoothed out by diffusion and which naturally leads to the dependence of the grating decay time on the grating period L. On the basis of the above analysis it was stressed in our paper [21] that in the experiments [6] the decay of the grating was, actually, due to the diffusive motion of polaritons in the bottle neck region rather than that of excitons. Naturally, the motion of polariton wavepackets was meant, so that the diffusion coefficient is D = $v_g \ell$, where ℓ is the mean free path of the polaritons and v_g is their group velocity.

This physical concept is now accepted and is already used to discuss TG experiments on anthracene at low temperatures (see [15,16]). However, further experiments have brought forward new theoretical problems, and it is these problems that we shall discuss in this paper.

Note that the mean free path ℓ and the group velocity v_g of polaritons in the bottle neck region can be taken from experiments if they are available or can be calculated if the complex dielectric tensor

$\epsilon_{ii}(\omega)$ in the vicinity of the lowest excitonic transition $\omega \approx \omega_b$ is known. For example, according to [20], in the frequency region $\omega \approx \omega_b$ and when $\vec{k} \| \vec{c}'$ the penetration depth (i.e. the length ℓ) of the light polarized along the crystallographic axis b is of the order 0.1 - 1 μm. At the same time, in the bottle neck region $v_g \approx 10^5$ cm/s, so that $D \sim 1$ - 10 cm^2/s. For $\vec{k} \| \vec{a}$ and the same light polarization the value of D can hardly differ appreciably from the results of the above estimation, and this was used in [21]. It became known afterwards (see [15]) that in the experiments [6] the TG vector was actually directed along the b crystallographic axis rather than along the a axis as was stated in [6]. This correction turned out to be rather important. It meant that in all experiments carried out until now the exciton gratings were generated by pulses of p-polarized radiation with the b axis in the plane of incidence. Therefore, in an anthracene crystal not s-polarized, but only p-polarized states could be excited, i.e. the states with the excitonic polarization vector parallel to the b axis. Here, for simplicity, we assume the anthracene crystal to be uniaxial with the optical axis parallel to the b axis. Such an approximation is justified since we are interested only in the frequency region $\omega \approx \omega_b$, where ω_b is the frequency of the excitation linearly polarized along the b axis. But, as is known for uniaxial crystals, the dispersion laws for s- and p-polarized waves can, generally speaking, be quite different (see also below).

In discussion of the decay of TG with the grating vector parallel to the b axis it seems attractive, first of all, to evaluate the diffusion coefficient D_{bb} taking into account the excitons (or polaritons) with wave vectors parallel to the b axis, $\vec{k} \| \vec{b}$, i.e. longitudinal waves. However, at low temperatures ($T \approx 1K$) these states are populated negligibly, as their energy is $E_L = \hbar(\omega_b + \Delta)$, $\Delta \approx 400$ cm^{-1}. At the same time, the most strongly populated states are those with energies $E \approx \hbar\omega_b$ in the region of the minimal exciton energies. Therefore, in evaluating the diffusion coefficient along the b axis it is necessary to take into account the states of a more general type, i.e. states whose wave vector $\vec{k}$ also has a component along the c' axis. The first evaluations of this kind were made in [22]. Below we revise these calculations and, in addition, we point out some new theoretical problems requiring special consideration.

P-POLARIZED WAVEGUIDE POLARITONS PROPAGATING ALONG THE b AXIS

In order to understand the nature of the phenomena observed in [6] it should be taken into account that the decay kinetics of exciton gratings was studied, as already noted, on thin anthracene films (1 - 2μm thick) rather than on bulk samples. Polaritons in thin films are waveguide polaritons, both radiative and nonradiative. At low temperatures radiationless relaxation of excited vibronic states populate waveguide polaritons of both types. In this sense the situation is analogous to that encountered within the framework of the polaritonic mechanism of fluorescence of bounded crystals, viz. as a result of the rapid relaxation of excited vibronic states not only bulk polaritons but also surface polaritons are excited and they can both contribute to the emission spectrum. It is known that surface polaritons in a lossless medium are "nonradiative" modes, i.e. they do not possess a radiative width. However, scattering by acoustic phonons or by surface roughness enables surface polaritons to become radiative, and this leads to the appearance of additional structure in polariton fluorescence spectra [23] (for experimental observations of surface polariton emission, see, e.g. [24]. There is no doubt that nonradiative waveguide modes can be converted into radiation due to scattering by phonons or defects. What is more important, the frequencies of these modes are, in particular, in the region $\omega \gtrsim \omega_b$, thus at low temperatures these states must be

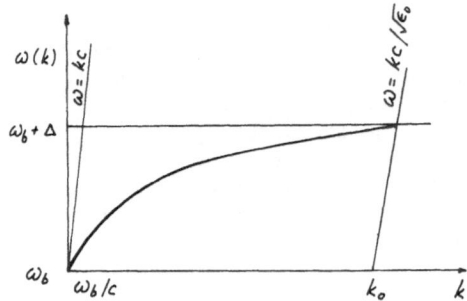

Fig. 1. Dispersion curves for surface polaritons along the b-axis.

populated. In Figs. 1-3 the dispersion curves are shown for p-polarized surface and waveguide polaritons propagating along the b axis in an anthracene film (note that the crystallographic a and b axes are in the plane of the film). The film thickness was assumed to equal d = 1.9 μm (as was the case in the experiments [16]). In calculations of the dispersion curves the dielectric tensor of the crystal was assumed to be diagonal rather than uniaxial. It was also assumed that $\epsilon_{c'c'} = \epsilon_0$,

$$\epsilon_{bb} \equiv \epsilon_b = \epsilon_0(1 + \frac{\Delta}{w_b - w}) \ ,$$

where $\epsilon_0 = 2.5$, $\Delta = 364$ cm^{-1}, $w_b = 25097$ cm^{-1}. The values of ϵ_0, Δ, and w_b were taken from [20]. We do not give here the value of the component ϵ_{aa} as it does not enter the dispersion relation for p-polarized waveguide polaritons propagating along the b axis. It was also assumed that the film was surrounded by vacuum. In this case all normal modes are either symmetric or antisymmetric. Their dispersion relations are

$$\tan\left[\frac{d\kappa}{2} \left(\frac{\epsilon_b}{\epsilon_0}\right)\right]^{1/2} = \epsilon_0 \frac{\kappa_0}{\kappa} \left(\frac{\epsilon_b}{\epsilon_0}\right)^{1/2}, \tag{1}$$

$$\cot\left[\frac{d\kappa}{2} \left(\frac{\epsilon_b}{\epsilon_0}\right)\right]^{1/2} = -\epsilon_0 \frac{\kappa_0}{\kappa} \left(\frac{\epsilon_b}{\epsilon_0}\right)^{1/2} \ , \tag{2}$$

where $\kappa_0 = \left(k^2 - \frac{w^2}{2}\right)^{1/2}$, $\kappa = \left(\epsilon_0 \frac{w^2}{3} - k^2\right)^{1/2}$, and $\vec{k}$ is the wave vector of the waveguide polariton propagating along the b axis (the x axis). Equations (1) and (2) determine the frequencies of the waveguide polaritons as functions of their wave vector k.

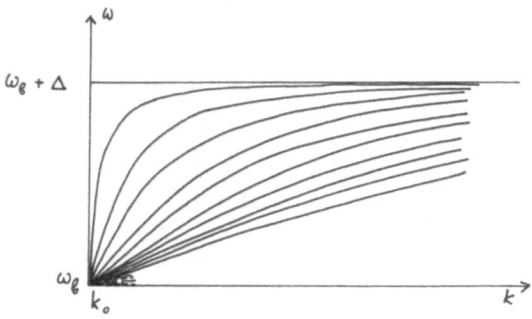

Fig. 2. Dispersion curves waveguide modes in the frequency region $w \gtrsim w_b$ propagating along the b axis.

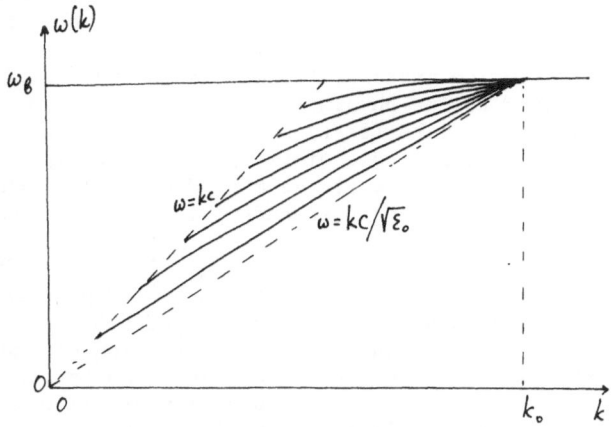

Fig. 3. Dispersion curves for waveguide modes in the frequency region $w \leq w_b$ propagating along the b axis.

First, let us consider the surface polariton states. In the case of the uniaxial crystal under discussion a surface polariton propagating along the optical axis exists in the whole frequency region where $\epsilon_b(w) < 0$, but its wave vector k varies within the interval $(w_b/c; \sqrt{\epsilon_0}\ w_b/c$ where κ and κ_0 are both real. For the lowest transition in anthracene $\frac{b}{c} \sim 1.5 \times 10^5$ cm^{-1}, thus kd >> 1 over the whole region of the surface polariton's existence. Therefore, coupling between the surface modes on the opposite faces of the film is negligible. Both surface modes are independent in this case and their dispersion curves coincide with that for surface polaritons propagating along the surface of a uniaxial crystal (see [26]).

Now we turn to a discussion of the properties of waveguide modes (see Fig. 2,3). In Fig. 2 are shown the dispersion curves for waveguide modes in the frequency region $w \gtrsim w_b$, while in Fig. 3 these curves are shown in the frequency region $w \lesssim w_b$. In view of the dispersion curves it follows that all normal modes can, roughly speaking, be separated into two groups of states. We refer to the first of them as photon-like states, i.e. states which have a small "exciton content" and have an almost photon character. These states, including also the surface polaritons, have large group velocities and, accordingly, long mean free paths. Propagation of wavepackets of the photon-like states on the scale of the grating periods cannot be described within the diffusion approximation, as the mean free path of these states is larger than the grating period. The states of the second group are, on the contrary, exciton-like. Their mean free path can be much shorter than the grating period, so that the propagation of the wavepackets can be considered in the diffusion approximation. In what follows it is namely these states which are of special interest. However, before we turn to a discussion of the properties of these states we shall make some remarks about the experimental results [6,15,16]. In the above-referenced works the decay of exciton gratings was studied on the nanosecond time scale. On this time scale the intensity of the diffraction beam was found to decrease exponentially I $\propto$ exp(-Kt) with a decay rate $K = \frac{2}{\tau_0} + 2D(2\pi/L)^2$, where $L = \lambda/\phi$, and π is the angle between the laser beams exciting the grating. For small angles ϕ the decay rate K is a linear function of ϕ^2. These two observations prove unambiguously that on the above time

149

scale the initial periodic distribution of the concentration of quasi-particles is smoothed out due to their diffusion, i.e. that for these quasiparticles the diffusion approximation is, actually, sufficiently good. If now we go back to the physical concept we are using here it should be noted that photon-like (fast) states can also make a certain contribution to the initial periodic distribution of quasiparticles. However, the contribution of these states must disappear on the time scale S/v_g, where S is the grating size and v_g is the group velocity. Assuming $S \approx 300\mu m$ (see [6.16]) and e.g. for $v_g \approx 10^8 - 10^{10}$ cm/s we conclude that this is very short, viz. of the order of picoseconds. Thus, it follows from the above that the exciton grating decay observed on the nanosecond time scale is mainly due to diffusion of short path (exciton-like) states with a mean free path ℓ much shorter than the grating period, $\ell \ll L$. Naturally, the separation of all waveguide polaritons into two strictly definite groups is rather approximate. In reality along with short path exciton-like states (for which $\ell \ll L$) there exist also such exciton-like states with longer mean free paths $\ell \lesssim L$ which can no longer be studied within the diffusion approximation. We shall return below to the discussion of a possible role of these states. But now we shall evaluate the diffusion coefficient for short path exciton-like waveguide polaritons.

GROUP VELOCITY, MEAN FREE PATH AND DIFFUSION COEFFICIENT OF SHORT PATH WAVEGUIDE MODES

As has been stressed more than once, polaritons at low temperatures $T \approx 1 - 4K$ accumulate in the bottle neck region, i.e. in the vicinity of the resonance frequency ω_b. This conclusion is, as is known, confirmed by observations of the spectra of polaritonic fluorescence and their temperature dependence [19,20], and now can be considered as established. In this frequency region the modulus of the functions on the right-hand side of the dispersion relations (1) and (2) is much larger than unity. Therefore, from Eqs. (1) and (2) simple size quantization of the z component of the wave vector follows:

$$\frac{d}{2} k_z = N \frac{\pi}{2}, \quad N = 0,1,2\ldots, \tag{3}$$

where $k_z = \kappa \sqrt{\epsilon_b/\epsilon_0}$. It was supposed in (3) that the z-axis is directed along the c' axis. The above expression for k_z is equivalent to the well-known dispersion relation for the extraordinary wave in a uniaxial crystal:

$$\frac{k_z^2}{\epsilon_b} + \frac{k^2}{\epsilon_0} = \frac{\omega^2}{c^2}. \tag{4}$$

The dielectric constant $\epsilon_b(\omega)$ is, generally speaking, complex:

$$\epsilon_b(\omega) = \epsilon_0 (1 + \frac{\Delta}{\omega_b - \omega - i\Gamma}) \tag{5}$$

The quantity Γ appearing in this expression leads to damping of polaritons and is a function of frequency. As the frequency ω_b is the minimal frequency of Coulomb excitons in anthracene, the function $\Gamma(\omega)$ drops sharply upon a transition from the frequencies $\omega \gtrsim \omega_b$ to the frequencies $\omega \lesssim \omega_b$ (the so called effect of the long-wavelength edge of the exciton absorption bands; see [27] and also [28] and references there). Therefore, the mean free times and mean free paths of waveguide

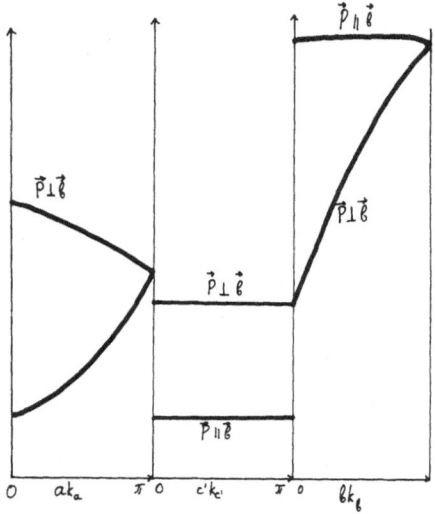

Fig. 4. Dispersion curves for the lowest exciton bands in anthracene ([30], dipole-dipole approximation).

polaritons for $\omega < \omega_b$ increase abruptly and, due to what was said above, we may neglect these modes in calculating the diffusion coefficient. For the states with $\omega > \omega_b |\omega - \omega_b| \approx 1 - 3 \text{ cm}^{-1}$) we assume $\Gamma = 1 \text{ cm}^{-1}$. This value of Γ corresponds to the mean free time $\tau_0 = 30\text{ps}$ obtained previously in experiments on the polariton fluorescence kinetics at $T = 1.8\text{K}$ in anthracene [20]. Besides, this value of Γ should be postulated when fitting the absorption spectra of anthracene with the use of Eq. (5) [29]. We also take into account spatial dispersion, i.e. we assume that the frequency of Coulomb excitons is

$$\omega_b(k) = \omega_b^0 + \frac{hk_z^2}{2M_z} + \frac{hk_x^2}{2M_x} \; ,$$
(6)

where $\omega_b^0 = 25097 \text{ cm}^{-1}$, and the effective mass M_x, according to [20], equals 500 m_0, m_0 with the free electron mass. We are not aware of any experimental data which would permit to evaluate the effective mass M_x. Theoretical estimations, always approximate, are rather diverse. Keeping in mind what has been said we turn, nevertheless, to the results of calculations [30] (Fig. 4). These calculations were made with the inclusion of only a dipole-dipole resonance interaction and are now considered as the most trustworthy. In Fig. 4 the dispersion curves of the lowest exciton bands in anthracene (the Davydov doublet $\mu = 1,2$) are shown for different directions of the wave vector ($\vec{k}$ is along the a, c', b axes). For each exciton branch $\mu = 1,2$ the directions of the transition dipole moment $\vec{P}$ as $\vec{k} \to 0$ are given. Because the contribution of higher multipoles was omitted in the calculations [30] their results shown in Fig. 4 can be used only for a qualitative analysis. In particular, it follows from this figure that the effective mass in the direction of the b axis (the x axis) for the exciton band $\mu = 1(\vec{P}\|\vec{b}]$ is rather large $M_x \sim 100m_0$, but smaller than the effective mass M_z in the direction of the c' axis. We shall show below that the comparison of our calculations with the experimental results on the TG decay can yield information about the value of the effective mass M_x.

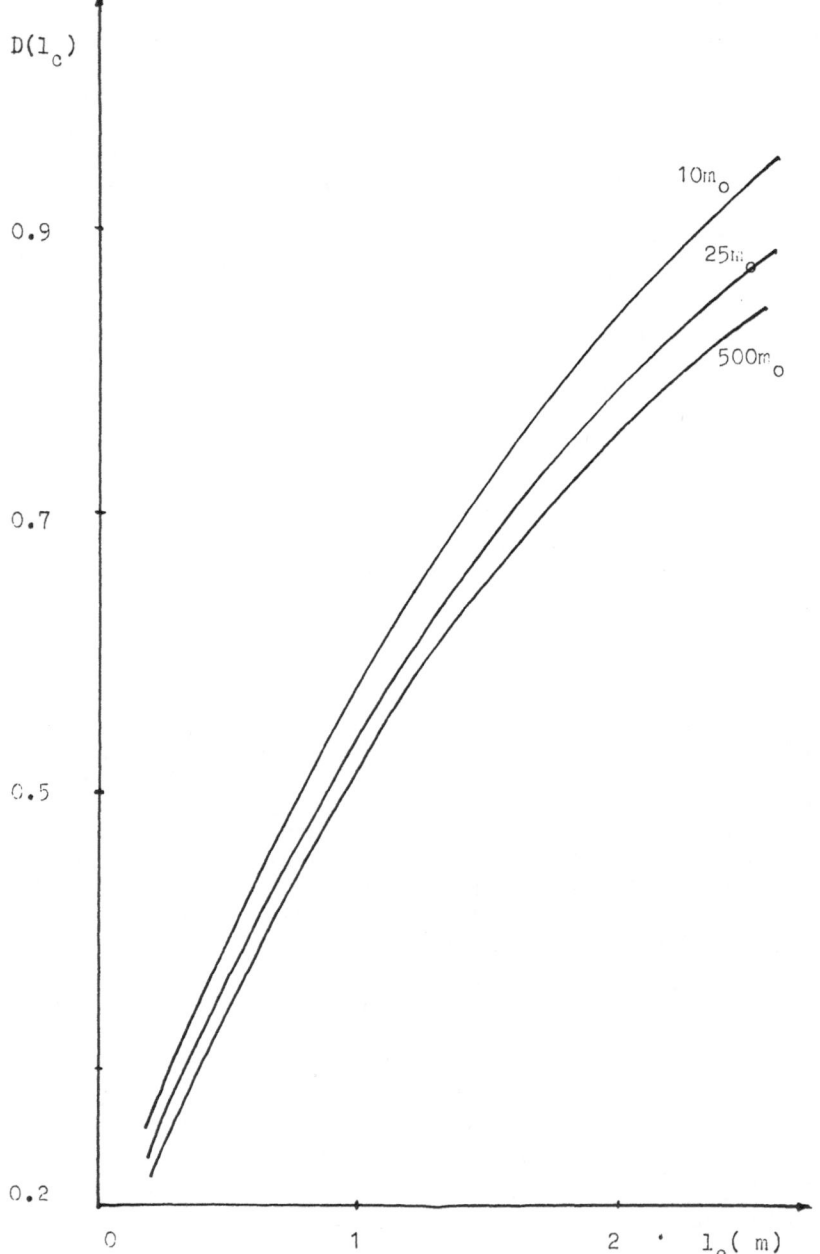

Fig. 5. The diffusion coefficient of waveguide modes versus ℓ_c for different values of M_x.

With regard for what has been said above we find from Eq. (5) the following expression for the frequencies of the waveguide modes existing in the frequency region $\omega \gtrsim \omega_b^0$:

$$\omega = \omega_b^0 + \frac{\Delta(k^2 - k_0^2)}{k_z^2 + k^2 - k_0^2 + a^2} + \frac{hk_x^2}{2M_z} + \frac{hk^2}{2M_x} \quad , \tag{7}$$

where $k_0^2 = \epsilon_0 (w_b^0)^2 / c^2$, $a^2 = 2\Delta k_0^2 / w_b^0$. Consequently, the group velocity is

$$v_g = \frac{\partial \omega(k, k_z)}{\partial k} = \frac{2\Delta k (k_z^2 + a^2)}{(k_z^2 + k^2 - k_0^2 + a^2)^2} + \frac{hk}{M_x} \quad . \tag{8}$$

As the diffusion character of the grating decay was observed in [6,15,16] for gratings whose period L fell within the interval 1 - 10μm, we calculate the diffusion coefficient of the waveguide modes averaging its value only over all short-path whose mean free path ($\ell(k, k_z)$ is, by definition, shorter than some length ℓ_c, $\ell_c < L$. Naturally, such a selection of the class of states has some arbitrariness. We shall verify below the importance of this arbitrariness for the evaluation of the diffusion coefficient sought.

As we are interested in the region of k for which $k \underset{\sim}{>} k_0$ (see Fig. 2), and the group velocity (8) increases with k, the mean free path $\ell(k, k_z) = v_g(k, k_z) \tau_0$ at $k = k_0$ is smaller than ℓ_c only when $k_z \underset{\sim}{>} bk_0$, where $b = (2\Delta \tau_0 / k_0 \ell_c)^{1/2}$. Thus, the above inequality determines the class of states over which the diffusion coefficient should be averaged. Assuming now that the exciton-like short-path waveguide modes are in thermodynamic equilibrium, we find that

$$D = D_0 \frac{A}{B} \quad , \tag{9}$$

where $D_0 = \dfrac{\ell_c^2}{\tau_0}$, where

$$A = b^4 \int_1^\infty \int_1^\infty \left[\frac{y^2 b^2 + a^2}{(y^2 b^2 + x^2 - 1 + a^2)^2} + \frac{\gamma T}{\Delta} \right]^2 x^2 \exp(-\Psi(x,y)) \, dy \, dx \quad , \tag{10}$$

$$B = \int_1^\infty \int_1^\infty \exp[-\Psi(x,y)] \, dy \, dx \quad , \tag{11}$$

where

$$\Psi(x,y) = h(\omega_N - \omega_b^0) = \frac{\beta(x^2 - 1)}{(y^2 b^2 + x^2 - 1 + a^2)^2} + \alpha y^2 b^2 + \gamma x^2 \quad ,$$

$$\beta = \frac{h\Delta}{T}, \quad \alpha = \frac{h^2 k_0^2}{2 M_z T}, \quad \gamma = \frac{h^2 k_0^2}{2 M_x T}$$

and T is the temperature measured in energy units. In the expressions for A and B the sums over the numbers of the waveguide modes is replaced by the integrals over $y = k_z / (k_0 b)$, since when $y > 1$ then $k_z \gg \pi/d$. In addition, as the integrals converge exponentially rapidly the integration is extended up to infinity.

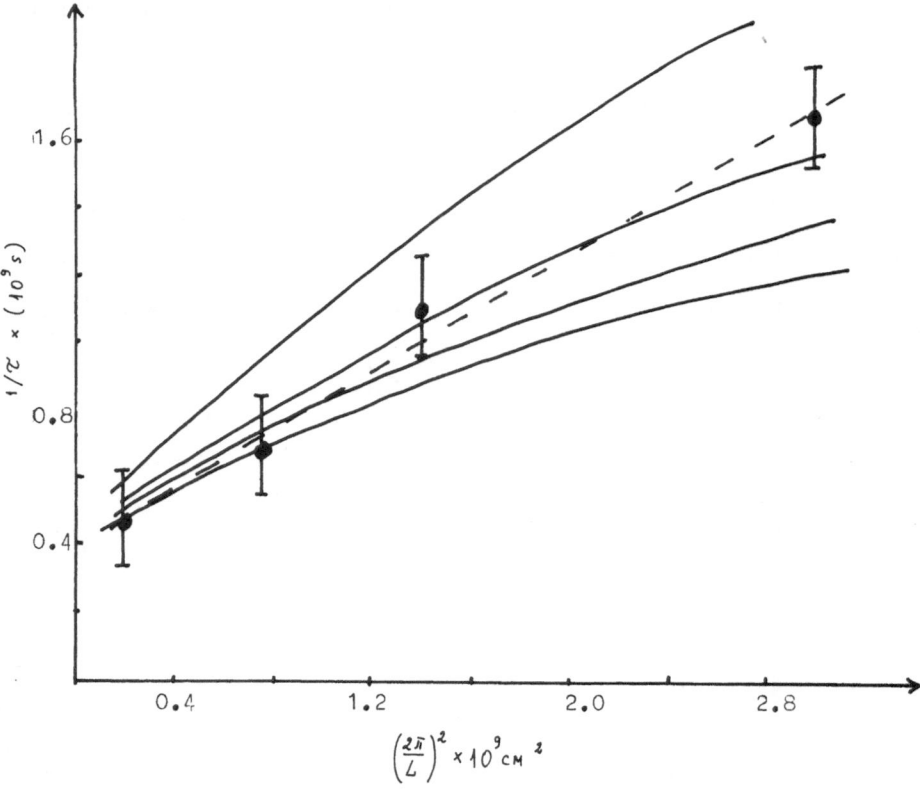

Fig. 6. The decay rate K as a function of $2\pi/L$ for different values
of r. The results of [16] are also shown (the dashed line; it
corresponds to r = 0.4).

DISCUSSION

The calculated diffusion coefficient D for several values of the
effective mass M_x (M_x = 500; 25; $15m_0$) is shown in Fig. 5 as a function
of the parameter ℓ_c. One's attention is immediately attracted by the
very weak dependence of the diffusion coefficient on the effective mass
M_x. More appreciable is the dependence of D on the parameter ℓ_c. Note
that the values of this parameter must satisfy the condition $\ell_c \ll L$, and
its choice restricts the class of short-path waveguide modes whose propagation can be described within the diffusion approximation. We assume
that the ratio ℓ_c/L = r is smaller than unity (r ~ 0.1 - 0.5) and independent of the grating period L. This ratio is the only parameter which
is introduced in our calculations and should be chosen from a comparison
with experimental data. It appears due to our use of the diffusion
approximation, which is inapplicable for the long-path waveguide modes
with $\ell \gtrsim L$. Therefore, the dependence of the diffusion coefficient D on
the ℓ_c is, in fact, its dependence on the grating period L (or on the
angle ϕ between the laser beams exciting the grating, L = λ/ϕ).
Consequently, the decay rate $K(\phi^2) = \frac{1}{\tau} + 2D(2\pi/L)^2$, must differ somewhat
from a linear function of ϕ^2. From the qualitative point of view this
result seems to be quite natural. Even if we had not used the diffusion
approximation, i.e. if within the framework of the kinetic Boltzman
equation we had also taken into account long-path waveguide modes the
dependence of $K(\phi^2)$ on ϕ^2 would have been different from a linear one.
Indeed, the long-path waveguide modes in the case under discussion should

154

play the same role as fluorescence photons at room temperature, e.g. in anthracene. These photons are known to lead to reabsorption. Although their mean free path ℓ is much longer than that of the exciton-like states, nevertheless, as was shown in [21], they can influence the decay rate of exciton gratings. In the case when $\ell \ll L$, i.e., when the motion of photons is diffusive, the decay rate $K(\phi^2)$ remains a linear function of ϕ^2; however, instead of D the effective diffusion coefficient D_{eff} appears, $D_{eff} = D+R$, where the quantity R is the contribution to the diffusion coefficient from reabsorption and is, practically, independent of L. If, however, $\ell \gg L$ (this is the case of interest to us) the quantity R begins to depend on L, but is small due to the parameter L/ℓ_c.

In the situation under consideration the long-path waveguide modes can influence the decay rate of exciton gratings due to processes analogous to reabsorption at room temperature because of phonon-induced scattering of long-path waveguide modes into short-path waveguide polaritons, and vice versa. The full inclusion of these processes requires, in fact, a theory of reabsorption in thin films with size quantization effects taken into account. However, even in the absence of such a theory we can state that, in view of what has been said above, the long-path waveguide modes ($\ell \gg L$) must not influence appreciably the decay kinetics of exciton gratings at long times ($t \sim \tau_0$). Therefore, we do not expect that further revision of the above calculations can change substantially the results obtained for the diffusion coefficient $D(\phi)$. To illustrate them in Fig. 6 the decay rate K is shown as a function of $(2\pi/L)^2$ for the value $\tau_0 = 2.5$ns obtained in [16]. By fitting this dependence to a straight line, as was done in [16], one obtains $D = 0.5$ cm^2/s for small values of $2\pi/L$, while for large $2\pi/L$, $D = 0.3$ cm^2/s. Thus, these results yield the mean value of the diffusion coefficient obtained in [16] and, in addition, permit understanding of such a great uncertainty of the value of the diffusion coefficient obtained in [16] $D = 0.43 \pm 0.2$ cm^2/s. It should be stressed that the use of the function $D(\phi^2)$ in fitting the experimental data can also change the value of τ_0, since the latter also depends on the way in which experimental data are processed. Since experimental investigations of the decay of exciton gratings in thin anthracene films are continuing we are planning in the future to go beyond the diffusion approximation and to return to a refinement of the above results.

ACKNOWLEDGEMENTS

In conclusion the authors express their gratitude to Professor M. D. Fayer for sending his results [16,17] before their publication.

REFERENCES

1. H. J. Eichler, P. Gunter, D. W. Pohl, in "Laser Induced Dynamic Gratings," ed. T. Tamir (Springer, Berlin, 1986); Special Issue on Dynamic Gratings and Four-Wave Mixing, IEEE J. Quantum Electron, QE-22 (1986).
2. J. R. Salcedo, A. E. Siegman, D. D. Dlott, M. D. Fayer, Phys. Rev. Lett. 41, 131 (1978).
3. K. A. Nelson, D. D. Dlott, M. D. Fayer, Chem. Phys. Lett. 64, 88 (1979).
4. K. A. Nelson, M. D. Fayer, J. Chem. Phys. 72, 5602 (1980).
5. M. D. Fayer, in "Excitation Dynamics and Spectroscopy of Condensed Molecular System," eds. V. M. Agranovich, R. M. Hochstrasser (North-Holland, Amsterdam, 1983).
6. T. S. Rose, R. Righini, M. D. Fayer, Chem. Phys. Lett. 106, 13 (1984).
7. V. M. Kenkre, in "Exciton Dynamics in Molecular Crystals and Aggregates," ed. G. Holer, (Springer, Berlin, 1982).

8. V. M. Kenkre, Phys. Rev. B18, 4064 (1978).
9. Y. M. Wong, V. M. Kenkre, Phys. Rev. B22, 3072 (1980).
10. V. M. Kenkre, V. Ern, A. Fort, Phys. Rev. B28, 598 (1983).
11. A. Fort, V. Ern, V. M. Kenkre, Chem. Phys. 80, 205 (1983).
12. V. M. Kenkre, D. Schmid, Phys. Rev. B31, 2430 (1985).
13. V. M. Kenkre, G. P. Tsironis, J. Lumin. 34, 107 (1985).
14. D. K. Garrity, J. L. Skinner, J. Chem. Phys. 82, 260 (1985).
15. T. S. Rose, J. V. Newell, J. S. Meth, M. D. Fayer, Chem. Phys. Lett.
    145, 475 (1988).
16. J. S. Meth, C. D. Marshall, M. D. Fayer, J. Lumin., submitted.
17. J. S. Meth, C. D. Marshall, M. D. Fayer, Solid State Commun.,
    accepted (1990).
18. V. M. Agranovich, "Theory of Excitons" (Nauka, Moscow, 1968) (in
    Russian).
19. V. M. Agranovich, M. D. Galanin, "Electronic Excitation Energy
    Transfer in Condensed Matter" (North-Holland, Amsterdam, 1982).
20. J. Aaviksoo, A. Freiberg, J. Lipmaa, T. Reinot, J. Lumin. 37, 313
    (1987).
21. V. M. Agranovich, A. M. Ratner, M. Salieva, Solid State Commun. 63,
    329 (1986).
22. V. M. Agranovich, T. A. Leskova, Solid State Commun. 68, 1029 (1988).
23. V. M. Agranovich, T. A. Leskova, Pis'ma Zh. Eksp. Teor. Fiz. 29, 151
    (1979).
24. V. V. Travnikov, Pis'ma Zh. Eksp. Teor. Fiz. 29, 151 (1979).
25. J. Zyss, D. S. Chemla, in "Nonlinear Optical Properties of Organic
    Molecules and Crystals" V.1, p. 23 (1987).
26. "Surface Polaritons," eds. V. M. Agranovich, D. L. Mills (North-
    Holland, Amsterdam, 1982).
27. V. M. Agranovich, Yu. V. Konobeev, Sov. Phys.- Solid State 3, 260
    (1961).
28. V. M. Agranovich, V, L. Ginzburg, "Crystaloptics with Spatial
    Dispersion and Excitons," (Springer, Berlin, 1984).
29. S. V. Marisova, E. N. Myasnikov, A. N. Lipovchenko, Phys. stat. sol.
    (b) 115, 649 (1983).
30. M. R. Philpott, J. Chem. Phys. 54, 111 (1971).

EXCITATION TRANSPORT IN POLYMERIC SOLIDS

M. D. Fayer

Department of Chemistry
Stanford University, Stanford, California 94305-5080

By attaching chromophores to polymer backbones, it is possible to make complex chromophore systems in which changes in the nature of the polymer or changes in the environment of the polymer have substantial influences on the nature and rate of electronic excitation transport. Chromophores bound to a polymer chain (tagged chain) or microphase separated domains of tagged chains are finite volume excitation transport systems with complex, nonrandom, distributions of chromophores. We have developed a general statistical mechanical theoretical approach for describing energy transport in complex, finite (or infinite) volume systems. We have also employed time resolved fluorescence depolarization experiments to test the theory and make the first detailed measurements of energy transport in complex systems.

In systems involving donor-donor excited state transport, the fundamental quantity of theoretical and experimental interest is $G^S(t)$; the ensemble averaged probability that an originally excited chromophore is excited at time t.[1] $G^S(t)$ contains contributions from excitations that return to the initially excited chromophores after one or more transfer events. $G^S(t)$ does not contain loss of excitation due to lifetime (fluorescence) events. The transition dipoles of the chromophores in a solid polymer matrix are randomly oriented. The initially excited ensemble is polarized along the direction of the excitation $\bar{E}$ field and gives rise to polarized fluorescence. Transport occurs into an ensemble of chromophores with randomly distributed dipole directions and the fluorescence becomes unpolarized. Thus $G^S(t)$ can be obtained from the time dependence of fluorescence depolarization.

The decay of $G^S(t)$ (the rate of excitation transport) is intimately related to the spatial distribution of the chromophores. For flexible polymers (freely jointed chains) the distribution of chromophores which are tagged on the chain in relatively low concentration is determined by the chain structure. Relatively low concentration means one or less chromophores per statistical segment of the chain. The chain structure and size is characterized by the ensemble averaged root-mean-squared radius of gyration, $\langle R_g^2 \rangle^{1/2}$.

The theory employed to analyze the data presented here has been described in detail elsewhere.[2] Random flight statistics with an appropriate statistical segment length,[2] are employed to describe the average chain conformation of the copolymer chains. This model has been applied successfully to polymer coils in solution. For a polymer with chromophores randomly distributed along the chain, the chromophore distribution function can be modeled by an appropriate pair correlation function which describes the probability that a chromophore (labeled 2) on any chain segment j is a distance r_{12} from a chromophore (labeled 1) on chain segment i. $\overline{N}$ and a are the number of statistical segments and the statistical segment length of the polymer, respectively.

Excited state transport has been described by many formalism.[1,3,4] Peterson and Fayer have applied the first order cumulant expansion to this problem.[2] The details of the calculations and demonstrations of its accuracy are reported in reference 2. The cumulant expansion is applied to $G^S{}_i(t)$, and then the average over i, the position of initial excitation, is performed exactly.

Using the chromophore distribution function, $P_i(r_{12})$, performing the cumulant expansion and then averaging over the possible positions of the initially excited chromophore, an expression for $G^S(t)$ is obtained[2]

$$G^S(t) = \frac{1}{\overline{N}} \sum_{i=1}^{\overline{N}} \exp\left[\frac{4\pi}{2} \int_0^\infty \left(1 - e^{-2\omega_{12}t}\right) P_i(r_{12}) r_{12}^2 dr_{12}\right] \qquad (1)$$

ω_{12} is the rate of excitation transport between two chromophores at a separation of r_{12}.

$$\omega_{12} = \gamma^2 \left(\frac{R_0}{r_{12}}\right) / \tau \qquad (2)$$

τ is the fluorescence lifetime, R_0 is the critical transfer radius, and $\gamma = 0.85$.[5]

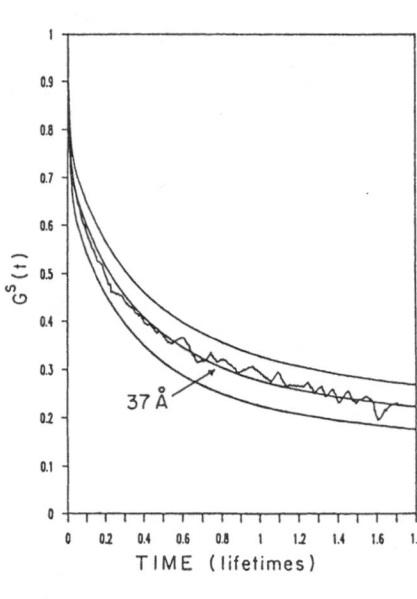

Fig. 1

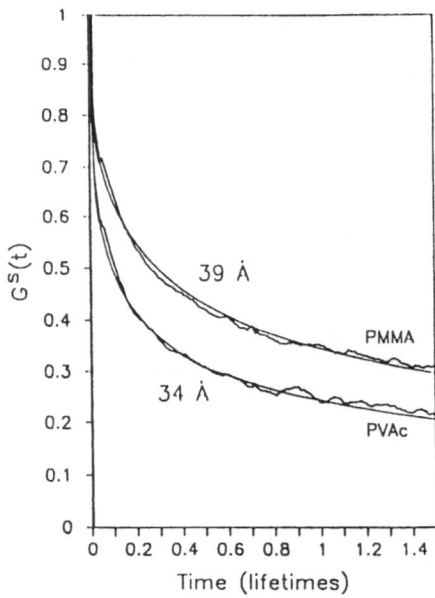

Fig. 2

Provided the molecular weight of the copolymer is known, Eq. (1) has only one adjustable parameter; the statistical segment length a. This is directly related to $\langle R_g{}^2\rangle^{1/2}$ by $\langle R_g{}^2\rangle = 1/6 \ (\bar{N} \ a^2)$. Thus, a fit of the experimentally determined $G^S(t)$ with a theoretically calculated $G^S(t)$ determined by adjusting the statistical segment length will give a measure of $\langle R_g{}^2\rangle^{1/2}$ for the copolymer.

Figure 1 displays data taken on a copolymer of methyl methacrylate (MMA) and 2-vinylnaphthalene (2-VN) in a PMMA host matrix.[6] The copolymer has a molecular weight of 23,400 and contains 20 naphthalenes (9% tagged). $G^S(t)$ is found using eq. (1). The theoretical curve through the data clearly has the correct functional form and yields a $\langle R_g{}^2\rangle^{1/2}$ of 37 Å. The calculated curves above and below the data are for 39 Å and 35 Å, respectively. This shows the incredible sensitivity of the rate of excitation transport to the size of the chromophore cluster. A 2 Å change in size is easily detectable given the signal to noise ratio (S/N).

Table I displays results taken on several copolymers. The first number (in the first column) is the percent of tags, and the second number is the molecular weight divided by 1000. Table I compares these results to those obtained by light scattering studies in θ liquid solvents.

TABLE I.   Summary of Results

Copolymer	$\langle R_g{}^2\rangle^{1/2}$ Determined by Excitation Transport	$\langle R_g{}^2\rangle^{1/2}$ for equivalent M_w θ-condition PMMA
4-23	38 ± 3	39 ± 4
6-22	39 ± 3	39 ± 4
9-23	37 ± 3	39 ± 4
9-60	61 ± 3	64 ± 7

Solid PMMA is a θ solvent for itself. The agreement between the light scattering determinations and these measurements is complete.

Figure 2 displays the effects of changing the medium on a tagged chain. Energy transport measurements on 23,000 M_w PMMA chain tagged with 9% naphthalene chromophores were made in two hosts, PMMA and Poly-vinylacetate (PVAc).[7] In PVAc the excitation transport is considerably faster because poor thermodynamic interactions have caused partial chain collapse. The theoretical fits to the data show that $\langle R_g{}^2\rangle^{1/2}$ has decreased from 39 Å to 34 Å in going from a PMMA host (θ solvent) to a PVAc host. This is 26% of the total chain collapse. This is the first measurement of chain collapse in an incompatible polymer blend.

At or near room temperature the rate of excitation transfer has been assumed to be independent of chromophore energy.[8] The spectral lines of the chromophores are taken to be homogeneously broadened; thermal fluctuations in the solvent broaden the spectral lines to an extent which is much greater than the inhomogeneous linewidth. No explicit consideration of how thermal energy is absorbed or emitted by the chromophores in conserving energy during the transfer step is necessary. In the case of donor-donor transfer, excitation transfer between two chemically identical species, the Förster theory predicts symmetric forward and back transfer rates.[8]

Situations in which the inhomogeneous broadening is larger than the thermal, homogeneous broadening of spectral lines are generally assumed only to occur at low temperatures. For a spatial distribution of chromophores with a distribution of electronic transition energies an excitation not only moves spatially among the chromophores, but it moves to lower energy as it is transferred. The excitation will tend to move to the red edge of the absorption line by a series of transfer steps to lower energy chromophores while emitting phonons (heat) to conserve energy. This is referred to as dispersive transport.

Recently, we reported the first observation of dispersive electronic excitation transport at and near room temperature by monitoring the time-resolved fluorescence depolarization from an ensemble of chromophores.[9,10] Figure 3 shows $G^S(t)$ at three different excitation wavelengths in the S_0-S_1 absorption of naphthyl chromophores tagged to PMMA polymer chains.[10] $G^S(t)$ depends on the excitation wavelength, i.e., the transport is dispersive.

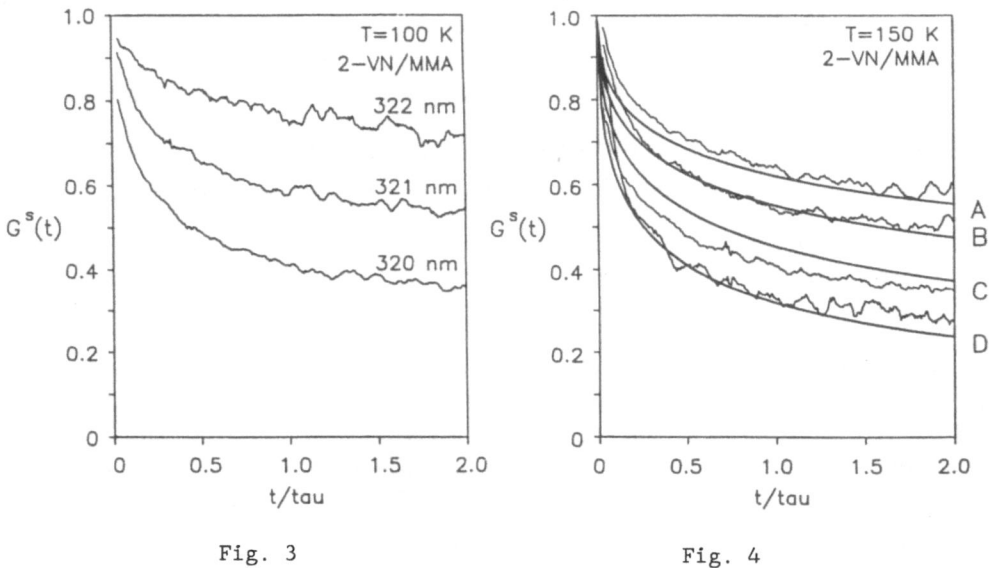

Fig. 3   Fig. 4

The sample is a solid mixture of 0.38% by weight 2-VN/MMA copolymers M_w = 23,000, containing 9% 2-VN, in PMMA. The sample temperature is 100 K. There is a large decrease in the rate of decay of $G^S(t)$ as the excitation wavelength is moved to the red. Data from the same sample at room temperature (298 K)[10] still depends on the excitation wavelength, but to a smaller extent than in fig. 3. This demonstrates that the Förster description of donor-donor transport is inadequate to describe the excitation transfer dynamics in this system, even at room temperature.

To understand dispersive excitation transport at or near room temperature it is necessary to calculate $G^S(t)$ for an ensemble of chromophores with a distribution of site energies. This situation applies to molecules for which the absorption linewidth of an electronic transition has a large contribution from inhomogeneous broadening. Excitation transport between two chromophores with differing transition energies between their ground and excited electronic states requires either absorption or emission of one or more phonons to conserve energy. The complications in calculating the pairwise transfer rate in situations such as this are due to the need to consider both the interaction of the electronic states of the naphthyl chromophores with the phonon bath and the phonon density of states. Since both of these quantities are difficult to obtain experimentally or theoretically, we have developed an alternative approach to this problem.[9,10]

The theory separates the dispersive transport problem into two parts.[9,10] The first part involves calculating the extent of dispersive transport that occurs in a system under given conditions. This is based on the two limits for an acceptor chromophore. It is either a donor itself, capable of back-transferring an excitation to the initially excited chromophore, or it is a trap, which cannot back-transfer. An effective concentration of chromophores and a scaling factor which accounts for the amount of dispersive transport in the system are calculated. Parameters involved in this calculation include the sample temperature and the inhomogeneous linewidth of the electronic transition. Since, experimentally a distribution of chromophore frequencies are excited initially by the laser light, the calculation is repeated for all possible starting chromophore transition energies.

The second part of the theory uses the results developed in the past for nondispersive electronic excitation transfer (see above). We combine the modified number density of chromophores available for energy transfer and the dispersive transport parameter calculated in the first part of the calculation, with the theory for a nondispersive system of the appropriate chromophore geometry. In this way, the powerful techniques developed over the past years to handle incoherent excitation transfer are applied to the problem of dispersive transport.

The important feature of the theory is that all of the input parameters necessary to calculate the wavelength dependence at a given temperature and the temperature dependence at a given wavelength are available from experiments. Figure 4 shows a comparison of theory with experiment for a tagged copolymer in PMMA. The parameters used in the calculations were the inhomogeneous line center (319.5 nm, 31298 cm^{-1}), the inhomogeneous line width (300 cm^{-1}), the root-mean-squared radius of gyration of the polymer chain (39 Å), the critical Förster radius R_0 (13 Å), the number of chromophores on the tagged chains (20), the temperature (150 K), and the laser excitation wavelength. The homogeneous linewidth is assumed to be kT. $G^s(t)$ is shown for excitation at (A) 322 nm (31056 cm^{-1}), (B) 321 nm (31153 cm^{-1}), (C) 320 nm (31250 cm^{-1}), and (D) 318 nm (31446 cm^{-1}). The calculation has no adjustable parameters. The agreement between theory and experiment, considering the complexity of the problem and the lack of adjustable parameter, is quite remarkable.

The detailed experiments and theory[9,10] which have been briefly discussed show that the basic assumptions about room temperature and near room temperature excitation transport are not necessarily applicable to all systems. There is then the question of why the experiments described above concerning polymer structure can be successfully interpreted with the Förster mechanism. As part of the theoretical analysis we discovered that there is a particular excitation wavelength which makes the effects of dispersive transport vanish for the $G^s(t)$ observable. We call this wavelength the "magic wavelength".

Figure 5 shows a calculation of the temperature dependence of 2-ethylnaphthalene in PMMA for excitation at the magic wavelength. The magic

wavelength, which depends on the spectroscopic parameters of the system, is
317.3 nm for 2-EN in PMMA. Note that $G^S(t)$ displays no temperature
dependence. This is in contrast to large changes with temperature
calculated and observed at other wavelengths, e.g., compare the data in
figs. 3 and 4. The curves not only lack temperature dependence but are
accurately described by the non-dispersive theory of $G^S(t)$.

Figure 6 shows an experimental verification of the existence of the
magic wavelength. These are data at three temperatures, 298 K, 200 K, and
100 K. The curves are identical. This is in contrast to curves taken on

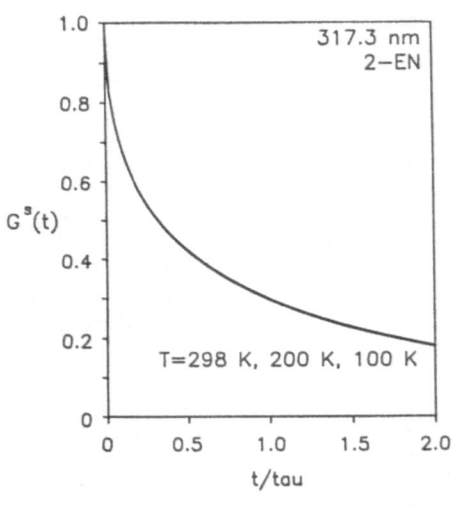

Fig. 5

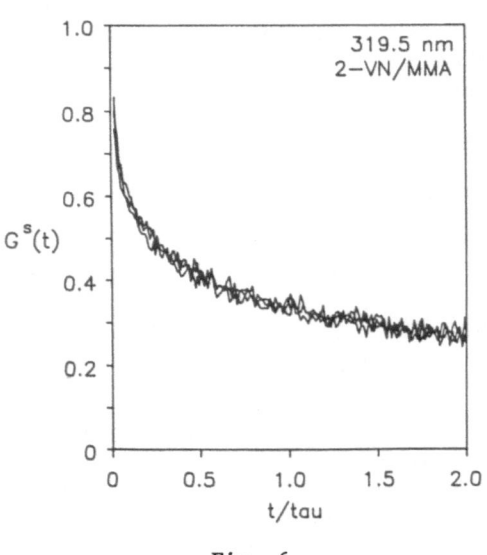

Fig. 6

the same sample but with other excitation wavelengths. (Compare curves in
figs. 3 and 4.) We theoretically predicted the magic wavelength to be
319 nm and found it at 319.5 nm. The experiments on polymer structure were
conducted at 320 nm, which is the peak of the absorption spectrum. 320 nm
is so close to the magic wavelength that dispersive transport effects did
not influence the measurements.

ACKNOWLEDGMENT

This work was supported by the Department of Energy, Office of Basic
Energy Sciences (DE-FG03-84ER-13251). Additional equipment support was
provided by the National Science Foundation, Division of Materials
Research (DMR87-18959). I would also like to thank the Stanford Center for
Materials Research Polymer Thrust Program for support of this research, and
Dr. K. A. Peterson and A. D. Stein for their contributions to the
experiments and theory (references 6, 7, 9, and 10).

REFERENCES

1. a) Gochanour, C. R.; Andersen, H. C.; Fayer, M. D. *J. Chem. Phys.*, **70**, 4254 (1979).
    b) Loring, R. F.; Andersen, H. C.; Fayer, M. D. *J. Chem. Phys.*, **76**, 2015 (1982).
2. Peterson, K. A.; Fayer, M. D. *J. Chem. Phys.*, **85**, 4702 (1986).
3. Haan, S. W.; Zwanzig, R. *J. Chem. Phys.*, **68**, 1979 (1978).
4. a) Huber, D. L. *Phys. Rev. B*, **20**, 2307 (1979).
    b) Huber, D. L. *Phys. Rev. B*, **20**, 5333 (1979).
5. Gochanour, C. R.; Fayer, M. D. *J. Phys. Chem.*, **85**, 1989 (1981).
6. Peterson, K. A.; Zimmt, M. B.; Linse, S.; Domingue, R. P.; Fayer, M. D. *Macromolecules*, **20**, 168 (1987).
7. Peterson, K. A.; Stein, A. D.; Fayer, M. D. *Macromolecules*, **23**, 111 (1989).
8. Förster, Th. *Ann. Phys.*, **2**, 55 (1948).
9. Stein, A. D.; Peterson, K. A.; Fayer, M. D. *Chem. Phys. Lett.*, **161**, 16 (1989).
10. Stein, A. D.; Peterson, K. A.; Fayer, M. D. *J. Chem. Phys.*, **92**, 5622 (1990).

VIBRON LIFETIMES IN MOLECULAR CRYSTALS

Salvatore Califano

European Laboratory for non Linear Spectroscopy
Largo Enrico Fermi 2, 50125 Florence, Italy

Typical phonon lifetimes in molecular crystals range from few picoseconds to some hundred picoseconds at the liquid He temperature. In some special cases, for very isolated phonon levels, these can even reach higher values, up to several nanoseconds.

The finite phonon lifetime[1,2] is a consequence of the fact that anharmonic terms of the crystal hamiltonian activate energy and phase relaxation channels involving the thermal bath phonons. As the temperature increases the lifetime decreases, since the number of decay channels increases with temperature.

Anharmonic phonon-phonon interactions in crystals give rise to two different relaxation mechanisms[1]. The first, due to the decay of the phonon energy into the thermal bath, produces a depopulation of the phonon state and is associated to a T_1 relaxation time. The second mechanism[1,3,4] is instead due to scattering processes with thermal bath phonons which randomize the phase of the collective excitation without changing the population of the state and is associated to a T_2^* relaxation time. The total relaxation time T_2 of a phonon state is then given by [1]

$$\frac{1}{T_2} = \frac{1}{2T_1} + \frac{1}{T_2^*} \qquad (1)$$

The phonon lifetime can be measured experimentally in the time domain by time-resolved spectroscopic techniques or in the frequency domain from the profile of the absorption or scattering bands. The two types of experiments are equivalent in principle, since the Fourier transform of the time domain relaxation curve gives the band profile in the frequency domain. When the relaxation time T_2 is truly exponential, the band profile is a Lorentzian, with full width γ at half maximum (FWHM in units of cm^{-1}) given by[1]

$$\gamma = \frac{1}{\pi c T_2} \qquad (2)$$

Infrared absorption and Raman scattering spectroscopy are widely utilized for lifetime measurements in the frequency domain. In most of the cases of interest, especially at low temperatures, high resolution is needed for accurate band profile measurements. Commercial infrared interferometers and Raman grating spectrometers have today resolution up to 0.001 cm⁻¹ and 0.01 cm⁻¹ respectively, largely sufficient for the determination of accurate band profiles in solids. For higher resolution in Raman spectroscopy, necessary for very long lived phonons, it is convenient to couple a Fabry-Perot interferometer to a Raman spectrometer [5], as shown schematically in Fig.1.

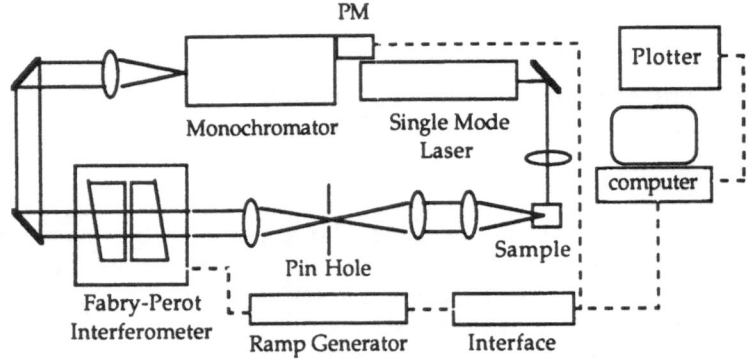

Figure 1.   Fabry- Perot and Raman Spectrometer Tandem instrument
for high resolution Raman Spectroscopy.

Another convenient technique for the determination of the band profiles of Raman active phonons is that of CARS (Coherent Antistokes Raman Spectroscopy)[6,7,8]. In a CARS experiment two laser beams of frequency ω_1 and ω_2, choosen so that $\omega_1 - \omega_2 = \omega_{phonon}$, are sent on a sample to coherently populate a phonon level. The same laser beam of frequency ω_1 (in some cases a third laser beam of different frequency ω_3 is used), stimulates the emission at the antistokes frequency $\omega_{as} = 2\omega_1 - \omega_2$, as shown schematically in the diagram of Fig. 2 . In this parametric process the sample plays the role of transferring photons between different modes of the field with energy and momentum conservation. The momentum conservation relation imposes that, if the beams ω_1 and ω_2, are incident

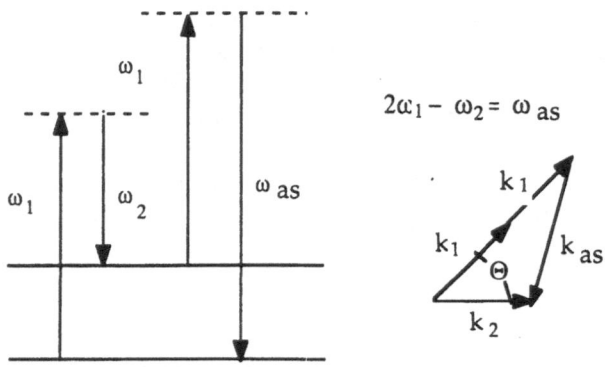

Figure 2. Scheme of a CARS process.

on the sample with an angle Θ, the new beam ω_{as} will be emitted in the direction defined by the triangle shown in the diagram.

A great advantage of the CARS technique is that it can be utilized for both frequency and time domain experiments.

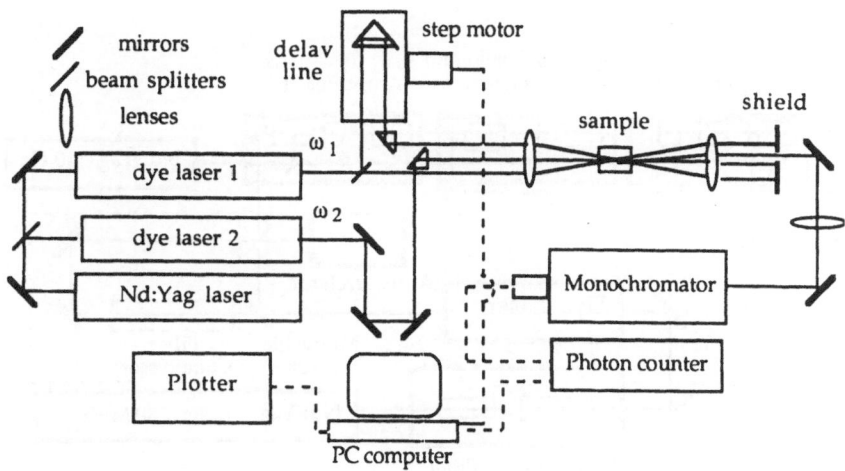

Figure 3. System for time resolved CARS Spectroscopy with ps pulses.·

Two systems will be described here for time resolved spectroscopy, both available in our laboratory. The first, shown schematically in Fig.3, is made of a mode locked, frequency doubled Nd. Yag laser synchronously pumping two dye lasers and has a time resolution of about 3 ps. The output of the pump laser is divided into two beams by means of a beam splitter. One beam is used to pump the first dye laser, tuned at the frequency ω_1, whereas the other beam pumps the second dye laser, tuned at the frequency ω_2. The output of the first laser is again divided into two, by means of another beam splitter. One of the two ω_1 beams is used, together with the beam ω_2 to coherently populate the phonon state. The second beam of frequency ω_1 is passed through a motorized delay line so that the pulses reach the sample with a variable delay, controlled by a PC computer. The CARS beam at frequency $\omega_{as} = 2\omega_1 - \omega_2$ is spatially filtered, collimated and sent to a double monochromator. The signal from the cooled photomultiplier is then processed by a photon counting system. By retarding the delayed ω_1 pulse with respect to the other two, one observes the decay of the CARS signal over several decades.

The second system is designed for shorter pulses (from 250 to 100 femtoseconds) and is shown in Fig 4.
The infrared pulses of a mode locked Nd. Yag laser of about 80 ps duration (80 MHz repetition rate) are compressed in an optical fiber and through four passes on a grating, to 4 ps. The pulses are then frequency doubled and used to pump a dye laser which emits pulses of 250 fs. These are amplified by a three stage amplifier pumped by a second Nd. Yag laser (10 Hz repetition rate) and the output beam is devided into two parts by a beam splitter. One beam is directly used whereas the second is sent into a

water cell to generate a 250 fs continuum. A frequency component of the continuum is then selected and amplified three times by a second amplifier, pumped by the same Nd. Yag laser. If pulses shorter than 250 fs are needed, the output of the first dye laser is again fiber compressed before being sent to the amplification system.

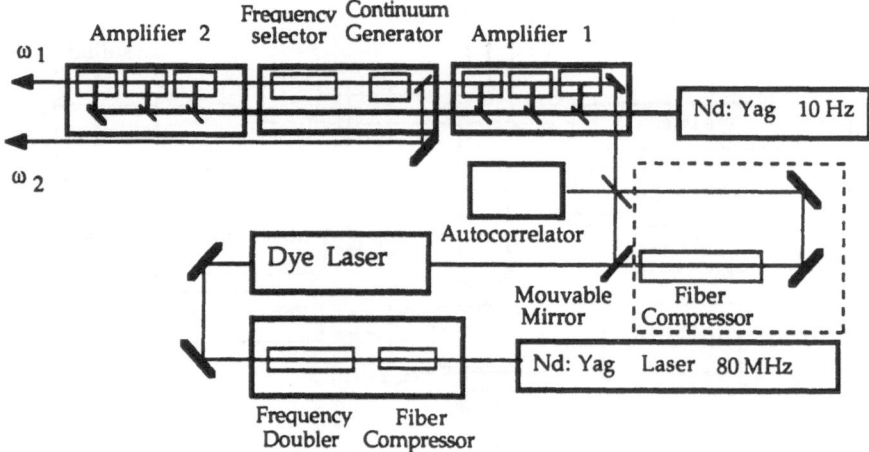

Figure 4. System for time resolved CARS Spectroscopy with fs pulses.

The dynamics of molecular crystals is most conveniently treated in terms of molecular normal coordinates (molecular translations, rotations and internal vibrations). From these, crystal normal coordinates are easily constructed and many-body perturbation techniques are then used to evaluate anharmonic phonon shifts and bandwidths. The perturbation expansion can be carried out in principle to any order, although in practice the inclusion of fourth order terms is already a major task. The theory is discussed in details in many books and review articles [1,2]. Here we present a summary of the theory, limiting the exposition to the basic parts, necessary for the interpretation of the experimental results.

First we expand the crystal potential V in powers of the crystal normal coordinates Q

$$V = \frac{1}{2}\sum_{lm} C_{lm}Q_lQ_m + \frac{1}{3!}\sum_{lmn}Q_lQ_mQ_n + \frac{1}{4!}\sum_{lmnp}C_{lmnp}Q_lQ_mQ_nQ_p + \ldots \quad (4)$$

where l,m,n,p are composite indices, comprehensive of both the phonon branch label j and of the phonon wavevector label k. Whenever necessary we shall specify these two labels separately. The coefficients C are derivatives at equilibrium of the potential with respect to normal coordinates [1,2]

$$C_{lm} = \left(\frac{\partial^2 V}{\partial Q_l \partial Q_m}\right)_0 \delta(\mathbf{k}_1 + \mathbf{k}_m) \quad (5a)$$

$$C_{lmn} = \left(\frac{\partial^3 V}{\partial Q_l \partial Q_m \partial Q_n}\right)_0 \delta(k_l + k_m + k_n) \tag{5b}$$

$$C_{lmnp} = \left(\frac{\partial^4 V}{\partial Q_l \partial Q_m \partial Q_n \partial Q_p}\right)_0 \delta(k_l + k_m + k_n + k_p) \tag{5c}$$

where the δ functions assure the momentum conservation condition. The C coefficients of eqs. 5a - 5c can be calculated from an analytical form of crystal potential. For instance the cubic coefficient C_{lmn} of interest for us is given by

$$C_{lmn} = C(j0, j_1 k, j_2 \text{-} k) = L^{-\frac{3}{2}} \sum_{abc} \sum_{j_1 j_2} \sum_{\rho\sigma\tau} D(^{abc}_{\rho\sigma\tau}, k) E(\rho a, j0) E(\sigma b, j_1 k) E(\tau c, j_2 \text{-} k) \tag{6}$$

where a,b,c count molecules in the unit cell, ρ, σ, τ label molecular internal or external coordinates, L is the number of unit cells and $E(\sigma b, j_1 k)$ is the eigenvector of the harmonic phonon involving the σ molecular coordinate of the b molecule and the j_1 dispersion branch with k wavevector. $D(^{abc}_{\rho\sigma\tau}, k)$ is an element of the third order dynamical matrix

$$D(^{abc}_{\rho\sigma\tau}) = \sum_{\beta\gamma} \left(\frac{d^3 V}{d\rho^{0a} d\sigma^{\beta b} d\tau^{\gamma c}}\right)_0 e^{k(r_\beta - r_\gamma)} \tag{7}$$

where 0 is a reference unit cell and β and γ count unit cells.

For the perturbation approach the crystal normal coordinates are conveniently expressed in terms of phonon creation Q_l^+ and annihilation Q_l^- operators

$$Q_l = \left[\frac{h}{2\omega_l}\right]^{\frac{1}{2}} (Q_l^- + Q_{-l}^+) = \left[\frac{h}{2\omega_l}\right]^{\frac{1}{2}} A_l \tag{8}$$

where the index -l stands for -kj, i.e. for reversal of the wavevector direction. In terms of the sum operators $A_l = (Q_l^- + Q_{-l}^+)$ the various terms of the hamiltonian can be rewritten in the form

$$H = H_0 + H_1 + H_2 + \cdots\cdots\cdots \tag{9}$$

$$H_0 = \sum_l h\omega_l^0 (Q_l^+ Q_l^- + \tfrac{1}{2}) = \sum_l h\omega_l^0 (n_l + \tfrac{1}{2})$$

$$H_1 = \sum_{lmn} B_{lmn} A_l A_m A_n$$

$$H_2 = \sum_{lmnp} B_{lmnp} A_l A_m A_n A_p$$

In these expressions ω_l^0 is the harmonic frequency of the l-th phonon, n_l is the corresponding phonon occupation number

$$n_1 = \left[\exp\left(\frac{\hbar\omega_1}{kT}\right) - 1\right]^{-1} \qquad (10)$$

and the B coefficients are defined as

$$B_{1mn} = \frac{1}{3!}\left[\frac{\hbar^3}{2^3\omega_1\omega_m\omega_n}\right]^{\frac{1}{2}} C_{1mn} \quad , \quad B_{1mnp} = \frac{1}{4!}\left[\frac{\hbar^4}{2^4\omega_1\omega_m\omega_n\omega_p}\right]^{\frac{1}{2}} C_{1mnp} \quad , \quad .... \quad (11)$$

The calculation of phonon anharmonic shifts and bandwidths is normally made using the Green's function method. The method is well-known in solid state physics and extensively discussed in many books and review articles [1,2,9]. We recall that it leads to the Dyson equation for the anharmonic phonon propagator

$$G_1(i\omega_m) = G_1^0(i\omega_m) + G_1^0(i\omega_m)\Sigma_{11'}(i\omega_m)G_1(i\omega_m) \qquad (12)$$

where $\Sigma_{11'}(i\omega_m)$ is the crystal self-energy, and $G_1(i\omega_m)$, $G_1^0(i\omega_m)$ are anharmonic and harmonic phonon propagators, respectively. The self-energy is the sum of a series of elementary contributions which are conveniently expressed in terms of diagrams. The diagrams of interest for our problem are those which describe the relaxation of optical phonons with k=0 wavevector, i.e. $Q_1 = Q_{j0}$. The lowest order diagrams, representing the lowest order processes, are shown below and involve two cubic coupling terms

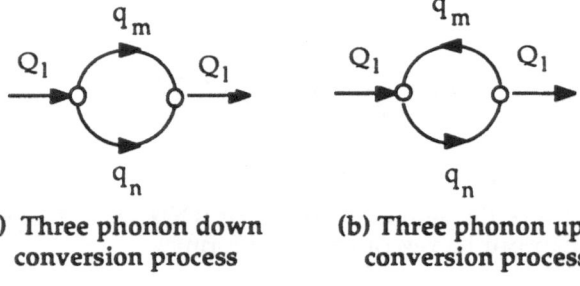

(a) Three phonon down
conversion process
$\omega_1 = \omega_m + \omega_n$

(b) Three phonon up
conversion process
$\omega_1 = \omega_m - \omega_n$

Diagram (a) represents an energy decay process in which the phonon ω_1 is annihilated, giving rise to two lower energy phonons ω_m and ω_n, or is created by the fusion of two phonons of the thermal bath. For momentum conservation the ω_m phonon has wavevector k and the ω_n phonon has -k wavevector. The process is called **down** since the phonon energy is sent down in the phonon manifold. In the same way diagram (b) represents a process in which the ω_1 phonon is fused with a thermal bath phonon ω_n to produce a higher energy phonon ω_m (**up** conversion) or is created from a decay process involving two bath phonons.

The calculation of the contribution of all these processes to the phonon bandwidth is given in many articles and books [1,2,9,10,11]. For our purposes it is sufficient to list only the contributions of the diagrams of interest to the phonon bandwidth. Those of diagrams (a) and (b) are

$$\gamma_{l(a)} = 32\pi h^{-2}\sum_{mn}|B_{lmn}|^2(n_m+n_n+1)\delta(\omega-\omega_m-\omega_n) \qquad (13)$$

$$\gamma_{l(b)} = 72\pi h^{-2}\sum_{mn}|B_{lmn}|^2(n_m-n_n)\delta(\omega+\omega_m-\omega_n) \qquad (14)$$

The next higher order diagrams involve either two quartic, or one quartic and two cubic, or four cubic coupling terms. Those with two quartic terms are [1,2]

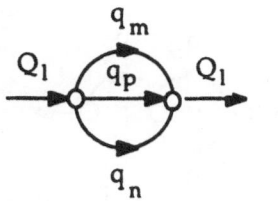

(c) Four phonon down
conversion process

$$\omega_l = \omega_m + \omega_n + \omega_p$$

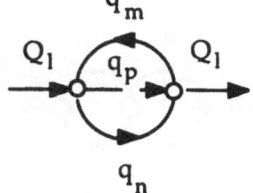

d) Four phonon up
conversion process '

$$\omega_l = \omega_m + \omega_p - \omega_n$$

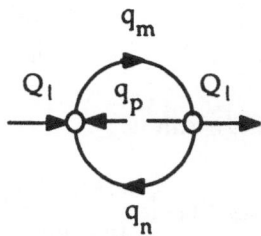

(e) Four phonon up
conversion process

$$\omega_l = \omega_m - \omega_n - \omega_p$$

$$\gamma_{l(c)} = 192\pi h^{-2}\sum_{mn}|B_{lmnp}|^2[(n_m+1)(n_n+1)(n_p+1)-n_m n_n n_p]\delta(\omega-\omega_m-\omega_n-\omega_p) \qquad (15)$$

$$\gamma_{l(d)} = 576\pi h^{-2}\sum_{mn}|B_{lmnp}|^2[n_n(n_m+1)(n_p+1)-n_m(n_n+1)n_p]\delta(\omega-\omega_m+\omega_n-\omega_p)$$

$$\gamma_{l(e)} = 576\pi h^{-2}\sum_{mn}|B_{lmnp}|^2[(n_m+1)n_n n_p - n_m(n_n+1)(n_p+1)]\delta(\omega-\omega_m+\omega_n+\omega_p)$$

Diagrams of the same order in the perturbation expansion, representing processes with one quartic and two cubic or with four cubic terms [1,2], are shown below

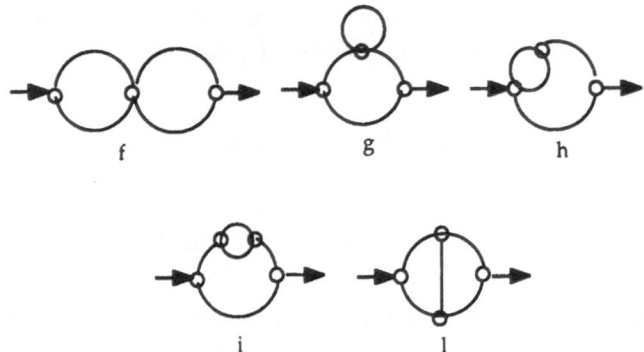

Each of the diagrams from f to l is representative of a whole series of diagrams with all possible combinations of creation and annihilation of phonons. For instance diagram f represents the four possible processes

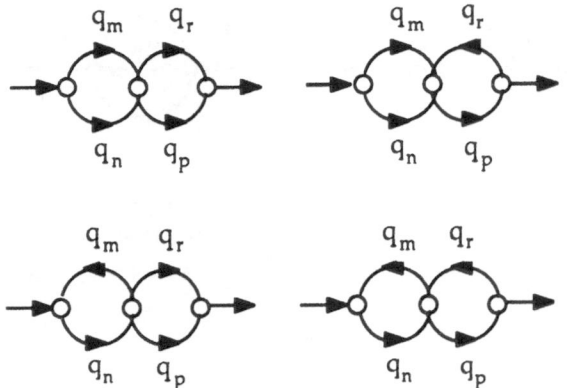

whose contribution to the bandwidth is:

$$\gamma_{1(f)} = 438\pi h^{-3} \sum_{mnpr} B_{lm-n} B_{-mnp-r} B_{l-pr} \left[\frac{(n_m+n_n+1)(n_p+n_r+1)}{(\omega_m+\omega_n-\omega_p-\omega_r)} \delta(\omega-\omega_m-\omega_n) \right. +$$

$$+ \frac{(n_m+n_n+1)(n_p-n_r)}{(\omega_m+\omega_n-\omega_p+\omega_r)} \delta(\omega-\omega_m-\omega_n) + \frac{(n_p+n_r+1)(n_m-n_n)}{(\omega_m-\omega_n-\omega_p-\omega_r)} \delta(\omega-\omega_m+\omega_n) +$$

$$+ \frac{(n_p-n_r)(n_m-n_n)}{(\omega_m-\omega_n+\omega_p-\omega_r)} \delta(\omega-\omega_m+\omega_n) \Big] \tag{16}$$

There is an important difference between third- and fourth-order contributions to the bandwidth. Those of third-order (diagrams a and b) depend on single occupation numbers (see eqs. 13 and 14). In the classical regime, when $\hbar\omega < kT$, the occupation numbers are proportional to T, as can be seen from the expansion of eq. (9) as a function of T [1,2] .

$$n = \left(\frac{k}{h\omega}\right)T - \frac{1}{2} + \frac{1}{12}\left(\frac{h\omega}{k}\right)\frac{1}{T} - \frac{1}{720}\left(\frac{h\omega}{k}\right)^3 \frac{1}{T^3} + \dots \tag{17}$$

The fourth order contributions depend instead on products of two occupation numbers and thus are proportional to T^2 in the classical regime. Plots of the phonon bandwidth as a function of T allow us therefore to decide whether or not fourth-order contributions are important.

In addition to the energy decay processes discussed above there are pure dephasing processes which also contribute to the phonon bandwidth. These processes randomize the phase of the phonons, without changing their occupation number. The simplest of these processes[1] is represented by diagram (m) below

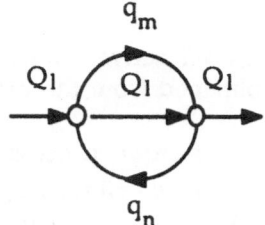

**(m) Energy exchange
dephasing process**

which is a special case of diagram (d), with $Q_p = Q_l$. In this process two phonons Q_m and Q_n of the thermal bath exchange energy and this affects the width of the Q_l phonon. The contribution[1] to the bandwidth of diagram (m) is easily obtained from that of diagram (d) and is

$$\gamma_{l(m)} = 576\pi h^{-2} \sum_{mn} |B_{lmnp}|^2 [n_n(n_m+1)]\delta(\omega_n - \omega_m) \qquad (18)$$

We discuss now the interpretation of experimental lifetime measurements for internal vibrons of crystalline benzene in terms of the theory discussed before. The interpretation of relaxation processes for vibrons is simpler than for lattice phonons for two reasons. The first is that these modes occur normally at relatively high frequencies, above 400 cm^{-1} . Their occupation numbers are then vanishingly small in the temperature range from liquid He to room temperature and play therefore no role in all equations from (13) to (18). The second reason is that the number of decay channels is small for high frequency vibrations.

Benzene crystallizes in the orthorombic system, space group D_{2h}^{15} with four molecules per unit cell on C_i sites. Each molecule has 30 internal modes, which classify in the symmetry species of the D_{6h} molecular group as 10 Raman active modes ($2a_g$, $1a_{2g}$, $2b_{2g}$, $2e_{1g}$, $3e_{2g}$) and 10 infrared active modes ($1a_{2u}$, $2b_{1u}$, $2 b_{2u}$, $3e_{1u}$, $2e_{2u}$), all e modes being doubly degenerate[12]. Each non-degenerate Raman active internal mode splits in the crystal into four components and each degenerate mode into eight components according to the scheme:

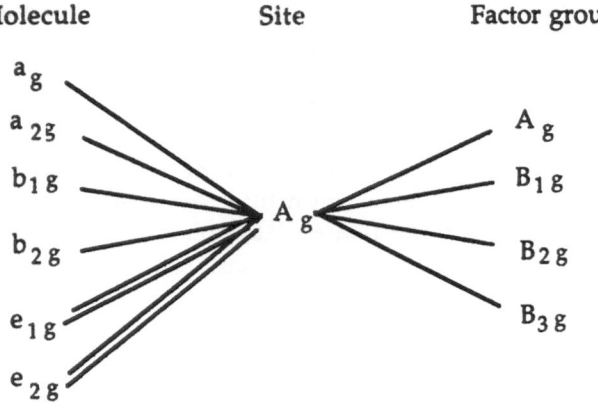

Molecule	Site	Factor group

The same occurs for the infrared active modes and the corresponding correlation diagram can be obtained from the previous one changing the labels from g to u.

Here we consider the relaxation processes[13] of the Ag crystal components of the four Raman-active vibrons $v_1(a_g)$ at 991 cm^{-1}, $v_6(e_{2g})$ at 606 cm^{-1}, $v_9(e_{2g})$ at 1174 cm^{-1} and $v_{10}(e_{1g})$ at 854 cm^{-1}.

The measurements[13] were made using the ps system shown in fig. 3 with the only difference that the Nd.Yag laser was replaced by a mode locked Ar laser. For an accurate determination of the decay signal we used a deconvolution procedure which takes into account the temporal shape of the laser pulses and the contribution of the non-resonant background to the signal. the theory shows that the total signal S(t) is the sum of six different contributions[14]

$$S(t) = |P_1|^2 + |P_2|^2 + 2P_1P_2 + |P_{nr}|^2 + 2(P_{1nr} + P_{2nr}) \qquad (19)$$

with

$$|P_1|^2 = \int_{-}^{-} dt\{|E_1(t-\tau)|^2 \int_{-}^{'} dt' E_1(t')E_2^*(t')R(t'-t)\int_{-}^{'} dt''E_1^*(t'')E_2(t'')R^*(t''-t)\}$$

$$|P_2|^2 = \int_{-}^{-} dt\{|E_1(t)|^2 \int_{-}^{'} dt'E_2^*(t')E_1(t'-\tau)R(t'-t)\int_{-}^{'} dt''E_1(t''-\tau)E_2^*(t'')R^*(t''-t)\}$$

$$P_1P_2 = Re\{\int_{-}^{-} dt[|E_1(t-\tau)E_1^*(t)\int_{-}^{'} dt'E_1(t')E_2^*(t')R(t'-t)\int_{-}^{'} dt''E_1^*(t'')E_2(t'')R^*(t''-t)]\}$$

$$|P_{nr}|^2 = \int_{-}^{-} dt\{|\chi_{nr}|^2 |E_1(t)|^2 |E_2(t)|^2 |E_1(t-\tau)|^2\} \qquad (20)$$

$$P_{1nr} + P_{2nr} = Re\chi_{nr}\int_{-}^{-} dt\{|E_1(t-\tau)E_1^*(t)E_2(t)\int_{-}^{'} dt'E_2^*(t')E_1^*(t')R(t'-t) +$$

$$+ |E_1(t)|^2E_1^*(t-\tau)E_2(t)\int_{-}^{'} dt'E_2^*(t')E_1(t'-\tau)R(t'-t)\}$$

where $|P_1|^2$ and $|P_2|^2$ describe the contributions to the signal when the role of probe beam is interchanged between the two ω_1 pulses, P_1P_2 is the cross term between the two, P_{nr} the contribution from the non resonant contributions.

Since we are interested in processes in which one internal vibron is coupled to a second internal vibron and one lattice phonon, and since the internal vibrons are well separated and do not mix appreciably, we have used the approximations [13]

$$E(\rho a, j0) = \delta_{\rho j} \ , \ E(\sigma b, j_1 k) = \delta_{\sigma j_1}, \ E(\tau c, j_2 - k) = \delta_{\tau j_2} \qquad (21)$$

and

$$|C(j0, j_1 k, j_2 - k)|^2 = |\overline{C(j0, j_1 k)}|^2 = \sum_{abc} \sum_{\tau} |D(^{abc}_{\rho\sigma\tau})|^2 \qquad (22)$$

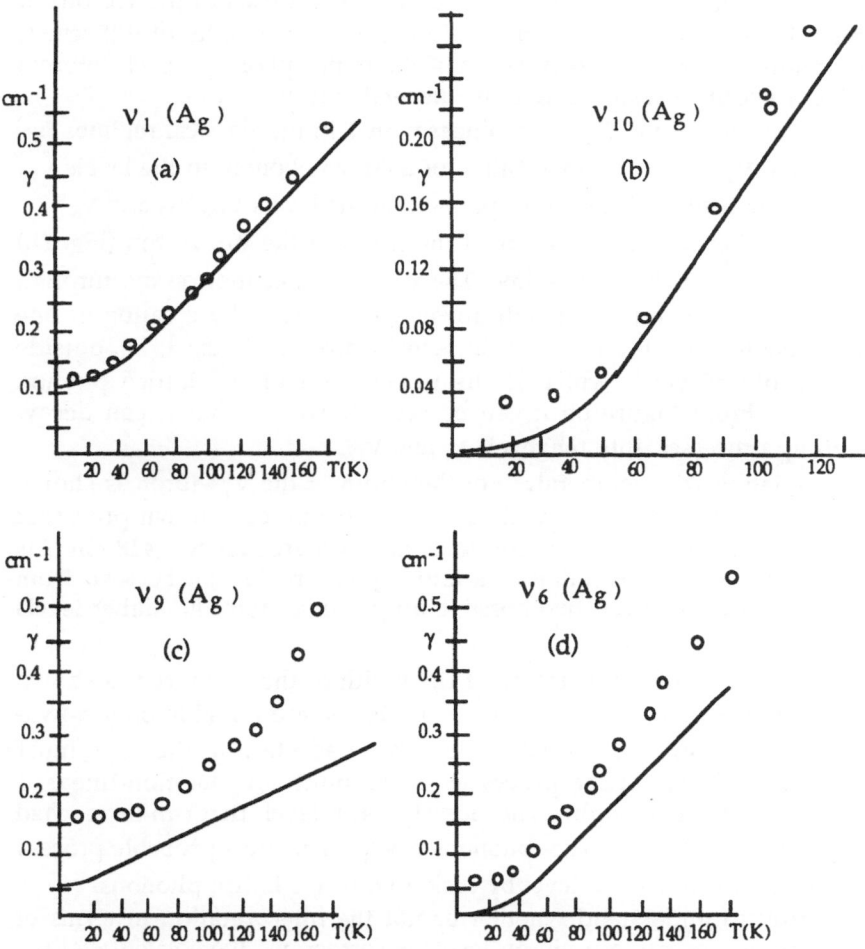

Figure 5. Experimental and calculated vibron bandwidths of benzene

where τ labels the external phonon involved in the decay process. The third order dynamical matrix elements were obtained from an intermolecular potential including atom-atom and quadrupole-quadrupole interactions that had been previously utilized for the resonant background and $P_{1nr} + P_{2nr}$ the cross terms between resonant and non resonant contributions.

The evolution of the bandwidth with temperature in the range 10-180 K of these four vibrons is shown in Fig.5. The experimental data[13] are represented by circles. The curves in the Figures are calculated according to the discussion that follows.

Inspection of Fig. 5 shows that the temperature variation of the bandwidth of ν_1, ν_6 and ν_{10} is linear in the classical regime, whereas that of ν_9 is clearly quadratic. This means that for the first three vibrons only third order processes are important whereas for the ν_9 vibron also fourth order processes must be considered.

In order to interpret the data, we consider the scheme of vibron levels of Fig. 6. The figure is devided in four parts, one for each of the vibrons of interest. The vertical arrows in the figure have a lenght of 135 cm^{-1}, corresponding to the frequency range of the lattice phonons. Each internal level has a width corresponding to its dispersion in the crystal.

The relaxation of ν_1 (Fig. 5a) is linear with T in the classical regime. Fig. 6 shows that ν_1 can decay, by creation of a lattice phonon, in the levels ν_{10} and ν_{17} or, by fusion with a lattice phonon, in the levels ν_{18}, ν_{12} and ν_5.

The variation with temperature of the width of the ν_{10} vibron (Fig. 5b) is also linear in the classical regime. The ν_{10} vibron cannot decay, through a three-phonon down-process, into lower vibron levels by creation of one lattice phonon, since the next vibron level occurs at 707 cm^{-1}, i.e. outside the range of 135 cm^{-1}, which is the upper limit of the lattice phonon frequencies. From Figure 6b it can be seen, however, that it can decay, through up processes, into the levels ν_1 and ν_{17}.

The variation with temperature of the width of the ν_6 vibron is shown in Fig. 5d and is also linear with T. Again in this case down-processes cannot occur, since the next vibron level at lower frequency (418 cm^{-1}) is not accessible through creation of a lattice phonon. As can be seen from Fig. 5d there are, however, two possible up-processes into the higher levels ν_6 and ν_{11}.

The variation with temperature of the width of the ν_9 vibron is shown in Fig. 5c and is not-linear with T. From Fig.6c are possible only down-processes in the ν_{15} and in the ν_{18} levels. In addition to these we must consider also fourth-order processes to account for the non-linear T dependence of the bandwidth. The closest vibron level that can be reached by fusion of ν_9 with two lattice phonons is ν_{14} . The most probable process is thus the decay in the ν_{14} level by fusion with two lattice phonons.

In order to understand whether or not the interpretation in terms of phonon-assisted decay into vibron levels is correct, we have calculated the contribution of three-phonon processes to the bandwidth using an intermolecular potential[15] and the harmonic eigenvectors[12] taken from previous papers from our laboratory.

According to eqs. 13 and 14 the bandwidth is determined by the phonon density of states and by the numerical value of the third order coupling coefficients. The calculation of the density of states is relatively simple and fast whereas the evaluation of the coupling coefficients requires heavy computing.

calculation of the bandwidth of ν_1 at low temperature[14]. the harmonic eigenvectors were taken from a previous calculation of the crystal normal modes of benzene[12].

The result of the calculations are shown in graphical form in figs. 5a to 5d. For a correct comparison of experimental and calculated data it is necessary to take into account the fact that the experimental data refer to crystalline benzene with natural isotopic composition whereas the curves were calculated for the pure ^{12}C species.

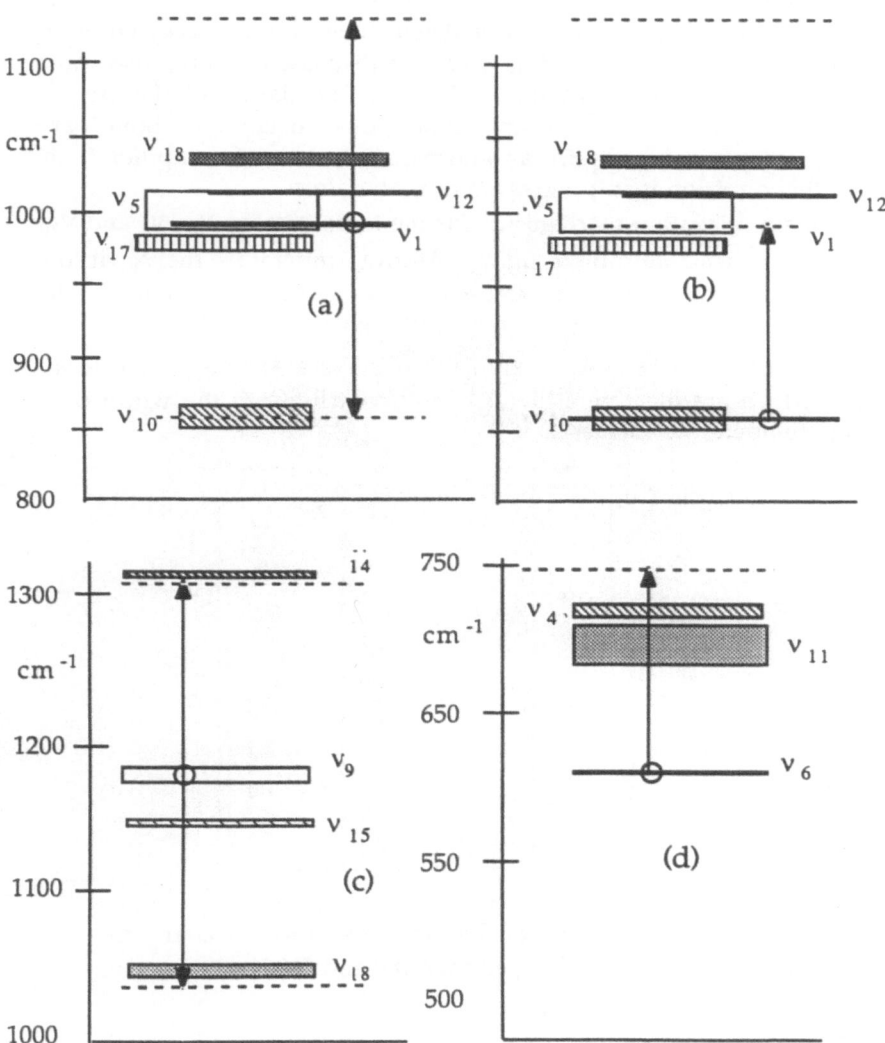

Figure 6. Schematic representation of the vibron levels of Benzene

An estimate of the effect of isotopic impurities can be made from the data of ref. 16, which show that the low temperature width of ν_6 and ν_{10} is practically zero for a pure ^{12}C benzene crystal, in perfect agreement with our result. This means that the difference between the experimental data and the calculated curves in figs. 5b and 5d is essentially due to a contribution from molecules containing ^{13}C atoms and thus that, in order to take this contribution into account, it is sufficient to shift the calculated curve for ν_6 of about 0.055 cm^{-1} and that of ν_{10} of about 0.035 cm^{-1}, respectively. In the case of ν_1 the agreement is very satisfactory even without this correction, in agreement with the fact that the isotopic effect is in this case vanishingly small. The authors of ref. 16 have in fact measured for this band a residual width at 4 K of 0.09 cm^{-1} for the pure isotopic species, close to our calculated value of 0.11 cm^{-1}.

In the case of the ν_9 vibron the calculated three phonon decay curve of fig.5c cannot fit the observed data since, as discussed before, also four-phonon processes are important in this case. Calculation of the quartic contributions, even with the approximations discussed above, are extremely complex and require an enormous amount of computer time. Work in this direction is in progress in our laboratory.

The main relaxation mechanisms for the three vibrons ν_6, ν_1 and ν_{10} are represented schematically in fig. 7 . At low temperature the ν_1 vibron, the only one whith a non-zero residual width, relaxes essentially in the ν_{10} level, whereas at 100 K the decay in the levels ν_{18} and ν_5 becomes important. For the ν_6 and for the ν_{10} vibrons there are only two decay channels which are inactive at T= 0 K and contribute to the width only when the temperature increases.

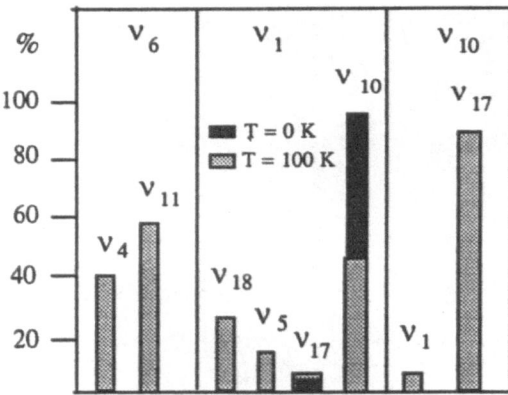

Fig. 7 . Schematic representation of the main relaxation mechanisms for the ν_6, ν_1 and ν_{10} vibrons at 0 and 100 K.

REFERENCES

1. S. Califano and V. Schettino, Int. Rev. in Phys. Chem. **7**, 19, 1988
2. S. Califano, V. Schettino and N. Neto, *Lattice Dynamics of Molecular Crystals*, Lecture Notes in Chemistry, Vol. **26**, Springer (Berlin) 1981
3. M.A. Ivanov, L.B. Kvashnina and M.A. Krivoglaz, Sov. Phys. Sol. State **7**, 1652 (1966)
4. C.B. Harris, R.M. Shelby and P.A. Cornelius, Phys. Rev. Letters, **38**, 1415 (1977)

5. P. Ranson, R. Ouillon and S. Califano, Chem. Phys. **86**, 115,1984
6. S. Velko and R. M. Hochstrasser, J. Phys. Chem. **89**, 2240, 1985
   S. Velko and R. M. Hochstrasser, J. Chem. Phys. **82**, 2180, 1985
7. D. D. Dlott, Ann. Rev. Phys. Chem. **37**, 157, 1986
8. W. Demtröder, *Laser Spectroscopy*, Springer Series in Chemical Physics, Vol.5, Springer (Berlin) 1982
   J. R. Shen, *The Principles of Nonlinear Optics*, Wiley (New York) 1984
9. R. F. and M. Balkansky, *Many-body Aspects of Solid State Spectroscopy*, North-Holland (Amsterdam), 1986
10. R. S. Tripathi and K. N. Patak, Nuovo Cimento, **21B**, 289, 1974
11. V. K. Jindal, R. Righini and S. Califano, Phys. Rev. B, **38**, 4259,1988
12. G. Taddei, H. Bonadeo, M. P. Marzocchi and S. Califano, J. Chem. Phys. **58**, 966, 1973
13. R. Torre, R. Righini, L. Angeloni and S. Califano, J. Chem. Phys. 1990, in press.
14. R. Torre, R. Righini, P. Foggi and L. Angeloni, Appl. Phys. B, in press.
15. R. Righini, Chem. Phys. **84**, 97 (1984)
16. J. Trout, S. Velsko, R. Bozio, P.L. Decola and R.M.Hochstrasser,J. Chem. Phys. **81**, 4746 (1985).

ANOMALIES OF THE ELASTIC LIGHT SCATTERING AT PHASE TRANSITIONS IN

CRYSTALS WITH POINT DEFECTS

A. P. Levanyuk

Institute of Crystallography
Academy of Sciences of the USSR
Moscow 117333, USSR

1. Introduction

The study of light scattering anomalies at second order phase transitions in solids started in the fifties with the observation of the "opalescence" at the $\alpha \rightleftarrows \beta$ transition in quartz[1]. After a decade it was shown that this anomaly has a static origin[2]. Then in the beginning of the seventies the observation of the temperature dependent central peak in neutron scattering near phase transitions attracted much attention. One of the proposed mechanisms for the central peak was defect-induced scattering (see e.g. Ref. (3)).

The reason for the increase of the elastic scattering near a second order phase transition seems to be fairly obvious. Indeed, among the disturbances of the crystal lattice caused by defects there is, in general, one corresponding to the order parameter (η), and some defects create a non-zero value of the order parameter at their sites even in the symmetrical phase. Such defects are conventionally called random field (RF) defects. The size of the region disturbed by an RF-defect is the correlation radius (ξ) of the order parameter. The increase of ξ on approaching the phase transitions leads to an increase in the cross-section for the defect-induced elastic scattering of neutrons or of light[5]. Thus, one can seemingly expect the maximum of the elastic scattering at the phase transition ($T = T_c$). This idea was used to treat the temperature dependence of elastic scattering at various phase transitions in crystals with various defects. A review of the results was given in Refs. (3,6).

Note that in the above explanation of the light scattering anomalies near phase transitions it has been implicitly assumed that in the nonsymmetrical (ordered) phase the crystal is single-domain. This assumption is, seemingly, reasonable because it was shown that RF-defects of small concentration don't remove the long-range order, i.e. the equilibrium state of the crystal in the nonsymmetrical phase is single-domain in spite of the presence of RF-defects (we don't take into account the long-range forces which may lead to domain formation in an ideal crystal). However, when the nonsymmetrical phase is quenched (as is almost always the case in a real experiment) the many-domain state, in spite of being a non-equilibrium one, may exist for a very long time in a crystal with defects.

The aim of this paper is to take into account the scattering from this non-equilibrium domain structure.

2. TEMPERATURE EVOLUTION OF THE DEFECT-INDUCED DISTORTIONS IN CRYSTALS WITH PHASE TRANSITIONS

In this section the results obtained by many authors are briefly reviewed. One can find many more details in Refs. (3,6). The discussion is confined to RF-defects and a one-component order parameter.

It is appropriate to use the continuum approximation and the Landau thermodynamic potential

$$F = \frac{1}{2} A\eta^2 + \frac{1}{4} B\eta^4 + \frac{1}{2} D\langle\nabla\eta\rangle^2 - h(r)\eta(r) \tag{1}$$

with

$$h(r) = \sum_i h_i \delta(r-r_i) , \tag{2}$$

where $h_i = \pm h_I$, both signs occur with the same probability, and the positions of the defects are completely random. The concentration of the defects (N) and the "defect strength" h_I are assumed to be small, the conditions of the smallness will be specified below. As to the temperature dependence of the coefficients A, B, D the classical (mean field) approximation will be used because taking nonclassical corrections into account would be to exceed the precision of the theory presented in Section III. Specifically, $A = A_o \tau \left(\tau = \frac{T-T_c}{T_c}\right)$, B, D = const.

For $T > T_c$ one can neglect the second term in Eq. (1) and find $\eta(r)$ from the linearized Euler equation. Within this approximation $\eta(r)$ is a superposition of contributions due to isolated defects:

$$\eta(r) = \sum_i \eta_i(r-r_i), \quad \eta_i(r) = \frac{h_I}{4\pi Dr} \exp(-\frac{r}{\xi}) , \tag{3}$$

where $\xi = (D/A)^{1/2}$. This approximation is valid if[3]:

$$\frac{h_I}{h_{at}} \ll 1, \quad N\xi^3 \left(\frac{h_I}{h_{at}}\right)^2 \ll 1 \tag{4}$$

where $h_{at} = 4\pi D^{3/2}/B^{1/2}$ has the meaning of the "atomic value" of the defect strength. One can see that the approximation becomes invalid close to the phase transition and almost nothing can be said about the form of $\eta(r)$ in this region.

For $T < T_c$ one has to remember first of all that if the defect concentration and strength are small long-range order persists in spite of the presence of defects (see e.g. Refs. (3,7). It means that the ground state is a single-domain one. Seemingly, for a small defect concentration one can use the same perturbation theory as for $T > T_c$ with the only difference that the zero approximation corresponds to the non-zero equilibrium value of η ($\eta_e \neq 0$). However, one can be sure that such an approach to the description of the nonsymmetrical phase is realistic only in the case when the crystal in the nonsymmetrical phase with defects is prepared in the following way: first the crystal is cooled to the nonsymmetrical phase and then the defects are introduced. In a real experiment the situation is, normally, the opposite: the crystal with defects is quenched from the symmetrical phase. Let us imagine that the temperature has been lowered instantly. At the first moment the domain structure is very fine because the sign of the order parameter at a given point is determined by the sign of the local field, just as in the symmetrical phase. These domains grow because the surface energy of the domain boundaries tends to be smaller. The growth is very rapid until the force tending to expand the domains (Gaussian pressure) becomes insufficient to overcome the barriers due to the defects. Then the expansion of the domain begins to proceed very slowly, the time dependence of the domain size being logarithmic. For an RF-defect in a three-dimensional system[7]

$$R = C_1 + C_2 \ln t . \tag{5}$$

This logarithmic growth has never been observed in experiment because, probably, the first term in Eq. (5) is generally greater than the second one for the typical times of the experiment.

Thus it is reasonable to assume that the domain size is given by the first term of Eq. (5). For RF-defects[7]

$$C_1 = R_{min} = \xi^2/\lambda , \tag{6}$$

where λ is the so-called fuzziness length

$$\lambda = Nh_I^2/\Sigma , \tag{7}$$

and Σ is the surface energy of the domain wall. Using the conventional estimates for the coefficients in Eq. (1) (see, e.g. Ref. 3) one obtains

$$\lambda = d\left(\frac{h_I}{h_{at}}\right)^2 \frac{Nd^3}{\tau^3} \psi^3 \tag{8}$$

where $\psi = 1$ and T_{at}/T_c for order-disorder and displacive systems respectively, d is the "atomic length," i.e. the interatomic distance, and T_{at} is the "atomic temperature" $(10^4 \div 10^5$ K).

It follows from Eq. (6) that $R_{min} \sim \tau^2$. This means that at some temperature $(\tau = \tau^*)$ R_{min} becomes comparable with ξ. This temperature is the boundary of applicability of Eq. (6) because the width of the domain wall is equal to ξ and, of course, R_{min} should exceed ξ. The condition $R_{min} \approx \xi$ or $\lambda \approx \xi$ may be considered also as the equation for the new (shifted due to defects) temperature of phase transitions to the ordered phase. Using Eq. (8) and taking into account that $\xi = d\psi^{1/2} \tau^{-1/2}$ one obtains

$$|\tau^*| = \left(\frac{h_I}{h_{at}}\right)^{4/5} (Nd^3)^{2/5} \psi. \tag{9}$$

Let us note that for displacive systems $(\psi \approx 10^2 \div 10^3)$ $|\tau^*| \approx 1$ at $h/h_{at} \approx 10^{-1}$ and $Nd^3 \approx 10^{-4}$, i.e. at a fairly small concentration of "moderately strong" defects.

3. LIGHT SCATTERING INTENSITY

The intensity of the scattered light per unit solid angle and unit intensity of the incident light is (see, e.g. Ref. 3):

$$I(q) = \left(\frac{V}{4\pi}\right)^2 q_o^4 < |\Delta\epsilon(q)|^2> \equiv VQ <|\Delta\epsilon(q)|^2> . \tag{10}$$

Here V is the scattering volume, ϵ is the dielectric constant for optical frequencies, $\Delta\epsilon(r)$ is the change in ϵ due to spatial inhomogeneity, q_o is the wave vector of the incident light, $\Delta\epsilon(q) = V^{-1}\int\Delta\epsilon(r)dr$ is the Fourier transform of $\Delta\epsilon(r)$, q is the difference between the wave vectors of the scattered and incident light.

In the case of a one-component order parameter $\Delta\epsilon = a\eta^2$. For $T > T_c$ and RF-defects one finds[3]:

$$I(q\approx0) = VQ \, a^2\Big(\frac{h_I}{h_{at}}\Big)^4 \, N\xi^2 \, \frac{4\pi^2 D^2}{B^2} \, (1 + 2\pi Nr_c^3) \, . \tag{11a}$$

At the boundary of applicability of the linear theory (Eq. (4)) one has:

$$I(q\approx0) = VQ(10^2 \div 10^3)d^3\Big(\frac{h_I}{h_{at}}\Big)^{2/3}(Nd^3)^{1/3} \equiv I_1. \tag{11b}$$

In obtaining Eq. (11b) we have assumed that $D = A_{at} d^2$, a $A_{at}/B \simeq 1$. These estimates are valid both for order-disorder and for displacive systems.

Let us now estimate the light scattering intensity in the non-symmetrical phase, assuming that the characteristic domain size is equal to R_{min}. Unfortunately nothing is known about the statistics of the domain sizes. Thus only very rough estimates of the scattering intensity are possible. The contribution to $\Delta\epsilon(r)$ from a domain wall can be estimated as $\Delta\epsilon(r) = \epsilon_o\delta(r-r_o)$ where r_o describes the position of the domain wall,

$$\epsilon_o \simeq a\eta_e^2\xi = a(|A|/B) \propto |\tau|^{1/2}.$$

We consider first the temperature region close to T_c where $q_oR_{min}<1$. Here one can consider as independent the scattering from the volume with characteristic dimension q_o^{-1}. It seems to be reasonable to assume that such a volume is divided into volumes R_{min}^3, and that half of the domain walls are absent, the locations of the domain wall vacancies being random. For the scattering intensity one obtains

$$I = VQ \, \epsilon_o^2 \, R_{min} \, . \tag{12}$$

Thus in the region of applicability of Eq. (12) $I \propto |\tau|^3$.

It is instructive to compare $I(\tau^*)$ and I_1 (see Eq. (11b)). One finds:

$$I_1/I(\tau^*) = \Big(\frac{h_I}{h_{at}}\Big)^{4/5}(Nd^3)^{2/15}(10^2 \div 10^3)^{2/5} \, . \tag{13}$$

We see that for very small concentrations and very "weak" defects $I_1 < I(\tau^*)$. However, because of the very rough estimate of $I(\tau^*)$ one cannot be sure that the exponents in Eq. (13) are reliable. Because they are fairly small it is quite possible that in reality I_1 and $I(\tau^*)$ differ by some numerical factor only. Thus one cannot conclude if the function $I(T)$ has a maximum somewhere in the vicinity of the phase transition temperature of the ideal crystal.

For the case $R_{min} > q_o^{-1}$ the probability to find a domain wall in the volume q_o^{-3} can be estimated as $(q_oR_{min})^{-1}$. For the scattering intensity one has

$$I = VQ \, \epsilon_o^2 q_o^{-2} R_{min}^{-1} \, . \tag{14}$$

It follows from Eq. (14) that in this temperature region $I \propto |\tau|^{-1}$.

Let us now summarize the results obtained. The temperature

dependence of the scattering intensity can be described as follows. On approaching T_c from the symmetrical phase the intensity increases. It is not clear if there is a maximum at the phase transition temperature of the ideal crystal, and the behavior of the intensity in the temperature region between the "old" and "new" phase transition temperatures is unknown. On lowering the temperature from the "new" (renormalized due to defects) T_c the intensity increases as long as the characteristic domain size is smaller than the wavelength of the light. In this temperature region the domain walls can be considered as particles whose volume is $R_{min}^2 \xi$. In spite of the fact that the number of these "particles" decreases as the temperature is decreased, the scattering cross-section increases very rapidly, and as a result, the scattering intensity increases also. In the temperature region where R_{min} is larger than the wavelength of the light the effective scattering cross-section is estimated as $q^{-2} \epsilon_0$ and increases as $|\tau|^{1/2}$, i.e. much more slowly. As the result the decrease of the probability for a domain wall to be located in the volume q^{-3} leads to the decrease of the scattering intensity.

Let us emphasize that for the entire temperature range of the non-symmetrical phase the scattering intensity due to domains is greater than I_1 and, consequently, is greater than the scattering intensity due to point defects in a single-domain crystal, because I_1 is the maximum value of the intensity.

Because of the different dependence of I on R_{min} in the two temperature regions for $T < T_c$ (see Eqs. (12) and (14)) one expects different time dependences of the intensity in these two regions if the crystal is kept at a fixed temperature. Close to T_c where $R_{min} < q_0^{-1}$ the intensity first increases and then decreases. Far from T_c, where $R_{min} > q_0^{-1}$, a monotonic decrease of the intensity is predicted.

Let us mention finally that it is the simplest case that was discussed in this report. To compare the theoretical consclusions with experimental results seems to be of importance to take into account the crystal anisotropy and, for proper and improper ferroelectrics and ferroelastics, the long range fields arising from the bending of the domain walls.

REFERENCES

1. I. A. Yakovlev, T. S. Velichkina, and L. F. Mikheeva, Kristallografia 1, 123 (1956).
2. S. M. Shapiro and H. Z. Cummins, Phys. Rev. Lett. 21, 1578 (1968).
3. A. P. Levanyuk and A. S. Sigov, Defects and Structural Phase Transitions, (Gordon and Breach, New York, 1988).
4. J. D. Axe and G. Shirane, Phys. Rev. B8, 1965 (1973).
5. A. P. Levanyuk, V. V. Osipov, and A. A. Sobyanin. In: Theory of Light Scattering in Condensed Matter, B. Bendow, J. L. Birman, and V. M. Agranovich, eds., (Plenum Press, New York, 1976), p. 517.
6. E. B. Kolomeisky, A. P. Levanyuk, and A. S. Sigov, Report given at the Seventh International Meeting on Ferroelectricity, to be published in Ferroelectrics.
7. T. Natterman and J. Villain, Phase Transitions, 11, 5 (1988); T. Natterman and P. Rujan, to be published in the International Journal of Modern Physics.

DYNAMICAL FLUCTUATIONS IN A DIPOLAR GLASS

K. B. Lyons and P. A. Fleury
AT&T Bell Laboratories, Murray Hill, NJ 07974

H. Chou, J. Kjems,* and S. Shapiro
Brookhaven National Labs, Upton, New York 11973

D. Rytz
Hughes Research Laboratories, Malibu, CA 90265

INTRODUCTION

Other reports in this symposium have focused on the static properties of defects near a structural phase transition. A different set of questions surrounds the dynamic behavior of such systems. $KTaO_3$:Nb, or KTN, is one such system of particular interest where, due to the availability of high quality materials, it may be possible to make quantitative contact between the behavior of defects in the dilute limit and the effects of similar defects at concentrations where they affect the macroscopic behavior of the order parameter. We have previously shown[1] that the average radius ξ of the ferroelectrically distorted microregions surrounding symmetry breaking defects in nominally pure $KTaO_3$ is directly related, as a function of T, to the phase diagram of $KTa_{(1-x)}Nb_xO_3$ (KTN) in the Tx plane. That is, the phase transition to long range ferroelectric order at high concentration x occurs when $\xi(T_c)=An^{0.33}$, where n is the number density of Nb ions, and A is a constant very close to unity. Thus, the structural phase transition occurs when the distorted regions overlap continuously.

Such a defect-driven phase transition, in a host system such as $KTaO_3$ which has no phase transition in its pure form, affords the opportunity for studying a structural phase transition at arbitrarily low T, if x is adjusted close to the value x_c where T_c extrapolates to zero. It may be expected in such a case that a crossover will occur between a true structural phase transition and glassy behavior on laboratory time scales, as the ionic relaxation slows down for $T_c{\rightarrow}0$.

PREVIOUS RESULTS

There is strong evidence that such a crossover occurs in KTN. Whereas Raman measurements[2] indicate a soft mode behavior for $x=0.009$, Brillouin scattering[2] and AC dielectric measurements[3] exhibit a transition temperature $T_c(\omega)$ with substantial ω dependence. In fact, the order parameter relaxation obeys a Vogel-Fulcher law[2] $\tau_c^{-1}=\nu_o\exp[-E_a/(T-T_o)]$ over nine decades in the relaxation time τ_c, with $T_o=3.0$ K. This dependence, which is more rapid than a simple activated single-ion model would predict, suggests a cooperative slowing down of defect motion as correlated defect clusters grow in size at low temperature.

This behavior may be explained as follows. As T decreases, the Nb hopping among the eight available equivalent sites slows down and the host lattice correlation length increases, thus causing the Nb-related dipolar clusters (polarization clouds surrounding one or more Nb moving

---

* Permanent Address: Risø National Laboratory, Roskilde, Denmark

cooperatively) to grow. As the clusters grow, their relaxation slows precipitously. For $x=0.009$, the mean Nb-Nb distance is about 5 unit cell diameters. The value of ξ reaches this distance at temperatures of order a few K. Hence, the competition between ordering and glassy freezing may be very delicately balanced in this system.

Additional evidence for glassy behavior has been found in the study of the polarization response of KTN ($x=0.009$). The response is essentially quadratic[4] at low temperature for $x=0.009$, even at exceedingly low fields, indicating a nonanalytic (e.g. bistable) response. This response can be modeled by imparting a switching behavior to the clusters mentioned above, but with a switching field (E_c) distribution which has a nonzero value at zero E_c. Such a distribution is a signature of the random nature of the low temperature state. Furthermore, when the step response is measured for times $\leq 10^3$ s a significant "waiting time effect" is observed,[5] which is the signature of a crossover between equilibrium and nonequilibrium relaxation.[6] This latter effect, reminiscent of analogous effects in spin glasses,[6] manifests the nonergodicity characteristic of a glassy system.

In an effort to characterize further the dynamics of the KTN system, we have studied a 1 cm^3 sample at a higher concentration, namely $x=0.012$, using both neutron and light scattering. The neutron scattering results[7] clearly show a soft mode and exhibit a clear *hardening* of the order parameter fluctuations below $T=20$ K, where the zone center optic mode frequency reaches a minimum value of 7 cm^{-1}. Thus, on a fast enough time scale the material behaves as if it undergoes a ferroelectric phase transition, although the finite minimum frequency observed remains to be understood, and may be related specifically to the glassy nature of the polar phase. Indeed, the long-time behavior indicates a dipolar glass behavior in this concentration range.[3] In the present manuscript, we report light scattering measurements on the same sample used for the neutron measurements. These measurements present a curious dichotomy between the high frequency and low frequency behaviors which suggests a phenomenon of cluster percolation not yet observed at $x=0.009$.

RESULTS

The purposes of the light scattering studies reported here were two-fold: (i) to probe via Brillouin spectroscopy the order parameter dynamics at a frequency significantly different from that accessible to neutron scattering and (ii) to check for inhomogeneity in the crystal, as evidenced in the Raman spectrum as a function of position in the crystal near T_c.

The Raman scattering was performed in a right angle geometry with a Spex 1402 double grating monochromator and photon counting electronics, using about 10 mW of 5145Å radiation as the excitation source, focused to a spot size of $\sim 50\mu m$. The collection optics employed were f/3.5. The entrance and exit slits were 40μm in width, with an active height of 1 cm except as noted. In the comparison of the vv and vh scattering intensities, a scrambler was employed to remove the polarization dependence of the holographic grating response. The spectrometer and temperature control apparatus were controlled by an LSI 11/02 computer which performed the data acquisition.

The spectra obtained are shown in Fig. 1a, out to a shift of 300 cm^{-1}. There are several noteworthy aspects of these data. First, as has often been noticed before, even in nominally pure material, the hard modes, which should become allowed by symmetry only below T_c, appear in the spectrum substantially above $T_c \cong 20$ K. Second, a soft mode is seen, with a frequency (~ 10 cm^{-1}) which reproduces accurately the neutron scattering observations. The mode frequencies shown in Fig. 1b were extracted from (i) the frequency position of the soft mode maximum and (ii) the center of gravity of the scattering observed in excess of that found well above T_c. It was possible to extend the latter technique with acceptable accuracy substantially above the nominal value of T_c. Over the entire range, the Raman soft mode data essentially reproduce the curve observed using neutron scattering. As expected, our ability to resolve the soft mode feature was substantially better than in the neutron scattering data in the immediate vicinity of T_c. Nevertheless, the results lie in quantitative agreement except very near the minimum, where the neutron scattering analysis was only qualitative in any event.

There is a third phenomenon which becomes most evident when the two-phonon background of the spectrum is examined. Since this scattering is allowed on both sides of the transition, it should exhibit a temperature dependence characteristic of normal two-phonon scattering. However, it does not do so. Rather it exhibits a precipitous drop in intensity quite precisely at the temperature where the soft mode frequency is a minimum. This drop in intensity

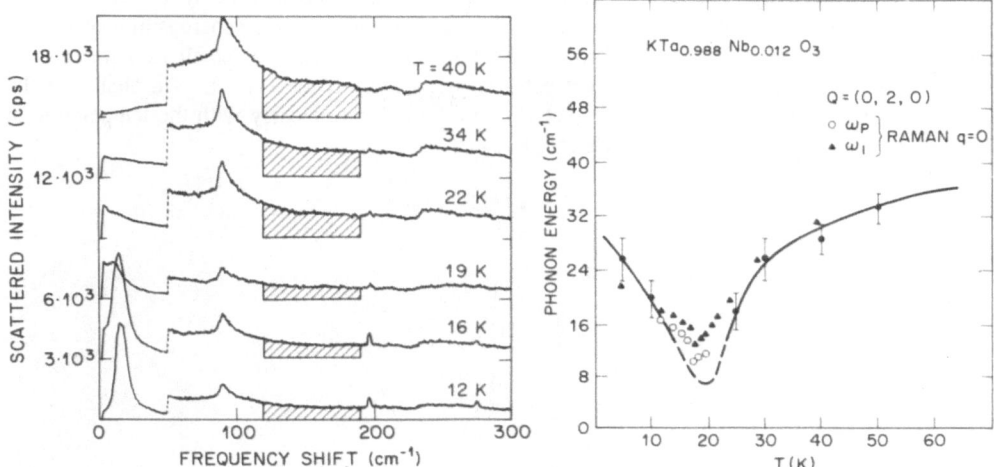

FIGURE 1. Raman spectra of KTN (x=0.012) are shown on the left. The scattering geometry is polarized $(x(zz)y)$, and an iodine cell is employed to reduce elastically scattered light. Integrals of the shaded regions are plotted in Fig. 2. The region below 50 cm^{-1} is plotted with gain reduced ×4. On the right we show the soft mode peak positions extracted by location of the spectral maxima (open circles) and by calculation of the first moment (solid triangles). The values of the soft mode frequency inferred by neutron scattering are shown by the solid dots and error bars, The line is a guide to the neutron data, taking into account qualitative indications of a minimum frequency near 7 cm^{-1}.[7]

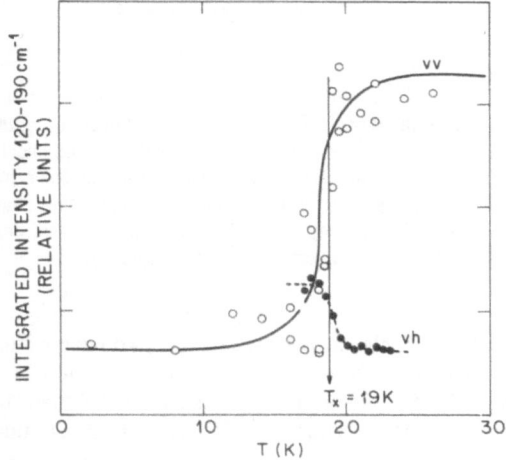

FIGURE 2. Temperature dependence of the two-phonon Raman intensity, integrated from 120 to 190 cm^{-1} (shaded regions in Fig. 1). A scrambling of the polarization is clearly evident near 19 K.

was studied quantitatively by integration of the spectral intensity over the range 120-190 cm^{-1}, where no sharp features interfere. Data were obtained for both vv and vh scattering geometries, using a scrambler on the spectrometer input. The results of this analysis are shown in Fig. 2. These data make it quite clear that the drop in intensity apparent in Fig. 1 is the result of polarization scrambling, affecting both the incident beam and the scattered light, and thus causing a fourfold drop in intensity for the predominantly vv two-phonon component below the apparent T_c. The drop occurs sharply, reproducibly, and reversibly, as evidenced by the superposition of several runs in Fig. 2. Moreover, in tests performed with a reduced slit height as small as 1 mm it was found that the behavior was also quite independent of position in the crystal. We shall call the temperature of this drop T_x, and we note that T_x corresponds quite closely with the temperature at which we observe the minimum soft mode frequency.

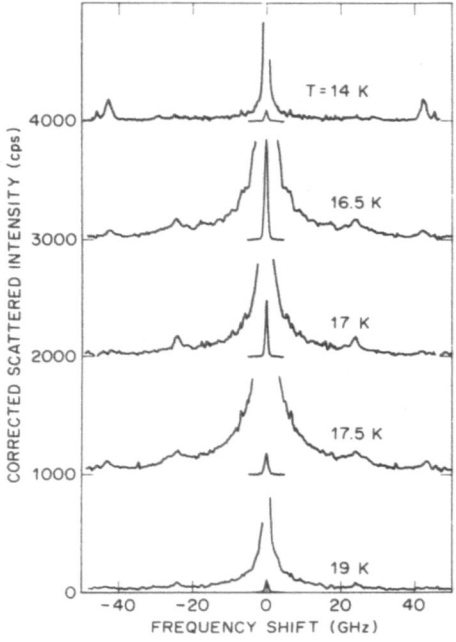

FIGURE 3. Fabry Perot spectra of the $x=0.012$ KTN sample, obtained in right angle vv geometry, using an iodine cell to remove the elastically scattered light. The spectral profiles are corrected for the iodine transmission function. The region near the laser frequency is plotted on a scale reduced 15x on the same baseline in order to reveal the behavior of the scattering below 1 GHz. The elastically scattered light, observed with the laser detuned from the I_2 absorption, shows no dependence on temperature over the same region.

If this were the only data available on the system, one would conclude that the sample enters a rather normal multi-domain state below a ferroelectric transition at $T_c=T_x$, with some evidence of precursor effects in the hard mode behavior (see summary of characteristic temperatures in Table I). Indeed, a previous neutron scattering study concluded as much.[8] However, the Brillouin spectra do not fit such a pattern. These spectra, displayed in Fig. 3, were obtained using a tandem pressure scanned Fabry Perot interferometer in conjunction with an iodine reabsorption cell to remove the elastically scattered component. The spectral resolution is 0.6 GHz (HWHM), and the same mounting, collection optics, and incident power were employed as for the Raman spectra above. In fact, the sample was not heated above 25 K between the two runs. The behavior of the polarization fluctuation peak centered at the laser frequency belies the simple interpretation indicated above for the Raman and neutron scattering data. First, the maximum intensity near 10 GHz occurs near 17.5 K, nearly two full degrees *below* T_x. We note that any contribution from the tail of an elastic component may be completely excluded at this frequency under the conditions of our experiment. Secondly, at still lower temperatures, there is a growth in a component with a width (HWHM) of 0.5 GHz or less, with a maximum intensity occurring near 16.5 K. As indicated in Table I, we interpret these two features as different manifestations of the same phenomenon, namely polarization fluctuation. In our spectra the narrow component is indistinguishable spectrally from purely elastic scattering. However, checks of the elastic scattered intensity, which were carried out before and after every scan of the apparatus indicated that the level of elastic scattering remained quite constant as a function of temperature and time. We therefore conclude that the increased intensity is the result of a dynamic component which lies outside the iodine absorption ($\Gamma_{I2}{\sim}200$ MHz, HWHM) but inside the Fabry Perot resolution (0.6 GHz). We may infer, moreover, that the reduction of that intensity on further cooling past 16.5 K results from a lowering of the relaxation frequency below the I_2 absorption width. On the basis of this qualitative argument, we make the last entry in the righthand column of Table I.

Table I

Characteristic Temperatures for Various Frequency Probes

METHOD	ω (Hz)	$x{=}0.009$	$x{=}0.012$
neutron scattering	$2.5{\cdot}10^{11}$	-	20 K
Raman scattering Hard mode onset Soft mode	$2{\cdot}10^{12}$ $2.5{\cdot}10^{11}$	18 K 12 K	22 K 20 K
Brillouin scattering P fluctuation (Γ_{12})	10^{10} $0.5{\cdot}10^{9}$ $0.3{\cdot}10^{9}$	11 K <10 K	17.5 K 16.5 K 16 K
$\varepsilon(\omega)$ (Ref. 3)	10^{6} 10^{2}	8 K 6 K	
Field step response	$10^{-3}{-}1$	5 K	

We note in passing the change in width and intensity of the LA and TA modes on cooling. Similar effects were seen[4] for x=0.009. It is likely that this behavior is a manifestation of the scattering of phonons by the same clusters responsible for the polarization dynamics.

The above discussion clearly leads to a simplified view of the actual behavior of the spectra in Fig. 3. There is no doubt a range of widths involved, not just two, with a weight function which is a function of temperature. Indeed, attempts to fit the spectra for x=0.012 as a superposition of two Lorentzian peaks yields good fits, but unreliable width parameters, since there is a strong correlation between the parameters extracted for the two components (widths $\Gamma_n{\ll}\Gamma_b$ and amplitudes $A_n{\gg}A_b$). However, if we treat this superposition as a *parametrization* of the spectra, we can then use it to extract measures of the integrated intensity $I{=}\int I(\omega)d\omega$ and first spectral moment $\Gamma_{av}{=}\int \omega I(\omega)d\omega$ in the form

$$I{\equiv}A_n\Gamma_n{+}A_b\Gamma_b \; ; \; \Gamma_{av}{\equiv}A_n\Gamma_n^2{+}A_b\Gamma_b^2 \; .$$

Performing the calculation in this manner excludes the effects of the phonon modes which also appear in the spectral data. The parameters I and Γ_{av} then exhibit a smooth variation with temperature, as shown in Fig. 4. They both approach an anomaly near 16 K, nearly 3 K lower than T_x. Power law fits to the respective data sets yield critical exponents $\gamma_I{=}1.5{\pm}0.5$ and $\gamma_\Gamma{=}0.6{\pm}0.3$, where the uncertainties stem mostly from the uncertainty in the proper value of $T_c{=}16.2{\pm}0.2$, and the small temperature range usable for the fit. Additional uncertainty results from the subtraction of a small background component from the narrow peak intensity in order to account for leakage through the iodine cell. Neither of these uncertainties influence the qualitative picture, though. Indeed any reading of the data makes it obvious that there is a substantial range of temperature over which the slowing down of the polarization fluctuations occurs, with two apparent anomalous points, quite clearly separated in temperature.

DISCUSSION

The juxtaposition of the Brillouin and Raman data above seems at first glance to pose a contradiction. Indeed, the observation of polarization scrambling requires the presence of distorted

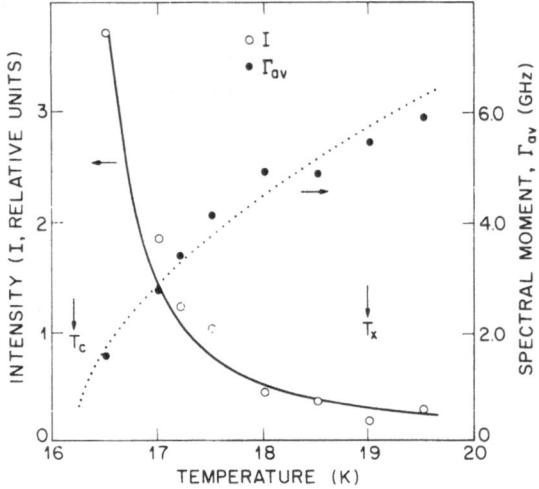

FIGURE 4. Integrated intensity (I) and average width (Γ_{av}) derived from the Brillouin spectra as described in the text. The lines represent power law fits to the two data sets with T_c=16.2 K, which yields the exponents γ_I=1.23±0.1 and γ_Γ=0.55±0.1.

regions with a dimension at least that of the exciting wavelength. Such large regions must be stationary on our experimental time scale. On the other hand, the Fabry Perot data show clearly the presence of a relaxation component which *continues to narrow* over a range of several degrees below T_x. This behavior is certainly not consistent with a normal transition at T_x to a multidomain ferroelectric state.

We have used laser ionization mass analysis[9] to check for possible variations of Nb content with 1 µm spatial resolution. Within a measurement accuracy of 0.2 at.%, no variation was seen for this sample. This provides additional evidence against any segregation of the Nb dopant which could cause smearing of a true phase transition. Hence it appears that the dual anomalies near 16 and 19 K are most likely due to intrinsic properties of the random mixed KTN system.

We can suggest two possible pictures to account for this behavior, one based on the percolation of a fractal cluster of correlated dipoles at the well defined temperature T_x, and the other based on the development of quadrupolar order at T_x, with dipolar order or freezing occurring at lower temperature, below 16 K. The first concept is a simple extension of ideas previously presented. A symmetry-breaking defect in the pure lattice is expected to generate a ferroelectric microregion with an extent of several lattice spacings, even at relatively high temperatures.[1] Each of the offcenter Nb ions should produce such a region. As the host lattice is cooled, approaching its incipient ferroelectric instability, the dielectric susceptibility increases and the characteristic size of the microregions grows. Thus, groups of closely spaced Nb ions will find their polarization clouds overlapping, and begin to execute strongly correlated motion.[2] These correlated clusters grow as the crystal is cooled until at some point a percolation occurs, forming a cluster with a macroscopic extent, and hence essentially no dynamic response. At this point, however, even the percolated cluster could remain fractal, in the sense that it occupies only a portion of the sample volume, leaving a significant portion of the crystal volume uncorrelated with this "static" cluster. It is possible then that the percolation occurs at T_x, where the polarization scrambling provides evidence for long-range static correlation in the birefringence, and that the dynamic slowing down observed in the Fabry Perot spectra below T_x relates to regions excluded from the percolated cluster.

We note in passing that such a correlated cluster does *not* necessarily possess the spontaneous polarization characteristic of the ferroelectric phase. For a very dilute system of dipoles such as this one, the local details of the Nb positions probably favor a very complicated dipolar arrangement, since the sign of their interaction will vary with their relative positions, in relation to the direction of the local polarization.[10] On any length scale, though, even in such a random system, there will remain a static polarization and a static birefringence which might be sufficient to cause the polarization scrambling observed.

The alternative picture for understanding the observed behavior is based on the fact that the distortion of the unit cell is independent of the *sign* of the Nb displacement (or of P_s). Moreover, the elastic deformation entails a long-range interaction which could lead to development of quadrupolar order, causing the birefringence and consequent polarization scrambling at T_x. The Nb ions would then be biased towards a choice of two possible positions, of the eight available equally above T_x, but they could continue to fluctuate between those two to lower temperatures. It is apparent that dipolar randomness, as measured by $<P_s(0)P_s(r)>$, and in the distribution function for $P_s(r)$, could still exist in such a case, depending on the relative dynamics of the freezing and ordering processes. X-ray measurements are in progress to study the possible development of quadrupolar order,* while acoustic and dielectric measurements are planned to elucidate the low frequency dynamics of the polarization. Until these experiments are complete, it is not possible to ascertain the exact nature of the glassy phase for $x=0.012$.

We emphasize that our percolated cluster picture is quite speculative. However, not only does it appear quite attractive as an explanation of the dynamic behavior observed but also it should be amenable to theoretical development to yield more quantitative comparison with experimental data. An important question surrounds the way in which such a picture evolves into a normal ferroelectric transition at higher Nb concentration. Further studies are planned in applied electric field, in order to check the sensitivity of the observations to small ordering fields, which may allow us to distinguish between the two suggested pictures. We anticipate, on the basis of the qualitative discussion given above, that the very occurrence of this cluster as well as the low frequency dynamics should be a strong function of applied field. On the other hand, the development of quadrupolar order may be quite unaffected by a small field.

The behavior of this sample relative to that of the $x=0.009$ sample previously studied[1-5] is summarized in Table I. In the near future we plan to carry out the experiments required to fill in the lower frequency data for the present ($x=0.012$) sample. For now, we simply note that the behavior appears similar as far as the temperature range over which the relaxation occurs; however, no analog of T_x has been observed for $x=0.009$. It will be necessary to observe the behavior for $x=0.012$ over a similarly broad frequency range before a reliable comparison can be made. In particular, we cannot yet draw any conclusion as to the applicability of a Vogel-Fulcher parametrization of the data. [The data in Fig. 4 do not yield sensible parameters if a VF fit is attempted. A fit to $\Gamma_{av}=\nu_o\exp[-E_a/k(T-T_o)]$ yields an attempt frequency $\Gamma_{av}\sim 10^{10}$ Hz, a freezing temperature $T_o\sim 15$ K, and an activation energy $E_a\sim 1$ cm^{-1}.]

The possibility remains that the polarization scrambling effect, not observed for the $x=0.009\%$ sample,[2] signals a fundamental difference in the order parameter behavior at this increased concentration, $x=0.012\cong 1.5x_c$. Indeed, early dielectric measurements in similar materials[12] found evidence of a change in the characteristic temperature dependence of $\varepsilon(T)$ over this range of composition. Due to the extreme dependence of material parameters on x near x_c it will be necessary to perform dielectric measurements on our own sample in order to carry out accurate comparison.

We also recognize an additional picture which could be consistent with our data, namely that the material is first entering a tetragonal phase upon cooling, as KTN does at much higher concentration. In this case, there would not be one site that predominates, but rather four equally, and fluctuations among these four could cause the dynamics we observe. However, no evidence has been found in previous studies for such a transition below $x=0.05$. Indeed, preliminary TEM[11] and x-ray[13] measurements indicate a lack of such distortion. Hence, we feel this explanation is an unlikely one.

In this work we have established some similarity of the dynamic behavior observed for a KTN sample substantially above the dipolar glass concentration threshold, x_c, in the frequency regime above ~1 GHz, when compared to that observed near x_c. We have pointed out a seemingly paradoxical anomaly in the polarization behavior of the scattered spectrum, and have speculated as to its origin. Although neutron and Raman data, as in earlier studies, seem consistent with a

---

* No evidence of quadrupolar order was seen in a previous x-ray study[11] at slightly higher concentration.

transition to a normal (multidomain) ferroelectric state, the lower frequency dynamics deviate strongly from such a picture. We have suggested two possible pictures which invoke (i) a percolated cluster of correlated dipoles which develops at a temperature substantially above that of the dipolar freezing or (ii) a quadrupolar ordering temperature $T_x=19$ K, substantially above the dipolar freezing/ordering temperature at about 16 K. The two models are not mutually exclusive since a random behavior in the dipolar order could develop within a quadrupolar domain. Future measurements are indicated at lower frequencies in order to elucidate further the nature of the low temperature state.

We are grateful to L. Heimbrook for carrying out the spatially resolved mass spectroscopic analysis of our sample. The work at Brookhaven National Labs is supported by the Division of Material Science, U. S. Department of Energy, under contract number DE-AC02-76CH00016.

REFERENCES

1. H. Uwe, K. B. Lyons, H. L. Carter, and P. A. Fleury, Phys. Rev. B **33**, 6436 (1986).
2. K. B. Lyons, P. A. Fleury, and D. Rytz, Phys. Rev. Lett. **57**, 2207 (1986).
3. G. A. Samara, Jpn. J. Appl. Phys. Suppl. **24**, Pt. 2, 80 (1985).
4. K. B. Lyons, P. A. Fleury, T. J. Negran, and H. L. Carter, Phys. Rev. B **36**, 2465 (1987).
5. K. B. Lyons and P. A. Fleury, in: *Time Dependent Effects in Disordered Materials*, Ed. by R. Pynn and R. Riste, Plenum, New York, 1987, pp. 297-300.
6. P. Svedlindh et al, Phys. Rev. B **35**, 268 (1987).
7. H. Chou, S. M. Shapiro, K. B. Lyons, D. Rytz, J. Kjems, to be published.
8. M. D. Fontana, W. Kress, G. Kugel, N. Lehner, and D. Rytz, Ferroelectrics **55**, 23 (1984).
9. L. Van Vaeck and R. Gijbels, Microbeam Analysis **17**, 1256 (1989).
10. B. E. Vugmeister and M. D. Glenchuk, Sol. St. Comm. **48**, 503 (1989).
11. P. Buffat and D. Rytz, unpublished data.
12. D. Rytz, U. T. Höchli and H. Bilz, Phys. Rev. B **22**, 359 (1980).
13. H. Chou, et al, to be published.

LOCALIZATION OF LIGHT IN RANDOM MEDIA

Ad Lagendijk

Natuurkundig Laboratorium, Universiteit van Amsterdam
Valckenierstraat 65, 1018 XE Amsterdam, THE NETHERLANDS

ABSTRACT

After reviewing shortly the issue of localization of light we will discuss recent developments. Spatial amplitude correlations build up in a strongly scattering medium due to interference effects. Theory predicts three different kinds of correlations: short range (F_1), long range (F_2,) and universal (F_3). These correlations have several observable consequences: enhanced backscattering and intensity fluctuations. We will discuss measurements of intensity fluctuations in three-dimensional and two-dimensional media. Both F_2 and F_1 have been observed separately. The results will be compared with theory.

QUANTUM OPTIC AND TRANSIENT EFFECTS OF EXCITONIC POLARITONS, AND

PROPERTIES OF PHONORITONS

Joseph L. Birman, M. Artoni, S.V. Branis, O. Martin and
Bing Shen Wang

Physics Department
City College of New York, C.U.N.Y.
New York, New York 10031

ABSTRACT

The following topics are briefly reviewed: Demonstration that polaritons are intrinsically squeezed, and methods to detect the squeezing:
a) use of the crystal-vacuum surface for homodyne processes, and b) two
scattering experiments. The Phonoriton - a generalization of the exciton
polariton and two non-linear methods to detect it: a) Non-Linear Resonance
Scattering and b) Non-Linear Reflectivity; an examination of possible
polariton-soliton formation and conclusion that previous treatments
omitted a highly singular term, making formation highly unlikely if not
impossible.

INTRODUCTION

In this paper we shall give a brief survey of some results our group
has obtained recently in the study of the family of exciton polaritons.
Special emphasis will be on a few topics of our group's work: some quantum
optic squeezing properties of these mixed modes; the response of the medium
to a transient optical pulse, especially the possible existence of a
polariton soliton; and certain non-linear effects related to the phonoriton, an excitation which has not yet been positively identified experimentally.

1. Polaritons

First recall some well known material about the excitonic polariton.
We may consider a system comprising exciton and photon to be described
by the following total Hamiltonian:

$$H = \sum_{k>0} [H_k + H_{-k}] + h.c.$$

Paper for Proceedings IVth Binational US-USSR Symposium "The Physics of
Optical Phenomena and Their Use as Probes of Matter" U. of California,
Irvine, 23-27 January 1990.

with

$$H_k = E_k^{ph} a_k^\dagger a_k + E_k^{exc} b_k^\dagger b_k + B_k' a_k a_{-k}$$

$$- C_k' b_k b_{-k} + i A_{1k} a_k b_k + i A_{2k} a_k^\dagger b_k \tag{1}$$

Here, the operators a_k, b_k are photon, exciton destruction operators. Justification for this starting Hamiltonian was given by Heller and Marcus [1951] and Hopfield [1958]. It can be derived from an extreme tight-binding (Frankel) model for the exciton. Terms included in H are bare exciton, bare photon, and bilinear exciton and photon interactions, which are usually kept in the simplified treatments generally ascribed to Hopfield as in standard texts: Kittel [1963]. Also included is the non-resonant exciton-exciton term, which can originate for example in dipolar exciton-exciton interaction. Noteworthy also is that this form of Hamiltonian is an element in the Lie Algebra $S_p(8,C)$; see also Kim and Birman [1988] for another dynamical algebraic treatment. To distinguish the usual "Hopfield Hamiltonian" from the case where added terms are present, we call the Hamiltonian (1) an <u>generalized</u> polariton Hamiltonian. It reduces to the Hopfield case when $A_{1k} = A_{2k}$ $C_k' = 0$ and $B_k' = B_k$.

Now we diagonalize H_k by use of the Bogolyubov method. Introduce the polariton operators

$$\eta_k = x\, a_k + y\, b_k + z\, a_{-k}^\dagger + w\, b_{-k}^\dagger \tag{2}$$

and determine the energy and the coefficients (x,y,z,w) from

$$[\eta_k, H_k] = \varepsilon_k\, \eta_k \tag{3}$$

This produces a biquadratic eigenvalue equation for ε_k

$$\varepsilon_k^4 - \varepsilon_k^2 f_{2k} + f_{0k} = 0 \tag{4}$$

where

$$f_{2k}\ (E,A,B,C) \quad \text{and} \quad f_{0k}\ (E,A,B,C)$$

are simple algebraic functions of the coefficients in H_k. We then determine a set of these coefficients from the observed dispersion $\varepsilon(k) = \hbar\omega(k)$ as obtained e.g. via RBS experiments reviewed by Koteles [1982]. These coefficients are then used to obtain (x,y,z,w).

The same transformation to polaritons can be achieved via a different route which brings out the relationship to squeezing. Define new mixed mode operators by

$$a_\pm = \alpha_k\, a_{\pm k} + e^{2i\chi} \beta_k\, b_{\pm k} \tag{5}$$

Then H_k can be written

$$H = \sum_{k>0} \{\Omega_k\, [a_+^\dagger a_+ + a_-^\dagger a_- + 1] + 2i\,\xi_k\, [a_+ a_- - a_-^\dagger a_+^\dagger]\}$$

$$\equiv H_0 + H_I \equiv \sum_k H_k^{sq} \tag{6}$$

where "sq" means "squeezing", if we identify coefficients as follows:

$$\alpha_k^2 \, \Omega_k = hck + 2B_k \qquad ; \qquad \Omega_k \, \beta_k^2 = hw_k - 2C_k$$

$$\xi_k \, \alpha_k^2 = B_k' \qquad ; \qquad \xi_k \, \beta_k^2 \, e^{4i\chi} = -C_k'$$

$$2 \, \xi_k \, \alpha_k \, \beta_k \, e^{2i\chi} = i \, A_{1k} \qquad ; \qquad \Omega_k \, \alpha_k \, \beta_k \, e^{2i\chi} = i \, A_{2k} . \tag{7}$$

Evidently Ω_k and ξ_k depend upon (E, A, B, C) in a well determined form and thus can be obtained. Of course diagonalization of H_k^{sq} proceeds as before

$$[\eta_k \, H_k^{sq}] = \varepsilon_k \, \eta_k \tag{8}$$

and produces the eigenvectors and eigenvalues as before. But now we can identify

$$\varepsilon_k^2 = \Omega_k^2 - 4\xi_k \tag{9}$$

and the coefficients are related as

$$\{x, y, z, w\} = F(\Omega_k, \, \xi_k, \, \chi_k, \, \varepsilon_k) \tag{10}$$

The same diagonalization can be carried out by use of a canonical transformation, which will exhibit the relationship to squeezing. We return to this below, after a brief digression about squeezing.

2. Squeezed States

In a quantum system, let $\hat{K}$ and $\hat{F}$ be two Hermitian operators for which $[\hat{K}, \hat{F}]_- = i \, M$, where $\hat{M}$ is Hermitian. Define

$$\Delta \hat{K} \equiv \hat{K} - \langle K \rangle_\psi \tag{11}$$

where $|\psi\rangle$ is a state. Then

$$\langle (\Delta \hat{K})^2 \rangle_\psi \cdot \langle (\Delta \hat{F})^2 \rangle_\psi \geq \langle \hat{M} \rangle_\psi^2 / 4 \tag{12}$$

For a system which is a simple harmonic oscillator, for example a single mode of the electromagnetic field, the canonical variables $\hat{p}$ and $\hat{q}$ obey

$$\langle (\Delta \hat{p}^2) \rangle_\psi \cdot \langle (\Delta \hat{q})^2 \rangle_\psi \geq 1/4 \tag{13}$$

where $\hbar = 1$. If we define the field "quadrature" operators $\hat{X}$ and $\hat{Y}$ by writing the electric field

$$\hat{E}(\vec{r}, t) = E_c (\hat{X} \cos \omega t + \hat{Y} \sin \omega t) \tag{14}$$

then the Heisenberg Uncertainty Principle gives

$$\langle (\Delta \hat{X})^2 \rangle_\psi \cdot \langle (\Delta \hat{Y})^2 \rangle_\psi \geq 1/16 \tag{15}$$

where

$$\hat{X} = \omega^{\frac{1}{2}} \, \hat{q} / \sqrt{2} \qquad ; \qquad \hat{Y} = \omega^{-\frac{1}{2}} \, \hat{p} / \sqrt{2} \tag{16}$$

Now if the system state $|\psi\rangle$ is a coherent state, denoted $|\alpha\rangle$ and defined as an eigenstate of the mode destruction operator a

$$\hat{a} \; |\alpha> \; = \; \alpha \; |\alpha> \tag{17}$$

then we find

$$<(\hat{\Delta X})^2>_\alpha \;\; = \;\; <(\hat{\Delta Y})^2>_\alpha \; = \; 1/4 \tag{18}$$

If the system state $|\psi>$ is prepared as a squeezed state $|>_r$ then

$$<(\hat{\Delta X})^2>_r \; = \; e^{-2r}/4, \text{ and } <(\hat{\Delta Y})^2>_r \; = \; e^{+2r}/4 \tag{19}$$

– the product of variances is of course fixed by the Heisenberg uncertainty principle. As discussed in standard references such as Schumaker [1985], the coherent and squeezed states can be generated from the vacuum state $|0>$, by defining a "displacement" operator $\hat{D}(\alpha)$ which has the property

$$\hat{D}(\alpha) \; |0> \; = \; |\alpha> \tag{20}$$

and a squeezing operator $\hat{S}(r)$

$$\hat{S}(r) \; |0> \; = \; |0>_r \tag{21a}$$

or squeezed coherent state

$$\hat{S}(r) \; D(\alpha) \; |0> \; = \; |\alpha>_r . \tag{21b}$$

3. Polaritons are Intrinsically Squeezed

Returning to the squeezed Hamiltonian H_k^{sq} which is expressed in terms of operators $a_\pm$, we define the (squeezing operator)

$$\hat{S}_\pm(r,\phi) \; = \; \exp[\rho \; \hat{a}_\pm^\dagger \; \hat{a}_\pm^\dagger \; - \; \rho^* \; \hat{a}_\pm \; \hat{a}_\pm] \tag{22}$$

where $\rho \equiv r \, e^{i\phi}$ is a complex squeeze factor. Then we establish that $\hat{S}_\pm(r,\phi)$ transforms the operator $a_\pm$ as

$$\hat{S}_\pm^\dagger \; (r,\phi) \; \hat{a}_\pm \; \hat{S}_\pm(r,\phi) \equiv \mu_\pm$$

$$= \hat{a}_k \cos \theta_k \cos h \, r + \hat{b}_k \, e^{2i\chi} \sin \theta_k \cos h \, r + \tag{23}$$

$$\hat{a}_{-k}^\dagger \, e^{2i\phi} \cos \theta_k \sin \hat{h} \, r + b_{-k}^\dagger \, e^{2i(\phi-\chi)} \sin \theta_k \sin h \, r.$$

We verify that

$$H_k = \epsilon_k \; (\hat{\mu}_+^\dagger \; \hat{\mu}_+ + 1/2) + \epsilon_k (+ \to -) \tag{24}$$

But

$$H_k = \epsilon_k (\hat{\eta}_k^\dagger \; \hat{\eta}_k + 1/2) \tag{25}$$

Hence we identify the transformed operators μ_+ as identical to the polariton operators $\mu_+ \leftrightarrow \eta_{+k}$. At the same time term-by-term comparison of the coefficients allows us to obtain (θ, r, ϕ) from (x,y,z,w). Finally we get

$$r = \tanh^{-1} \frac{2\xi_k}{\varepsilon_k + \Omega_k} \quad \text{taking} \quad \phi = 0.$$

Thus we obtained the squeeze factor ($\omega(k)$) or $r(\varepsilon(k))$ giving the amount of squeezing, on each branch of polariton, as a function of frequency or energy: Artoni and Birman [1989]. This demonstrates that polariton formation produces a squeezed Bose entity. Otherwise put the transformation from the virtural intermediate mixed modes $a_\pm$ to the polariton modes $\hat{\eta}_\pm = \mu_\pm$ is a squeezing transformation. We can and do identify the quadrature operators $\hat{X}$ and $\hat{Y}$ in this mixed exciton-photon system and verify a variety of properties: in the time-independent stationary squeezed state the equality Eq.(18) holds. A variety of interesting behavior is found in the time-dependent cases. For example, when an initially coherent state with respect to the partial Hamiltonian H_0 (see eqn.(6)) at time t=0 begins evolution under influence of the interaction term H_I, time-dependent squeezing occurs. Several important oscillatory effects may then be observed. These are discussed in detail elsewhere in Artoni and Birman [1990a].

4. Measuring Polariton Squeezing

In order to confirm the existence of squeezing, measurements are needed. A second motivation is to distinguish between the generalized and the usual truncated Hopfield Hamiltonians: the magnitude of predicted squeezing is much greater in the former case than in the latter, and this is discussed in Artoni and Birman [1990b]. Here we shall describe the essence of the proposed experiments. Our proposed experiments modify the type of experiment which was successfully used to detect squeezing of photons in a cavity, or cavity plus atom, detection scheme. An excellent reference is Knight and Loudon [1988].

Two proposed experiments involve scattering of a particle (incident from vacuum to the crystal) in a properly prepared coherent superposition state from a standing wave volume polariton. By analysis of the cross-section for production of the scattered particle, one detects the presence of squeezing. In one case if squeezing is present it will be possible to tune the scattering so that the cross-section vanishes. In the second case absence or presence of squeezing is revealed by whether or not the cross-section is sensitive to a relative phase χ which is adjustable by tuning a phase shifter and a mirror in the proposed set-up. These experiments are variations on a proposal of Yurke [1998], see Artoni and Birman [1989], [1990a].

For a completely different detection scheme we take advantage of using the "leaky" volume polariton. Assuming that the squeezed(statistical) properties of the volume polariton are preserved at the surface of a bounded sample, we take advantage of the external evanescent part of the volume polariton. It is of course well known that a volume wave in a medium, i.e. electromagnetic, or acoustic, when extended to the boundary, will join onto an evanescent portion, exponentially decaying into the vacuum. Then with an incident local oscillator coherent beam, mix at the surface. Then, in a suitably constructed geometrical configuration as given in Artoni and Birman [1990b] the local oscillator due to a beam-split external (control) laser will interfere with a copropagating leaky polariton wave (both propagate with wave vector parallel to the bounding plane). The resultant signal is obtained via homodyne detection. In fact, using a "balanced" homodyne detection system, it can be shown that by a ratio measurement of two detectors the amount of squeezing r can be directly measured. For details of these proposed experiments see the cited references.

5. Phonoriton: A Non-Linear "Generalized" Polariton

When a semiconductor is irradiated by intense electromagnetic (optical) excitation with frequency near the exciton resonance one can produce a polariton state which is effectively macroscopically occupied. The polaritons in this state then can scatter with (Stokes or Antistokes) resonant phonons to reconstruct the dispersion and give a quasiparticle called <u>phonoriton</u> by Keldysh, Ivanov, Tikhodeev [1983] and collaborators. This entity is predicted to have some quite remarkably changed properties by comparison to the parent polariton. For example, the dispersion curve can be strongly modified and new gaps can open for example on lower (or upper) branches: LPB and UPB respectively. Elsewhere Wang and Birman [1990a,b] have shown that, to good approximation the reconstructed phonoriton dispersion near the new gap will take the form

$$\omega_{1,2}^{\pm}(p) = \frac{1}{2}\left[\omega(p) \pm \Omega_{k_0-p} - \omega_{k_0} + i\,\Gamma\right]$$

$$+ (-)\,(1/2)\left[\omega(p) \mp \Omega_{k_0-p} + \omega_{k_0} + i\gamma)^2 \pm \Psi_{ex}(p)\,Q^2\right]^{\frac{1}{2}}$$

Here Γ is an effective lifetime (= phonon + polariton), γ = (phonon-polariton), $\omega(p)$ is the parent polariton dispersion $\Psi(p)$ the exciton "strength function" and Q is the phonon-polariton coupling which depends upon $N_0^{\frac{1}{2}}$ where N_0 is the laser intensity. A number of approximations were made in deriving this expression - these are discussed in Wang and Birman [1990a,b], as are the experiments mentioned below.

Since (to date) no experiments have conclusively shown that the phonoriton exists in any real medium, we proposed two new experiments for direct confirmation. Both of these are also generalizations of the usual experiments carried out for polaritons: 1) Non-Linear (Intensity Dependent) Resonant Brillouin Scattering can map the entire reconstructed dispersion curve of the parent polariton. Quantitative calculations for CdS and GaAs show that the Brillouin Shift, line width and cross-section become intensity-dependent and indeed reveal details of the new quasiparticle dispersion, especially the existence of a new gap and change in group velocity near the gap region; 2) A Non-Linear Reflectivity experiment shows a striking new anomaly in the dispersion wave due to the new dispersion - NLR should be defectable by modulation spectroscopy. We look for experimental tests of these NLRBS and NLR effects - which we expect can conclusively establish existence of the phonoriton.

6. Polariton - Solitons?

The discovery of self-induced transparency (SIT) by McCall and Hahn [1967] opened a fruitful line of study of the pulse propagation in local dispersive media. A quite remarkable prediction along this direction was made later by Akimoto and Ikeda [1977], that in local dielectric media described by the "single resonant-oscillator model" of a dielectric function a mixed-mode polariton soliton could propagate if a suitably prepared pulse were incident. According to their results both SIT pulses <u>and</u> also steady pulse solutions can exist in resonant semiconductors if the pulse width is much longer than the reciprocal of the polariton gap frequency. According to their calculation, a long pulse (relatively sharply peaked in frequency outside the gap region) behaves as a polariton-soliton, while a pulse with peaked frequency inside the gap propagates slowly as a standing wave of non-linear polariton. They reported that for typical semiconductor parameters, a weak pulse of duration of the order of picosecond behaves as a polariton-soliton while a strong, shorter pulse will exhibit SIT. They in essence found a continuum of such possible solutions taking the pulse width as an arbitrary adjustable parameter and solving the relevant equations. This result was a starting point of our investigation.

Making the usual physical assumptions that the bounded dielectric comprises a system of non-interacting (i.e. local) two-level atoms, and the electromagnetic field is a circularly polarized plane wave with varying amplitude and phase, the system is governed by a set of five coupled non-linear differential equations: the Maxwell-Bloch equations (MBE). The variables are: magnitude of the pulse envelope E, phase modulation ϕ, real and imaginary components of the macroscopic polarization per unit volume (u and v) and the population difference w between ground and excited states of the two-level system. Akimoto and Ikeda examined the solutions of these equations beyond the slowly varying envelope approximation (SVEA) and by obtaining corrections in the form of power series expansion. The expansion ε is related to pulse width, and detuning and is so chosen to be $\varepsilon \ll 1$ in case of long or short pulses.

A key Ansatz of their work is that the corrections to the SVEA are obtained from series expansions such as:

$$E = \varepsilon^{\nu} (E_0 + \varepsilon E_1 + \varepsilon^2 E_2 + ...) + \quad \text{(Remainder)}$$

We reexamined the solutions of the MBE and found that the remainder term has a singular behavior like $\exp(-\Gamma f(\Delta)/2\varepsilon)$ where Δ is the detuning parameter, Γ a constant and $f_0(\Delta)$ a regular function of Δ. This led to reexamination of the totality of the solutions of Akimoto-Ikeda. The work has been carried out by a combination of analytical and numerical methods [Branis, Martin, and Birman (1990), and Branis (1990)]. Our conclusion is that polariton-soliton propagation is not possible. We find that solitary waves can occur only for isolated values of parameters: detuning and pulse width. Very careful adjustment of the physical system would be needed to observe such solutions experimentally.

Acknowledgement

This work was supported in part by contract from NASC #N00019-87-C-0251 and by a grant from PSC-BHE CUNY FRAP.

References

1. Akimoto, O and Ikeda, K (1977) Journal of Physics A10, 425.
2. Artoni, M. and Birman, J.L. (1989) Quantum Optics 1, 95 (1990,a,b) in preparation.
3. Branis, S.M. (1990), Ph.D. Dissertation, Dept. of Physics. City University of New York.
4. Branis, S.V., Martin, O. and Birman J.L. (1990) in preparation.
5. Heller, W.R. and Marcus, A. (1951), Phys. Rev. 84, 809.
6. Hopfield, J.J. (1958), Phys. Rev. 112, 1555.
7. Keldysh, L.V., Ivanov, A.L. (1983) Sov. Phys. JETP 57, 234.
8. Kim, S.K. and Birman, J.L. (1988) Phys. Rev. B38, 4291.
9. Kittel, C. (1963), "Quantum Theory of Solids" Chap.3.
10. Knight, P. and Loudon, R. (1987), Journal of Modern Optics 34, 708.
11. Koteles, E. (1982) in "Exitons" Chap.3, ed. E. Rashba and M.B. Sturge, North-Holland Press.
12. Schumaker, B.L. (1985) Phys. Rep. 135, 317.
13. Wang, B.S. and Birman, J.L. (1990a) sumbitted to Phys. Rev. Lett. (1990b) in preparation.
14. Yurke, C. (1963), in "Quantum Theory of Solids", Chap.3.
15. McCall, S. and Hahn, E.L. (1967), Phys. Rev. Lett. 18, 908.

NONCLASSICAL FIELD CORRELATIONS IN QUANTUM OPTICS

H. Jeff Kimble

Norman Bridge Laboratory of Physics 12-33
California Institute of Technology, Pasadena, CA 91125

ABSTRACT
    The manifestly quantum or nonclassical character of the electro-
magnetic field is investigated in several optical experiments. Nonclassical correlations as manifest in measurements of photon statistics
and of quadrature phase amplitudes offer possibilities for detection
strategies with sensitivity beyond the vacuum-state limit and for
quantitative studies of quantum state reduction in a dissipative setting.

PHASE-CONJUGATED WAVE ENHANCED

BY WEAK LOCALIZATIONS OF EXCITON-POLARITONS

Eiichi Hanamura

Department of Applied Physics
University of Tokyo
7-3-1 Hongo, Bunkyo-ku, Tokyo 113, Japan

INTRODUCTION

An exciton has mesoscopic or macroscopic transition dipolemoment depending upon the system. This enhancement is limited by the size of the microcrystallite for the excitons, e.g., in CuCl microcrystallites embedded in insulators or glasses. As the first effect of the mesoscopic transition dipolemoment, the excitons in CuCl microcrystallie can show rapid superradiative decay[1] : $(T_1)^{-1} \equiv 2\gamma = 64\pi(R/a_B)^3\gamma_s$, where γ_s is a radiative decay rate due to band-to-band transition. The magnitude as well as R-dependence of T_1 were confirmed experimentally[2] in quantitative agreement with the theoretical prediction[1]. As the second effect of the mesoscopic transition dipolemoment, the third-order optical susceptibility $\chi^{(3)}(\omega; -\omega, \omega, -\omega)$ was shown[3] to be enhanced under nearly resonant pumping of these excitons as shown later. The size-dependence and the absolute magnitude of $\chi^{(3)}$ were also observed in agreement with the theory for the same system as in [2].[4]

The exciton in 2D and 3D crystals has a macrosopic transiton dipolemoment in an ideal limit[1]. The real 2D and 3D excitons, however, have the finite coherent length L^* due to their collisions with impurities and phonons: $L^* = \sqrt{2\hbar/M L}$, Γ is a half spectrum width of the exciton and M is an effective mass for center-of-mass motion of the exciton. The enhancement of the transition dipolemoment P is limited by this coherent length[5,6]: $P \propto \mu(L^*/a)^{d/2}$, where μ is a band-to-band transition dipolemoment, a is an exciton Bohr radius and d is a dimension of the system. The third-order optical susceptibility $\chi^{(3)}(\omega; -\omega, \omega, -\omega)$ is enhanced by a factor $(L^*/a)^d$ through the coherent nature of the excitons. Therefore the sharper exciton (with the smaller secptrum width Γ) has the more rapid radiative decay and the much larger $\chi^{(3)}$. These facts have been observed for the 2D excitons localized at the top surface of anthracene crystal.[7,8]

Exciton-polaritons are formed in the bulk crystal due to this strong dipolar interaction with the radiation field and the conservation law of the wave vector coming from translational symmetry of the crystal. If we don't care of the slow response, we may expect the large optical nonlinearity due to these 3D excitons or exciton-polaritons. In addition to the dipolemoment enhancement, we have another enhancement of phase-conjugated wave due to weak localization of exciton-polaritons.[9] Origin of this enhancement will be fully discussed from the physical point of view.

EXCITONIC ENHANCEMENT OF OPTICAL NONLINEARITY

Excitons and exciton-polaritons may be considered as boson particles in good approximation. Ideal bosons, i.e., harmonic oscillators usually cannot show any non-linearity by themselves. We will answer why and how these excitons can show finite, and sometimes enhanced optical nonlinearity under nealy resonant pumping of these exctions. This will be related to the question on the physical origins of enhancement of the phase-conjugated wave due to weak localization of excitons.[9] The generation of the phase-conjugated wave and the degenerate four-wave mixing are described in terms of the third-order-optical susceptibility $\chi^{(3)}(\omega; -\omega, \omega, -\omega)$. Therefore we will discuss how this $\chi^{(3)}$ becomes finite and is enhanced under nearly resonant pumping of these excitons in this section.

We consider such nearly resonant pumping of optically allowed excitons as contributions from other levels are negligible. Then the electronic system can be described in terms of boson operators of excitons $(b_n, b_n^\dagger)$. Here n means a quantum number for a center-of-mass motion of the 1S exciton with the lowest electron-hole relative motion in the case of microcrystallites. For two- or three-dimensional crystals, n means both a wave-number state $\mathbf{K}$ for a center-of-mass motion and a discrete quantization level of the electron-hole relative motion. The Hamiltonian of this electonic system is

$$H_0 = \sum_n \hbar\omega_n b_n^\dagger b_n + \sum_{n,m} \hbar\omega_{int}^{nm} b_n^\dagger b_m^\dagger b_m b_n, \tag{1}$$

where the exciton-exciton interaction ω_{int} is effective only for the microcrystallite and negligible for 2D- and 3D-excitons. This system is pumped by external radiation field $E(t) = E\exp(-i\omega t) + c.c.$, which is treated as a classical variable. When the exciton has a transition dipolemoment $P = \sum(P_n b_n^\dagger + P_n^* b_n)$, interaction Hamiltonian H' of this electronic system with the radiation field $E(t)$ is in a dipolar approximation:

$$H' = -P \cdot \{E\exp(-i\omega t) + E^*\exp(i\omega t)\}. \tag{2}$$

Here the mesoscopic or macroscopic enhancement of the transition dipolemoment is taken into account in the expression of P_n and P_n^*. This exciton interacts with three kinds of reservoirs: (1) radiation vacuum responsible to radiative decay of the exciton, (2) phonons and (3) impurities distributed randomly. These interactions are described, respectively, as

$$V_1 = \sum_{n,\mathbf{k}} V_{n\mathbf{k}}^{(1)} a_\mathbf{k}^\dagger b_n + h.c., \tag{3a}$$

$$V_2 = \sum_{m,n,\mathbf{q}} V_{mn\mathbf{q}}^{(2)}(c_\mathbf{q} + c_{-\mathbf{q}}^\dagger) b_{m\mathbf{k}+\mathbf{q}}^\dagger b_{n\mathbf{k}}, \tag{3b}$$

$$V_3 = \sum_{imnq} V_\mathbf{q}^{(3)} e^{i\mathbf{q}\cdot\mathbf{R}_i} b_{m\mathbf{k}+\mathbf{q}}^\dagger b_{n\mathbf{k}}. \tag{3c}$$

Here $(a_\mathbf{k}, a_\mathbf{k}^\dagger)$ and $(c_\mathbf{q}, c_\mathbf{q}^\dagger)$ denote annihilation and creation operators of photons and phonons, respectively. The elastic scattering centers of excitons are located randomly at $\{\mathbf{R}_i\}$. In Eqs.(3b) and (3c), the wave-number vectors are written explicitly for the center-of-mass motion of excitons in 2D and 3D crystals. Hereafter we use the notations $H = H_0 + V_1 + V_2 + V_3$ and $H_t = H + H'$.

Under the nearly resonant pumping of the lowest exciton, the electronic polarization of the system with time dependence $\exp(-i\omega t)$ is expanded in the amplitude of external radiation field E, E^* as follows:

$$\langle P \rangle = \chi^{(1)}(\omega; -\omega)E\exp(-i\omega t) + \chi^{(3)}(\omega; -\omega, \omega, -\omega)|E|^2 E\exp(-i\omega t) + \cdots.$$
$$\langle P \rangle \equiv \mathrm{Tr_{SB}}\{P\rho(t)\}/\mathrm{Tr_{SB}}\{\rho(t)\} \quad, \tag{4}$$

Here
$\rho(t) = \exp(-iH_t t)\rho_0 \exp(iH_t t)$, and $\chi^{(3)}(\omega; -\omega, \omega, -\omega)$ means the third-order optical susceptibility. Here and hereafter we take $\hbar = 1$. The latter three arguments ω and

$-\omega$ in $\chi^{(3)}$ corresponds, respectively, to $E^* \exp(i\omega t)$ and $E \exp(-i\omega t)$ components in $E(t)$ but the order does not mean choronological orders. The initial state density $\rho_0 = |0\rangle \rho_B \langle 0|$ means that the relevant electronic system is in a ground state $|0\rangle$, the radiation field besides the external field $E(t)$ is in a vacuum $|0\rangle_{photon} \langle 0|$ and other reservoir such as phonons is in a thermal equilibrium. As will be shown in the next section, we can eliminate the reservoir coordinates by introducing the projection operators so as to justify the stochastic model and give the microscopic expressions to the dephasing rate γ' and the decay rate γ.

It is noted first that $\mathrm{Tr_S}\langle \rho(t)\rangle_B \neq \mathrm{Tr_{SB}}|0\rangle \rho_B \langle 0| = 1$ for the stochastic model, when we consider the nonlinear optical response. Here $\langle \rho(t)\rangle_B$ means that the effects of the reservoirs are taken into account by the stochastic model. Therefore we must take account of the renormalization of the state density to the corresponding order of $E(t)$ as given by Eq.(4). Note also that the second-order polarization $P^{(2)}$ is missing for a system with an inversion symmetry. We confine ourselves to such a case. In order to obtain the third-order optical susceptibility $\chi^{(3)}(\omega; -\omega, \omega, -\omega)$, we expand $\mathrm{Tr_S}\{P\langle \rho(t)\rangle_B\}$ and $\mathrm{Tr_S}\langle \rho(t)\rangle_B$ in terms of H' Eq.(2):

$$\mathrm{Tr_S}\{P\langle \rho(t)\rangle_B\} = P^{(1)} \exp(-i\omega t) + P^{(3)} \exp(-i\omega t) + \cdots,$$

where

$$P^{(1)}e^{-i\omega t} = \frac{1}{i\hbar} \int_{-\infty}^{t} dt_1 \mathrm{Tr_S}\{Pe^{-i\bar{H}(t-t_1)} H' e^{-i\bar{H}t_1} \bar{\rho}_0 e^{i\bar{H}t}\},$$

$$P^{(3)}e^{-i\omega t} = \frac{1}{(i\hbar)^3} \int_{-\infty}^{t} dt_1 \int_{-\infty}^{t_1} dt_2 \int_{-\infty}^{t_2} dt_3$$
$$\times \mathrm{Tr_S}\{P(t)[H'(t_1), [H'(t_2), [H'(t_3), \bar{\rho}_0]]]\}\} \tag{5}$$

where $\bar{H}$ describe the Hamiltonian H_0 of the relevant electronic system modified by the decay γ and the dephasing γ' depending on the state density of the relevant system, and $\bar{\rho}_0 = |0\rangle\langle 0|$ means the electronic ground state. The denominator $\mathrm{Tr_{SB}}\rho(t) = \mathrm{Tr_S}\langle \rho(t)\rangle_B$ in the stochastic model is also expanded in H' as follows:

$$\mathrm{Tr_S}\langle \rho(t)\rangle_B = 1 + \frac{1}{(i\hbar)^2} \int_{-\infty}^{t} dt_1 \int_{-\infty}^{t_1} dt_2 \mathrm{Tr_S}\{[H'(t_1), [H'(t_2), \bar{\rho}_0]]\} + \cdots$$
$$\equiv 1 + \rho^{(2)} + \cdots. \tag{6}$$

The third-order polarization which is proportional to $|E|^2 E e^{-i\omega t}$ is given by[10] $\chi^{(3)}$ $(\omega; -\omega, \omega, -\omega) |E|^2 E e^{-i\omega t} \equiv P^{(3)} e^{-i\omega t} - P^{(1)} e^{-i\omega t} \rho^{(2)}$. The second term was obtained by expanding Eq.(6) in $|E|^2$. The contributions of the first term $P^{(3)}$ are expressed by eight terms in Fig.1, and the second term comes from the product of linear polarization Fig.2(a) and the sum of four diagrams of Figs.2(b) which correspond to $\rho^{(2)}$ in Eq.(6).

In the stochastic theory, we introduce the decay rate 2γ for an exciton population, e.g., for time intervals between t_2 and t_1 in the diagrams of Figs.1(2) and (3), and the relaxation $\gamma+\gamma'$ for an exciton polarization, e.g., for time intervals between t_3 and t_2 in the above diagram, where γ' is a pure dephasing. This means that excitons suffer from very state-dependent relaxations and decays in discussing nonlinear optical response which involve interband transitions many times. On the other hand, excitons may behave as ideal bosons to the good approximation for discussing the quasi-equilibrium of excitons.

MICROSCOPIC DERIVATION OF DEPHASING AND DECAY

In evaluating the higher-order optical response, we must calculate the following time-development of the state density for the time interval (0 and t) between the

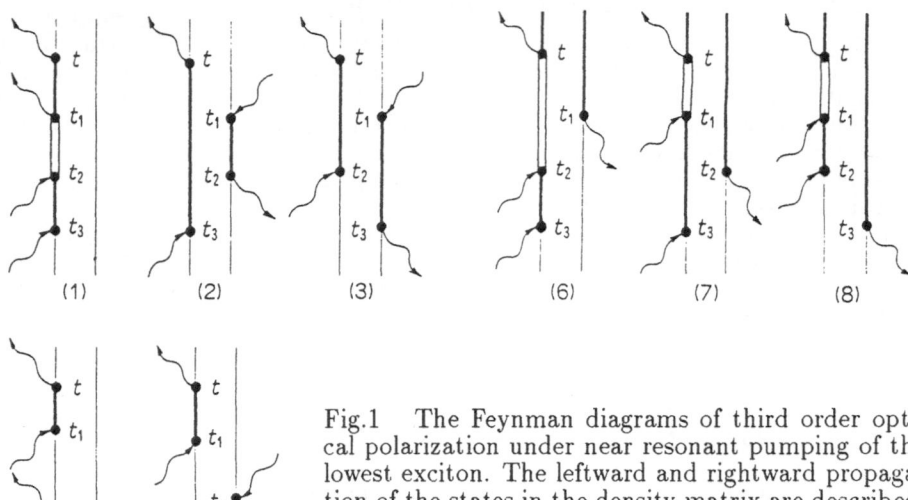

Fig.1　The Feynman diagrams of third order optical polarization under near resonant pumping of the lowest exciton. The leftward and rightward propagation of the states in the density matrix are described, respectively, by left and right two lines with upward time propagation. Single and double solid lines describe a single exciton and double excitons, respectively, and thin lines that of the ground state.

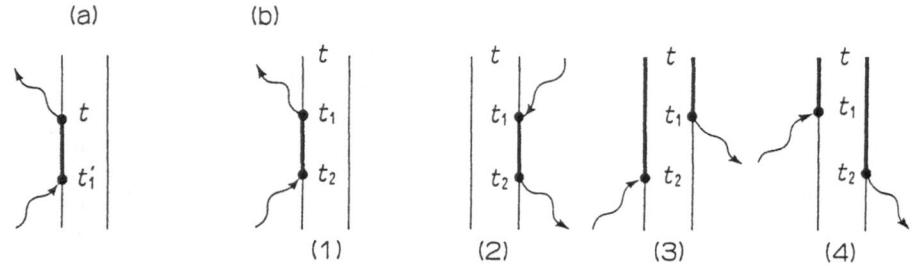

Fig.2　(a) The Feynman diagram of the linear optical polarization under nearly resonant pumping of the lowest exciton. (b) Contributions to the normalization of the state density in the second-order of the radiation field. The meanings of two thin lines and solid line are the same as in Fig.1.

successive external excitations and/or de-excitations of excitons after expanding $\rho(t)$ in H' in Eq.(4) :

$$\langle\!\langle mn|G(t)|mn\rangle\!\rangle \equiv \mathrm{Tr_S}\langle\!\langle m|e^{-iHt}|m\rangle\rho_{\mathrm{B}}\langle n|e^{iHt}|n\rangle\!\rangle. \tag{7}$$

Here the outside bra-ket means taking the projection operator $\hat{P}A \equiv \rho_{\mathrm{B}}\mathrm{Tr_B}A$ over the reservoir variables or taking the ensemble average over impurity distributions. When the m or n excitons are distributed over many states, $|m\rangle\langle n| \equiv |m_1, m_2, \cdots, m_i, \cdots\rangle$ $\langle n_1, n_2, \cdots, n_i, \cdots|$ with $\sum m_i = m$ and $\sum n_i = n$. The Hamiltonian $H = H_0 + V_1 + V_2 + V_3$. We also assume that the exciton-reservoir interactions are Markoffian and that the reservoir is always in thermal equilibrium ρ_{B}. This Hamiltonian H was replaced in the stochastic model by $\bar{H}$, i.e., H_0 and state-dependent decay γ and relaxation $\gamma + \gamma'$.

We derive microscopic expressions of decay and relaxation operators induced by interactions of excitons with radiation vacuum, impurities and phonons, respectively. The Hamiltonian $H = H_0 + V$ here describe a relevant electronic system and reservoir, and their interaction V. This V is chosen to be V_1, V_2 or V_3 depending upon which mechanism we discuss. The time-development of state density Eq.(7) is described in terms of cumulant expansions[12]. First, the reservoir of the radiation field is considered to be a vacuum $|0\rangle_{photon}\langle 0|$ because we consider a system of excitons which have excitation energy larger than 1eV = 11,604 K. We apply also the projection operator onto the radiation vacuum $\rho_B \equiv |0\rangle_{photon}\langle 0|$:

$$\rho_B \mathrm{Tr}_B\{\langle m|e^{-iHt}|m\rangle \rho_B \langle n|e^{iHt}|n\rangle\} \equiv \exp[-i(H_0^\times - i\Gamma^\dagger)t]|m\rangle \rho_B \langle n|$$

$$= \exp[-i(H_0 - i\sum_{\mathbf{k}} \gamma_{\mathbf{k}} b_{\mathbf{k}}^\dagger b_{\mathbf{k}})t]|m\rangle \rho_B \langle n| \exp[i(H_0 + i\sum_{\mathbf{k}} \gamma_{\mathbf{k}} b_{\mathbf{k}}^\dagger b_{\mathbf{k}})t] \tag{8}$$

where

$$\gamma_{\mathbf{k}} = \pi \sum_{\mu} \delta(\omega_\mu - \omega_{\mathbf{k}})|V_{\mathbf{k}\mu}^{(1)}|^2 = \pi N_{photon}(\omega_{\mathbf{k}})|V_{\mathbf{k}}^{(1)}|^2. \tag{9}$$

In the case of 3D excitons, $\mathbf{k}$ and μ in Eq.(9) should be equal wave-number vectors of the exciton center-of-mass motion. Here we have exciton-polaritons rather than decay for the 3D excitons. On the other hand, the wave-number vector conservation law is violated for the exciton in microcrystallites and in the direction perpendicular to the surface for 2D excitons. To these systems, Eq.(9) is applicable and $V_{\mathbf{k}}^{(1)}$ means an average over possible directions of radiative decay. Second we derive phase-relaxation of an exciton due to impurity scatterings Eq.(3c) with $V = V_3$. Here the scattering of excitons over wave number vector states for the center-of-mass motion plays key roles in phase relaxations of 2D and 3D excitons at low temperature. The even-order perturbational terms in V's in Eq.(7) remain finite and the ensemble average of 2n-th order perturbations is well approximated by n products of the ensemble average of a pair of V's:

$$\langle V_{\mathbf{qq'}} V_{\mathbf{kk'}}\rangle = \langle \sum_i e^{i(\mathbf{q}-\mathbf{q'})\cdot \mathbf{R}_i} V(\mathbf{q}-\mathbf{q'}) b_{\mathbf{q}}^\dagger b_{\mathbf{q'}} \sum_j e^{i(\mathbf{k}-\mathbf{k'})\cdot \mathbf{R}_j} V(\mathbf{k}-\mathbf{k'}) b_{\mathbf{k}}^\dagger b_{\mathbf{k'}}\rangle$$

$$= n_i |V(\mathbf{q}-\mathbf{q'})|^2 b_{\mathbf{q}}^\dagger b_{\mathbf{q'}} b_{\mathbf{k}}^\dagger b_{\mathbf{k'}} \delta_{\mathbf{q}-\mathbf{q'},\mathbf{k'}-\mathbf{k}}. \tag{10}$$

The leading contributions from second- and fourth-order perturbations due to impurity- and phonon-scatterings of excitons are confirmed to have the following forms :

$$\langle\langle m|e^{-iHt}|m\rangle \rho_B \langle n|e^{iHt}|n\rangle\rangle = \exp[-iH_0^\times t - \Gamma^\times t]|m\rangle \rho_B \langle n|, \tag{11a}$$

and

$$\Gamma^\times A \equiv (\gamma_2' + \gamma_3') \sum_{\mathbf{k},\mathbf{k'}} [b_{\mathbf{k}}^\dagger b_{\mathbf{k}}, [b_{\mathbf{k'}}^\dagger b_{\mathbf{k'}}, A]]. \tag{11b}$$

Here γ_2' and γ_3' are sums of contributions to the exciton dephasings from 2nd and higher even-order connected scatterings of excitons with phonons and impurities, respectively. The higher order interference effects between phonon- and impurity-scatterings and the exciton populations at the intermediate higher states of excitons were assumed to be negligible. The time-convolutionless formula[13] and the projection method[11] give the same result as long as we assume the Markoffian approximations. If the exciton states in $|m\rangle$ and $\langle n|$ are degenerate in energy and connected by impurity- and phonon-scatterings, the dephasing $\gamma' \equiv \gamma_2' + \gamma_3'$ works depending upon the square of number difference $(m-n)^2$ between the left $|m\rangle$ and the right state $\langle n|$ in the state density at t = 0. On the other hand, when these exciton-states are not connected by impurity- and phonon-scatterings, we have the dephasing $\sum_i \gamma'(m_i - n_i)^2$. Thus in any cases, the phase-relaxation is sensitive to the left- and right- states in the state-density of excitons.

These state-dependent operators $\Gamma^\times$ and $\Gamma^\dagger$ make the commutators of boson operators in Eq.(5) remain finite and state-dependent. This point is very important to obtain the enhanced $\chi^{(3)}$ as discussed in the former sections.

First, these state-dependent decay Eq.(8) and relaxations Eqs.(11b) make three commutators in the integrand in Eqs.(5) to remain finite and result in enhanced $\chi^{(3)}$. Note also that the decay and relaxation operators give no decay nor relaxation on the

electronic ground state. It is noted that $\mathrm{Tr}_S\langle\rho(t)\rangle_B$ in the stochastic model with the state-dependent decay and relaxations have finite contributions of even-orders in $E(t)$. After taking into account all these contributions from the diagrams of Fig.1 and Fig.2 to $\chi^{(3)}$, mesoscopic or macroscopic enhancement of $\chi^{(3)}(\omega; -\omega, \omega, -\omega)$ is confirmed theoretically for the following two cases:
(1) In the case of $w_{int} \gg |\omega - \omega_0|$, excitons in semiconductor microcrystallites with such a radius R as $a_B \ll R < \lambda$ show the mesoscopic enhancement $(R/a_B)^3$ and the dephasing enhancement (γ'/γ) of $\chi^{(3)}$:

$$\chi^{(3)} = \frac{2N_c|P_0|^4}{\hbar^3(\omega - \omega_0 + i\Gamma)^2(\omega - \omega_0 - i\Gamma)}(1 + \frac{\gamma'}{\gamma}) \tag{12}$$

$$\propto (\frac{R}{a_B})^3, \qquad (\omega_{int} \gg |\omega - \omega_0|).$$

Here we assumed a constant volume fraction r of semiconductor microcrystallites over the matrix, i.e., $N_c \equiv 3r/(4\pi R^3)$. Under the condition $w_{int} \gg |\omega - \omega_0|$, semiconductor microcrystallites are considered as two-level atomic system with mesoscopic transition dipolemoment $(2\sqrt{2}/\pi)\mu_{cv}(R/a_B)^{3/2}$. Thus we can understand the expression Eq.(12) quite well. The exciton-exciton interaction $\hbar w_{int} = (13\pi/3)E^b_{exc}a^3_B/(4\pi R^3/3)$ comes from the electron exchange process between two excitons with the same spin structure, and otherwise $\hbar w_{int}$ is almost negligible. Therefore when we take account of electron spins, the probability in which the second exciton has the same spin structure as the first one in the CuCl microcrystallite is half so that half the microcrystallites have the contribution of Eq.(12) to $\chi^{(3)}$ and the other half contribution is evaluated assuming that two different kinds of excitons are excited in Fig.1 with $\hbar w_{int} = 0$.
(2) When w_{int} is negligible in comparison to Γ, e.g., in 2D and 3D crystals, $\chi^{(3)}$ is given by the following equation : $(|\omega_1 - \omega_0| > |\omega - \omega_0|, \quad \Gamma \gg w_{int})$,

$$\chi^{(3)} = \frac{2N_c|P_0|^4(\Gamma + \gamma')i}{\hbar^3(\omega - \omega_0 + i\Gamma)^2(\omega - \omega_0 - i\Gamma)(\omega - \omega_0 + i(\Gamma + 2\gamma))}. \tag{13}$$

This expression is obtained by fully taking account of the facts that the two kinds of two excitons are created but that the exciton-exciton interaction $\hbar w_{int}$ is negligible in comparison to Γ and $|\omega - \omega_0|$. Here we have also the size-dependent $\chi^{(3)} \sim (R/a_B)^3$ for the microcrystallites. If the coherent length L^* of the exciton center-of-mass motion is long enough, the transition dipolemoment P_0 has an enhancement factor of (L^*/a_B) for 2D excitons and $(L^*/a_B)^{3/2}$ for 3D excitons. In this case, N_c is of an order of $(L^*)^{-d}$ so that $\chi^{(3)}$ is proportional to $(L^*)^d$ under the off-resonant excitation $|\omega - \omega_0| > \Gamma$ and is strongly dependent on L^* under the resonant excitation $|\omega - \omega_0| < \Gamma$ also through Γ as L^* is determined by Γ, where d is a dimension of the crystal.

Second, let us consider why these phase relaxations bring about finite $\chi^{(3)}$ or enhanced $\chi^{(3)}$. (1) First, the microscopic Hamiltonian of the exciton-phonon interaction is intrinsically anharmonic as it is described as three products of boson operators, i.e., two exciton- and one phonon- operators (Eq.(3b)). Operating the projection operators so as to eliminate the phonon operators, we have obtained phase relaxation operators written in terms of four products of exciton boson operators. In the case of exciton-impurities scatterings, similar anharmonic relaxation operators in Eq.(10) are obtained after taking ensemble average of a pair of V_3's over spatial distribution of impurities. These anharmonic operators of exciton-bosons work succsessively on excitons and induce nonlinear optical response. Therefore this effect works as a factor (γ'/γ), which means how many times these relaxations work in the lifetime $T_1(= 1/2\gamma)$. (2) Excitons may be treated as good boson particles for the quasi-stationary phenomena in 2D and 3D systems. In the case of nonlinear optical response, however, e.g., for four-wave mixing under nearly resonant pumping of excitons, abrupt changes of way of relaxations and decays are induced whenever the excitons are created or annihilated. These state-dependent relaxations and decays make the commutators of exciton-boson operators nonvanishing in the integrand of Eqs.(5). This makes $\chi^{(3)}$ to be finite and works so as to make the best use of mesoscopic or macroscopic transition dipolemoment of the exciton in enhancing $\chi^{(3)}$. (3) The exciton-exciton interaction is clearly anharmonic in exciton operators. This make $\chi^{(3)}$ finite but it was size-independent in the damped harmonic oscillators with a state-independent decay. Therefore only ω_{int} is not enough to produce the mesoscopic and macroscopic enhancement of $\chi^{(3)}$. These mesoscopic or macroscopic enhancement of $\chi^{(3)}$ is possible only under such a resonant pumping of the lowest exciton ω_0 as $|\omega_1 - \omega_0| > |\omega - \omega_0|$ in which the lowest exciton with the largest oscillator strength contributes dominantly to $\chi^{(3)}(\omega; -\omega, \omega, -\omega)$. Here $\hbar|\omega_1 - \omega_0|$ is the energy separation of the second lowest exciton $\hbar\omega_1$ from the lowest one $\hbar\omega_0$. We cannot extrapolate this enhancement of $\chi^{(3)}$ to the off-resonance case, as had been mentioned clearly in the first papers[1,2,6]. The energy separation $\hbar|\omega_1 - \omega_0|$ is a few tens of meV for the CuCl microcrystallites with the radius 30 A $\sim$ 80 A, and $\hbar\Gamma$ is estimated to be of an order of 0.1 meV.

ENHANCEMENT OF PHASE-CONJUGATED WAVE

The enhacement of phase-conjugated wave generation due to weak localization of exciton polarions[9] is discussed in the relation to the dephasing process Eqs.(10) and (11) due to impurities scattering of excitons. Specified scattering processes in Eq.(10) were chosen in the lowest order phase-relaxation process because we were interested in the diagonal components of exciton states in Eq.(7). On the other hand, this restriction is relaxed in the weak localization of excitons or exciton-polaritons. We choose a series of the scattering processes of Eq.(10) in which the difference $\mathbf{k'} - \mathbf{q'} = \mathbf{k} - \mathbf{q}$ is conserved for the ladder diagrams (Fig.3(b)) and those in which the sum $\mathbf{k} + \mathbf{q} = \mathbf{k'} + \mathbf{q'}$ is conserved for the maximally crossed diagrams (Fig.3(c)). It is noted here that these scattering effects are also described by anharmonic terms of bosons as Eq.(10) shows. As a result, we can get a physical picture of enhancement (γ'/γ) of phase-conjugated wave through the weak localization of excitons or exciton-polaritons. Here the enhancement of the phase-conjugated wave is also shown to come from the successive operations of exciton-anharmonic operators Eq.(10) on the nonlinear optical processes.

We consider such a bulk crystal as the effect of exciton-radiation field interaction (of an order of longitudinal-transverse splitting of the exciton) is much larger than the exciton-impurity scattering rate. We also assume the localized state of the excitons at impurities to be negligible in the present work. In this case, the largest interaction between the exciton and the photon is diagonalized and the exciton-polariton described by d_k and $d_k^\dagger$ is formed at first. Then the material system is described by the Hamiltonian :

$$H = H_0 + V$$
$$= \sum_{\mathbf{k}} \omega(\mathbf{k}) d_{\mathbf{k}}^\dagger d_{\mathbf{k}} + \sum_{\mathbf{k},\mathbf{q},i} V_0(\mathbf{q}) e^{i\mathbf{q}\cdot\mathbf{R}_i} d_{\mathbf{k}+\mathbf{q}}^\dagger d_{\mathbf{k}},$$

where V_0 is described by the impurity potential and the exciton-polariton transformation matrix elements. Then this exciton-polariton is scattered by impurities through the excitonic component of the polaritons. The radiation field is divided into two parts inside and outside the crystal. The former was already taken into account in forming the polaritons. The external radiation fields $\mathbf{E}_j$ excite the polaritons at the crystal surface. This process is described by the interaction Hamiltonian H' :

$$H' = \sum_{\alpha,j,\mathbf{k}} (d_{\mathbf{k}\alpha} + d_{\mathbf{k}\alpha}^\dagger) \boldsymbol{\mu}_\alpha \cdot \mathbf{E}_j,$$

where $\boldsymbol{\mu}_\alpha$ is the polarization vector of the polarion and is evaluated in terms of the exciton dipolemoment and the transformation matrix between the exciton and the polariton. The generation of the phase-conjugated wave is descirbed by the diagrams of

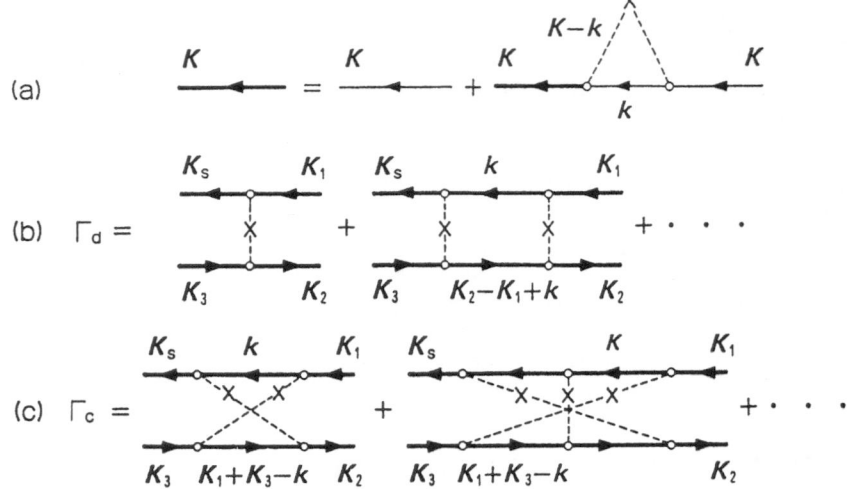

(a)

(b) $\Gamma_d =$

(c) $\Gamma_c =$

Fig. 3   (a) Diagram describing the lowest-order contribution to the self-energy of polariton $\mathbf{K}$. The crosses and dotted lines mean scattering centers and elastic scattering interactions of the polariton with an impurity. (b) Ladder-like scatterings of the leftward and rightward polaritons. (c) Maximally crossed scattering of polaritons at impurity centers.

Fig.1(1~3) and 1(6~8). Other contributions of Fig.1 and 2 vanish in the present problem. This is purely nonlinear optical phenomena in two respects. First, normal process of phase-conjugated wave even due to excitons is in itself four-wave mixing because the population gratings are formed by two kinds of radiation fields associated with the dephasing and decay of excitons. These are described by the diagrams of Fig.3(a) and the first terms of Fig.3(b). The third wave is scattered by this grating and the signal of the phase-conjuated wave is produced. It is noted that this nonlinear response is induced by the association of decay and dephasing of the exciton-polaritons. Second, this phase-conjugated wave was shown to be enhanced furthermore by the weak localization of exciton-polaritons, i.e., by constructive interference between multiple elastic scatterings of the polariton through its exciton-component and its time-reversed processes, which is described by Fig.3(c).

A single scattering process can be described by effective anharmonic exciton operators of Eq.(10). It is noted here that a weak localization of excion-polariton is described by successive operations of anharmonic interactions Eq.(10), resulting in enhancing the phase-conjugated wave. The amplitude of signal polariton $(\mathbf{K}_s, w_s)$ is expressed in the following form[9] :

$$\langle d_{\mathbf{K}_s}(\omega_s)\rangle \doteq 2\pi A\delta(\omega_s - \omega_0 - \omega_0' + \omega_i)$$
$$\times \frac{1}{\gamma_0(\gamma_0 + \gamma)^4} \frac{2(\gamma_0 + \gamma)U_0}{2\gamma_0 - i(\omega_0 - \omega_0') + D(\mathbf{K}_0 + \mathbf{K}_0')^2}, \tag{14}$$

where $A = \langle 0|d_{\mathbf{K}_s}|w_s\mathbf{K}_s\rangle\langle\omega_0\mathbf{K}_0|d_{\mathbf{K}_0}^\dagger\boldsymbol{\mu}\cdot\mathbf{E}(\mathbf{K}_0)|0\rangle \langle 0|d_{\mathbf{K}_i}\boldsymbol{\mu}\cdot\mathbf{E}(\mathbf{K}_i)^*|\omega_i\mathbf{K}_i\rangle\langle\omega_0\mathbf{K}_0'| d_{\mathbf{K}_0'}^\dagger\boldsymbol{\mu}\cdot$ $\mathbf{E}(\mathbf{K}_0')|0\rangle$, $2\gamma_0$ is a sum of leakage rate through the crystal surfaces and inelastic phonon scattering rate of the polariton, $\gamma = \pi N(\omega)n_i|V_0(\mathbf{q} = 0)|^2$ and $D \equiv v_g^2/6(\gamma_0 + \gamma)$ the diffusion constant. Here $N(\omega)$ is the state density of polariton and n_i the impurity concentration. The last factor of Eq.(14) comes from the maximally crossed diagrams Fig.3(c) so that the phase-conjugated wave $(\mathbf{K}_s, w_s)$ is enhanced by the factor $(\gamma_0 + \gamma)/\gamma_0$ when the forward $(\omega_0\mathbf{K}_0)$ and backward $(\omega_0'\mathbf{K}_0')$ pump waves are degenerate $\omega_0 = \omega_0'$ and $\mathbf{K}_0 = -\mathbf{K}_0'$ in comparison to the ordinary processes neglected here. The ladder diagrams Fig.3(b) and other diagrams result in ordinary contribution without the last factor of Eq.(14). Only the constructive interference between the multiple scattering and its time-reversed process enhance the formation of the population grating and the reflection of the third wave. We will list some characteristics of the phase-conjugated wave generated by the present mechanism. Observation of these characteristics will also serve to check the present theory. The phase conjugation was observed already near the exciton-frequency region of CuCl[14] and the polaritons of this material have been studied very well[15], so we will especially study the case of CuCl. The group velocity v_g of the polariton ranges from $\hbar k/M \sim 1.2 \times 10^6$ cm/sec to $c/(\epsilon_\infty)^{1/2} = 1.3 \times 10^{10}$ cm/sec. As far as the incident frequency is limited below the upper polariton and above the polariton bottleneck, the group velocity changes from 0.2×10^6 to 3.0×10^6 cm/sec[16]. The value of $\hbar(\gamma_0 + \gamma)$ depends also on the incident frequency as well as the samples, e.g., the concentration of impurities. This is observed as the spectrum half-width at half maximum of the intensity of the hyper-Raman scattering leaving the polariton in the final state[17]. It ranges from $\hbar(\gamma_0 + \gamma) = 0.05$ to 2.0 meV. The inelastic scattering rate γ_0 was observed as a function of temperature and γ_0 may almost be determined by the polariton transmission through surfaces and nonradiatibe decay below 40 K. The lifetime $\tau_0 = (2\gamma_0)^{-1}$ is tentatively set to be of the order of a nanosecond, i.e., $\hbar\gamma_0 \sim 0.03$ meV[18]. Therefore $D \sim v_g^2/6(\gamma_0 + \gamma)$ is estimated to be of the order of 1 cm^2/sec so that $D(\mathbf{K}_0 - \mathbf{K}_0')^2$ is much larger than $2\gamma_0$ as long as the deviation of the angle between $\mathbf{K}_0$ and $\mathbf{K}_0'$ from π is larger than 2.2×10^{-2} rad.

(1) The first characteristic of the present theory is the singular dependence of $\langle d_{\mathbf{K}_s}(\omega_s)\rangle$ through Γ_c on $\omega_0 - \omega_0'$ and $\mathbf{K}_0 + \mathbf{K}_0'$. For the degenerate four-wave mixing $\omega_0 = \omega_i = \omega_s$, the intensity $|\langle d_{\mathbf{K}_s}(\omega_s)\rangle|^2$ of the phase-conjugated wave is very sensitive to an angle $(\pi - \theta)$ between the two pump fields $\mathbf{K}_0$ and $\mathbf{K}_0'$. As far as $\theta > 2.2 \times 10^{-2}$ rad, the signal intensity is proportional to $(1 - \cos\theta)^{-2} \doteq 4\theta^{-4}$.

(2) On the other hand, when we use nondegenerate four-wave mixing, we will be able to determine the dispersion of the collective mode from the dependence

$$|\langle d_{\mathbf{K}_s}(\omega_s)\rangle|^2 \propto \{(\omega_0 - \omega_0')^2 + [2\gamma_0 + D(\mathbf{K}_0 + \mathbf{K}_0')^2]^2\}^{-1} \qquad (15)$$

(3) The diffusion constant $D \equiv v_g^2/6(\gamma_0 + \gamma)$ can be varied by changing the input frequency ω_0 over several orders of magnitude as the group velocity v_g depends sensitively on the input frequency over the polariton dispersion. This brings about the incident frequency dependence of the phase-conjugated wave intensity even for the degenerate case with the fixed θ.

(4) The cutoff frequency $2\gamma_0$ in Eq.(15) is sensitive to the lattice temperature $T > 40K$.

The formation of polariton gratings and reflection of one pump wave into the conjugated wave are enhanced in poportion to the square of impurity concentration n_i through $2(\gamma_0 + \gamma)U_0$. However, the intensity of the phase-conjugated wave depends inversely on the impurity concentration n_i for the case of $\gamma \gg \gamma_0$ as follows :

$$\langle d_{\mathbf{K}_s}(\omega_s)\rangle \sim n_i^{-2} \quad \text{for} \quad D(\mathbf{K}_0 + \mathbf{K}_0')^2 \ll 2\gamma_0, \qquad (16a)$$

$$\sim n_i^{-1} \quad \text{for} \quad D(\mathbf{K}_0 + \mathbf{K}_0')^2 \gg 2\gamma_0, \qquad (16b)$$

This is because the propagation of input as well as signal polarions is incoherently scattered by impurities although the grating formation as well as the reflection due to the grating are enhanced by the impurity scatterings through Γ_c. Note that we cannot extrapolate the results of Eqs.(16) to the limit of n_i zero as these are restricted to the case $\gamma \equiv \pi N(\omega_0) n_i |V_0(0)|^2 \gg \gamma_0$. Of course $\langle d_{\mathbf{K}_s}(\omega_s)\rangle$ vanishes in the limit of $n_i |V_0(0)|^2 \equiv U_0 \to 0$ as Eq.(14) shows. Therefore we have the optimum impurity concentration n_i to get the largest phase-conjugated wave. It is noted that only the phase-conjugated wave is enhanced by this weak localization in comparison to other four-wave mixing signals.

CONCLUSION AND DISCUSSION

As a conclusion, we have understood the roles of phase-relaxation and decay of excitons as well as the exciton-exciton interaction in enhancing the optical nonlinear response due to excitons. First, whenever the excitons are created or de-excited by the external field for the nonlinear optical processes, the numbers of the excitons in the left- and/or right states in the state-density of the system change so that the way of the phase-relaxations and decays is changed. These abrupt changes result in finite third-order optical polarization and the excitonic enhancement of $\chi^{(3)}(\omega; -\omega, \omega, -\omega)$ under nearly resonant excitations. Second, the multiple scatterings of exciton-polaritons and their time-reversed process interfere constructively and enhance furthermore the generation of the phase-conjugated wave in comparison to other four-wave mixing. Both the generation of the phase-conjugated wave and the weak localization of exciton-polaritons come from the time-reversal symmetry of the system.

These anharmonic terms ω_{int}, γ' and γ are much smaller than the exciton frequency ω_0 but we can make the best use of these anharmonic effects under such a resonant pumping of the excitons as $|\omega - \omega_0| \sim \omega_{int}, \gamma'$ or/and γ. Therefore it is noted that these enhancements of $\chi^{(3)}$ are limited to the nearly resonant pumping of the lowest exciton with the largest oscillator strength.

References

[1] E. Hanamura: Phys. Rev. **B38**, 1228(1988).
[2] T. Itoh, F. Jin, Y. Iwabuchi, and T. Ikehara: Springer Proceedings in Physics **36** (Nonlinear Optics of Organics and Semiconductors) ed. by T. Kobayashi, (1989) pp.76, and J. Lumminescence (Proceedings of DPC'89) (1989) to appear.
[3] E. Hanamura: Phys. Rev. **B37**, 1273(1988).
[4] Y. Masumoto, M. Yamazaki, and H. Sugawara: Appl. Phys. Lett. **53**, 1527(1988).
[5] J. Feldmann, G. Peter, E. O. Göbel, P. Dawson, K. Moore, C. Faxon, and R. J. Elliott: Phys. Rev. Lett. **59**, 2337(1987).
[6] E. Hanamura: in Optical Switching in Low Dimensional Systems ed. by H. Haug and Y. Banyai, p (Plenum 1988).
[7] Ya. Aaviksoo, Ya. Lippma, and T. Reinot: Opt. Spectrosc. (USSR) **62**, 247 (1987).
[8] M. Kuwata: J. Lummin. **38**, 247(1987).
[9] E. Hanamura: Solid State Commun. **67**, 1039(1988); ibid. Phys. Rev. **B39**, 1152(1989).
[10] E. Hanamura, M. Kuwata-Gonokami and H. Ezaki: to appear in Solid State Commun.
[11] For example, see F. Haake: Springer Tracts in Modern Physics **66**, (Springer, Heidelbery, 1973) pp.98.
[12] R. Kubo: J. Phys. Soc. Jpn. **17**, 1100(1962).
[13] F. Shibata and T. Arimitsu: J. Phys. Soc. Jpn. **49**, 891(1980).
[14] G. Mizutani and N. Nagaosa: J. Phys. Soc. Jpn. **52**, 2251(1983).
[15] M. Ueta, H. Kanzaki, K. Kobayashi, Y. Toyozawa and E. Hanamura: Excitonic Processes in Solids (Springer-Verlag, Berlin, 1986), Chaps. 2 and 3.
[16] Y. Masumoto, Y. Unuma, Y. Tanaka and S. Shionoya: J. Phys. Soc. Jpn. **47**, 1844(1979).
[17] T. Itoh, T. Katohno, T. Kirihara and M. Ueta: J. Phys. Soc. Jpn. **53**, 854(1984).
[18] M. Kuwata: J. Phys. Soc. Jpn. **53**, 4456(1984).

THE HISTORICAL RELATIONSHIP BETWEEN NONLINEAR OPTICS AND CONDENSED MATTER

Nicolaas Bloembergen

Department of Physics, Harvard University, Cambridge, MA 02138, U.S.A.

ABSTRACT

The first operating laser, developed by Maiman in 1960, utilized the energy levels of the Cr^{3+} ion in ruby. The high light intensity available in focused ruby laser pulses led to the creation of a new subfield of physics. This field, known as nonlinear optics, concerns itself with the optical properties of matter at high light intensity. The pioneering experiment by Franken and coworkers in 1961, utilizing a quartz crystal, demonstrated second harmonic generation of light. In the same year, two-photon optical absorption of ruby light was demonstrated by Garrett and Kaiser in a crystal of CaF_2 with Eu^{2+} ions. Many other nonlinear optical phenomena were also demonstrated first in condensed matter. The history of the generalization of the optical laws of reflection (Hero of Alexandria), refraction (Snell), and Fresnel's laws to the nonlinear regime will be reviewed. The phenomenon of conical refraction, characteristic of biaxial optical crystals, also has an interesting nonlinear counterpart.

OPTICAL NONLINEARITIES ENHANCED BY CARRIER TRANSPORT

Elsa Garmire

Center for Laser Studies
University of Southern California
Los Angeles, CA 90089-1112, USA

INTRODUCTION

A non-local nonlinearity results when optically excited carriers move within the internal electric fields of depletion regions, causing field-dependent changes in absorption and/or refractive index. This paper will review some recent experimental results which use these concepts in heter-Schottky barriers and hetero-nipi structures. It will be shown that large, extremely sensitive nonlinearities result.

Three examples which have been demonstrated in our laboratory will be discussed [1,2]: 1) Lengthening of carrier lifetimes due to separation of optically-induced charges within internal fields in hetero-Schottky barriers; 2) Using optically induced modulation of the Quantum Confined Stark Effect (QCSE) within a n-i-p-i structure to increase the index change per induced carrier density over that measured from band-filling; 3) Enhanced photo-refractive effect observed using electro-refraction and operating near the bandedge in semi-insulating semiconductors.

DEVICE IMPLICATIONS FOR OPTICAL NONLINEARITIES

In order to compare optical nonlinearities in semiconductors for practical applications it is necessary to understand that the optical nonlinearity is due to an induced carrier density which saturates. How the nonlinearity affects device performance can be seen from the threshold for switching in a Nonlinear Fabry-Perot which is filled with a saturating nonlinear medium [3]:

$$I_{sw}(\lambda)\tau = \frac{N_S 2h\nu (1-R_F) f^4}{|\Delta n_S(\lambda)k| - \alpha(\lambda)f},\tag{1}$$

where τ is the carrier lifetime, N_S is the saturation carrier density, $h\nu$ is the photon energy, R_F is the front mirror reflectivity, Δn_S is the saturation value of the optically induced index change, k is the free space wave vector, α is the absorption per unit length and f is a factor which is approximately one and describes device performance (f is one-third the square root of the contrast ratio).

There is an absolute requirement on the materials to have nonlinear switching [4],

$$|\Delta n_s k/\alpha| > f,\tag{2}$$

which is always be satisfied if operation is sufficiently far from the bandedge, since the absorption falls off exponentially [5] and the index change falls off hyperbolicly [6]. However, if operation takes place too far from the bandedge, the switching intensity increases because $|\Delta n_s|$ becomes small. For a given set of parameters, there is a particular wavelength at which the switching intensity is minimum, expressed in terms of $\Delta n(1)$ and $\alpha(1)$, the values measured at one URbach tail from the resonance peak. The minimum switching intensity is given by

$$I_{min}\tau = \frac{N_s}{|\Delta n(1)|} (hc/\pi)(1-R_F)f^4 \frac{X_m^2}{X_m-1},\tag{3}$$

with X_m given (in units of the number of Urbach parameters) by

$$fX_m^2 \exp(1-X_m) = k|\Delta n(1)|/\alpha(1).\tag{4}$$

Thus, figures of merit for nonlinear materials are:

$$M_1 = |\Delta n(1)|/\alpha(1) > f, \quad \text{which determines } X_m.\tag{5}$$

$$M = |\Delta n(1)|/N_s, \text{ as large as possible.}\tag{6}$$

We have compared $\Delta n_s/N_s$ in QWs of differing thicknesses [7], resulting in the data shown in Fig. 1. At the peak of the exciton resonance, $|\Delta n|/\alpha \approx 0.1$, essentially independent of well-thickness. The values at one Urbach parameter from the resonance are $\Delta n(1) = 0.5\Delta n$ and $\alpha(1) = 0.37\alpha$. This means $k|\Delta n(1)|/\alpha(1) = 1.0$, independent of well thickness. Inserting numbers into (3) and (4), $X_m = 3.4$, and the switching energy is 5.5 $\mu j/cm^2$.

Since $|\Delta n_s|$ decreases below the bandedge, the length of material required to produce sufficient optical phase change for switching increases:

$$L > \frac{(1-R_F)f}{|\Delta n_s k| - \alpha}.\tag{7}$$

There is a trade-off between switching intensity and carrier lifetime, as shown by (3). This means that for a material with given figures of merit, the CW switching intensity can be lowered by lengthing carrier lifetime. Long carrier lifetime is an important reason for the high sensitivity of the carrier-transport enhanced optical nonlinearities.

HETERO-SCHOTTKY BARRIERS

Although bandfilling provides a large optical nonlinearity in semiconductors, estimations from (3) show that the required switching intensity is on the order of 300 W/cm^2 when carrier lifetimes τ are 20 nsec. This is too high to be practical in applications where high degrees of parallelism (as much as 10^6) are required, but longer carrier lifetimes reduce the threshold.

In hetero-Schottky-barriers we have demonstrated $\tau = 4.3$ μs, a factor of 200 over typical bulk values. Estimated threshold for optical bistability, using bulk $\Delta n_s/N_s$, is 13 W/cm^2.

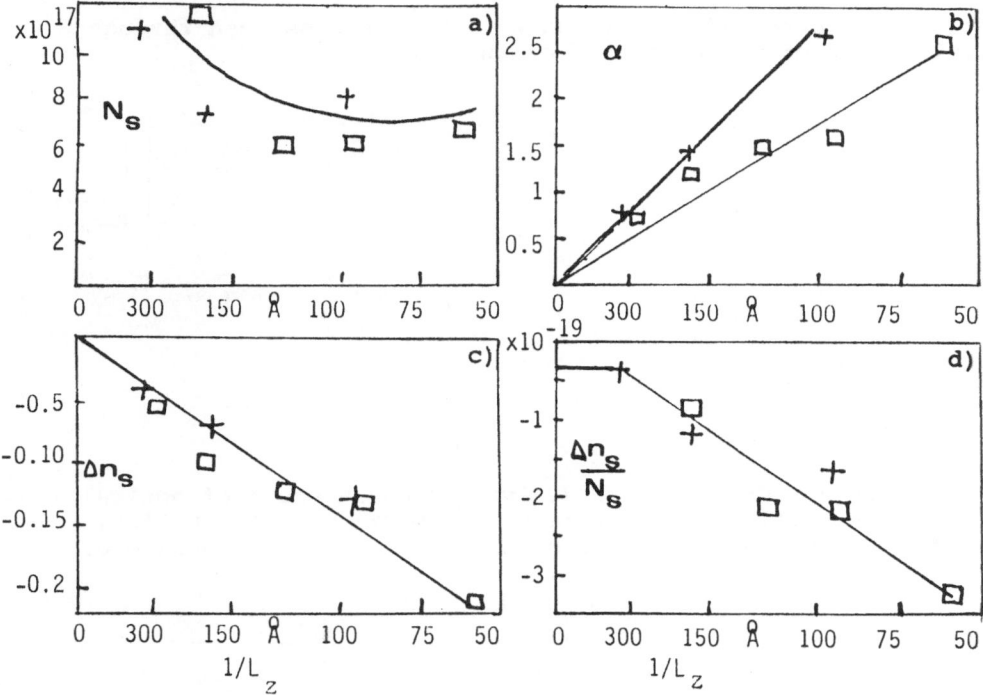

Figure 1. Measurements made at the peak of the exciton resonance in MQW as a function of inverse well thickness (thickness in angstroms is shown). Boxes are our data, crosses are those of [10]. Quantities shown are a) saturation carrier density N_S ($\times 10^{17}$ carriers/cc); b) absorption loss α (inverse microns); c) saturation index change; and d) saturation index change per carrier density ($\times 10^{-19}$ cc).

The experiment consisted of a single layer of n-InGaAsP with a bandgap near 1.06 μm, grown on a transparent semi-insulating substrate of InP [8]. A gold Schottky barrier was illuminated in reflection through the substrate with a YAG laser in the geometry shown in Fig. 2. Typical experimental results (Fig. 3) show the transmission as a function of time, when step-function illumination is applied. The data are fit to time constant of 4.3 μs. Figure 3b shows the cw transmission as a function of incident intensity, and a comparison with the theory of band-filling (plus some electro-absorption), assuming the measured time constant and no free parameters. An understanding of the role of carrier transport can be seen by observing in Fig. 2b that optically induced electrons will be confined within the hetero-barrier, but optically-induced holes will float out into the semi-insulating substrate and into the Schottky barrier.

SELF-MODULATION OF INTERNAL FIELDS

Improvements in the switching energy can only occur by increases in $|\Delta n_s/N_s|$. This is possible by using transport of optically-induced carriers to change internal electric fields in media whose refractive index depends on electric field (electro-

optic or electro-refractive media). This can be seen through a
simple "back-of-the-envelope" calculation.

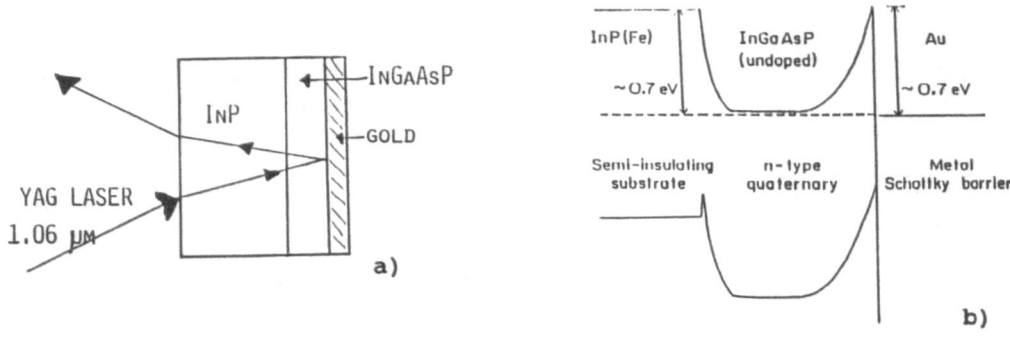

Figure 2. Experimental design for measurement of optical
nonlinearities in hetero-Schottky barriers. a) Optical
measurement method; b) Band diagram for LPE-grown sample.

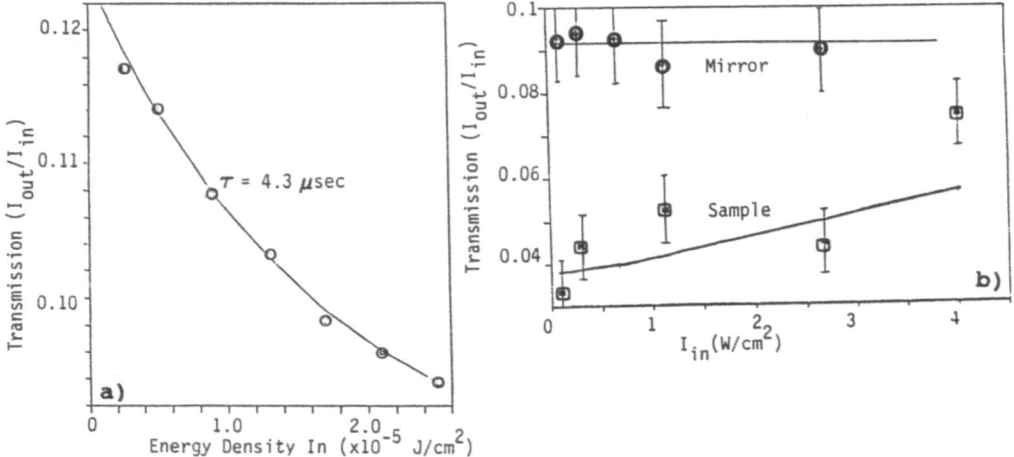

Figure 3. Experimental results of saturable transmission in
hetero-Schottky barrier. a) Transmission as a function of
energy density (time) for a step-function input; b)
Experimental transmission as a function of intensity for cw input
(boxes) and comparison with theory.

The magnitude of the electric field within an abrupt
semiconductor junction of width d is given by E = [e/ε]Nx, where
x is measured from the beginning of the depletion region to the
junction and N is the uncovered carrier density. The depletion
width is related to the junction voltage V through

$$d^2 = [2\epsilon/e]V/N. \qquad (8)$$

Below the bandedge the field-induced index change due to the
Franz-Keldysh effect (FKE) is quadratic in field, $|\Delta n| = bE^2$, so
that the average index change in the depletion width is

$$<|\Delta n|> = b[d^2/3][e/\epsilon]^2 N^2. \qquad (9)$$

The figure of merit can be calculated using (8) and (9):

$$<|\Delta n/N|> = [2/3][e/\epsilon]bV. \tag{10}$$

Equation (10) is an essentially "universal" figure of merit, depending only on the magnitude of the field-induced index change and the internal voltage (assuming that the material is fully depleted). However, the FKE b cannot be determined until the wavelength of operation is chosen, as required by (4). Inserting numbers into the FKE theory in GaAs [9], X_m = 3 (20 meV) and b(3) = 4 x 10^{-13} (cm/V)^2 (at E = 50 kV/cm). For N = 1X10^{16}/cc, d = 0.3 μm; with a band discontinuity of 1.43 V, this gives an average internal field of 12 kV. The figure of merit (at X_m = 3) is

$$|\Delta n/N| (FKE) = 6x10^{-20} \text{ cc.} \tag{11}$$

If the bulk band-filling nonlinearity [10] were used, X_m = 3.4 and $|\Delta n/N|$ = 0.6 x 10^{-20} cc. Thus, carrier transport is ten times better than band-filling.

If carrier transport nonlinearities are utilized in MQW, even larger effects are expected, using QCSE. Measurements of Weiner and Miller [11] at 20 meV (three Urbach tails) away from the bandedge give $\Delta n(3)$ = 0.005 for V = 65 kV. This gives b(3) = 12 x 10^{-13} (cm/V)^2, where is three times the FKE value. Thus

$$|\Delta n(3)/N_s| (QCSE) = 20 \times 10^{-20} \text{ cc,} \tag{12}$$

compared to state-filling value of $\Delta n(3)/N_s$ = 5.6 x 10^{-20} cc, a four times improvement over the state-filling nonlinearity.

Carrier transport nonlinearities can use external or internal electric fields. One convenient way of providing internal fields is the use of n-i-p-i structures [12]. Hetero-structures and/or MQW within n-i-p-i structures provide large optical nonlinearities due to carrier transport. We have investigated optically induced index changes in two cases: hetero-npnp materials and MQW-hetero-nipi materials.

OPTICAL NONLINEARITIES IN HETERO-NPNP STRUCTURES

The geometry for the hetero-npnp is shown in Figure 4a, and consists of thin GaAlAs p-regions acting as transparent sinks for carriers sandwiched between thicker GaAs n-regions, which act as electro-absorptive media. Illumination generates free carriers which move so as to oppose the internal fields. The space charge resulting from carrier motion flattens the bands and the absorption and refractive index changes due to the FKE. Figure 4b shows the measured absorption change and the calculated FKE (assuming idealized parabolic bands) in a field of 7.2 kV/cm. In reality, the electric field varies throughout the device, from 150 kV/cm at one edge of the 0.093 μm thick depletion region to zero, with a 63% fill factor and an average field of 47 kV/cm. The difference of almost a factor of 7 in predicted internal fields is due to the simplicity of the model. While the hetero-npnp structures are very sensitive (changes were measured with less than a W/cm^2), the 150 cm^{-1} absorption change was only a few percent of the total absorption. Larger effects can be found by including MQW in the high-field regions.

MULTIPLE QUANTUM WELL HETERO-NIPI MATERIALS

In order to increase the nonlinearity, we inserted GaAs QWs in the i regions of a GaAlAs n-i-p-i structure [13], in the geometry shown in Fig. 5a. The (saturated) differential absorption spectrum introduced by a pump of 375 mW/cm^2 (wavelength 808 nm), is shown in Fig. 5b, along with the index change calculated by the Kramers Kronig relation. The spectral shape of the absorption change agrees well with the measurements of QCSE [11]. However, the overall absorption change was roughly half of that observed by Weiner and Miller for a comparable field. The fact that the absorption change was smaller than that produced in an electro-absorptive modulator indicates that the optical excitation is unable to reduce the internal electric field more than a factor of two. We attribute this to the fact that the carriers become trapped in the QW's when the internal field lowers sufficiently.

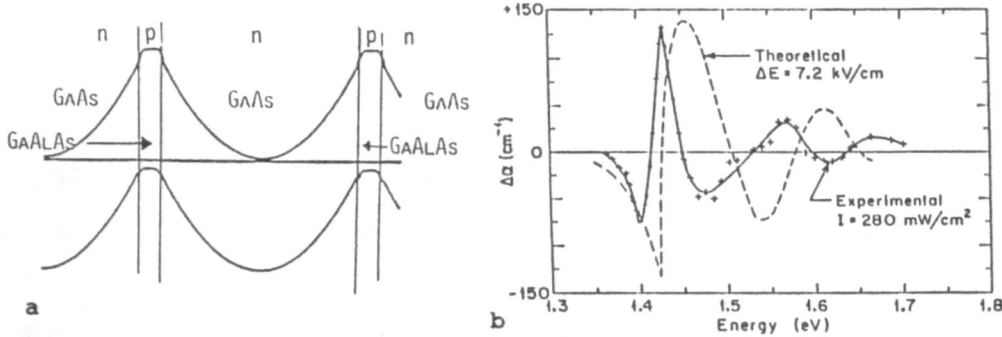

Figure 4. Nonlinear absorption in a hetero-nipi sample. a) Bandgap diagram of MOCVD-grown sample and b) experimental and theoretical results for changes in absorption of a broad-band probe due to excitation with a 100 mW/cm^2 probe [12].

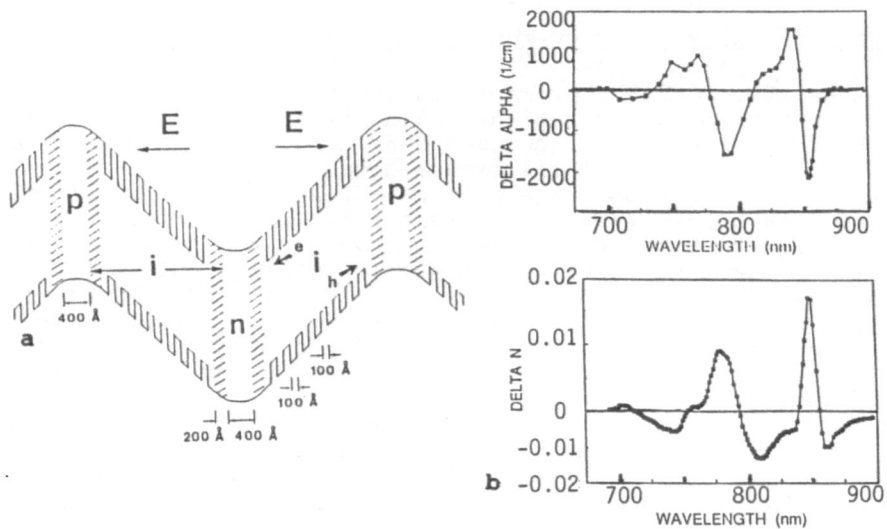

Figure 5. Nonlinear transmission in MQW H-nipi. a) Energy-band diagram. b) Absorption change in a broad-band probe due to 375 mW/cm^2 pump along with index change calculated by Kramers-Kronig.

The intensity-dependence of this nonlinearity is shown in Fig. 6a [1]. The extreme sensitivity is apparent; the absorption change saturates abruptly at a 50 mW/cm^2. Exact modelling of the intensity-dependence of this nonlinearity is difficult because the carrier lifetime is a function of intensity. However, heuristically modelled by a simple saturation equation, the saturation intensity I_s = 700 μW/cm^2. In a pump-probe experiment we measured a carrier lifetime τ = 20 μsec at I = 30 mW/cm^2 [14], lengthening as the optically induced carrier concentration decreases. Using the fact that <N> = <α>(I/hv)τ, and <α> = 0.84 μm^{-1}, we obtain N = 1.7x10^{16}/cc and N$_s$ = 0.85x10^{16}/cc. The saturation value of the index change at three Urbach tails is Δn(3) = -.0025, so that Δn(3)/N = 3x10^{-19} cc. This is close to the theoretical estimate of 2x10^{-19} cc for the index change per carrier density expected for a QCSE carrier transport nonlinearity (13).

Another way to calculate a similar result is to compare with the experimentally measured QCSE, under the conditions that the apparent field change is 35 kV/cm (which occurs for a depletion region thickness of 0.19 μm). For an optically-induced space charge to cause this much field change, a net charge must be accumulated of N = (ϵ/e)E/d = 1 x 10^{16}/cc, and N$_s$ = 5 x 10^{15}/cc. This gives for a Δn(3)/N = 5 x 10^{-19} cc.

We have shown experimentally that using MQW in a h-nipi structure gives Δn/N which is ten times better than state filling. This occurs at a very low intensity, due to enhanced carrier lifetimes. An important factor in understanding these materials is their intensity-dependent lifetimes. Our estimate of lifetimes based on the internal field-dependence of the excitonic resonance [14] is shown in Fig. (6b). The discrepancy of the experimental results with estimates based on thermionic emission recombination indicates that other recombination mechanisms (such as at surface defects) dominate.

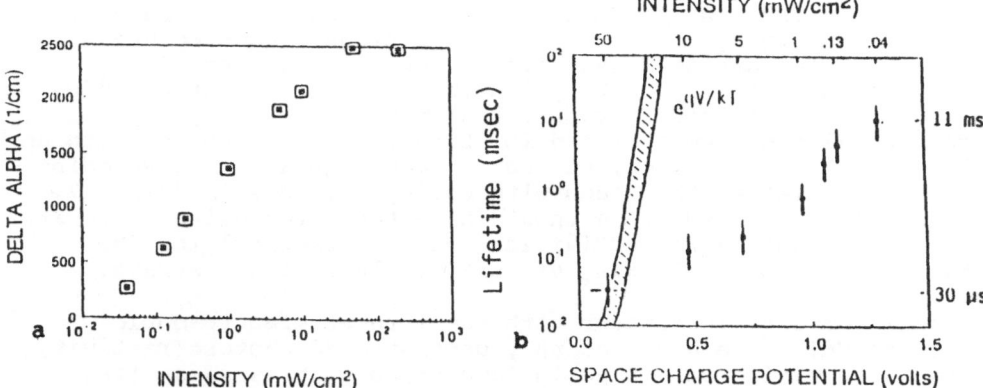

Figure 6. Intensity-dependent measurements in MQW-H-nipi. a) Pump-induced change in probe absorption as a function of pump intensity; b) Carrier lifetime inferred from measurements of the energy-shift of the exciton resonance due to QCSE, along with theoretical estimates from simple thermionic recombination.

ELECTRO-REFRACTIVE PHOTOREFRACTIVE EFFECT

In two-wave mixing, the photorefractive effect uses carrier transport across an optically induced sinusoidal space charge. Because this causes a phase shifted index grating, it is possible to couple light from one beam to another. The challenge is to have photorefractive coupling with gain, opening up a variety of applications in optical switching, signal processing and phase conjugation. We have investigated the photorefractive effect near the bandedge in semiconductors in order to use electro-refraction (the quadratic electro-optic effect resulting from FKE) rather than the linear electro-optic effect [2]. We have observed two-wave mixing in GaAs with a net gain of 13 cm^{-1} at a wavelength about 20 nm below the bandedge. This gain is comparable to BaTiO$_3$ and means that infrared phase conjugation is practical for the first time in GaAs, InP and CdTe. At near-infrared wavelengths GaAs is 10^6 times more sensitive than BaTiO$_3$ and locking of semiconductor laser arrays is practical for the first time. In fact, the optimum wavelength for bandedge effects is close to that of the newly popular strained layer semiconductor lasers [15].

In electro-refractive photo-refractivity (ERPR) the index change depends on the magnitude of the electric field and not on its sign. Thus, a grating in the space-charge electric field will cause an index grating at twice the frequency, eliminating the possibility of two-wave mixing. However, when a positive electric field is applied, by ensuring that the total field is always positive, an index grating is created which is now out of phase with the sinusoidal intensity profile and two-wave mixing can occur. With a negative voltage the phase shift is in the opposite direction. This means that the sign of two-wave mixing is opposite and thus ERPR provides voltage-control to the switching direction.

By appropriately choosing the crystal axes, pure ERPR or a combination of ERPR and electro-optic photo-refractivity (EOPR) can be measured. In combination, gains of 7.6 cm^{-1} with DC field and 16.3 cm^{-1} by adding a moving grating have been observed in the presence of a loss of 3 cm^{-1} [2].

The high gains are important for applications involving self-pumped phase conjugation, but moving gratings cannot be used since the beams are self-generated. However, we have demonstrated that large ERPR gains are possible in InP by using temperature-stabilization rather than moving gratings, a technique already demonstrated for EOPR [16]. At 960 nm, with an applied field of 10 kV/cm, at 290° K we have measured ERPR gains of 8.5 cm^{-1}, larger than seen with moving gratings in GaAs [17]. When combined with EOPR, we should have the large gains necessary for phase conjugate reflectivities, making practical for the first time the phase-locking of strained-layer laser arrays.

Finally, the semiconductor with the largest electro-optic effect is CdTe. We have recently demonstrated photorefractivity by two-wave mixing at 1.5 μm in CdTe doped with Vanadiam [18]. This is the longest wavelength at which two-wave mixing has been reported; a gain of 0.6 cm^{-1} was observed with only 1 mW/cm^2 incident power. By using the moving grating technique, gain coefficients of 2.4 cm^{-1} were obtained with a background absorption of 2 cm^{-1}. In the future, by alloying with HgCdTe, these results indicate that it may be possible to achieve photorefraction throughout the infrared.

In conclusion, by operating near the bandedge, one can take advantage of the Franz-Keldysh electro-refractive effect in photorefractive materials to achieve beam coupling energy transfer. The new ERPR mechanism differs from conventional EOPR in that the direction of energy transfer is dictated by the direction of an externally applied field. By combining ERPR and EOPR and using a moving grating or thermal stabilization, very large gains are possible.

Many researchers contributed to this work: Dr. Alan Kost, students Nan Marie Jokerst, Masa Kawase, Afshin Partovi, James Millerd, Drs. M. Klein and G. Valley at Hughes Research Laboratories, Prof. D. D. Dapkus and his students H. C. Lee and A. Danner, and Prof. W. Steier and his student M. Zairi. The research was supported in part by the National Science Foundation, the Air Force under its URI on Optical Computing, JSEP, Hughes Research Laboratories and Northrop Corporation.

References

1. E. Garmire, N. M. Jokerst, A. Kost, A. DAnner, P. D. Dapkus, J. Opt. Soc. Am. B6, 579 (1989)
2. A. Partovi, A. Kost, E. Garmire, G. Valley, M.Klein, Appl. Phys. Lett. March, 1990.
3. E. Garmire, to be published.
4. A variation of this analysis appeared as "Criteria for Optical Bistability in a Lossy Saturating Fabry-Perot" E. Garmire, IEEE J. Quantum Electr. QE-25, 289 (1989)
5. A. Kost "Bandedge absorption coefficients from photoluminescence in semiconductor MQW" Appl. Phys. Lett. 54, 1356 (1989)
6. E. Garmire, to be published.
7. H. C. Lee, A. Kost, M. Kawase, A. Hariz, P. D. Dapkus, E. M. Garmire, IEEE J. Quantum Electr QE-24, 1370 (1988)
8. N. M. Jokerst and E. Garmire, Appl. Phys. Lett. 53, 897 (1988). See also, N. M. Jokerst, PhD Dissertation, USC, 1989.
9. B. O. Seraphin and N. Bottka, Phys. Rev. A239, 560 (1965)
10. S. H. Park, J. F. Morhange, A. D. Jeffery, R. A. Morgan, A. Chavez-Pirson, H. M. Gibbs, S. W. Koch, N. Peyghambarian, M. Derstine, A. C. Gossard, J. H. English, W. Wiegmann, Ppl. Phys. LEtt. 52, 1201 (1988)
11. J. S. Weiner, D. A. B. Miller, D. S: Chemla, Appl. Phys. Lett. 50, 842 (1987)
12. A. D. Danner, P. D. Dapkus, A. Kost, E. GArmire, J. Appl. Phys. 64, 5206 (1988)
13. A. Kost, E. Garmire, A. D. Danner, P. D. Dapkus, Appl. Phys. Lett. 52, 637 (1988)
14. A. Kost, M. Kawase, E. Garmire, A. D. Danner, H. C. Lee, P. D. Dapkus. Proc. SPIE 943, 114 (1988)
15. D. P. Bour, P. Stabille, A. Rosen, W. Janton, L. Elbaum, D. J. Holmes, Appl. Phys. Lett. 54, 2637 (1989)
16. G. Picoli, P. Gravey, C. Ozkul, V. Vieus, J. Appl. Phys. Sept. 15, 1989
17. A. Partovi, E. Garmire, G. Valley, M. Klein to be published
18. A. Partovi, J. Millerd, E. Garmire, M. Zairi, W. Steier, etc. to be published.

ORGANIC NONLINEAR OPTICAL MATERIALS AND DEVICES FOR OPTOELECTRONICS

Gary C. Bjorklund

IBM Almaden Research Center, Department K95/801
650 Harry Road, San Jose, CA 95120, U.S.A.

ABSTRACT

Light has several unique properties that are potentially of great
use for storage, transmission, processing, and display of information.
Among these are action at a distance (focusability), massive 2-D
parallelism, fast access by beam deflection, and a huge usable bandwidth.
For many applications, the fact that photons are bosons without charge or
rest mass is an advantage, since beams of light can usually pass freely
through each other without interaction or cross-talk. However, for
advanced opto-electronics applications, where light is being harnessed
to perform a function previously done exclusively by electronics, it is
often necessary to find ways to rapidly control beams of light by electric fields or by other beams of light. Nonlinear optical materials
provide a means of doing this

Organic nonlinear optical materials have high figures of merit for a
variety of device applications, including frequency doubling, electro-
optic modulation and switching, and real time holography using the photorefractive effect. Molecular engineering approaches can be used to
produce organic NLO materials with specialized device properties and with
chemical and mechanical properties that permit low cost device fabrication. Our research activities on synthesis and crystal growth of
candidate materials for frequency doubling , and on the synthesis,
fabrication, and poling of novel covalently functionalized linear and
crosslinked polymers will be reviewed. Prospects for organic photorefractive materials will be discussed.

NONLINEAR OPTICAL SUSCEPTIBILITIES OF SUBSURFACE LAYERS OF METALS AND

SUPER- AND SEMICONDUCTORS RELATED TO ELECTRONIC STRUCTURE AND CRYSTAL

SYMMETRY

S. V. Govorkov, N. I. Koroteev, I. L. Shumay, and V. V. Yakovlev

R. V. Khokhlov Laboratory of Nonlinear Optics
Moscow State University
Moscow 119899, USSR

INTRODUCTION

Numerous experimental data have been accumulated in recent years on the use of nonlinear optical techniques for studying surfaces, interfaces, and adsorbates. These techniques are based on the generation of optical harmonics (second and third) and sum and difference frequencies in reflection. The main advantage of these methods, besides the nondestructive nature of the probing and their high temporal resolution, consists in the fact that nonlinear optical response governed by the corresponding nonlinear optical susceptibility tensor reflects the symmetry of the interface layers of solids and adsorbed molecules[1-4].

Recently, we have observed a strong anisotropy of the second harmonic generation (SHG) in reflection from an $A\ell$ single crystal[5]. This fact was interpreted in terms of the anisotropy of the nonlinear optical response of the 3s- and 3p-electrons.

We report here on a theoretical and experimental investigation of the influence of the crystalline structure on the nonlinear optical response of metal single crystals. It is shown that lattice disordering leads to the decrease of both the anisotropy and the absolute value of the quadrupole second order nonlinear optical susceptibility as revealed by SHG in reflection from various $A\ell$ samples.

Optical SHG provides valuable information on the structure of high-T_c superconducting films.

We report also the observation of the primary stages of GaAs surface melting under femtosecond laser excitation using SHG in reflection. A numerical fit of the experimental data yielded the characteristic time delay between excitation and lattice melting of about 75 fs.

THE INFLUENCE OF DISORDERING ON SHG FROM METAL SURFACES

It was discovered recently that metal single crystals, for example Cu [6], Ag [7], Ni [8], and Al [5], demonstrate a significant anisotropy of the nonlinear optical response as revealed by SHG in reflection.

Hydrodynamic models have mostly been used to account for the nonlinear optical response of metals so far[9]. Those models deal with the plasma of free electrons and do not take into consideration the electric field of the lattice. The hydrodynamic model of free electrons provides satisfactory results in the case of nearly amorphous metal films However, when analyzing the nonlinear optical response of metal single crystals one should take into account the anisotropy of the lattice and its influence on the nonlinear optical response of the electrons.

The SHG in reflection from a surface of a solid is governed by the nonlinear polarization at the second harmonic frequency (2ω) given by

$$P_i(2\omega) = \chi_{ijk}^{(2)D} E_j(\omega)E_k(\omega) + \chi_{ijk}^{(2)S} E_j(\omega)E_k(\omega) + \chi_{ijk\ell}^{(2)Q} E_j(\omega)\nabla_k E_\ell(\omega).$$
(1)

The first term in (1), corresponding to the bulk electric dipole contribution, vanishes in media with inversion symmetry (an Aℓ single crystal, for example) due to spatial symmetry selection rules. The second term is the surface dipole contribution due to the lack of inversion symmetry in a few interface layers of atoms. This contribution is of importance only in the case of atomically clean surfaces prepared under high vacuum[10]. The third term corresponds to the electric quadrupole contribution. The quantity

$$\xi = \chi_{1111}^{(2)Q} - \left(\chi_{1122}^{(2)Q} + \chi_{1212}^{(2)Q} + \chi_{1221}^{(2)Q}\right)$$
(2)

is known[11] to describe the degree of anisotropy of the second order nonlinear optical susceptibility tensor and thus the anisotropy of the nonlinear optical response ($\xi = 0$ in isotropic media). We have shown that (see Refs. 8 and 12):

$$\chi_{ijk\ell}^{(2)Q}(\hbar\omega) = \frac{1}{2}\left|\frac{e}{m\omega}\right|^3 \sum_{b,b',b''} P(b,b',b'') \int_{BZ} \frac{d^3k}{4\pi^3} \times$$

$$\times \left\{R(\omega,\vec{k})\right\}\left\{p_{bb'}^i(\vec{k})p_{b'b''}^j(\vec{k})q_{b''b}^{k\ell}(\vec{k})\right\}.$$
(3)

Here $p_{bb'}^i(\vec{k})$ and $q_{bb'}^{ij}(\vec{k})$ are transition dipole and quadrupole matrix elements, respectively. $R(\omega,k)$ is a resonant factor of the form

$$\left[(\mathcal{E}_b(\vec{k}) - \mathcal{E}_{b'}(\vec{k}) - \hbar\omega)(\mathcal{E}_{b'}(\vec{k}) - \mathcal{E}_{b''}(\vec{k}) - 2\hbar\omega)\right]^{-1},$$
(4)

$P(b,b',b'')$ is a permutation operator, and $\mathcal{E}_b(\vec{k})$ is the band energy. The major difficulty is the direct calculation of the band structure and of the wave functions. To do it we have used the pseudopotential method also used in Ref. 13. However, we have taken into account the long-range order of the lattice. The plane wave representation of the real crystal wavefunctions was used (see, for example, Ref. 13). The Schrödinger equation enables one to derive the following equation used for the band structure calculation:

$$\det||T_{ij}\delta_{ij} + v_{ij}|| = 0 .$$
(4)

Here v_{ij} is the pseudopotential matrix, and T_{ij} depends on the degree of crystal disorder, which can be described by an order parameter α ($\alpha = 0$ in the case of an ideal crystal and $\alpha = 0.5$ corresponds to a complete loss of short-range order)[14]:

$$T_{ij}^{-1} = \frac{1}{(\sqrt{\pi}\ a\ g_i)^3} \exp\left\{-[\vec{k}+\vec{g}_i - \vec{q}]^2/[\alpha\ g_i]^2\right\} \left\{\mathcal{E} - \vec{q}^2\frac{\hbar^2}{2m}\right\}^{-2} d^3q .$$
(5)

Here $\vec{g}_i$ - is a reciprocal wavevector, and $\mathcal{E}(\vec{k}) = \mathcal{E}_1(\vec{k}) + i\Gamma(\vec{k})$ is a complex energy which in a simple form reflects the appearance of the new energy states in the case of disordering[13]. The inverse value of α is in fact the characteristic number of unit cells which maintain a crystalline order. In the case of an ideal crystal (5) has the usual form of

$$T_{ij} = \frac{h^2}{2m} (\vec{k} - \vec{g}_i)^2 - \mathcal{E} \qquad (6)$$

where $\mathcal{E}$ is real. Typical phase transitions to the amorphous state occur when α is close to 0.1[14].

It is clearly seen that the energy levels spread with the increase of the parameter α. This fact greatly simplifies integration over the Brillouin zone since it removes the singularity of the form $(\mathcal{E}_b(\vec{k}) - \mathcal{E}_{b'}(\vec{k}) - \hbar\omega)$ in the denominator.

We have calculated both the linear and nonlinear optical parameters of Al. The main conclusions are as follows: i) the linear optical susceptibility of Al (both crystalline and randomized) is governed by the Drude formula remarkably well with an error less than 20% (note that the characteristic feature near the photon energy $\hbar\omega = 1.5$ eV corresponds to the direct transition at the W point of the Brillouin zone); ii) the interband electronic transitions provide the major contribution to the nonlinear optical susceptibility of Al (this contribution contributes up to 80% of its value).

The quantities $\chi_{1212}^{(2)Q}$ and ξ which determine the intensity of SHG and the anisotropy of the SH rotational dependence are shown in Fig. 1 as functions of the disorder parameter α in the case of the fundamental wavelength $\lambda = 1.06 \mu m$. The maximum values of $\chi_{1212}^{(2)Q}$ and ξ are 7.8×10^{-13} CGSE, and 6.5×10^{-13} CGSE, respectively. It is clearly seen from Fig. 1 that ξ decreases faster than $\chi_{1212}^{(2)Q}$. Thus, SGH from amorphous Al ($\alpha > 0.1$) should be nonvanishing but isotropic, since ξ is close to 0.

The SH intensity and its dependence on the angle of sample rotation about its surface normal have been measured for three Al samples. The first one was an Al film with a thickness of about several microns deposited on a glass substrate by vacuum vapor deposition (sample No. 1). The second one was an Al(111) single crystal mechanically polished with a $1\mu m$ diamond powder (sample No. 2). The third sample was the same as No. 2, additionally electrochemically etched to remove a disordered layer (sample No. 3).

All measurements were made in the normal atmosphere at room temperature so that the samples had a native oxide layer on the surfaces. A passively mode-locked Nd^{3+}:YAG laser with a pulse duration of about 30 ps was used as the source of probe radiation. The laser fluence at the sample surface ($< 10^{-4}$ J cm^{-2}) was at least an order of magnitude lower than the damage threshold. The SH radiation was recorded at the angle of specular reflection (about 45°) after being filtered out by a set of spatial and spectral glass filters and a monochromator. The SH signal from the sample was normalized to that from a GaAs reference sample.

The corresponding SH rotational dependences (i.e. the dependence of the SH intensity on the angle of sample rotation about the surface normal) for the three Al samples listed above are shown in Figs. 2 a,b. The p- and s-polarized SH components have been measured for the electrochemically etched Al single crystal (No. 3) with a p-polarized pump laser beam. Only the p-polarized component has been measured for the other samples (No. 1 and No. 2), since the s-polarized component is nearly zero

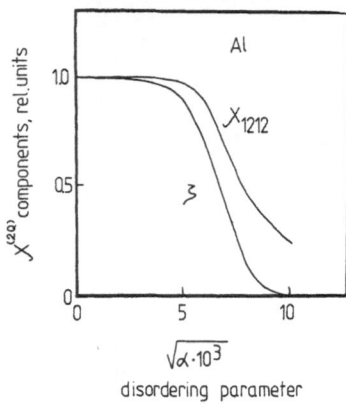

Fig. 1. The decrease of anisotropy (ξ/ξ_{cryst}) and absolute values $(\chi^{(2)Q}_{1212}/\chi^{cryst}_{1212})$ of the second order quadrupole optical susceptibility of Al with increasing lattice disorder (theory).

in this case. The clearly seen three-fold symmetry in Figs. 2a,b reflects the crystalline symmetry of Al(111) and shows that the single crystal surface layer gives the main contribution to the SH intensity. The relatively large amplitudes of the third and sixth Fourier components in the p → p experimental geometry are consistent with the theoretical SH rotational dependence calculated assuming that the electric quadrupole contribution dominates in the second order nonlinear optical susceptibility $\chi^{(2)}$ which governs the SHG in materials with cubic spatial symmetry[11]:

$$I^{p-p}_{SH}(\psi) \sim |\gamma \cos 3\psi + \varsigma|^2 . \tag{7}$$

In the p → s experimental geometry the SH rotational dependence for the Al(111) single crystal is given by

$$I^{p-s}_{SH}(\psi) \sim \varsigma^2 \sin^2 3\psi . \tag{8}$$

Consequently the third and sixth Fourier components are related to the symmetry of the sample surface symmetry. The solid lines in Fig. 2a,b are drawn in accordance with Eqs. (7)-(8) using Fourier amplitudes calculated numerically from experimental data. The degree of SH anisotropy $|\xi/\varsigma^{(2)Q}_{1212}|^{exp.} = 0.5$ derived from the numerical fit to the experimental data is in a good agreement with the theoretical value of 0.7. Here the results of Ref. 11, where the problem of calculating the rotational dependence of the SH intensity was solved, were taken into account.

The rotational dependences of the SH intensity for samples No. 1 and 2 were found to be isotropic, which showed the lack of any crystalline order in the surface layer of those samples. In fact, the surface layer, whose thickness is about the size of the polishing powder grains, is known to be randomized as a result of the polishing procedure. In addition, the glass substrate was amorphous, so there should not be any preferable directions in film deposition unless a special geometry of film deposition is used[15].

Note also that the absolute values of the SH intensity are different for all three Al samples studied. This difference can be attributed neither to the differences in the linear optical constants of the samples, which are practically identical, nor to the macroscopic structure of the

TABLE 1

No.	Composition	annealing	substrate	SH intensity a.u.	TH intensity a.u.
1	Y-Ba-Cu-O	yes	sapphire	0.22	$< 10^{-3}$
2	Y-Ba-Cu-O	no	sapphire	2.06	$< 10^{-3}$
3	La-Sr-Cu-O	yes	sapphire	0.18	$< 10^{-3}$
4	La-Sr-Cu-O	no	sapphire	11.6	$< 10^{-3}$
5	La-Ba-Cu-O	no	quartz	0.5	3×10^{-3}
6	La-Ba-Cu-O	no	quartz	14.4	$< 10^{-3}$
7	$YBa_2Cu_3O_{7-x}$	yes	$SrTiO_3$	0.25	-
8	$YBa_2Cu_3O_{7-x}$	single crystal		0.4	-

surface layer. Actually, the SH escape depth is smaller than the characteristic thickness of the disordered layer of both the mechanically
polished $A\ell$ single crystal and the amorphous $A\ell$ film (several μm).
Nevertheless, the two samples clearly demonstrate different SHG
efficiencies. Thus, we come to the conclusion that the nonlinear optical
susceptibility $\chi^{(2)}$ which governs the SH intensity depends on the crystal
structure of the surface layer.

The experimental observations reported above should be compared with
the theoretical predictions. Let us compare the anisotropy and the
absolute value of the SH intensity in the case of an amorphous $A\ell$ film
and an electrochemically etched $A\ell$ single crystal. According to theory,
SH anisotropy and SH intensity should decrease with the increase of
lattice disorder (see Fig. 1). There is no noticeable third Fourier
amplitude in the SH rotational dependence obtained in the case of an
amorphous $A\ell$ film. In addition, by comparing the isotropic contributions
to the SH rotational dependence for an $A\ell$ film and an $A\ell$ single crystal,
one finds that

$$|\chi_{1212}^{(2)Q,3}/\chi_{1212}^{(2)Q,1}| = 3.5$$

(note that the $\chi_{1212}^{(2)Q}$ component of the nonlinear optical susceptibility
tensor governs the isotropic contribution to the SH intensity). Thus,
the $A\ell$ film can be characterized by a disorder parameter value of about
0.1 (see Fig. 1).

The case of a mechanically polished $A\ell$ single crystal is much more
complicated. The surface layer is known to consist of monocrystalline
grains of various sizes dependent on the polishing powder. Thus, various
grains contribute to the total nonlinear optical response of a surface
layer. Moreover, the characteristic value of α can be found only if one
knows the grain size distribution. However, since the value of $\chi_{1212}^{(2)Q}$
decreases with the increase of the parameter α, we can estimate the upper
limit for $\chi_{1212}^{(2)Q}$ by averaging its tensor components over all grain
orientations, assuming the characteristic grain size is of the order of
10^{-6} cm. We have

$$\chi_{1111}^{aver} = 0.6 \; \chi_{1111}^{(2)Q} + 0.4 \left(\chi_{1212}^{(2)Q} + 2 \; \chi_{1122}^{(2)Q} \right)$$

$$\chi_{1212}^{aver} = 0.2 \; \chi_{1111}^{(2)Q} + 0.8 \; \chi_{1212}^{(2)Q} - 0.4 \; \chi_{1122}^{(2)Q} \tag{8}$$

$$\chi_{1122}^{aver} = 0.2 \; \chi_{1111}^{(2)Q} - 0.2 \; \chi_{1212}^{(2)Q} + 0.6 \; \chi_{1122}^{(2)Q} \; .$$

Naturally, the anisotropy parameter $\xi^{aver} = 0$. The calculated effective value of χ_{1212}^{aver} is equal to $0.7 \; \chi_{1212}^{(2)Q}$ (keeping in mind the calculated tensor components for a pure Al single crystal). This value is consistent with the experimental value of the ratio of the zero Fourier components of the SH rotational dependences for the two samples, found to be 2.3. This fact, probably, means that the surface layer of the mechanically polished sample consists of grains with α of about 0.03.

Finally, let us compare our theoretical value of $\chi_{1212}^{(2)Q}$ with the one obtained from hydrodynamic models. It seems reasonable that the absolute value of $\chi_{1212}^{(2)Q}$, obtained from our calculations for the case of totally amorphous Al should be of the same order of magnitude as that found from the hydrodynamic models because in this case crystal order and hence its

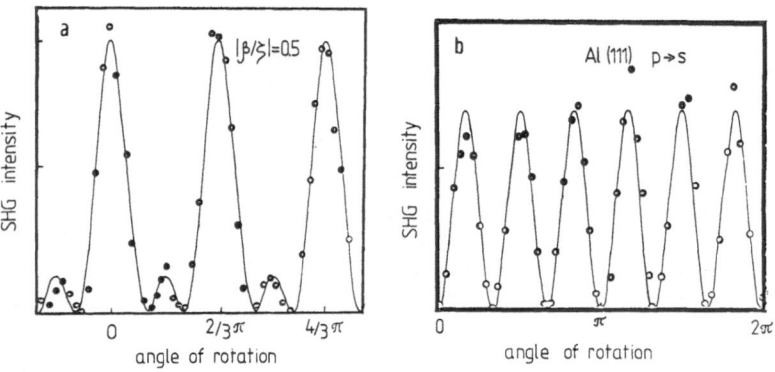

Fig. 2. SH rotational dependences for a mechanically polished electrochemically etched Al single crystal. The p- and s-polarized SH components are shown in Figs. 2a and 2b, respectively.

influence on the electrons is lost. Our calculated absolute value $\chi_{1212}^{(2)} (\alpha = 0.1) = 2.5 \times 10^{-13}$ CGSE is close to the one obtained by J. E. Sipe et al.[9] from the hydrodynamic model (free electron gas).

NONLINEAR OPTICAL STUDIES OF HIGH-T_c SUPERCONDUCTING FILMS AND SINGLE CRYSTALS

The microscopic structure of the new high-T_c superconductors is of great interest. It was found by the use of x-ray and particle beam scattering that those compounds have a perovskite-like structure. In particular, the structure of $YBa_2Cu_3O_{7-x}$ depends significantly on the number of oxygen atoms and on the distribution of oxygen vacancies. The "x" value determines the temperature of the transition from the orthorhombic to the tetragonal phase. Sometimes the so-called "green" phase can appear with a structure completely different from that of $YBa_2Cu_3O_{7-x}$.

We have studied SHG in reflection from high-T_c superconducting films of different composition deposited on quartz, sapphire or $SrTiO_3$ substrates. Some of the films have been annealed (see Table 1).

It is interesting to note that a relatively high SH signal was observed in some cases as compared with that from a Si wafer. It should be noted that Si has a strong electric quadrupole nonlinearity described by the second order nonlinear optical susceptibility $\chi^{(2)Q}$ at the fundamental Nd^{3+}:YAG laser frequency, since the latter is close to resonance with a Si interband transition. Thus, we have also measured the third harmonic intensity for the same samples to make sure that there is no influence of resonant effects on the SH intensity. However, the third harmonic intensity was found to be of the same level in all cases. Consequently, the anomalously high SH intensity can be a manifestation of the presence of grains with noncentrosymmetric symmetry possessing an electric dipole contribution to the second order optical nonlinearity. It is worth noting that high-quality samples with the correct composition (both thin films and single crystals) demonstrate a low SHG efficiency.

We made an attempt to establish the point symmetry of the high-T_c superconducting samples using the SH rotational dependence. The latter was found to be essentially isotropic in all cases. Thus, the $\chi^{(2)}$ tensor was found to have the following components:

$$
\chi^{(2)}_{ijk} = \begin{pmatrix}
0 & 0 & 0 & \chi^{(2)}_{123} & \chi^{(2)}_{113} & 0 \\
0 & 0 & 0 & \chi^{(2)}_{113} & -\chi^{(2)}_{123} & 0 \\
\chi^{(2)}_{311} & \chi^{(2)}_{311} & \chi^{(2)}_{333} & 0 & 0 & 0
\end{pmatrix}.
$$

This symmetry of the $\chi^{(2)}_{ijk}$ tensor corresponds to the point groups 4, 4mm and mm2. The latter can be compared with the symmetry groups obtained from electron diffraction experiments[16] (mmm, mm2, 4/mmm, 42m and $\bar{4}$mm).

Thus the nonlinear optical diagnostic technique provides valuable information on the structure of the high-T_c superconductors.

FEMTOSECOND LASER-INDUCED MELTING OF GaAs PROBED BY OPTICAL SHG

The primary steps in the melting (disordering) of GaAs surface layers following femtosecond laser pulsed excitation have been detected by using second harmonic generation (SHG) in reflection. A numerical fit of the experimental data yielded the characteristic time delay between laser excitation and lattice melting of about 75 fs.

Laser-induced phase transitions at the surface of semiconductors (e.g. melting, evaporation, amorphization, etc.) have been of great interest since the discovery of the pulsed-laser annealing effect[17]. A complete understanding of these phenomena requires knowledge of the nature of the melting dynamics and the fundamental electron-lattice energy transfer time. Therefore, experiments with ultrashort laser pulses which allow one to study directly the dynamics of surface melting with femtosecond temporal resolution are of great importance.

So far several experiments of this kind have been carried out. Shank et al.[18] have measured laser induced modification of Si reflectivity. Kash et al.[19] have used spontaneous Raman spectroscopy to estimate the electron-phonon energy transfer time in GaAs. Shank et al.[2] have studied the onset of the Si surface layer melting under femtosecond pulsed-laser excitation by using second harmonic generation in reflection. Recently Tom et al.[2] have reported the disappearance of the anisotropy of SHG on a time scale of the order of 100 fs in Si. They used a pump-and-probe technique first introduced by Shank et al.[2] to study the subpicosecond dynamics of the laser-induced melting of Si.

Here we report on the subpicosecond dynamics of the laser-induced melting of a GaAs single crystal studied by SHG in reflection by using the "self-action" technique proposed by Bloembergen et al.[20] The latter consists of studying the SH energy $E_{2\omega}$ as a function of the incident pulse energy E_ω. GaAs is a noncentrosymmetric material and exhibits a large second-order nonlinear optical susceptibility. This allows strong SHG in the bulk in the electric dipole approximation. In contrast, the melted material is an isotropic liquid in which SHG is forbidden in the electric dipole approximation due to symmetry selection rules. Thus the deviation of the SH intensity as a function of the pump-laser intensity from the square law unambiguously indicates a modification of the spatial symmetry of the surface layer, which can be attributed to surface melting. The "self-action" technique has several obvious advantages over the pump-probe technique, since the former does not require the special procedure of spatial and temporal overlap of the pump and probe pulses. However, computer simulation is needed to obtain quantitative information on the dynamics of laser-induced melting and to estimate the electron-phonon energy transfer time.

We used 50 μJ of energy in the pulses from a fs CPM dye laser amplified in four-gain stages[21]. The pulses were coupled out before recompression and had a duration of 260 fs. The sample was moved after each laser shot (a repetition rate $\approx$1 Hz) to ensure equal excitation conditions in each pulse, i.e. to avoid accumulation effects.

The measured dependence of the SH pulse energy versus the incident pulse energy is shown in Fig. 3. The arrow indicates the onset of amorphization. As can be clearly seen, the slope of the curve shows a quadratic behavior at low laser intensities. The slope decreases at an intensity corresponding to the threshold for laser-induced surface amorphization which we attribute to surface melting. This means that the laser fluence in the central part of the laser spot is sufficient to melt the surface layer of the GaAs sample during the laser pulse. In this manner the melted area gives no contribution to the SH signal. Thus, with further increase of the pump laser intensity the SH signal value is

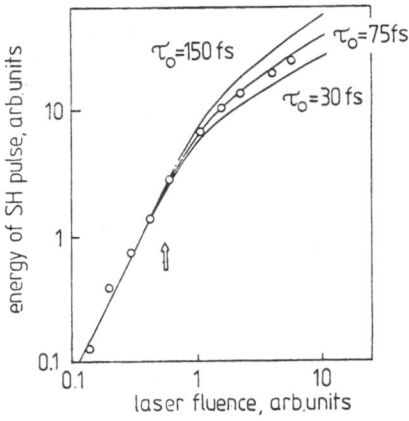

Fig. 3. The dependence of SH intensity of laser fluence. Experimental dots; numerical modeling—solid curves.

governed by the two competing processes: i) the increase of the SH signal from the unmelted part of the laser spot on the sample surface, and ii) the decrease of the energy of the SH pulses due to growth of the melted area at the surface. A correct estimate of the influence of these two factors requires a numerical treatment of the laser-induced phase transition.

The energy of the absorbed light pulse is stored first in the electron-hole sub-system of the semiconductor crystal and is transferred to the lattice, causing its melting, within a characteristic time τ_o. Thus the energy deposited in the electron-hole plasma at the moment t' flows to the lattice so that the energy density $Q_L(t)$ stored in the lattice (i.e. phonon sub-system) up to the moment t can be approximated by

$$Q_L(t) = A \int_{-\infty}^{t} I_\omega(t')\Big(1-\exp(-(t-t')/\tau_o)\Big)dt'. \tag{9}$$

Here A is the factor which takes into account the sample reflectivity, absorption, and the difference between the energy of the laser quantum and the fundamental band gap. When Q_L exceeds the threshold value Q_o the lattice melts, so that the second order nonlinear optical susceptibility $\chi^{(2)}(t)$ can be approximated by

$$\chi^{(2)}(t) = \begin{vmatrix} \chi_o^{(2)} & \text{if } Q_L(t) < Q_o \\ 0 & \text{if } Q_L(t) \geq Q_o \, . \end{vmatrix} \tag{10}$$

The total energy of the SH is proportional to the integral

$$E_{2\omega} \int_{-\infty}^{+\infty} |\chi^{(2)}(t)|^2 \, I_\omega^2(t)dt. \tag{11}$$

The value of τ_o determines the decrease of the slope above the threshold. Calculated dependences $E_{2\omega}(E_\omega)$ for various values of τ_o are shown in Fig. 3 by solid curves. The laser pulse shape was assumed to be $I_\omega = I_o ch^{-2} (1.76t/\tau_L)$ with $\tau_L = 260$ fs. The value of the SH signal was averaged over the laser beam cross-section with a Gaussian intensity distribution. The best fit was achieved for $\tau_o = 75$ fs.

The time τ_o is basically the effective time delay between the moment at which the laser pulse deposits energy in the electron sub-system and melting. This time is connected with the electron-phonon energy transfer time and with the characteristic time scales of various processes leading to the phase transition from a crystal lattice to a disordered medium, which is usually identified with a melt. It is known that the characteristic time of energy thermalization in the phonon system of a semiconductor can be of the order of several ps[22]. Thus, one can conclude that the melting of the lattice occurs before the lattice is in thermal equilibrium. Moreover, the derived time τ_o is shorter than the time for complete energy transfer to long wavelength LO-phonons, measured by Kash et al.[19] Assuming that in polar semiconductors the electrons and holes are coupled with LO phonons near the center of the Brillouin zone[23], one comes to the conclusion that this result gives additional evidence for the hypothesis of "cold melting" introduced by Tom[2].

In conclusion, we observed[23] ultrafast melting (disordering) of a GaAs surface layer during 260-fs laser pulses using SHG. From a numerical computer simulation and a fit to the experimental data we have estimated the characteristic time delay between laser energy deposition and lattice melting to be about 75 fs, which is shorter than that found from theoretical considerations.

CONCLUSIONS

We have reported the most interesting applications of nonlinear optical diagnostics of surface electronic structure and crystal symmetry. We have shown both theoretically and experimentally that the second order nonlinear optical response of metal surfaces is sensitive to the crystal structure of the surface layer. The absolute value and the anisotropy of the second order quadrupole optical susceptibility depend on the degree of lattice order. The application of SHG in reflection for studying the electronic structure of high-T_c superconductors seems to be even more attractive when using high temporal resolution.

REFERENCES

1. H. W. K. Tom, T. F. Heinz, Y. R. Shen, Phys. Rev. Lett. 51, 1983 (1983).
2. C. V. Shank, R. Yen, C. Hirlimann, Phys. Rev. Lett. 51, 900 (1983); H. W. K. Tom, G. D. Aumiller, C. H. Brito-Crutz, Phys. Rev. Lett. 60, 1458 (1988).
3. S. V. Govorkov, V. I. Emelyanov, N. I. Koroteev, G. I. Petrov, I. L. Shumay, V. V. Yakovlev, J. Opt. Soc. Am. B6, 1117 (1989).
4. S. A. Akhmanov, N. I. Koroteev, I. L. Shumay, in Nonlinear Optical Diagnostics of Laser-Excited Semiconductor Surfaces, V. S. Letokhov, C. V. Shank, Y. R. Shen, H. Walther eds. (Harwood Academic Publishers, 1989).
5. H. W. K. Tom, G. D. Aumiller, Phys. Rev. B33, 8818 (1986).
6. V. L. Shannon, D. A. Koos, G. L. Richmond, Appl. Opt. 26, 3579 (1987).
7. R. J. M. Anderson, J. C. Hamilton, Phys. Rev. B38, 8451 (1988).
8. S. A. Akhmanov, S. V. Govorkov, N. I. Koroteev, G. I. Petrov, I. L. Shumay, V. V. Yakovlev, Sov. Pisma v ZhETF 49, 527 (1989).
9. J. E. Sipe, G. I. Stegeman, in Surface Polaritons, V. M. Agranovich and D. L. Mills, eds. (North-Holland, Amsterdam, 1982), 661.; O. Keller, J. Opt. Soc. Am. B2, 367 (1985).
10. T. F. Heinz, M. M. T. Loy, W. A. Thompson, Phys. Rev. Lett. 54, 63 (1985).
11. J. E. Sipe, D. J. Moss, H. M. Van Driel, Phys. Rev. B35, 1129 (1986).
12. S. S. Jha, S. C. Warke, Phys. Rev. 153, 751 (1967).
13. V. Heine, in The Pseudopotential Concept, H. Ehrenreich, F. Seitz, D. Turnbull, eds., Vol. 24 of ser. Solid State Physics - Advances in Research and Applications (Academic Press, New York-London, 1970).
14. B. Kramer, phys. stat. solidi (B)47, 501 (1971).
15. V. Mizrahi, F. Suits, J. E. Sipe, U.J. Gibson, G. I. Stegeman, Appl. Phys. Lett. 51, 427 (1987).
16. F. Beech, S. Miraglia, A. Santoro, R. J. Angel, Phys. Rev. B35, 8778 (1987).
17. See, for example, Energy-Beam Solids Interactions and Transient Thermal Processing, D. K. Biegelsen, G. A. Rosgonyi, C. V. Shank eds. (Pittsburgh, 1985).
18. C. V. Shank, R. Yen, C. Hirlimann, Phys. Rev. Lett. 50, 454 (1983).
19. J. A. Kash, J. C. Tsang, J. M. Hyam, Phys. Rev. Lett. 54, 2151 (1985).
20. N. Bloembergen, A. M. Malvezzi, J. M. Liu, Appl. Phys. Lett. 45, 1019 (1984).
21. W. Dietel, E. Dopel, V. Petrov, C. Rempel, W. Rudolph, B. Wilhelmi, G. Marowsky, F. P. Schafer, Appl. Phys. B46, 183 (1988).

22. See, for example, A. Laubereau in Semiconductors Probed by Ultrafast Laser Spectroscopy, R. R. Alfano, ed. (Academic Press, New York, 1984).

23. A. Compaan, J. of Luminescence, $\underline{30}$, 425 (1985).

24. W. Rudolph, T. Shroeder, S. V. Govorkov, I. L. Shumay, Appl. Phys. A, accepted for publication.

SECOND-HARMONIC GENERATION IN OPTICAL FIBERS

Ulf Österberg

Thayer School of Engineering
Dartmouth College
Hanover, N.H. 03755

INTRODUCTION

Second-harmonic generation (SHG) in optical fibers was first reported in 1981 [1]. Since optical fibers are made out of glass, which has inversion symmetry, the SH light was believed to be caused by higher order effects such as electric quadrupole and magnetic dipole interactions. The low conversion efficiency ($\approx 10^{-5}$) reported in these early experiments made SHG in optical fibers a not very useful or interesting effect. This was, however, changed in October of 1985, when efficient SHG was observed in commercial single-mode telecommunication fibers (conversion efficiency $\approx 5\%$, input peak power ≈ 20 kW) [2]. The amount of green light generated from frequency-doubling a Nd:YAG laser in a 1-m long piece of fiber was sufficient to pump a dye laser [3]. Since then, even higher conversion efficiency has been reported [4].

The most noticeable characteristic of efficient SHG in optical fibers is that it does not appear immediately upon illumination, but grows exponentially with time during 1-10 hours depending on how the fiber is illuminated. This period of time is referred to as the preparation time. Typically, the initially very weak SH light grows 8-10 orders of magnitude before it saturates. The preparation of the fiber can be done in many different ways: one, by using intense fundamental light only, internal seeding ($P_{th} > 0.5$ kW, $T_{prep} \approx 2$-10 hours) [3]; two, using fundamental and harmonic light, external seeding ($P_{th} > 100$ W, $T_{prep} \approx 1$-2 hours) [5]; and three, using fundamental light and a dc-electric poling field [6]. In this paper, we will focus on the first two methods of preparation. The preparation alters the properties of the glass so that efficient SHG can take place. This alteration is not completely permanent; if the fibers are left in darkness for about 6 months the conversion efficiency drops by a factor of two. However, it takes only minutes of preparation to regain the loss in conversion efficiency.

It is the aim of this paper to present some of the various models put forward to date explaining the highly surprising observation of efficient SHG in optical fibers. The paper is organized in the following way: first, three different physical models for explaining a growing second-order susceptibility in an optical glass fiber will be discussed. Polarization and modal properties of the induced SH light is then treated, followed by some new ideas about the anomalous length dependence of the frequency-doubled light. Finally, we are also mentioning the unexpected observation of growing third-harmonic light in an optical fiber.

MODELS FOR A GROWING SECOND-ORDER SUSCEPTIBILITY

The first model proposed [7] is based upon a weak non-phase-matched SH signal, created via the quadrupole polarization, to initiate the self-writing of an axially periodic pattern of colour centres which are then postulated to lead to the growth of a $\chi^{(2)}$ grating. Phasematching, in other words, is obtained through a periodic $\chi^{(2)}$. The preparation time is accounted for by the slow formation of the colour centres at 532 nm. The concept of aligning colour centres of defects in glass by intense polarized light has been experimentally observed [8] and therefore this part of the model is plausible. However, the phase-matching scheme is not valid. The reason is that the non-phase-matched SH light has a spatial modulation $I_{SH} \approx \cos^{(2)}\Delta kz$, $\Delta k = k(2\omega) - 2k(\omega)$, leading to an induced $\chi^{(2)} \approx \exp[i(\Delta k/2)z]$, which will give no net transfer of energy from the fundamental light to the SH light over the length of the fiber.

This problem was overcome in the next model proposed [5]. Here it is proposed that the alignment of defects be accomplished by a dc-field induced through the interaction between fundamental and SH light in the fiber via the third-order nonlinear susceptibility.

$$P_{dc} \approx \chi^{(3)}(0,\omega,\omega,-2\omega)\, E_\omega E_\omega E_{2\omega}^* \cos(\Delta kz) \tag{1}$$

It was then postulated that this dc-field would break the inversion symmetry of the glass and produce an effective second-order susceptibility,

$$\chi^{(2)} \approx \gamma P_{dc} \tag{2}$$

where γ is an imaginary constant. This $\chi^{(2)}$ will produce quasi phase-matched SH light in the fiber. The main problem with this model was soon realized to be the magnitude of the induced dc-field [9]. For most experiments the induced dc-field would be on the order of 1-1000 mV/cm. These dc-fields exert smaller forces on the glass molecules than the randomizing forces from thermal fluctuations at room temperatures. Direct experimental tests of this model [10] later confirmed that it could not explain growth of SHG in optical fibers.

The last model we will describe is based upon photo-induced charge redistribution [11]. The hypothesis for this model assumes that the nonlinear polarization, producing SH light, arises from a third-order nonlinear susceptibility mixing the pump field with a static electric field E_{dc}.

$$P_{2\omega} \approx \chi^{(3)}(2\omega,\omega,\omega,0)\, E_\omega E_\omega E_{dc} \tag{3}$$

The dc-field is produced by the migration of electrical charges as defects in the glass are ionized by the intense light in the fiber. In ref. 11 an expression for the induced dc-field is obtained by using the "hopping" model, well known in photorefractive work [12],

$$E_{dc} = -\frac{kT}{e}\frac{\nabla I}{I} \tag{4}$$

where I is the total light intensity in the fiber. DC-fields as large as 1000 V/cm have been predicted in the fiber core from this model. The dc-field does not have this magnitude from the beginning, however, but grows with time, as more light pulses pass through the fiber displacing more and more electrical charges. So far, no experiments have been performed to test all aspects of this new model.

Common for all these models is the assumption of defects being present in the fiber. Electron spin resonance (ESR) measurements in fibers before and after preparation indicate that a defect called Ge E´ is important for the growth of the SH signal [13]. An interesting observation in these ESR measurements was that Ge E´ defects already present in the fiber before illumination (due to mechanical and chemical processes) did not contribute to the SH signal, but only those defects generated by the light itself.

POLARIZATION AND MODAL PROPERTIES

Polarization measurements of the intrinsic and induced SH light are important for determining the tensor elements of the involved second-order nonlinear susceptibilities. In performing these measurements it is important to remember that even if care is taken to use a fiber that preserves the polarization well and the incident light is linearly polarized, the light propagating in the fiber core will inevitably have three polarizations due to the properties of the waveguide. Typically, if x is the direction of the incident light, one finds that the ratio E_x/E_y can vary between 0.0001 and 0.01, and E_x/E_z between 0.01 and 0.1. In addition to this, there are intensity-dependent contributions to the various polarizations. Taking all this into account we find, e.g. [14],

$$\frac{\chi^{(2)}_{xxx}}{\chi^{(2)}_{yxx}} = 30 \pm 10 \quad \text{and} \quad \frac{\Gamma^{(2)}_{xxxx}}{\Gamma^{(2)}_{yxyx}} = 2.4 \pm 1.2 \tag{5}$$

where $\chi^{(2)}$ refers to the induced second-order susceptibility and $\Gamma^{(2)}$ to the intrinsic. For the induced $\chi^{(2)}$ the fiber was prepared with both the fundamental photons in the x-direction. $\Gamma^{(2)}$ originates from a non-local nonlinearity, phenomenologically described by a term like:

$$P_i = \Gamma^{(2)}_{ijkl} E_j \nabla_k E_l \tag{6}$$

There are a couple of interesting observations to be seen in eq. 5. The ratios of the induced $\chi^{(2)}$ are obtained for many different pieces of fiber and at different power levels of preparation. No correlation was seen between the measured $\chi^{(2)}$ ratio and whether the fiber had been prepared at a high or low power level or if the fiber was highly polarization preserving or not (could vary over two orders of magnitude). Furthermore, for these particular fibers prepared in the x-direction (being polarization preserving to better than 1000:1), the SH light in the x-direction grew 8 orders of magnitude and at the same time the SH light in the y-direction grew just less than 7 orders of magnitude, showing the complex tensor nature of the growth process. During these measurements it was also shown that a fiber could be prepared at first in the x-direction and thereafter in the y-direction (with the same conversion efficiency for the SH light in the y-direction as in the x-direction) without affecting the conversion efficiency in the x-direction. The conclusion from this is that saturation of the overall conversion efficiency for SH light in an optical fiber is not due to lack of defects to interact with during the preparation process.

The ratios between the different tensor elements in eq. 5 cannot be measured directly. The reason is that the SH light can be in different modes for different polarization directions, which affects the conversion efficiency. The most general expression for the ratio between two tensor elements is

$$\frac{\chi^{(2)}_{ijk}}{\chi^{(2)}_{lmn}} = \frac{f_{lmn}}{f_{ijk}} \left[\frac{I^{\omega}_m \cdot I^{\omega}_n}{I^{\omega}_j \cdot I^{\omega}_k} \right]^{1/2} \left[\frac{I^{2\omega}_i}{I^{2w}_l} \right]^{1/2} \tag{7}$$

where

$$f_{ijk} = \frac{1}{N} \int_{A_{\infty}} P^{2\omega} \times H^{2\omega} \, dxdy \tag{8}$$

and H is the magnetic field for the SH light. So, the overlap integral f has to be calculated for each ratio to be determined. Significant for SHG in optical fibers is that in most reported cases the saturated SH light is not in the same mode as the initial SH light.

The particular mode the SH light is in, as can be seen from eq. 8, gives information about what type of nonlinearity could have caused it. All fibers reported so far have had the initial SH light in an asymmetric mode, either LP31 or LP11. This means that by examining the angular part of the overlap integral in eq. 8, the initial SH light has to come from either a quadrupole interaction (bulk and interface) or an electric dipole interaction (interface). It is, however, extremely difficult to separate the quadrupole contribution from the electric-dipole contribution [15]. Experiments on fibers with different compositions indicate that the intrinsic SH light is from quadrupole interactions at the core-cladding interface [14]. For the saturated SH light, it has been reported to occur in asymmetric as well as symmetric modes. This can only be explained by electric-dipole interactions being the source for the saturated SH light.

LENGTH DEPENDENCE-ABSORPTION

The fact that the mode for the SH light changes during the growth process indicates that the refracting index is being altered. We propose that this is related to a significant change in the absorption for the SH light.

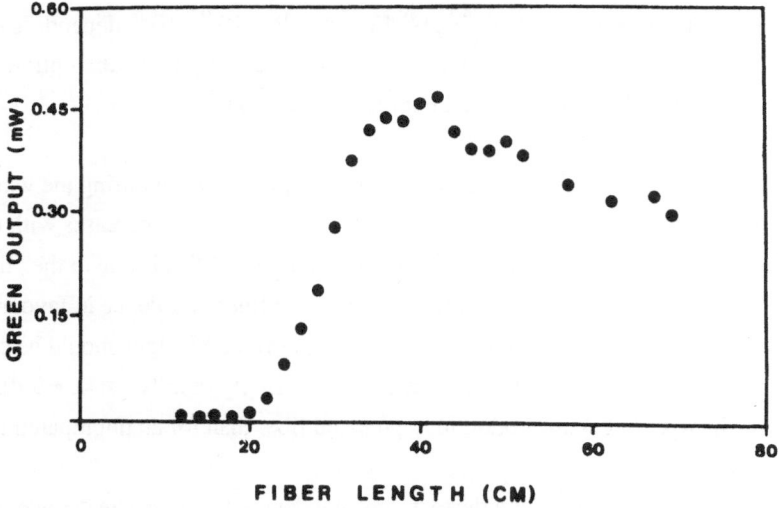

Fig. 1. Second-harmonic light as a function of distance into the fiber [16].

The reason for incorporating absorption into the analysis of SHG in optical fibers is related to the anomalous length dependence, fig. 1 [16]. The curve in fig. 1 was measured for the saturated SH light after ~10 hours of internal seeding. The late onset (~23 cm) of strong growth of the SH signal is characteristic for internal seeding. A thorough study of the length dependence [17] shows that the onset of strong growth of the SH signal is a function of the amount of SH light used to prepare the optical fiber. The more SH light used in the preparation, the earlier the induced SH light starts to grow strongly. All these observations can be accounted for by incorporating two-photon absorption into the equation for the length-dependence,

$$\frac{dE_{2\omega}^{(z)}}{dz} = i \cdot \gamma_1 \cdot \chi_{eff}^{(2)} \cdot E_\omega \cdot E_\omega - \gamma_2 \chi^{(3)} |E\omega|^2 E_{2\omega} \qquad (9)$$

γ_1 and γ_2 are well-known constants for second- and third-order processes, $\chi_{eff}^{(2)}$ has the overlap integral factor incorporated into it, and $\chi^{(3)}$ is the imaginary part of the third-order nonlinear susceptibility.

The origin of the two-photon absorption is surmised to be due to a resonance in the GeO molecules in the core of the fiber. These molecules have a resonant transition at 364 nm ($\omega + 2\omega \rightarrow 355$ nm) which is two-photon allowed [18]. Integrating eq. 9 we can obtain an expression for the distance it takes for the SH light to grow to 1/10 of its maximum value,

$$Z_{1/10} = \frac{0.38}{A} - B \cdot E_{2\omega}(0) \qquad (10)$$

where z is measured in meters and $A = \frac{3\pi}{4\lambda n} \chi^{(3)} |E\omega|^2$ and $B \approx 10^{-7} - 10^{-8}$, depending on the fiber. From fig. 1, where $z_{1/10} \approx 23$ cm, we can calculate the two-photon absorption coefficient $\beta \approx 0.3$ cm/GW. This is a small but realistic value [19].

Experimentally, what we find is that the distance $z_{1/10}$ is constant during the whole growth process, but that the slope of the exponentially growing region increases with the growth of the SH light [17]. As mentioned earlier we suggest that this is due to the "linear" absorption increasing with increasing visible SH light in the fiber. Evidence in favour of this can be seen from fig. 1. For the saturated regime ($z \gtrsim 40$ cm) the SH light should be constant ($\alpha_{loss} \approx 0.016$ dB/m) but is instead seen to decrease surprisingly rapidly ($\alpha_{loss} \approx 9$ dB/m), indicating a much larger linear absorption for a prepared fiber than for an unprepared fiber.

In order to match our model with the experimental data in fig. 1 we can deduce that the linear absorption coefficient α has to depend on the SH light intensity to the power of 3 $\left(\alpha_{2\omega} \propto I_{2\omega}^3\right)$. The validity of this has to be studied more extensively.

In eq. 9 we have written down the length dependence as if it were phase-matched. The justification for this is that the model is based upon a two-photon resonant transition between the fundamental and harmonic light, which will lock their phases together [20]. A consequence of this is that the coherence length is no longer important; instead the optimum length for our nonlinear process is a function of the induced absorption.

The assumption that the resonant transition provides full or partial locking of all the involved phases does not rule out a periodic $\chi^{(2)}$ for additional phase-matching. As a matter of fact, it has been experimentally shown that indeed a periodic $\chi^{(2)}$ is induced in the fiber [21].

CONCLUSION

We have tried to give a brief account for the models proposed to date of explaining SHG in optical glass fibers. Based on experiments and calculations, we have argued that SHG in optical fibers starts from electric quadrupole interactions at the core-cladding interface but that the induced SH light is from electric dipole interactions. We have also proposed a new model, based on absorption, for explaining the anomalous length dependence and mode-coupling. The question of what makes the $\chi^{(2)}$ grow in an optical glass fiber still remains to be properly explained, though.

To further confuse the issue, we would like to end our conclusion by mentioning the recent discovery of a growing third-harmonic signal in an optical fiber [22]. It was observed that the third-harmonic signal could grow exponentially with time, at the expense of the SH signal, almost 2 orders of magnitude.

It is obvious that the observations of growing harmonic signals in optical fibers is very interesting and that the physics behind it is far from understood.

REFERENCES

1. Y. Sasaki and Y. Ohmori, Phase-matched sum-frequency light generation in optical fibers, Appl. Phys. Lett. 39:466 (1981).
2. U. Österberg and W. Margulis, Efficient second harmonic generation in an optical fiber, Conference on Laser and Electro-optics, San Francisco, 1986. Paper WBB2.
3. U. Österberg and W. Margulis, Dye laser pumped by Nd:YAG laser pulses frequency doubled in an optical fiber, Opt. Lett. 11:516 (1986).
4. M.C. Farries, Laser Focus 24:12 (1988).

5. R. H. Stolen and H.W.K. Tom, Self-organized phase-matched harmonic generation in optical fibers, Opt. Lett. 12:585 (1987).

6. M.V. Bergot, M.C. Farries, M.E. Fermann, L. Li, L.J. Poyntz-Wright, P. St. J. Russell and A. Smithson, Generation of permanent optically induced second-order nonlinearities in optical fibers by poling, Opt. Lett. 13:592 (1988).

7. M.C. Farries, P. St. J. Russell, M.E. Fermann and D.N. Payne, Second harmonic generation in an optical fiber by self-written $\chi^{(2)}$ grating, Electron. Lett. 23:322 (1987).

8. J.H. Stathis, Selective generation of oriented defects in glasses: Applications to SiO_2, Phys. Rev. Lett. 58:1448 (1987).

9. V. Mizrahi, U. Österberg, J.E. Sipe and G.I. Stegeman, Test of a model of efficient second-harmonic generation in glass optical fibers, Opt. Lett. 13:279 (1988).

10. V. Mizrahi, U. Österberg, C. Krautschik, G.I. Stegeman, J.E. Sipe and T.F. Morse, Direct test of a model of efficient second-harmonic generation in glass optical fibers, Appl. Phys. Lett. 53:557 (1988).

11. D.Z. Anderson, Efficient second-harmonic generation in glass fibers: The possible role of photoinduced charge redistribution, Proc. SPIE 1148:paper 20 (1989).

12. P. Günter and J-P. Huignard, *Photoretractive Materials and their Applications I: Fundamental Phenomena*, Springer-Verlag, Heidelberg (1988).

13. T.E. Tsai, M.A. Saifi, E.J. Friebele, D.L. Griscom and U. Österberg, Correlation of defect centers with second-harmonic generation in Ge- and Ge-P-doped silica core single-mode fibers, Opt. Lett. 14:1023 (1989).

14. U. Österberg and R.I. Lawconnell, Tensor properties of intrinsic and photoinduced second-order nonlinear susceptibilities in glass optical fibers, submitted to JOSA B.

15. T.F. Heinz, Second-order nonlinear optical effects at surfaces and interfaces, in: *Nonlinear Electromagnetic Phenomena at Surfaces*, H. Ponath and G.I. Stegeman, eds., Elsevier, Amsterdam (1989).

16. U. Österberg and W. Margulis, Experimental studies on efficient frequency-doubling in glass optical fibers, Opt. Lett. 12:57 (1987).

17. B. Batdorf, C. Krautschik, U. Österberg, G.I. Stegeman, J.W. Leitch, J. R. Rotgé and T.F. Morse, Study of the length dependence of frequency-doubled light in optical fibers, Opt. Comm. 73:393 (1989).

18. M.J. Yuen, Ultraviolet absorption studies of germanosilicate glasses, Appl. Opt. 21:136 (1982).

19. V. Mizrahi, K.W. De Long, G.I. Stegeman, M.J. Andrejco and M.A. Saifi, Two-photon absorption as a limitation to all-optical switching, Opt. Lett. 14:1140 (1989).

20. A.I. Maimistov, L.R. Malov and E.A. Manykin, Harmonic generation under two-photon resonance conditions, Sov. J. Quant. El. 5:375 (1975).

21. M.C. Farries, Efficient second-harmonic generation in an optical fibre, Proc. Coll. on Non-linear Optical Waveguides, London, IEEE, 1988/88.

22. U. Österberg, Growth of the third-harmonic signal in an optical fiber, submitted to Electronics Letters.

NONLINEAR OPTICAL PROBES OF GLASSY POLYMERS

M. G. Kuzyk*, L. A. King, and K. D. Singer

AT&T Bell Laboratories
P. O. Box 900, Princeton, NJ, USA 08540

ABSTRACT

Theory and experiments show that the second order nonlinear optical susceptibility is a sensitive probe of polar orientational order. By incorporating optically nonlinear molecules in glassy polymers and orienting the dipole moments using an external field, it is possible to deduce the molecular orientational distribution from measurements of the nonlinear optical susceptibility. Presented here are results of such measurements of a poled dye-doped polymer film under external uniaxial stress measured using polarized second harmonic generation. The ratio of the tensor components of the second order susceptibility, $\chi_{113}^{(2)}/\chi_{333}^{(2)}(\equiv a)$, is shown to vary with poling field and stress. Tensor ratios of $a=0.33$ and $a=0.7$ are obtained for stresses of zero and $3.71\times10^7\,dyne/cm^2$, respectively, and fall into a regime of order unlike other material classes such as poled liquid crystals and Langmuir-Blodgett films.

INTRODUCTION

The large electronic second order nonlinear optical susceptibilities of organic molecules can be built into bulk systems by imparting polar orientational order to an ensemble of nonlinear optical dopants. Noncentrosymmetric orientational order is required for second order nonlinear optical processes and has been demonstrated in noncrystalline materials, such as molecule-doped liquid crystals and polymer glasses, using electric field poling.[1] [2] [3] [4] A relationship between the second order molecular tensor susceptibility and the bulk tensor susceptibility of a polymer doped with optically nonlinear molecules has been calculated in the mean field approximation for poled materials under the influence of other bulk ordering forces.[5] Also, the effect of internal ordering through intermolecular interactions, as well as electric field ordering, has previously been reviewed. [6] Here, the effect of externally applied uniaxial stress is considered. The tensor properties of the second order susceptibility of a film poled under uniaxial stress in the direction of the poling field are determined and compared with those attainable in other classes of oriented materials, such as Langmuir-Blodgett films and poled liquid crystal systems. Findings suggest that uniaxially stressed materials can access tensor regimes not normally observed in other oriented materials. The tensor ratios are used to deduce the first four nontrivial order parameters, $<P_1>$, $<P_2>$, $<P_3>$, and $<P_4>$, thereby determining the orientational distribution function of a poled polymer film under stress. This method thus represents a powerful probe of structure/property relationships in poled guest-host polymer nonlinear optical materials.

---

*   Present address, Dept. of Physics, Washington State University, Pullman, WA 99164-2814

THEORY

In the dipole approximation and for instantaneous response, the bulk polarization, **P**, is defined as the dipole moment per unit volume and may be expressed as a power series in the electric field, **E**:

$$P_i(t) = \chi_i^{(0)} + \chi_{ij}^{(1)}(t)E_j(t) + \chi_{ijk}^{(2)}E_j(t)E_k(t) + \dots \tag{1}$$

Summation over repeated indices is implied and the tensor quantity $\chi^{(n)}$ is the *nth* order susceptibility. The molecular polarization is similar to the bulk polarization in that it too may be expanded in a power series in the electric field,

$$p_I = \mu_I^0 + \alpha_{IJ}(t)F_j(t) + \beta_{IJK}(t)F_J(t)F_K(t) + \dots, \tag{2}$$

where F_m is the local electric field, μ_I^0 is the molecular ground state dipole moment, α_{IJ} is the linear polarizability, and β_{IJK} is the molecular second order nonlinear optical susceptibility or hyperpolarizability. For weakly interacting molecules, the microscopic polarizability can be related to the macroscopic polarizability through the thermodynamic average of the microscopic polarization, $<p_I(t)>$,

$$P_i(t) = N<p_I(t)>_i, \tag{3}$$

where N is the number density of molecules. Under the assumptions leading to Eq. (3), bulk measurements can be used to determine microscopic properties. A bulk centrosymmetric material will not produce second harmonic light in the dipole approximation; however, a polar aligned material is noncentrosymmetric and will produce second harmonic light. The magnitude of the bulk nonlinearity depends on the degree of alignment and the magnitude of the molecular optical nonlinearity.

The thermodynamic average in Eq. (3) requires knowledge of the orientational distribution function, G. This distribution is a function of three Euler angles and yields the probability density of finding a molecule oriented in a given direction. For a material in the presence of an electric field, $\mathbf{E}_p$, and other orientational forces as described by a mean field potential, U_T, the distribution function can be expressed as,

$$G(\Omega) = \frac{\exp\left[-\dfrac{U_T}{kT}\right]}{\int d\Omega \exp\left[-\dfrac{U_T}{kT}\right]}, \tag{4}$$

where $U_T = U_E + U_p$, U_E is the potential due to the poling field, and U_p is the potential which accounts for axial forces. Assuming a one dimensional molecule and bulk azimuthal symmetry, the orientational distribution function, G, can be expressed as a function of only the polar angle θ. Expressed in terms of measurable parameters that describe the bulk order and as a power series in complete and orthogonal functions, the orientational distribution function is,

$$G(\theta) = \sum_{l=0}^{\infty} \frac{2l+1}{2} A_l P_l(\cos\theta), \tag{5}$$

where the A_l's are the order parameters and where the orthogonal functions are chosen to be the Legendre polynomials. The orthogonality of the Legendre polynomials allows the order parameters to be expressed as

$$<P_l> \equiv A_l = \int_{-1}^{+1} d(\cos\theta)G(\theta, \mathbf{E}_p)P_l(\cos\theta), \tag{6}$$

where $P_l(\cos\theta)$ is the *lth* Legendre polynomial.

The bulk second order susceptibility in the laboratory reference frame, $\chi_{ijk}^{(2)}$, can be related to the molecular hyperpolarizability in the molecular reference frame, β_{IJK}, through Eqs. (1), (2) and (3),

$$\chi_{ijk}^{(2)} = N<\beta_{IJK}^*>_{ijk}, \tag{7}$$

where β_{IJK}^* is the local field-corrected hyperpolarizability. Thus, the bulk susceptibility in Eq. (7) is dependent only on the number density, molecular hyperpolarizability and the orientational order as described by the distribution function.

The lowest four nontrivial order parameters, $<P_1>$ through $<P_4>$, are evaluated by expanding the distribution function while successively taking into account various contributions to the total potential. First, assuming the potential is undetermined, the bulk second order susceptibility tensor for an optically nonlinear one dimensional molecule embedded in an optically linear glassy polymer matrix is related to the first and third order parameters of the dopant molecules by[6] [7]

$$\chi^{(2)}_{333} = N\beta^*_{zzz}\left[\frac{3}{5}<P_1> + \frac{2}{5}<P_3>\right], \tag{8}$$

and

$$\chi^{(2)}_{113} = \chi^{(2)}_{131} = \chi^{(2)}_{311} = N\beta^*_{zzz}\left[\frac{1}{5}<P_1> - \frac{1}{5}<P_3>\right], \tag{9}$$

where β^*_{zzz} is the only nonvanishing local field-corrected hyperpolarizability. The second order susceptibility depends only on the first two odd order parameters given by Eqs. (8) and (9), and second harmonic generation measures these parameters directly. Second, if the molecules possess axial order above the glass transition temperature and are free to rotate, electric field poling results in a bulk second order susceptibility that is described by the second and fourth order parameters,[5]

$$\chi^{(2)}_{333} = N\beta^*_{zzz}\frac{m^*_z E_p}{kT}\left[\frac{1}{5} + \frac{4}{7}<P_2> + \frac{8}{35}<P_4>\right], \tag{10}$$

and

$$\chi^{(2)}_{113} = \chi^{(2)}_{131} = \chi^{(2)}_{311} = N\beta^*_{zzz}\frac{m^*_z E_p}{kT}\left[\frac{1}{15} + \frac{1}{21}<P_2> - \frac{4}{35}<P_4>\right], \tag{11}$$

where $1/kT$ is the Boltzmann factor at the poling temperature, E_p is the magnitude of the poling field, and m^*_z is the magnitude of the only non-vanishing local field corrected molecular dipole moment. Eqs. (10) and (11) show that the zero-field even order parameters will affect the odd order parameters when subjected to electric field poling. If the energy of the electric field is small compared to $1/kT$, as in this work, then the even order parameters will not change appreciably with poling and the even order parameters of the poled material will be given by the zero-field values.[8] When a material is poled above its glass transition temperature, T_g, the nonlinear optical susceptibility given to the material can be "frozen-in" by cooling below T_g in the presence of the field; this nonlinearity will be given by Eqs. (10) and (11), substituting T_g for T. Last, the order parameters may be expressed in terms of both the known electric field potential and the known stress potential. Since polarized second harmonic measurements determine the ratio of $\chi^{(2)}_{113}/\chi^{(2)}_{333}$, or a, second harmonic measurements can be used, along with Eqs. (8) and (9), to calculate the ratio of the third to the first order parameters,

$$\frac{<P_3>}{<P_1>} = \frac{1-3a}{1+2a}. \tag{12}$$

A film in the presence of only a poling potential results in $a=1/3$; therefore, $<P_3>$ vanishes in accordance with previous work.[9] The remaining order parameters, expressed in terms of $<P_1>$ by using Eqs. (8)-(11), are

$$<P_2> = \frac{1}{2}\left[\frac{<P_1>}{\left[\dfrac{m^*E_p}{3kT}\right]} - 1\right], \tag{13}$$

and

$$<P_4> = \frac{1}{4}\left[\frac{<P_3>}{\left[\dfrac{m^*E_p}{7kT}\right]} - 3<P_2>\right]. \tag{14}$$

Because the uniaxial stress felt by the material is much greater than the poling force, the even order parameters are the values that result from the independently acting stress, while the odd order parameter values result from poling. Knowledge of $<P_1>$ and $m^* E_p/kT$ is not sufficient to determine the values of the first four nontrivial order parameters, one must also assume the form of the stress potential, the force acting on the molecules from the uniaxial stress in terms of a potential. The simplest choice for the stress potential, U_p, is,

$$U_p = b(T_s)P_2(\cos\theta), \tag{15}$$

where $b(T_s)$ is a function of stress and $P_2(\cos\theta)$ is the second Legendre Polynomial. If $b(T_s)$ is chosen to be an adjustable parameter at fixed stress, Eqs. (6), (8)-(11), and (15) can be used to determine the values of $b(T_s)$ and $<P_1>$ through $<P_4>$. This choice of potential has been shown to result in order parameters that approximate the distribution function of a stressed polymer.[10] The assumed form for the stress potential is convenient in that it describes both poled polymers and liquid crystals under strain; therefore, comparisons between the two materials classes can be made directly through the sign of b.

Under the assumption that the poling potential in this work is small relative to thermal energies, the distribution function of a poled system, described by Eq. (4), can be expanded to first order in the poling field. Conversely, the stress applied to the film is large enough to require that the distribution function be expanded to fifth order in the stress potential. The order parameters can be determined by substituting the expansion of the distribution function in the stress potential into Eq. (4) and setting the result equal to Eq. (5). Use of another relation obtained when Eq. (15) is substituted in the same expansion allows the ratio of the third to the first order parameter, α, to be given by,

$$\alpha \equiv \frac{<P_3>}{<P_1>} = \frac{3}{5}\frac{a_3}{a_1} \tag{16}$$

where

$$a_3 = -\frac{3}{5}\left[\frac{b}{kT}\right] + \frac{1}{5}\left[\frac{b}{kT}\right]^2 - \frac{7}{110}\left[\frac{b}{kT}\right]^3 + \frac{19}{1430}\left[\frac{b}{kT}\right]^4 - \frac{41}{17160}\left[\frac{b}{kT}\right]^5 \tag{17}$$

and

$$a_1 = 1 - \frac{2}{5}\left[\frac{b}{kT}\right] + \frac{11}{70}\left[\frac{b}{kT}\right]^2 - \frac{4}{105}\left[\frac{b}{kT}\right]^3 + \frac{73}{9240}\left[\frac{b}{kT}\right]^4 - \frac{79}{60060}\left[\frac{b}{kT}\right]^5. \tag{18}$$

From Eq. (16), we see that α depends only on the stress potential through b and vanishes under zero stress. Second harmonic generation measurements can be used to determine α, and α can then be used to determine the coefficient in the stress potential. Once the parameter b/kT is determined, Eqs. (4), (5), and (18) can be used to deduce the first order parameter,

$$<P_1> = \frac{2}{3}\frac{m^* E_p}{kT}\frac{a_1}{A_n}, \tag{19}$$

where A_n, the normalization constant, is given as,

$$A_n = \left[\frac{2\pi kT}{3b}\right]^{1/2} \exp\left[\frac{b}{2kT}\left\{1 - \frac{1}{3}\left[\frac{m^* E_p}{b}\right]^2\right\}\right]$$
$$\times \left[g\left[\left[\frac{3b}{kT}\right]^{1/2}\left\{\frac{m^* E_p}{3b} - 1\right\}\right] - g\left[\left[\frac{3b}{kT}\right]^{1/2}\left\{\frac{m^* E_p}{3b} + 1\right\}\right]\right], \tag{20}$$

where the function $g(x)$ is the area under the normal distribution between x and infinity and is of the form, [11]

$$g(x) = \frac{1}{(2\pi)^{1/2}}\int_x^\infty \exp\left[-x'^2/2\right]dx'. \tag{21}$$

Eqs. (12)-(14), (19), and (20) can then be used to deduce the remaining order parameters, $<P_2>$ through $<P_4>$. Once the order parameters are determined, they can be used to generate an approximation of the distribution function of the nonlinear species.

In order to relate the second harmonic generation results to the second order susceptibility, it is essential to know the birefringence of the film poled under stress. The birefringence values listed in Table 1 for the measured films include only the ordered dopant contribution since the contribution from the polymer is comparable to that of the dopant and thus, may be ignored when calculating approximate values of the coherence lengths.

Table 1. Results of second harmonic measurements. Film 1 is poled without stress and film 2 is poled under stress.

Film #	a	$\dfrac{\mu^* E_p}{kT}$	$\dfrac{b}{kT}$	$<P_1>$	$<P_2>$	$<P_3>$	$<P_4>$	Δn
1	0.33±0.05	0.50	0.00±0.02	0.16±0.02	-0.020±0.003	0.00±0.02	0.015±0.006	0.0014±0.0002
2	0.7±0.1	0.20	1.2±0.3	0.041±0.004	-0.20±0.03	-0.019±0.001	-0.02±0.01	-0.0043±0.0007

EXPERIMENTAL

Measurements of second harmonic light generated from a material are used to determine the ratio of the third to the first order parameters, α. The angular dependence of the second harmonic is measured as a function of incident angle for both $\hat{s}$- and $\hat{p}$-polarized fundamental and $\hat{p}$-polarized second harmonic.[10] The ratio of the second harmonic power for the two input polarizations can be used to determine a, the tensor ratio of the second order susceptibility, and is related to the order parameter ratio, α, through Eq. (12).

The nonlinear material used in these experiments is Disperse Red 1 (DR1) dye dissolved in the glassy polymer poly(methyl methacrylate) (PMMA). Films are fabricated by spin coating the material on glass slides appropriately patterned with an Indium Tin Oxide (ITO) coating to serve as the transparent electrodes. A sandwich structure is formed by placing two such slides together with their surfaces in contact and applying uniaxial pressure to the structure at temperatures well above the glass transition temperature of the doped system ($T_g \sim 95^{\circ}C$). Processing parameters of the measured films are shown in Table 2.

Table 2. Film processing parameters for PMMA doped with Disperse Red 1 dye.

Film #	Dye Number Density N ($10^{20}/cm^3$)	Thickness (μm)	Poling Field $E_p(MV/cm)$	Stress $\sigma(dyne/cm^2 \times 10^7)$
1	2.42	4.0	0.60	0.00
2	1.08	4.9	0.25	3.71

Film 1 is not a stressed structure; it consists of DR1/PMMA coated on a single substrate and has been previously discussed. Film 2 is a stressed structure formed as described above. The temperature is raised well above the transition temperature to $150^{\circ}C$ to allow bonding between the two film surfaces and the electric field is applied while the temperature is ramped down. It is assumed that the effects of the electric field and stress are frozen in below the transition temperature. Both the electric field and stress are removed after the sample has cooled to room temperature.

RESULTS AND DISCUSSION

Table 1 shows the results of the second harmonic measurements for the two films. The uncertainty in the measurements of second harmonic was found to be less then 4%, and the uncertainties of the derived quantities are determined from the uncertainty of the power ratio, which is assumed to be ±5%. The stressed film's tensor ratio, a, is substantially larger than that of the stress-free film. Tensor ratios of a=0.33±0.05 and a=0.7±0.1 for films 1 and 2 result in stress parameters of b/kT=0.0±0.2 and b/kT=1.2±0.3 respectively. The order parameters were calculated from this data and used to determine the orientational distribution function. Figure 1 shows a polar plot of the resulting distribution functions. Film 1, the stress-free poled film, shows a skewed distribution as expected for an electric field-induced polar oriented ensemble. Film 2, the film formed under stress, shows a depletion in oriented molecules in the poling direction, relative to the stress-free film, while still possessing the same degree of asymmetry necessary to exhibit second order nonlinear optical processes. The large value of the χ_{113} component relative to the χ_{333} component is reasonable considering the depletion of the orientational distribution function perpendicular to the film plane.

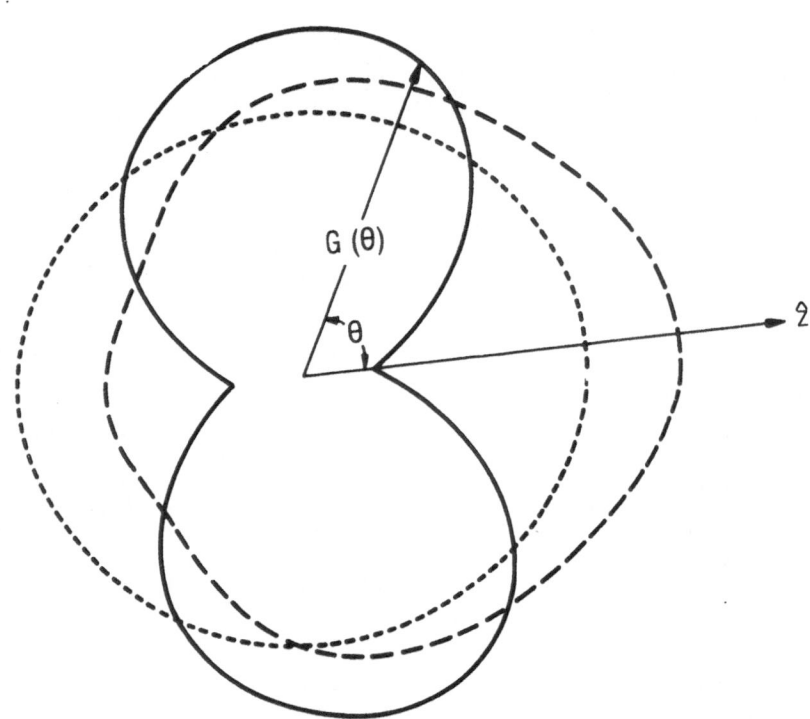

Figure 1. Polar plot of the distribution function, $G(\theta)$, as a function of the polar angle, θ. The length of the vector from the origin to the curve shows the magnitude of the distribution function in that direction. The curves shown are for the isotropic film (dotted curve), the poled film with no stress (dashed curve) and the poled film under stress (solid curve).

      Microwave scattering studies have been previously applied to polymers that are stressed by stretching to determine the orientational distribution function.[12] This technique results in a distribution that is similar to the one obtained with polarized second harmonic generation of poled and stressed films when the effect of the electric field is subtracted. The agreement between the nonpolar ordering of the two measurements shows that the nonpolar stress potential is adequately described. Although systematic errors can contribute to the uncertainties listed in Table 1, the shape of the orientational distribution is not very sensitive to such uncertainties.

      Figure 2 shows an overview of the tensor ranges of Langmuir-Blodgett films, poled liquid crystals, and poled polymer films under stress as calculated from Eqs. (8) through (11) and as determined from second harmonic generation. Most of the tensor ratios of the Langmuir-Blodgett films and the poled liquid crystalline systems measured fall within the range of $0<a<0.33$, whereas the tensor ratio of the poled polymer films under stress can theoretically achieve any ratio between $0.33<a<\infty$. Notice that film 2 falls within this range and that poled polymer films under stress can thus possess tensor ratios not normally achieved in other ordered systems such as Langmuir-Blodgett films and poled liquid crystals.

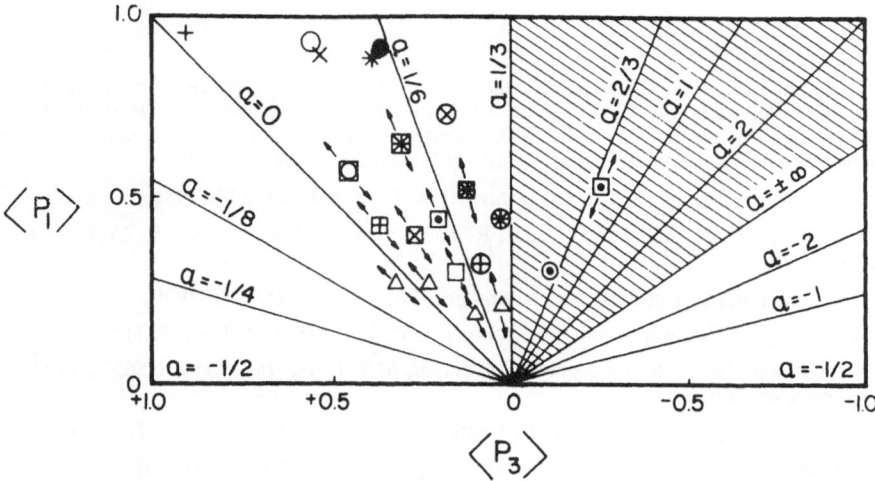

Figure 2. Survey of material order parameters $\langle P_1 \rangle$ and $\langle P_3 \rangle$. Lines represent constant values of a. The mesogenic systems are: 5CB as measured by two photon dichroism (solid point in square), MBBA as measured by Raman scattering (triangles) and third harmonic (asterisk in square), BBCA-MBBA as measured by Raman scattering (open circle in square), 40.8 as measured by Raman scattering in Smectic B phase (plus sign in square), Smectic A phase (square), and Nematic phase (cross in square).[6] Note that arrows show the direction of the poling field dependence ($a=constant$). The Langmuir-Blodgett systems are monolayers of SDNS (solid point), MC (plus sign), HC (cross), NS (asterisk), and HC/NS (open circle) and multilayers of COOH (open circle with plus), SO_2NH_2 (open circle with cross), $SO_2N(C_2H_5)_2$ (open circle with asterisk), and PS (open circle with dot).[6] Shaded area is region attainable with poled films under stress.

CONCLUSION

A method for controlling the ratio of the tensor components of the second order susceptibility, *a*, with stress was demonstrated and confirmed. The first four nontrivial order parameters were estimated and used to determine the orientational distribution function, which was consistent with the expected compression of molecules oriented in the film plane normal to the direction of the applied uniaxial stress. The tensor ratio was found to lie in a region not normally achieved by other ordered systems such as poled liquid crystals and Langmuir-Blodgett films.

ACKNOWLEDGEMENTS

We thank R. Moore for helpful discussions, W. Holland for refractive index measurements, J. Sohn and D. Fish for films, and H. Zahn for poling samples.

REFERENCES

1. G. R. Meredith, J. G. Vandusen, and D. J. Williams, "Characterization of Liquid Crystalline Polymers for Electro-optic Applications" in *Nonlinear Optical Properties of Organic and Polymeric Materials*, D. J. Williams, ed., ACS Symposium Series No. 233 (American Chemical Society: Washington, D.C., 1983), p 109.

2. K. D. Singer, J. E. Sohn, and S. J. Lalama, *Appl. Phys. Lett.* **49**, 248 (1986).

3. H. Ringsdorf, H. W. Schmidt, G. Baur, R. Kiefer, and F. Windscheid, *Liq Cryst. (GB)* **1**, 319 (1986).

4. E. E. Havinga and P. van Pelt, *Ber. Bunsenges. Phys. Chem.* **83**, 816 (1979).

5. K. D. Singer, M. G. Kuzyk, and J. E. Sohn, *J. Opt. Soc. Am. B* **4**, 968 (1987).

6. J. D. LeGrange, M. G. Kuzyk, and K. D. Singer, *Mol. Cryst. Liq. Cryst.* **150b**, 567 (1987) and references therein.

7. T. Rasing, Y. R. Shen, M. W. Kim, P. Valint, and J. Bock, *Phys. Rev. A* **31**, 537 (1985).

8. C. P. J. M. van der Vorst and S. J. Picken, *J. Opt. Soc. Am. B* **7**, 324 (1990).

9. K. D. Singer, M. G. Kuzyk, and J. E. Sohn, "Orientationally Ordered Electrooptic Materials" in *Nonlinear Optical and Electroactive Polymers*, P. N. Prasad and D. R. Ulrich eds. (Plenum: New York, 1988), p. 189.

10. M. G. Kuzyk, K. D. Singer, and L. A. King, *J. Opt. Soc. Am. B* **6**, 42 (1989) and references therein.

11. G. E. P. Box, W. G. Hunter, and J. S. Hunter, *Statistics for Experimenters* (Wiley: New York, 1978), p. 630.

12. S. Osaki, *Polym. Journal* **19**, 821 (1987).

PHOTOLUMINESCENCE OF HOT ELECTRONS AND SCATTERING PROCESSES IN QUANTUM-

WELL STRUCTURES

B. P. Zakharchenya, P. S. Kop'ev, D. N. Mirlin, I. I. Reshina,
V. F. Sapega, and A. A. Sirenko

A. F. Ioffe Physico-Technical Institute, Academy of Sciences of
the USSR, Politekhnicheskaja 26, Leningrad 194021, USSR

In earlier work we have observed the optical alignment of 2D-
electrons in quantum-well structures. which results in the polarization
of hot photoluminescence (HPL) [1]. It was emphasized that the peculiarity of the 2D-motion of electrons manifested itself in the strong
dependence of the degree of HPL polarization ρ on the energy of the 2D-
motion (such a dependence did not exist in the 3D case). Taking into
account the quantum confinement of electrons (but not holes) and ignoring
the valence band warping, we have obtained an approximate equation which
describes the energy dependence of ρ:

$$\rho = \frac{(E_o/E_1)^2}{8(1+E_o/E_1)+2(E_o/E_1)^2} , \tag{1}$$

where E_o is the kinetic energy of the 2D-motion (x,y) of electrons at the
point of photocreation and E_1 is the confinement energy. In deriving Eq.
(1) the z-component of the momentum was put equal to $p_z = (2mE_1)^{1/2}$.

In the present work the energy dependence $\rho(\epsilon)$ has been studied in
detail. It has been found that Eq. (1) reflects the measured dependence
rather well. Magnetic depolarization of HPL in a wide energy range was
used to measure the times of electron scattering from the point of
photocreation due to the intra-subband and inter-subband transitions, as
well as the transitions into the 3D-continuum.

1. Experimental

In this work we studied the spectra and the linear polarization of
the photoluminescence of hot electrons on the acceptor levels (1e-A^o
transitions), as well as the influence of the applied magnetic field on
polarization. The investigations were carried out onGaAs/Al$_x$Ga$_{1-x}$As
multiple quantum well structures with x close to 0.3. These structures
were grown by the MBE method on (100)-oriented substrates. The width of
the wells varied in the range 70-100 Å, and the barrier width was about
100 Å. The concentration of the acceptors (carbon) in the structures
used for the study of HPL 1e-A^o transitions was in the range 5×10^{15} -
2×10^{16} cm^{-3}. One of the structures was doped with Be up to

3×10^{18} cm^{-3}. The amount of Al in the barrier, x, was determined from Raman spectra, and the barrier height for the electrons, ΔE_c, was calculated from the equation $\Delta E_c = 0.64 \Delta E_g$, where ΔE_g is the difference in gaps between the barrier and well materials.

The luminescence was excited at 2K by the lines of Kr and He-Ne lasers at 1.65, 1.83, 1.92 and 1.96 eV. The registration of luminescence was made in the back-scattering configuration. The magnetic field was applied both perpendicular to the plane of the quantum wells, i.e. along the exciting beam (Faraday configuration), and in the plane (Voigt configuration). The pumping power did not exceed 10^2 W/cm^2, which corresponded a the concentration of photoexcited carriers close to 10^{10} cm^{-2}.

2.   The Energy Dependence of the Degree of Initial Polarization

The energy dependence of $\rho(E_o/E_1)$ for two structures with different E_1 is shown in Fig. 1 (E_1 is the confinement energy in the le subband, E_o is the initial kinetic energy in the le subband for electrons excited from the lhh subband).

One can see that Eq. (1) correctly reflects the trend of the $\rho(E_o/E_1)$ dependence and is in good agreement with the measured ρ values. The last circumstance is apparently accidental, if one takes into account the approximations made in deriving Eq. (1).

The increase in ρ with the increasing kinetic energy of the electrons, E_o, is due to the peculiarities of the 2D-motion. For small E_o ($E_o < E_1$) the electrons are "moving" nearly normal to the plane of the structure, and if observation is along the z-axis (normal to the plane of the structure) the recombination radiation is almost unpolarized. When $E_o \gg E_1$ the electrons in the first subband are moving almost in the plane of the quantum well, and their recombination radiation observed along z is strongly polarized perpendicular to the preferential direction of electron momenta. It should be noted that the increase of ρ with E_o is also observed when the electron energy exceeds the barrier height (shown in Fig. 1 by an arrow on the energy axis). Thus, by means of polarization characteristics we can observe the manifestation of the 2D-spectrum in the above-barrier region up to an electron energy that exceeds the barrier height by 100 meV.

3.   Polarization of HPL under Magnetic Field and Determination of Scattering Times

3.1.   Faraday Configuration

As in the 3D-case [2], the measurement of the depolarization of hot photoluminescence in a magnetic field allows the determination of characteristic scattering times of hot electrons. In the Faraday configuration, when the magnetic field is directed along the light beam, the polarization of luminescence is described by

$$\rho(\beta)/\rho(0) = (1 + 4\omega_c^2 \tau_o^2)^{-1}, \qquad (2)$$

where $\omega_c = eB/m_c^* c$ is the cyclotron frequency, and τ_o is the lifetime of electrons at the point of photocreation.

In calculating ω_c the energy dependence of m_c^* for GaAs, i.e. the nonparabolicity of the conduction band, has been taken into account in accordance with the results given in Ref. 3. In the 2D-case, it has been assumed that $m_c^*(E_o)$ corresponds to the 3D-value of $m_c^*(\epsilon)$ at $\epsilon = E_o + E_1$. At the moderate doping level and the pumping power used in this work τ_o is

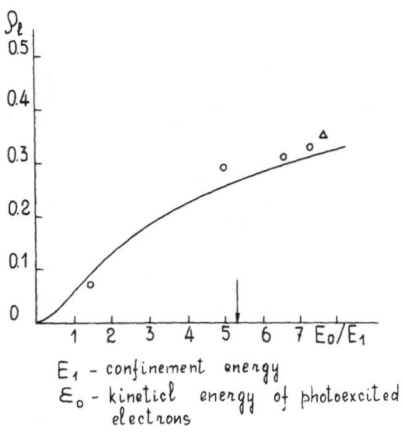

E_1 - confinement energy
E_o - kinetical energy of photoexcited electrons

Fig. 1. The energy dependence of the degree of the initial polarizaiton for two structures. The circles correspond to the structure with the well width L_z = 70 Å (E_1 = 45 meV), the triangles to the one with L_z = 100 Å ($E_1 \cong$ 30 meV).

identical to the LO phonon emission time τ_{po}. Note, that a four-fold increase of pumping power, i.e. up to the value 5×10^2 W/cm^2, produced no changes in the value of τ_o within the limits of measurement accuracy. On these grounds we could ignore in the interpretation of results the influence of screening and phonon heating, i.e. the factors which led to the spread of results in measurement of τ_{po} in a number of earlier works (see, for instance, the review, Ref. 4).

The measured value of τ_{po} for the 70 Å/100 Å structure was found to be 160 ± 10 fs at E_o = 60 meV. This energy is smaller than the energy of the bottom of the second subband, 2e. Therefore the measured scattering time corresponds to the intraband scattering in the 1e subband. The measured value of τ_{po} was found to be smaller by 20% than the calculated one for the scattering of 2D-electrons from bulk phonons. This is in agreement with the predictions of Ref. 5, where the rate of polar scattering of 2D-electrons in a thin slab of ionic crystal has been calculated.

The method described can be used in a wide range of energies to study the kinetics of intra- and intersubband scattering. The results of scattering rate measurements for values of E_o is shown in Fig. 2. Shown by the solid curve is the calculated dependence of the polar scattering of 2D-electrons from bulk phonons. In the same figure the schemes of possible scattering processes are presented. From the comparison of scattering rates at E_o = 60 meV and E_o = 215 meV we can estimate the intersubband scattering time $\tau_{1e \to 2e} = \left[\tau_o^{-1}(215 \text{ meV}) - \tau_o^{-1}(60 \text{ meV}) \right]^{-1}$. Such estimation becomes possible because in the studied structures at 215 meV the scattering of 2D-electrons is due both to the intra- (1e → 1e) and inter-subband transitions. In this way we have obtained $\tau_{1e \to 2e}$ = 0.75±0.25 ps. This value is somewhat smaller than the one we have calculated for the scattering from bulk phonons in a 70Å/100Å structure (1.1 ps).

The decrease of scattering time with increasing E_o for electrons excited into the above-barrier region (the dotted part of the curve in

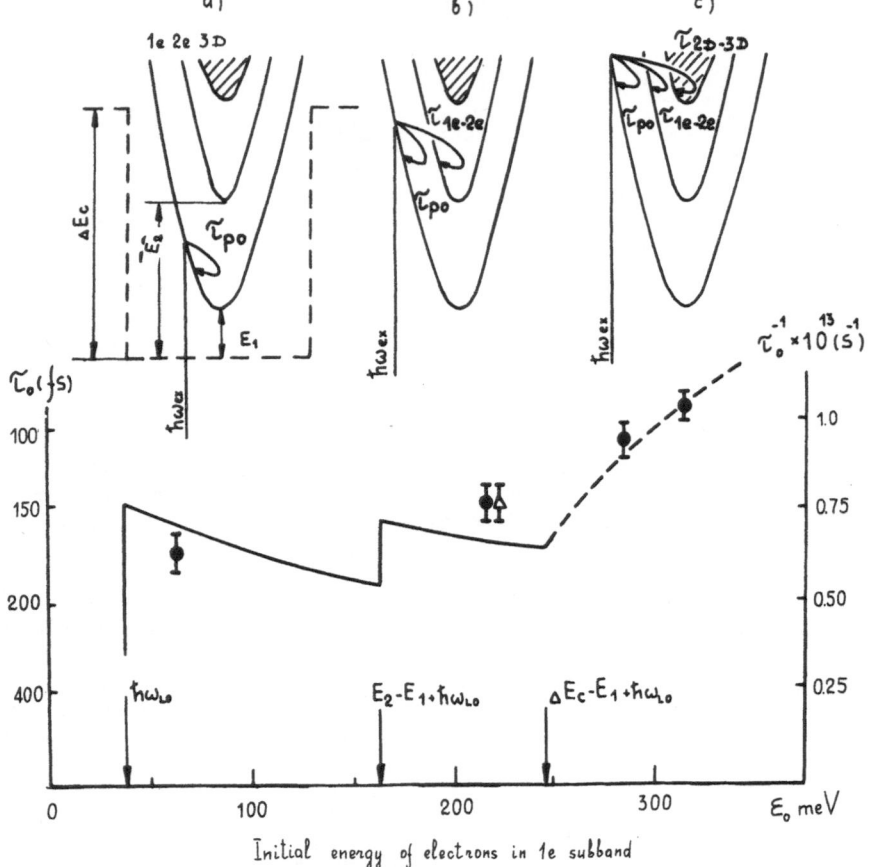

Fig. 2. The energy dependence of the scattering rate τ_o^{-1}. (a) the excitation below the bottom of the second, 2e, subband $E_o < E_2 - E_1 + \hbar\omega_{LO}$ (b) the excitation above the bottom of the 2e subband $E_o > E_2 - E_1 + \hbar\omega_{LO}$; (c) the excitation into the above-barrier region $E_o > \Delta E_c - E_1 + \hbar\omega_{LO}$

Fig. 2) can be explained by the addition of the processes of electron scattering from optical phonons in the region of 3D-states. When the electron energy surpasses the barrier energy by 100 meV the time of the 2D → 3D transition is close to 350 fs. One can expect that the time of such 2D → 3D transition decreases as E_o increases, due to the increase in the density of finite states in the 3D continuum.

In a heavily-doped structure ($N_{Be} \gtrsim 3 \times 10^{18}$ cm^{-3}) the scattering of impurities plays a significant role in the relaxation processes. This leads, in particular, to the decrease of the degree of polarization (ρ_o'), because the processes of <u>pure elastic</u> scattering become important. In this case $\rho_o' = \rho_o \dfrac{\tau_p}{\tau_p + \tau_o}$, where ρ_o is the degree of polarization at moderate doping (2).

In the structure studied the polarization at $E_o \simeq 200$ meV was found to be two times smaller than in a moderately doped structure, i.e. $\tau_p = \tau_o \equiv \left[\tau_{po}^{-1} + \tau_{imp}^{-1} \right]^{-1}$, where τ^{-1} is the rate of elastic scattering by impurities. From an analysis of experimental results we obtained $\tau_o = \tau_p = 80$ fs, $\tau_{imp} = 210$ fs.

3.2. Voigt Configuration

The peculiarity of the 2D-motion revealed itself in experiments on HPL depolarization in a magnetic field performed in the Voigt configuration. In bulk samples of GaAs the HPL depolarization has been observed both in the Faraday configuration and in the Voigt configuration (for $\vec{B} \perp \vec{e}$, where $\vec{e}$ is the polarization vector of the exciting light), although the ratio $\rho(B)/\rho(0)$ is different for these two configurations. In both cases the depolarization is due to the rotation and isotropization of the distribution function of electron momenta occurring in a magnetic field.

In the quantum-well structures studied by us the HPL depolarization in the Voigt geometry was not observed up to fields of 7 T, corresponding to $\omega_c \tau_o \geq 2$, both for the electrons excited within the wells and above the barriers. The depolarization was absent for $\vec{B} \perp \vec{e}$ as well as for $\vec{B} \| \vec{e}$ This result can be understood if one takes into account that the magnetic field can change the form of the initial momentum distribution of the electrons only if the size of the cyclotron orbit is smaller than the quantum well width L_z, i.e. if $\ell_B \ll L_z$, where ℓ_B is the magnetic length. In the structures studied, at $B = 7T$ $\ell_B \simeq 100$ Å, i.e. it practically equals L_z. Therefore the momentum distribution maintains its 2D-character.

The absence of depolarization in the Voigt geometry at the point E_0 in the case of above-barrier excitation, as well as the increase of the initial polarization with the increase of E_0 (the region $e_0 > E - E_1$), indicate that in this case, too, the hot photoluminescence is due to the recombination of exactly two-dimensional electrons. The electrons which scattered into the three-dimensional spectrum and after that were again trapped into the quantum well, do not make a contribution in the radiation from the point of photocreation. Thus, the branch of 2D states is revealed in luminescence in a rather wide energy range.

REFERENCES

1. B. P. Zakharchenya, P. S. Kop'ev, D. N. Mirlin, D. G. Polyakov, I. I. Reshina, V. P. Sapega, and A. A. Sirenko, Sol. State Comms. 69, 203 (1989).
2. B. P. Zakharchenya, D. N. Mirlin, V. I. Perel, and I. I. Reshina, Sov. Phys. Usp. 25, 243 (1982).
3. G. Ambrazevicius, M. Cardona, and R. Merlin, Phys. Rev. Lett. 59, 700 (1987).
4. S. A. Lyon, J. Lumin. 35, 121 (1986).
5. R. A. Riddoch and B. K. Ridley, Physica 134 B, 342 (1985).

HIGH RESOLUTION NONLINEAR LASER SPECTROSCOPY MEASUREMENTS
OF EXCITON DYNAMICS IN GaAs QUANTUM WELL STRUCTURES

Duncan G. Steel, Hailin Wang, Jeffrey T. Remillard, Min Jiang

Departments of Electrical Engineering and Physics
Randall Laboratory of Physics
University of Michigan
Ann Arbor, Michigan

The optical properties of GaAs/$Al_xGa_{1-x}As$ multiple quantum well structures are dominated by strong sharp excitonic resonances near the band edge which are observable in both absorption and luminescence spectra[1,2]. The quasi 2-dimensional excitons are confined in the GaAs layer by the $Al_xGa_{1-x}As$ barriers. The principal properties of the confined exciton include a binding energy which increases with decreasing well width and a blue shift in the exciton transition energy. Indeed the increase in binding energy due to confinement explains the clear observation of these resonances even at room temperature. These materials are grown by molecular beam epitaxy (MBE) methods and are important for application in high speed electronic and opto-electronic devices[3]. Moreover, the ability to fabricate crystals with dimensions controllable at the atomic level provides an excellent opportunity to study the basic physics giving rise to relaxation of the exciton with reduced dimensionality through the interaction of the exciton with the crystal lattice.

Optical resonant excitation of the exciton with quasi-monochromatic light with energy E leads to an optical induced coherence (the polarization) as well as a population of the excitons with energy between E and E+ΔE. Hence the decay of the coherent excitation must be characterized by decay of the coherence or polarization and the decay of the population at energy E. At room temperature, it is well known that LO phonons ($E_{LO} \sim 37$ meV) ionize the exciton (binding energy ~ 9 meV) on a time scale of a few hundred femtoseconds[4]. The decay rate of the coherence measured by the homogeneous linewidth or dephasing rate is then dominated by this ionization rate while the relaxation of the energy is then typically determined by electron-hole dynamics. At low temperature, the exciton is stable against phonon ionization and the predominant decay of the energy would be by recombination of the electron-hole pair of the exciton. In a perfect crystal at low temperature excitons are delocalized and described by Bloch type wave functions. The decay of the coherence is then expected to be due predominantly to elastic scattering by acoustic phonons along with contributions due to decay of the exciton by recombination.

However, in a quantum well structure, the problem can become more complicated[5,6]. Nonideal growth conditions result in interface roughness between the GaAs well region and the AlGaAs barrier region. From transport measurements[7,8] and chemical lattice imaging methods[9], it is known that the island regions which form during growth are typically one monolayer high and of order 50 Å in lateral extent. Since the exciton has a Bohr radius of order 65 Å for a 100 Å well, the exciton experiences a shift in the transition energy due to this interface roughness. These shifts lead to inhomogeneous broadening of the exciton absorption resonance[5]. In the low energy region of the absorption spectrum, the excitons are considered to be spatially localized by the island like structures. At low temperature (<10K), it is expected that excitons can then migrate among the islands by emitting or absorbing acoustic phonons leading to decay of the excitons at energy E and dephasing of the induced coherence. At higher temperatures, the excitons (at energy E) will experience an additional contribution to their decay by thermal activation to higher lying quasi-delocalized states. The general process of energy migration is designated spectral diffusion. Above line center, the excitons are believed to be quasi-delocalized. The dephasing of the induced polarization is due to scattering along the 2-D dispersion curve by acoustic phonons and disorder due to interface roughness. Evidence for this transition region designated the exciton mobility edge has been reported in the pioneering work of Hegarty and Sturge[10].

In this paper we describe the use of new precision frequency domain nonlinear laser spectroscopy methods for the general study of exciton dynamics[11,12]. The objective is to experimentally determine the completeness of the above description of exciton relaxation. The frequency domain methods are particularly well suited to this problem because the narrow bandwidth of the excitation permits improved resolution over the usual time domain measurements and it is straight forward to observe in a single measurement time scales ranging over twelve orders of magnitude associated with exciton dynamics. In addition the frequency domain methods allow us to eliminate contributions from inhomogeneous broadening and to obtain homogeneous line shapes along with information related to spectral diffusion kernels.

A complete analytical discussion of the basis for precision spectroscopy based on frequency domain four-wave mixing (FWM) in simple systems has been presented by us elsewhere[13,14]. However, the physical basis for interpretation of FWM lineshapes is understood by considering the basic details of the resonant nearly-degenerate FWM response in a semiconductor. The experiments are based on the backward FWM interaction shown in Fig. 1.

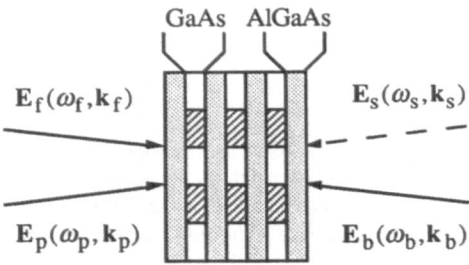

Figure 1: A schematic representation of the experimental configuration for frequency domain four-wave mixing spectroscopy in GaAs/AlGaAs quantum well structures. The cross hatched region represents the region of optically excited excitons which are confined by the AlGaAs barriers.

The three input fields (designated with subscript f, b, p representing forward, backward and probe fields, respectively) interact via the third order nearly degenerate resonant (i.e.,$\omega_i \approx \omega_j \approx \omega_o$) nonlinear susceptibility $\chi^{(3)}$ ($\omega_s = \omega_f - \omega_p + \omega_b$) to produce a coherent signal field, E_s. Physically, the signal field arises from a coherent scattering of the p-polarized backward field from the spatial and temporal modulation of the optical absorption and dispersion created by the interference of the s-polarized forward pump and probe fields, given by $E_f \cdot E_p^*$ with a time and space dependent phase given by $(\omega_f - \omega_p)t - (k_f - k_p) \cdot x$. (At low exciton densities, experiments show the absence of any tensor grating formed by $E_b E_p^*$)

Near resonance, it is well known now that the nearly degenerate third order susceptibility arises from many body effects due to the creation of excitons[15]. The effects include band filling, screening and exchange which modify the oscillator strength, the linewidth, and the resonance frequency. In a phenomenological description, these effects can be quantified by replacing the optical constants ξ_o, representing f_o, ω_o, and Γ_o, corresponding to the oscillator strength, resonant frequency, and linewidth, respectively in the linear susceptibility $\chi = \int d\omega_0 P(\omega_0) f_0 / [(\omega - \omega_0) - i\Gamma_0]$, with $\xi_0 \rightarrow \xi = \xi_0(1 + \xi_e n_e + \xi_h n_h)$ where n_i is the corresponding carrier density and expanding χ, keeping the terms first order in n_i[16]. The n_i are determined from their corresponding rate equations. At low temperature, the exciton is stable against phonon ionization and the parameters are modified to reflect the corresponding exciton density. The function $P(\omega_o)$ represents the distribution of resonant frequencies and gives rise to inhomogeneous broadening if the width of P is large compared to Γ_o.

Based on this discussion in the rate equation limit where energy relaxations are small compared to dephasing rate, Γ_o, (a reasonable approximation for many of the measurements discussed below, except where noted) it is easily seen that measuring the signal strength as function of $\omega_f - \omega_p$ while holding ω_f (or ω_p) = ω_b (designated the FWMp or FWMf response) provides a measure of the relaxation rate of the spatial modulation.

More specifically, the interference of E_f and E_p produces a traveling wave modulation of the excitation when $\omega_f \neq \omega_p$. When $|\omega_f - \omega_p|$ is larger than the spatial modulation decay rate, the signal intensity then decreases with increasing $|\omega_f - \omega_p|$. The lineshape function associated with this measurement is given by

$$L_{p(f)} = L \frac{1}{(\omega_f - \omega_p)^2 + (\gamma + D|\mathbf{k}_f - \mathbf{k}_p|^2)^2} \qquad (1)$$

where γ is the energy relaxation rate and $D |\mathbf{k}_f - \mathbf{k}_p|^2$ is the <u>spatial</u> diffusion rate which accounts for the fact that the excitation may diffuse in space, the rate being determined by diffusion coefficient and the reciprocal grating spacing $|\mathbf{k}_f - \mathbf{k}_p|^2 = (16\pi^2 \sin^2 \theta / 2) / \lambda^2$.

If the system is homogeneously broadened (i.e., $P(\omega_o)$ is a δ-function), tuning ω_b, designated the FWMb response, results in a lineshape closely related to the linear absorption profile. However, if the resonance is inhomogeneously broadened as is the contributions from localized excitons, then the linear absorption spectrum has little relation to the homogeneous lineshape. In this case, holding $\omega_f - \omega_p << \Gamma$ fixed results in creating excitons with energy in the region of $\hbar\omega_f$ ($\omega_f \sim \omega_p$). The energy spread is given by $\hbar\Gamma$; i.e., the interference of E_f and E_p produces a spatial modulation of a spectral hole in the inhomogeneous distribution. E_b only scatters from this spatial modulation when ω_b is tuned within the spectral hole. If $\chi(\omega - \omega_o)$ is the complex linear susceptibility associated with an excitation at a specific frequency ω_o, then in a simple hole burning picture, it is easily shown[14] that the lineshape function associated with the FWMb response in an inhomogeneously broadened system is given by:

$$L_b(\omega_b - \omega) = K \left| \int d\omega_o P(\omega_o) \chi(\omega_b - \omega_o) \operatorname{Im} \chi(\omega - \omega_o) \right|^2 \qquad (2)$$

where $\omega = \omega_f \sim \omega_p$ and $P(\omega_o)$ is the distribution function associated with the inhomogeneous broadening. Spectral diffusion effects are not included in this discussion. In this case, it can be shown that because χ must be analytic, even if $\chi(\omega - \omega_o)$ is asymmetric with respect to the maximum value at $\chi(\omega_o)$, $L_b(\omega_b - \omega) = L_b(\omega - \omega_b)$; i.e., L_b must be symmetric. In the data below, we see L_b is asymmetric at low temperature and is the result expected in the presence of spectral diffusion.

The experimental configuration for these experiments has been described in detail elsewhere[11], but is summarized here. A backward FWM geometry is used with two counterpropagating pump fields, described by $E_f(\omega_f, k_f)$ and $E_b(\omega_b, k_b)$, and the probe field, described by $E_p(\omega_p, k_p)$. Phase matching conditions result in a signal field, $E_S(\omega_S, k_S)$ which is counter propagating with respect to the probe field. The forward pump and probe fields are s-polarized while the backward pump and signal fields are p-polarized. The forward pump beam is chopped at a low frequency and the corresponding signal is phase sensitively detected and repeated scans of the signal as a function of frequency is averaged in a computer. In these experiments, two of the frequencies are held fixed (usually degenerate) while the third frequency is tuned. One frequency stabilized cw dye laser laser is used to provide the set of fixed frequencies while a second tunable frequency stabilized laser is used for scanning the remaining frequency. Two acousto-optic modulators driven by two phase locked digital frequency synthesizers are used to provide fixed frequency offsets between two different beams or to provide tuning in the $1\text{-}10^6$ Hertz region for high resolution measurements.

The sample consists of 65 periods of 96Å GaAs wells and 98Å $Al_{0.3}Ga_{0.7}As$ barriers grown by MBE methods and then mounted on a sapphire disk (c-axis normal to avoid birefringence problems and polarization mixing.)

The first line shapes we discuss were obtained at room temperature. As described above, the exciton is quickly ionized, producing an e-h plasma. Hence, the FWMb response provides little information of additional spectroscopic interest over the linear absorption spectrum. However, the FWMp response provides information on the e-h plasma dynamics, including e-h recombination and ambipolar diffusion, and is described by the line shape function L_p, given above. The result for a fixed angle between the forward pump and probe is shown in Fig. 2. The solid line is a least squares fit of L_p, showing a classical Lorentzian line shape, as expected. The presence of ambipolar diffusion of the spatially modulated e-h plasma results in an angle dependent decay of the grating. Figure 3 shows the measured angle dependence, varying as expected as $\sin^2\theta/2$. From the $\theta=0$ intercept, we obtain the e-h recombination time (5 nsec), and from fitting the quadratic portion of the curve, we obtain the ambipolar diffusion coefficient (18cm^2/sec). These measurements are in good agreement with those reported earlier[16].

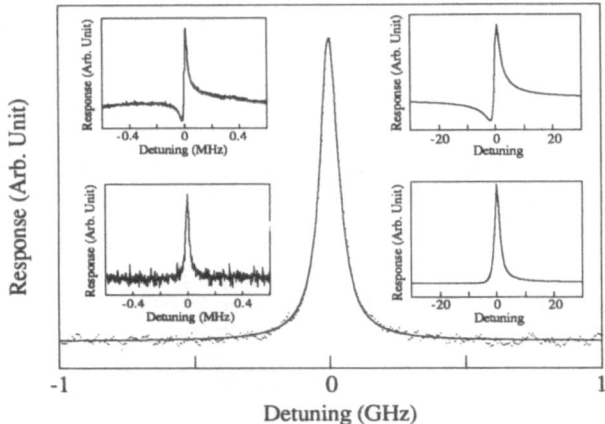

Figure 2: The FWMp response in the room temperature GaAs multiple quantum well. The solid line is a Lorentzian fit. The decay rate which determines the linewidth is due to recombination and ambipolar diffusion. The upper left inset is a high resolution scan of the FWMp response using the method of correlated optical fields. The interference dip is believed due to an excitation induced shift in the exciton resonance energy. The upper right inset is the theoretical FWMp response based on a phenomenological model. The curves in the lower insets represent experiment and theory when the back pump beam is detuned far from the exciton resonance energy, showing the disappearance of the interference effect.

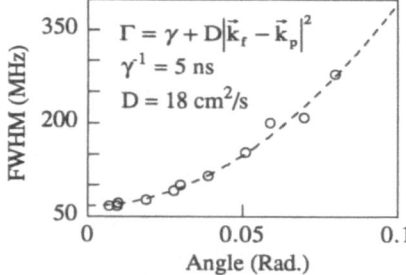

Figure 3: The FWMp response as a function of angle between the forward pump and probe beam in the room temperature GaAs multiple quantum well. The quadratic dependence on the sin of the angle is due to ambipolar diffusion of the carriers. The diffusion rate is determined by the fringe spacing which is angle dependent. The zero degree intercept is due to carrier recombination.

To determine the nature of dynamics corresponding to time scales below 1 MHz (determined by interlaser jitter), we used the method of correlated optical fields which enables us to obtain precision line shapes by tuning one acousto-optic modulator with respect to the fixed modulator[17]. For this experiment, one laser is used, but ω_f an ω_b are provided by the first order Bragg deflected beam of one fixed ao modulator while ω_p is provided by tuning the first order beam from the other modulator. The upper left inset in Fig. 2 shows resultant line shape. From the bandwidth of this resonance, it is clear that there exists a component of the excitation characterized by a decay on the order of 10 μsec. Even more striking is the presence of a classical interference dip on the left side of the resonance. Using the phenomenological description for many body effects given above, we considered the possibility that the structure contained e- or h-traps that were long lived. Inserting this effect into simple rate equations for the electron and holes including only band filling effects on the oscillator strength and collisional broadening on the line width, we easily obtained a narrow resonance due to trap dynamics in addition to the ordinary resonance due to free e-h carrier dynamics. However, no interference effect is obtained. In contrast, when a small shift in the exciton resonance due to many body effects is included in the calculation, the interference effect is observed in the scattering FWM signal, as shown in the upper right inset of Fig. 2. In addition, the simple theory showed that the interference effect would vanish if ω_b was tuned far away from ω_o, the exciton resonant frequency. Under this condition, the lower insets show agreement between experiment and the simple model. It is important to note that the time scale and signal magnitude is also roughly consistent with a temperature induced shift in ω_o, and we have calculated that such a shift would also give rise to this kind of profile. However, the thermal shift is red and would give a profile *reversed* from that in Fig. 2.

At low temperature, the linear absorption spectrum shows the clearly resolved HH1 and LH1 exciton peaks. HH1 has a width of order 2 meV and is inhomogeneously broadened. The luminescence shows a single emission peak corresponding to HH1 which is Stokes shifted by an amount varying between 10 and 20 cm^{-1}, depending on the sample. The degenerate FWM (DFWM) spectrum shows a single strong resonance also Stokes shifted by an amount comparable to the shift in the luminescence peak with a much weaker nonlinear signal being obtained at higher energies near the HH1 and LH1 absorption resonances. The Stokes shift in the luminescence and the DFWM response suggests that the cw low intensity nonlinear response in these samples is dominated by excitons localized by disorder. In contrast to many earlier measurements, the exciton density for these experiments was kept very low, near 10^7 excitons/cm^2/layer.

In a simple picture of the exciton dynamics, we would anticipate the FWMp response would be characterized by a simple Lorentzian line shape corresponding to L_p above with a width determined by the exciton recombination time (1-2 nsec for these structures[18]) along with contributions from phonon scattering on the psec time scale[19]. In fact, Fig 4 shows a much more complicated structure which was obtained at 5K. The spectrum shown in Fig. 4 is angle *independent*, suggesting that any spatial diffusion corresponds to a diffusion rate less than 1 cm^2/sec. The spectrum clearly indicates two obvious time scales corresponding to 15 nsec and 100 psec and clearly do not correspond to the expected 1-2 nsec time scale for recombination. The origin of the 15 nsec structure is currently under study and further discussion will be presented elsewhere. In the current paper, we will emphasize discussion of the 100 psec structure.

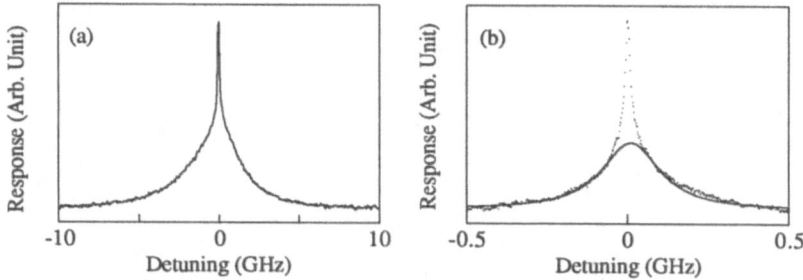

Figure 4a,b: The FWMp response in the GaAs multiple quantum well at 5K. (a) shows two major components to the spectrum corresponding to decay times of order 15nsec (the narrow component) and 100 psec (the broad component). A high resolution display of the central portion of the response is shown in 4b where a component corresponding to a 1.5 nsec decay is observed. The 100 psec component is believed due to spectral diffusion and the 1.5 nsec component is the radiative recombination time.

Earlier time domain work by Hegarty and Sturge[20] also showed the presence of fast decay components which they interpreted as arising from spectral diffusion, i.e., the scattering of the exciton at energy E to some energy E' outside the bandwidth of the spectral hole created by the forward pump and probe. Indeed, there has been recent work by Takagahara[21] suggesting that at low temperature, excitons localized by disorder spectrally diffuse to different localization sites by absorption and emission of acoustic phonons, changing their energy on the order of 0.01-0.1meV per scattering event. The process is identified as phonon assisted tunneling. In such a picture, we can imagine E_f and E_p create a spatial modulation of the exciton population at energy E in a narrow spectral hole with width $\Delta E = \hbar \Gamma$ within the inhomogeneous broadening profile. The narrow spectral hole decays at the scattering rate from E to E'. Since there is also scattering from E' to E, an equilibrium is established which describes the resultant distribution in energy space of the exciton population. This equilibrium distribution also contributes to the scattering of E_b but is characterized by a decay time of the total exciton population, i.e., the recombination time. Such dynamics are described by an equation of the form[22]

$$\dot{\rho}(E) = -(\gamma_{rec} + \Gamma_{SD})\rho(E) + \Gamma_{SD}\int f(E' \to E)\rho(E')dE' \qquad (3)$$

where $\rho(E)$ is the density of excitons at energy E, γ_{rec} is the recombination rate, Γ_{SD} is the integrated spectral diffusion rate or the rate at which excitons scatter from energy $E \to E' = E + \delta E$ and $f(E \to E')$ is the probability of scattering from E to E'. Such a model is analogous with the hard sphere collision model for velocity changing collisions in the gas phase. Analytical solutions assuming f to be independent of E' are easily obtained for standard distributions and show that the FWMp response has an additional component with a width given by γ_{rec} as anticipated based on physical arguments. The higher resolution experiment, shown in Figure 4b shows the expected component with a width corresponding to an inverse decay time of 1.5 nsec.

A more complete description of spectral diffusion is had, however, by making a direct measurement of the exciton scattered from energy E to E'. This measurement is made by scanning the backward pump. In the absence of spectral diffusion, the FWMb response is determined purely by the homogeneous linewidth associated with the spectral hole produced by the forward pump and probe. Recall from above (Eq. 2) that even if the hole were asymmetric, the FWMb response would be symmetric. However, excitons produced at energy E by the forward pump and probe and scattered to $E'=E+\delta E$ will result in a scattering of the backward pump beam when $\hbar(\omega_f - \omega_b)=\delta E$, producing a lineshape broader than the homogeneous lineshape and lacking symmetry if the distribution function of scattered states, f, is asymmetric. Figure 5a shows the FWMb response in the low energy energy

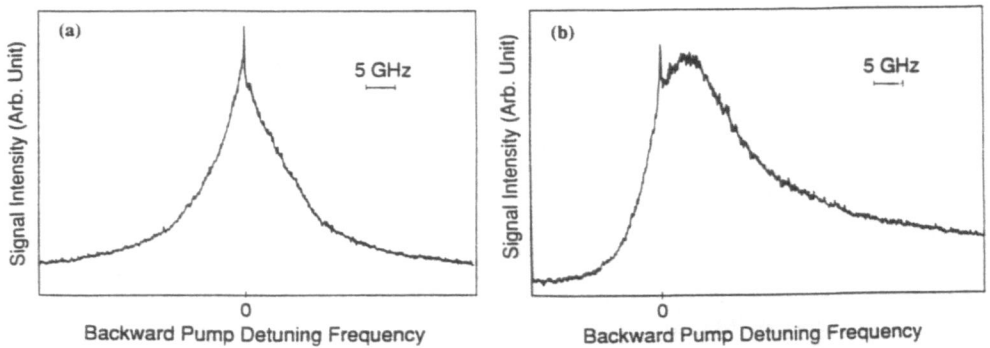

Figure 5a,b The FWMb response obtained at two different excitation energies. (a) is obtained 28cm[-1] below absorption line center. (b) is obtained 5 cm[-1] below line center.

region of the absorption spectrum, obtained 28cm[-1] below absorption line enter. The line shape is characterized by a slight skew to the high energy side of the spectrum as clearly seen by the discontinuity just to the right of line center, but no large degree of scattering was observed. However, the degree of asymmetry is greatly increased when the measurement is made near absorption line center as seen in Fig. 5b, corresponding to greatly increased scattering leading to spectral diffusion. This data shows that indeed the excitons are scattering to different energy states, and that the degree of scattering depends strongly on the excitation energy.

Further confirmation of the mechanism of exciton spectral diffusion is had by comparing the analytical prediction of Takagahara with the experimentally determined tunneling rate as a function of temperature. Figure 6 shows the FWMf response as a function of temperature for two different excitation energies. (The FWMf response rather than the FWMp response was used to reduce contributions to the response from the homogeneous line width, an effect which must be considered when the dephasing rate is comparable to the excitation decay rate. The interference creates an uncertainty of at most a factor of 2 in the overall rate constants which are measured[13].) As predicted, the tunneling rate varies as $\exp(BT^{1.6})$ as seen by the solid line fit of this function to the data, where B depends on the excitation energy. The inset shows a rapid increase of the tunneling rate when the exciton energy approaches the absorption line center. The behavior is consistent with the assignment of the absorption line

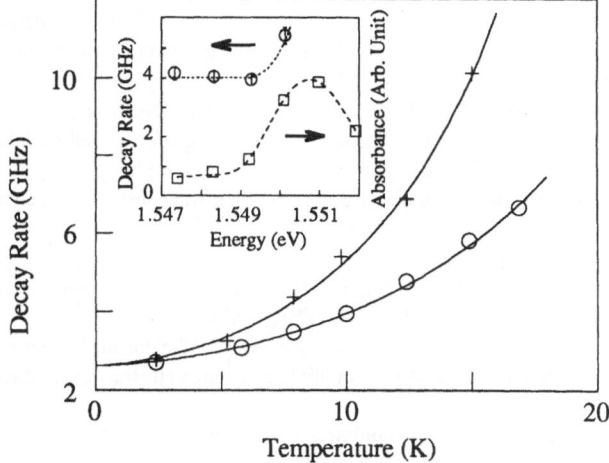

Figure 6 The measured exciton spectral diffusion rate as a function of temperature for two different excitation energies. Circles and crosses are data obtained at 1.5 meV and 0.6 meV below line center respectively. The solid curves are a fit of the theory for phonon assisted tunneling. The inset shows the energy dependence of the exciton decay rate at 10 K. The dash line is only a guide to the eye.

center as the exciton mobility edge. At higher temperatures, above 20K, we show in independent measurements made in the time domain that thermal activation dominates phonon assisted tunneling, in agreement with expectations.

In summary, we have examined the dynamics of excitons in room temperature and low temperature GaAs multiple quantum wells. Using a new kind of frequency domain nonlinear laser spectroscopy we have been able to make the first demonstration of phonon assisted tunneling in these structures. In addition, using the precision frequency domain capability of this method, we have also been able to make a direct measurement of the excitation lineshape which provide information on the redistribution kernel associated with spectral diffusion.

This work has been supported by the U.S. Army Research Office and the Air Force Office of Scientific Research.

REFERENCES

1. R. Dingle, W. Wiegmann and C.H. Henry, "Quantum states of confined carriers in very thin AlGaAs-GaAs-AlGaAs heterostructures," Phys. Rev. Lett. 33, 827(1974).
2. R.C. Miller and D.A. Kleinman, "Excitons in GaAs quantum wells," J. Lumin. 30, 520(1985).
3. D.S. Chemla and D.A.B. Miller, "Physics and applications of excitons confined in semiconductor quantum wells," in Heterojunctions: Band discontinuities and device applications, F. Capasso and G. Margaritondo, eds., (North-Holland, Amsterdam 1987).
4. W.H. Knox, R.L. Fork, M.C. Downer, D.A.B. Miller, D.S. Chemla, C.V. Shank, A.C. Gossard and W. Wiegmann, "Femtosecond, dynamics of resonantly excited excitons in room-temperature GaAs quantum wells," Phys. Rev. Lett. 54, 1306(1985).
5. C. Weisbuch, R. Dingle, A.C. Gossard and W. Wiegmann, "Optical characterization of interface disorder in GaAs-GaAlAs multi-quantum well structures," Solid State Comm.. 38, 709 (1981).
6. T. Takagahara, "Excitonic relaxation processes in quantum well structures," J. Lumin. 44, 347(1989).
7. R. Gottinger, A. Gold, G. Abstreiter, G. Weiman and W. Schlapp, "Interface roughness scattering and electron mobilities in thin GaAs quantum wells," Europhs. Lett., 6, 183 (1988).
8. H. Sakaki, T. Noda, K. Hirakawa and T. Matsusue, "Interface roughness scattering in GaAs/AlAs quantumwells," Appl. Phys. Lett. 51, 1934(1987).
9. A. Ourmazd, D.W. Taylor, J. Cunningham, and C.W. Tu, "Chemical mapping of semiconductor interfaces at near-atomic-resolution," Phys. Rev. Lett. 62, 933 (1989).
10. J. Hegarty, L. Goldner and M.D. Sturge, "Localized and delocalized two-dimensional excitons in GaAs-AlGaAs multiple quantum well structures," Phys. Rev. B30, 7346(1984).

11. J.T. Remillard, H. Wang, D.G. Steel, J. Oh, J. Pamulapati and P.K. Bhattacharya, "High resolution nonlinear laser spectroscopy of heavy hole excitons in GaAs/AlGaAs quantum well structures: a direct measure of the exciton line shape," Phys. Rev. Lett. **62**, 2861(1989).

12. J.T. Remillard, H. Wang, M.D. Webb, D.G. Steel, J. Oh, J. Pamulapati and P.K. Bhattacharya, "High resolution nonlinear laser spectroscopy of room temperature GaAs quantum well structures: observation of interference effects," Opt. Lett. **14**, 1131(1989).

13. D.G. Steel and J.T. Remillard, "Resonant nearly degenerate four wave mixing in open and closed systems," Phys. Rev. A **36** 4330(1987).

14. H. Wang, and D.G. Steel, to be published.

15. S. Schmitt-Rink, D.S. Chemla and D.A.B. Miller, "Theory of transient excitonic optical nonlinearities in semiconductor quantum-well structures," Phys. Rev. B32, 6601(1985).

16. D.S. Chemla, D.A.B. Miller, P.W. Smith, A.C. Gossard and W. Wiegmann, "Room temperature excitonic nonlinear absorption and refraction in GaAs/AlGaAs multiple quantum well structures," IEEE J. Quant. Elec. **20**, 265(1984).

17. D.G. Steel and S.C. Rand, "Ultranarrow nonlinear optical resonances in solids," Phys. Rev. Lett. **55**, 2285(1985).

18. J. Feldmann, G. Peter, E.O. Gobel, P. Dawson, K. Moore, C. Foxon, and R.J. Elliott, "Linewidth dependence of radiative exciton lifetimes in quantum wells," Phys. Rev. Lett. **59**, 2337 (1987).

19. L. Schultheis, A. Honold, J. Kuhl, K. Kohler and C.W. Tu, "Optical dephasing of homogeneously broadened two dimensional exciton transition in GaAs quantum wells," Phys. Rev. B34, 9027(1986).

20. J. Hegarty and M.D. Sturge, "studies of exciton localization in quantum-well structures by nonlinear optical techniques," J. Opt. Soc. Am. B2, 1143(1985).

21. T. Takagahara, "Localization and homogeneous dephasing relaxation of quasi-two-dimensional excitons in quantum-well heterostructures," Phys. Rev. B32 7013(1985).

22. P.R. Berman, "Validity conditions for the optical Bloch equations," J. Opt. Soc. Am. B3, 564(1986).

OPTICAL SPECTROSCOPY IN THE REGIME OF THE

FRACTIONAL QUANTUM HALL EFFECT

A. Pinczuk,[1], D. Heiman,[2] B. B. Goldberg,[3]
J. P. Valladares,[1] L. N. Pfeiffer[1] and K. W. West[1]

(1) AT&T Bell Labs, Murray Hill, NJ 07974.
(2) M.I.T. National Magnet Lab, Cambridge, MA 02139.
(3) Physics Dept., Boston University, Boston MA 02215

ABSTRACT

We present a brief review of recent optical spectroscopy experiments carried out in the regime of the fractional quantum Hall effect. Magneto-optics, optical absorption and emission, and inelastic light scattering measurements in high mobility GaAs-AlGaAs heterostructures are considered.

INTRODUCTION

The fractional quantum Hall effect (FQHE) was discovered in 1982 in magneto-transport studies of the high mobility 2D electron gas in GaAs-AlGaAs heterojunctions under a perpendicular magnetic field.[1] The effect occurs at low temperatures when there is partial occupation of a spin-split Landau level. It appears as a quantization of the Hall resistivity to values $\rho_{xy} = e^2/hi$. $i = p/q$ is a rational fraction that coincides with the Landau level filling factor $\nu = 2\pi n l_0^2$, where n is the free electron density and $l_0 = (\hbar c/eB)^{1/2}$ is the magnetic length. In the FQHE there is also vanishingly small longitudinal magnetoresistivity $\rho_{xx} \sim \exp(-\xi/2\,kT)$, where ξ is the activation energy.[2-4] Quantization at fractional values of ν, when the Fermi energy is within a partially populated Landau level, is evidence of unexpected new phenomena that could only arise from electron-electron interactions. The effect is interpreted as a manifestation of highly correlated many-body ground states.[5-8] Fractional quantization is a consequence of the formation of an energy gap that separates the new ground states from its excited states. These higher lying states are characterized by quasiparticle excitations with fractional charge and populations that are functions of filling factor and temperature.

The FQHE is one of the most remarkable novel phenomena discovered in free electron systems of reduced dimensionality. Magneto-optics experiments in the regime of the FQHE are expected to reveal new physics that may not be accessible in magneto-transport. Near bandgap optical recombination was reported in silicon metal-insulator-semiconductor transistors and in modulation doped GaAs-AlGaAs quantum wells. The silicon work was carried out at

T = 1.5K and at filling factors 7/3 and 8/3.[9-10] The measurements in GaAs quantum wells were at significantly lower temperatures, as low as 0.35K, and filling factor 2/3.[11-13]

In the interpretation of optical recombination in silicon MIS transistors the changes in spectral positions with changes in temperature and filling factor were used to extract values for the gaps associated with quasiparticle excitations of the FQHE.[9,10,14] Kukushkin and Timofeev found that the gaps determined from optical spectroscopy are close to activation energies ξ estimated from the temperature dependence of the magnetoresistance.

A very different approach was proposed for the interpretation of the results from modulation doped quantum wells.[12] First, we highlighted the well known fact that the presence of a valence band hole is a large perturbation in optical recombination of the electron gas.[15-20] We pointed out that the response of the free electrons to the Coulomb potential enhances the electron density near the positive charge of the hole. The increased electron-hole overlap associated with this response leads to changes in the recombination energy and in the matrix elements for the transitions. Because of the major role played by the valence band hole, the gap of the FQHE cannot be extracted in a simple manner from measured energy shifts. The response of the electron gas to the Coulomb potential of the hole, may be similar to the cases of charged impurities and disorder.[21-23] In this picture, the changes observed in optical recombination could reveal novel behavior in the FQHE that is not evident in magneto-transport experiments.

Recent work has shown that for a 2D electron gas in a perpendicular magnetic field inelastic light scattering gives access to elementary excitations with in-plane wavevectors $q \gtrsim q_0 = 1/l_0$.[24,25] This wavevector range, $q \gtrsim 10^6$ cm^{-1}, is the most relevant in the magneto-roton theories of the fractional quantum Hall effect.[26,27] In a partially occupied level the lowest excitations are related to intra-Landau level transitions. In the regime of the FQHE their dispersions show characteristic roton minima at finite wavevectors close to q_0. The "magneto-roton" minima are interpreted as the energy gaps of the FQHE and light scattering measurements could reveal new features of the FQHE.

In this paper we present a brief review of optical spectroscopy of modulation doped GaAs quantum wells in the regime of the fractional quantum Hall effect. In the next section we consider emission and absorption. In the subsequent sections we discuss inelastic light scattering and present our concluding remarks.

OPTICAL EMISSION AND ABSORPTION IN THE REGIME OF THE FQHE

We consider here magneto-optical experiments in GaAs quantum wells under conditions where simultaneous transport measurements display the characteristic signatures of the FQHE.[11-13,28] Magneto-optic experiments at larger filling factors, $\nu > 2$, have been reported in Refs. 29-33.

The samples consist of modulation-doped GaAs-AlGaAs quantum wells with free electron areal densities in the range $1.4 \times 10^{11} < n < 6 \times 10^{11}$ cm^{-2}, and mobilities $10^5 < \mu < 5 \times 10^5$ cm^2/Vsec. GaAs substrates were removed for optical transmission measurements and the ultra-thin samples were cemented on a thin AlGaAs epi-layer.[33] A fiber-optic apparatus[34,35] was used to carry the excitation light to the sample and return the luminescence or transmitted light from the sample to a spectrometer with optical multichannel detection. The apparatus was placed in a ^{3}He cryostat in which temperatures as low as 0.35K

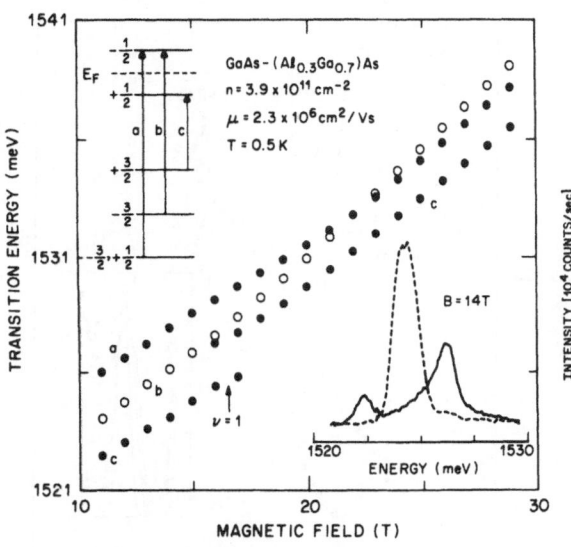

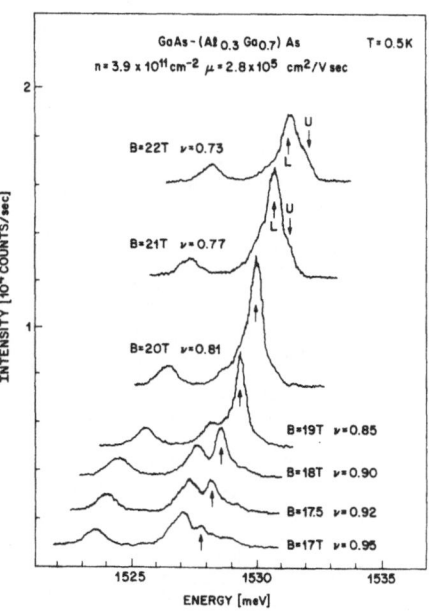

Fig. 1. Energies of lowest optical absorptions for a sample with n = 3.9 × 10^{11} cm^{-2}. Lower inset: dashed (solid) lines are absorption spectra in σ^+ (σ^-) polarization. After Ref. 28.

Fig. 2. PL spectra at several fields 0.95 < ν < 0.73. The arrows indicate the new peak appearing for ν < 1. After Ref. 12.

could be reached. Constant magnetic fields of up to 30 tesla could be achieved in hybrid (superconductor plus Bitter solenoid) magnets. The power of incident light was kept below 1 mW/cm^2. At these power levels we found no changes in magneto-transport, indicating absence of significant effects in temperature, density and mobility of the electron gas.

The upper inset to Fig. 1 indicates the three lowest optical transitions between states in the valence and conduction subbands measured at high magnetic fields. The states are Landau levels labeled by their components of angular momenta along the normal to the plane.[33] We show the two spin-split states of the lowest conduction Landau level and the three higher valence states. At $\nu = 1$ the Fermi level (E$_F$) is in the middle of the gap between the spin-split conduction Landau levels. The lowest Landau level has spin +1/2 because the g-factor is negative. In the valence band there is extensive coupling among Landau levels with mixing of different terms of angular momentum components m$_j$.[36-38] At fields B ~ 16T the highest valence state is mostly m$_j$ = 3/2 with some admixture of m$_j$ = -1/2. The second state has m$_j$ = -3/2 and the third state has a considerable admixture of -1/2 and 3/2 components.[33]

Figure 1 shows the positions of the three lowest absorption transitions in the magnetic field range of 1.5 > ν > 0.56.[28] The absorption spectra shown in the lower inset indicate that measured circular polarizations are consistent with the transition assignments of the upper inset. Transitions a and c are observed in σ^- polarization and transition b in σ^+. The observation of transitions c for $\nu > 1$ is unexpected at 0.5K since the the lowest spin-split level is fully occupied. The observation could be a manifestation of effects of residual disorder that cause the appearance of spin +1/2 states above the Fermi level.[28]

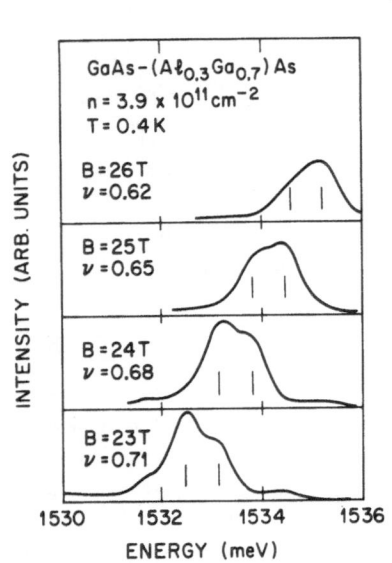

Fig. 3. PL spectra in the regime of the FQHE at $\nu = 2/3$. After Ref. 12.

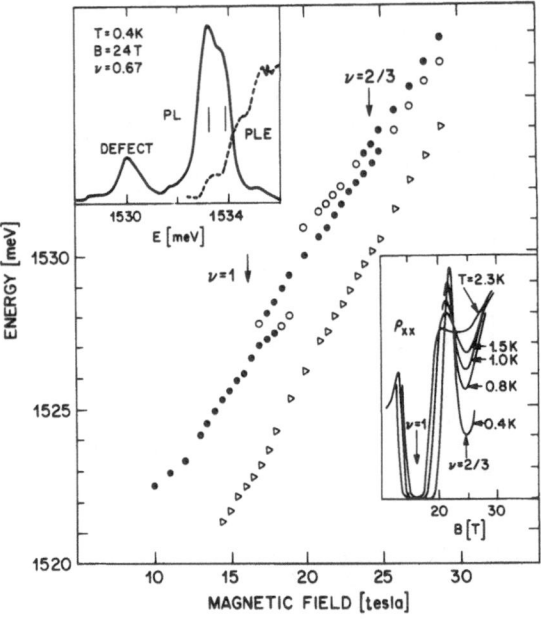

Fig. 4. Peak energies in PL spectra. The intrinsic peaks are represented by filled (open) circles for strong (weak) intensities and the triangles are for extrinsic (defect) peaks. After Ref. 12.

Figure 2 shows photoluminescence (PL) spectra measured at T = 0.4K in the range $0.95 > \nu > 0.73$ and Fig. 3 shows PL spectra in the range $.7 > \nu > .56$.[12] The higher energy features in these spectra are in the range of absorption due to transitions c. The weaker band appearing at 3-4 meV below is assigned to recombination at a defect. The energies of the PL peaks are plotted in Fig. 4.[13] The magnetic fields at filling factors $\nu = 1$ and $\nu = 2/3$, indicated by arrows, were obtained from ρ_{xx} traces as shown in the lower inset to Fig. 4. The results displayed in Figs. 2-4 indicate remarkable changes in optical emission near filling factors 1 and 2/3.

In the range of $\nu < 1$ we find that a new recombination line emerges at higher energy as indicated by arrows in Fig. 2. The line is absent below 16.2T($\nu=1$) and its intensity grows with increasing field. For fields above 18T($\nu<0.9$) the new emission line becomes the dominant feature of the spectrum. Its position is nearly coincidental with the strongest absorption peak due to transitions c. This reveals the intrinsic character of the new recombination. The lower energy emission, strong for fields below 17T($\nu>0.95$), could arise from localized states in the lowest Landau level.

The rapid growth of the intensity of intrinsic recombination occurs when the Fermi level shifts from the region of localized states into that of the extended states close to the center of the lowest, spin split, Landau level. To interpret this observation we have proposed that optical recombination in the magnetic quantum limit ($\nu<1$) is strongly affected by the response of the 2D electron gas to the hole

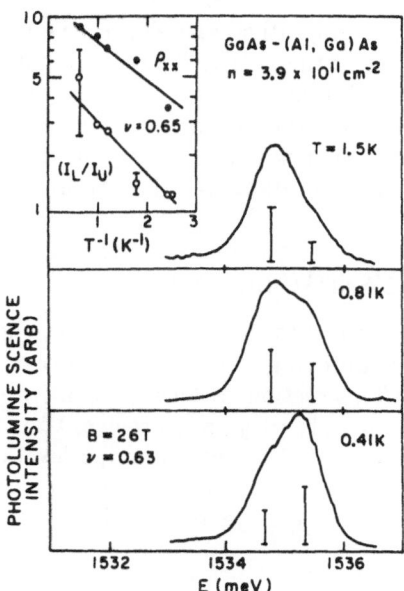

Fig. 5.   Temperature dependence of optical emission in the regime of the FQHE. Inset: $\log(I_L/I_U)$ and $\log(\rho_{xx})$ vs. $1/T$. After Ref. 12.

in the valence band.[12] When the Fermi level moves into the region of extended states there is an enhancement in the ability of the electron gas to screen the Coulomb potential of the hole. Such a "screening response" results in a build-up of electron density near the positive charge of the hole. The larger electron-hole overlap explains the observed increase in the strength of optical recombination.

When $\nu < 0.8(B > 20T)$ a new luminescence peak appears. It is labeled U in Fig. 2, where it is seen as a weak high energy shoulder on the main emission peak (labeled L). The new high energy emission increases in intensity with increasing field until, as seen in Fig. 3, the recombination is a well developed doublet for $\nu = 2/3$. At higher fields the higher energy component becomes dominant. Figure 4 displays the positions of the peaks in low temperature (0.4K) optical recombination.

The temperature dependence of optical recombination in the regime of the FQHE is very different from that at $\nu \lesssim 1$. The spectral lineshapes and intensities at $\nu \lesssim 1$ undergo only mirror changes in the range $.4 < T < 3K$. In the regime of the FQHE the temperature dependence is striking, as shown in Fig. 5 for filling factors 0.65 and 0.63. In this regime small variations in sample temperature result in large changes of the relative intensities in the doublet.

The temperature dependence in the regime of the FQHE is remarkable because the overall temperature variation, $1.5K = 0.13$ meV, is much smaller than the splitting between the two components of the doublet (0.5-0.7 meV). The inset to Fig. 5 compares the temperature dependence of the intensities ratio I_L/I_U of the two peaks with that of the FQHE. The results for ρ_{xx} at $\nu = 2/3$ shown in the lower inset to Fig. 4 have an activated behavior with $(\xi/2) = 0.44K = 0.04$ meV. The temperature dependence of I_L/I_U is similar. At $T = 2K$, where the FQHE minimum in ρ_{xx} is very weak the optical recombination becomes a singlet. These results suggest that the emergence of the high energy emission with intensity I_U is related to the states of the FQHE.

The "anomalies" observed in optical emission in the range $0.8 > \nu > 0.56$ suggest that also in the regime of the FQHE the recombination transitions are strongly altered by the interaction between the electron gas and the hole in the valence band. The many-body interactions are complex effects in which localization due to residual disorder might play a significant role. Current research in higher mobility systems could give clues to understand these intriguing phenomena.[39]

LIGHT SCATTERING BY LANDAU LEVEL EXCITATIONS

The changes in electron-electron interactions due to the appearance of the states of the FQHE should manifest in the elementary excitations of transitions between Landau levels.[40,41] This section considers recent results which indicate that the wavevector range $q > 1/l_0 = q_0$, most relevant to the FQHE,[26,27] is accessible in resonant light scattering by inter-Landau level excitations. These experiments have been carried out at integral values of ν.[24,25] Measurements in the fractional regime are expected to yield spectra of collective excitations of states of the FQHE.[42]

The inter-Landau level excitations, magnetoplasmons and spin-density excitations, have energies[43-46]

$$\omega(q) = \omega_c + \Delta(q,B) \tag{1}$$

where ω_c is the cyclotron energy. Hartree-Fock calculations of the dispersions $\Delta(q,B)$ show distinctive roton minima at finite wavevectors.[44,45] The roton is caused by the reduction at $q > q_0$ of the excitonic attraction between the hole in the ground state and the electron in the excited state.[47,48] These interactions also play a leading role in the theories of collective excitations in the regime of the FQHE.

Evidence of roton structure in the dispersions of inter-Landau level excitations was observed in the resonant inelastic light scattering spectra.[24,25] The measurements were carried out in high electron mobility GaAs-AlGaAs quantum wells and single heterojunctions. A massive breakdown of wavevector conservation, attributed to residual disorder, allows the observation of modes with large wavevectors $q > q_0 = 10^6$ cm^{-1}. The spectra were interpreted in terms of the critical points of calculated mode dispersions where $(\partial \omega / \partial q) = 0$.

Figure 6a presents results obtained at $\nu = 2$ in a single quantum well of width 250Å. The spectra consist of continua with well defined peaks. The relative intensities of the structures in the spectra depend on incident photon energy $\hbar\omega_L$. This is typical of light scattering spectra measured under strong and sharp resonant enhancements as shown in the inset. Figure 6b shows calculated mode dispersions based on results of Cheng.[49] The calculations are within the Hartree-Fock approximation. They consider finite-thickness of the 2D electron gas and coupling between modes at finite wavevector. The modes below ω_c are spin-density inter-Landau level excitations and those above are magnetoplasmons (charge-density inter-Landau level excitations). In Fig. 6 we find excellent agreement between the positions of the peaks in the spectra and the energies of the critical points in the calculated mode dispersion. Coupling to intersubband excitations at energies above 25 meV can be ignored.

Further work could unravel intriguing behaviors at non-integer filling factors. Thus far our experiments in the ultra-high mobility systems have been successful only near integer values of ν. As the Landau level filling factor is

278

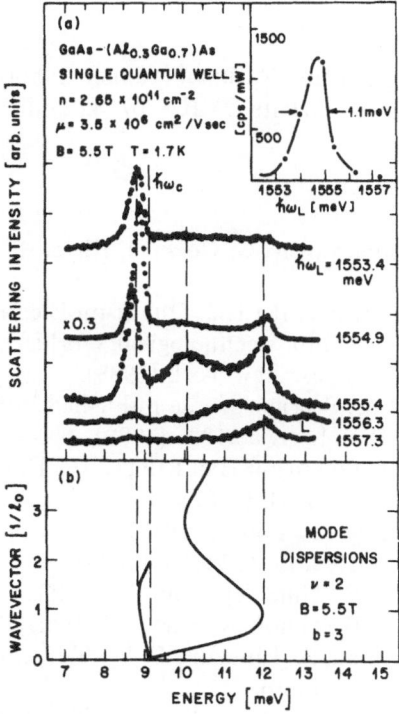

Fig. 6.   (a) Light scattering spectra at four incident photon energies.
Inset: profile of resonant enhancement at 8.8 meV.
(b) Calculated mode dispersions. After Ref. 25.

changed away from integer values there is a marked decrease in the light
scattering intensities. This indicates that the disorder potential required to relax
wavevector conservation is effectively screened by the 2D electron gas at non-
integer values of the filling factor. Current experiments at filling factors 2/3 and
1/3 could allow the observation of elementary excitations associated with the
FQHE.

CONCLUDING REMARKS

We have shown that optical absorption and emission measurements are
sensitive to electron-electron interactions within the 2D electron gas in the extreme
magnetic quantum limit. For the magnetic field range $\nu < 1$ we have proposed
that changes in optical emission are related to the response of the electron gas to
the Coulomb potential of the hole in the valence subband. This is supported by
recent calculations showing that at near integer filling factors the Coulomb hole
term for the valence subband hole dominate the effects of electron-electron
interactions in optical emission.[50,51] Extension of these theories to the regime of
the FQHE would provide a conceptual framework to interpret the intriguing
behaviors of optical emission near $\nu = 2/3$. Light scattering in the regime of the
FQHE is in its infancy. Screening of the disorder potential could be greatly
reduced due to formation of the gap associated with the states of the FQHE. The
concomitant loss of translational invariance may give access to the elementary
excitations of the FQHE.

ACKNOWLEDGEMENTS

The work at the Francis Bitter National Magnet Laboratory was supported by National Science Foundation Grants DMR-8807682 and DMR-8813164.

REFERENCES

1. D. C. Tsui, H. L. Stormer and A. C. Gossard, Phys. Rev. Lett. 48:1559 (1982).
2. D. C. Tsui and H. L. Stormer, IEEE J. Quantum Electron. 22:1711 (1986).
3. H. L. Stormer, in "Physics in a Technological World," A. P. French ed., American Institute of Physics, New York (1988).
4. J. P. Eisenstein and H. L. Stormer, to be published in Science.
5. R. B. Laughlin, Phys. Rev. Lett. 50:1395 (1983).
6. R. B. Laughlin, in "The Quantum Hall Effect," R. E. Prange and S. M. Girvin, eds., Springer, New York (1987).
7. B. I. Halperin, Helv. Phys. Acta 56:75 (1983).
8. F. D. M. Haldane, Phys. Rev. Lett. 51:605 (1983).
9. I. V. Kukushkin and V. B. Timofeev, Sov. Phys. JETP Lett. 44:228 (1986).
10. I. V. Kukushkin and V. B. Timofeev, Surf. Sci. 196:196 (1988).
11. B. B. Goldberg, D. Heiman, A. Pinczuk, C. W. Tu, A. C. Gossard and J. H. English, Surf. Sci. 196:209 (1988).
12. D. Heiman, B. B. Goldberg, A. Pinczuk, C. W. Tu, A. C. Gossard and J. H. English, Phys. Rev. Lett. 61:605 (1988).
13. B. B. Goldberg, D. Heiman, M. J. Graf, D. A. Broido, A. Pinczuk, C. W. Tu, A. C. Gossard and J. H. English, in "Proc. of the 19th Int. Conference of Semiconductors," W. Zawadski ed., Nauka, Warsaw (1989).
14. I. V. Kukushkin, V. B. Timofeev, K. von Klitzing and K. Ploog, in "Advances in Solid State Physics 28," U. Roessler ed., Vieweg, Braunschweig (1988).
15. A. Pinczuk, J. Shah, R. C. Miller, A. C. Gossard and W. Wiegman, Solid State Commun. 50:735 (1984).
16. G. E. W. Bauer and T. Ando, Phys. Rev. B31:8321 (1985).
17. Y. C. Chang and G. D. Sanders, Phys. Rev. B32:5521 (1985).
18. A. E. Ruckenstein, S. Schmitt-Rink and R. C. Miller, Phys. Rev. Lett., 56:504 (1986).
19. R. Sooryakumar, A. Pinczuk, A. C. Gossard, D. S. Chemla and L. J. Sham, Phys. Rev. Lett., 58:1150 (1987).
20. M. S. Skolnick, J. M. Rorison, K. J. Nash, D. J. Mowbraz, P. R. Tapster, S. J. Bass and A. D. Pitt, Phys. Rev. Lett. 58:2130 (1987).
21. F. C. Zhang, V. Z. Vulovic, Y. Guo and S. Das Sarma, Phys. Rev. B32:6920 (1985).
22. E. H. Rezayi and F. D. M. Haldane, Phys. Rev. 'B32:6924 (1985).
23. A. H. MacDonald, K. L. Liu, S. M. Girvin and P. M. Platzman, Phys. Rev. B33:4014 (1986).
24. A. Pinczuk, J. P. Valladares, D. Heiman, A. C. Gossard, J. H. English, C. W. Tu, L. Pfeiffer and K. West, Phys. Rev. Lett. 61:2701 (1988).
25. A. Pinczuk, J. P. Valladares, D. Heiman, L. N. Pfeiffer and K. W. West, to be published in Surface Science.
26. S. M. Girvin, A. H. MacDonald and P. M. Platzman, Phys. Rev. Lett. 54:581 (1985).

27. S. M. Girvin, A. H. MacDonald and P. M. Platzman, Phys. Rev. B33:7481 (1986).

28. B. B. Goldberg, D. Heiman and A. Pinczuk, Phys. Rev. Lett. 63:1102 (1989).

29. M. C. Smith, A. Petrou, C. H. Perry, J. M. Worlock and R. L. Aggarwal, in "Proc. of the 17th Int. Conf. on the Physics of Semiconductors," D. J. Chadi and W. A. Harrison eds., Springer, New York (1985).

30. C. H. Perry, J. M. Worlock, M. C. Smith and A. Petrou, in "High Magnetic Fields in Semiconductor Physics," G. Landwehr ed., Springer, New York (1987).

31. F. Meseguer, J. C. Maan and K. Ploog, Phys. Rev. B35:2505 (1987).

32. J. Sanchez-Dehesa, F. Meseguer, F. Borondo and J. C. Maan, Phys. Rev. 36:5070 (1987).

33. B. B. Goldberg, D. Heiman, M. J. Graf, D. A. Broido, A. Pinczuk, C. W. Tu, J. H. English and A. C. Gossard, Phys. Rev. B38:10131 (1988).

34. D. Heiman, Rev. Sci. Instrum. 56:684 (1985).

35. E. D. Isaacs and D. Heiman, Rev. Sci. Instrum. 58:1672 (1987).

36. T. Ando, J. Phys. Soc. Japan, 54:1528 (1985).

37. G. Bastard and J. A. Brum, IEEE J. Quantum Electron., 22:1625 (1986).

38. D. A. Broido and L. J. Sham, Phys. Rev. B31:888 (1985).

39. B. B. Goldberg, D. Heiman, A. Pinczuk, L. N. Pfeiffer and K. W. West, submitted for publication.

40. A. H. MacDonald, H. C. A. Oji and S. M. Girvin, Phys. Rev. Lett. 55:2208 (1985).

41. P. Pietiläinen and T. Chakraborty, Europhys. Lett., 5:157 (1988).

42. D. Heiman, A. Pinczuk, J. P. Valladares, L. N. Pfeiffer and K. W. West, work in progress.

43. K. W. Chiu and J. J. Quinn, Phys. Rev. B9:4724 (1974).

44. C. Kallin and B. I. Halperin, Phys. Rev. B30:5655 (1984).

45. A. H. MacDonald, J. Phys. C., 18:1003 (1985).

46. H. C. A. Oji and A. H. MacDonald, Phys. Rev. B33:3810 (1986).

47. I. V. Lerner and Yu. E. Lozovik, Sov. Phys. JETP 51:588 (1980).

48. Yu, A. Bychkov, S. V. Iordanskii and G. M. Eliashberg, Sov. Phys. JETP Lett. 33:143 (1981).

49. S. C. Cheng, PhD Thesis, Yale University (1989).

50. S. Katayama and T. Ando, Solid State Commun., 70:97 (1989).

51. T. Uenoyama and L. J. Sham, Phys. Rev. B39:11044 (1989).

GEMINATE RECOMBINATION IN MQW STRUCTURES IN A MAGNETIC FIELD

D. N. Mirlin, P. S. Kop'ev, V. F. Sapega, and A. A. Sirenko

A. F Ioffe Physico-Technical Institute
Academy of Sciences of the USSR
Leningrad 194021, USSR

INTRODUCTION

In studies of hot electron photoluminescence in multiple quantum well (MQW) structures [1] we have observed an unusual emission close to the excitation line. This luminescence arises in a magnetic field directed perpendicular to the plane of the quantum well. In what follows this luminescence will be called near-resonance luminescence.

We have studied MQW structures of the type of GaAs/A$\ell_{0.3}$Ga$_{0.7}$As with the width of the quantum wells in the range 50÷100Å and the width of the barriers of 100 Å. These structures have been grown by the molecular-beam epitaxy method onto <100> oriented GaAs substrates. The concentration of residual carbon acceptors was in the range of $10^{15} \div 10^{16}$ cm^{-3}.

The luminescence was excited by the 1.65, 1.83 and 1.92 eV lines of a Kr$^+$ laser. In this case the electron-hole pairs were excited inside the quantum wells. The exciting laser beam was directed perpendicular to the surface of the structure. The luminescence was observed in the back-scattering configuration. The magnetic field of up to 8T was applied perpendicular to the plane of the structure along the light beam (namely in the Faraday geometry).

Figure 1 shows emission spectra near the excitation line in one of the structures studied for an energy of excitation of 1.83 eV. The initial energy of the electrons created in the 1e subband was close to 215 meV. When the magnetic field exceeds 4T, near-resonance luminescence arises in the Stokes region. Its intensity increases rapidly with the increase of the magnetic field. The intensity decreases exponentially away from the excitation energy: $I(\epsilon) = I_0 \exp(-\epsilon/\epsilon_0)$, where ϵ_0 is close to 2 meV. If the exciting light was linearly-polarized the luminescence was also linearly polarized. The degree of polarization increased with the magnetic field. At B = 8T and for a Stokes shift of 1 meV the degree of polarization exceeds 60%. With the increase of the magnetic field a rotation of the plane of polarization was also observed, of the order of 3 degree per Tesla.

In a wide range of pumping intensities up to 5×10^2 W/cm^2 the luminescence intensity increased with the pumping linearly as is shown in the inset of Fig. 1. For higher pumping intensities the increase slows down.

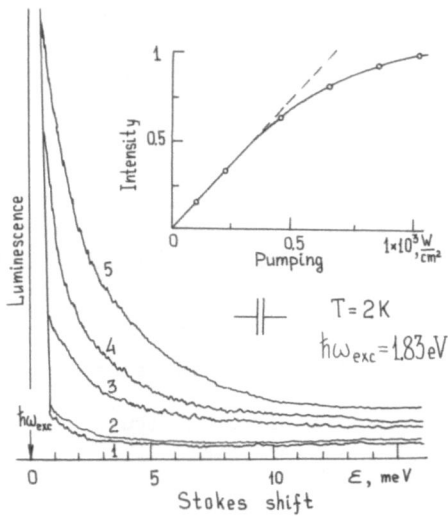

Fig. 1. Spectra of near-resonance luminescence of a 70/100
GaAs/A$\ell_{0.3}$Ga$_{0.7}$AS MQW structure in a magnetic field: 1 - B = 0,
2 - B = 5.5T, 3 - B = 6T, 4 - B = 6.5T, 5 - B = 7T, for the energy
of the initial electrons E_o = 215 meV. The notation a/b denotes the
well/barrier thickness in Å. The inset shows the dependence of the
near-resonance luminescence intensity on pumping.

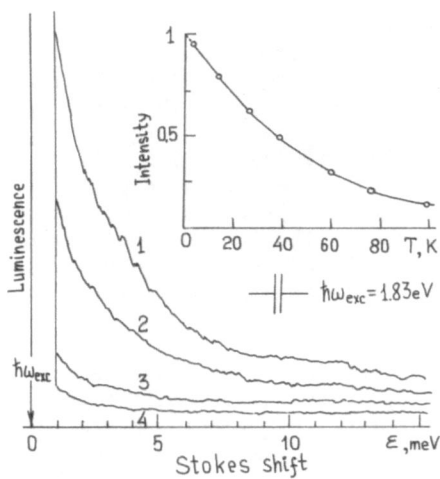

Fig. 2. Spectra of near-resonance luminescence (in a 50/100 MQW
structure) at different temperatures: 1 - T = 2K, 2 - 40K, 3 - T =
100K and magnetic field B = 7T, 4 - T = 2K, B = 0 and with
excitation at hω_{exc} = 1.833 eV. The inset illustrates the
temperature dependence of the intensity at ϵ = 1 meV and B = 7T.

Increasing the temperature leads to the disappearance of
luminescence as is shown in Fig. 2.

The second and third phonon replicas of the near-resonance
luminescence band studied were observed in the form of background
intensity near the LO-phonon Raman scattering lines (see Fig. 3).

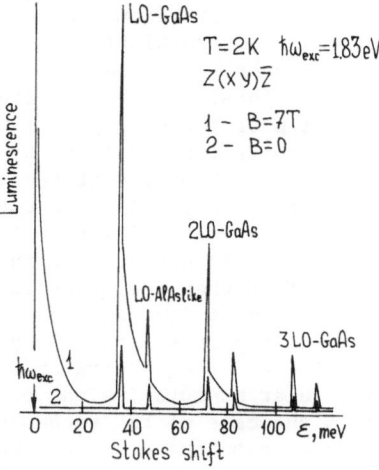

Fig. 3. Raman and near resonance luminescence spectra and its LO-phonon replicas at T = 2K and magnetic field: 1 - B = 7T, 2 -B = 0. Excitation with $h\omega_{exc}$ = 1.833 eV. The same structure as in Fig. 1.

One would like to note, that the behavior of the LO-phonon Raman scattering lines had very much in common with that of the near-resonance luminescence (see Fig. 4). An increase of Raman intensity with the magnetic field as well as the rotation of the plane of polarization have been observed. Also, as was the case for near-resonance luminescence, the magnetic field effects vanished as the temperature increased above 70K. The increase of Raman scattering intensity in quantum wells in a magnetic field has been described earlier [2].

In the Voigt geometry, i.e. when the magnetic field is in the plane of the quantum well, all described effects of the magnetic field in the secondary emission are absent.

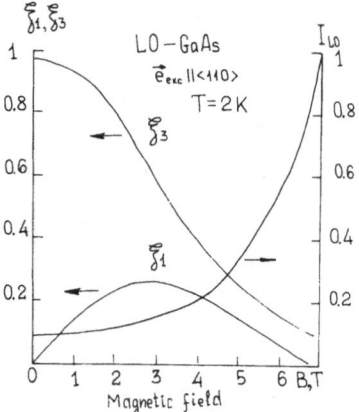

Fig. 4. Dependence of LO-phonon Raman scattering intensity in Z(XX+XY)Z geometry and Stokes parameters ξ_1 and ξ_3 on magnetic field with excitation at $h\omega_{exc}$ = 1.65 eV. The angle of rotation of the polarization plane is equal to $(1/2)\text{arctg}(\xi_1/\xi_3)$. The energy of the initial electrons E_0 - 60 meV (70/100 MQW structure).

We believe that the observed near-resonance luminescence is due to geminate recombination, namely the recombination of an electron and hole which have been created in the same act. The magnetic field restricts the spatial motion of two dimensional carriers in the plane of the quantum well, increasing the overlap of the wave functions of the electron and hole created in the same generation act (i.e. by one quantum of light), therefore increasing the probability of their recombination. A noticeable effect of the magnetic field can be expected when the product of the cyclotron frequency and the scattering time is larger than unity ($w_c\tau > 1$). If the electron energy is higher than the optical phonon energy this condition is satisfied when the magnetic field exceeds 4T. Near-resonance luminescence arises at just such a field. In the framework of this interpretation one can in a natural way explain the linear dependence on pumping intensity and the appearance of linear polarization due to correlation of the angular momenta (spins) of the electron and hole created in one act. In the case of the recombination of hot electrons and holes which have lost coherence, one would expect a square-law dependence of the luminescence intensity on pumping intensity, as has been observed in bulk GaAs [3].

The band-width of near-resonance luminescence and its LO-phonon replica are obviously due to the scattering of geminate electron-hole pairs from acoustic and optical phonons, respectively.

In the proposed model the increase of the intensity of the LO-phonon Raman lines in a magnetic field can also be explained if one assumes that geminate electron-holes pairs are an intermediate state in the scattering process.

When the electrons are excited with an energy lower than the optical phonon energy the scattering time is determined by the acoustic phonon scattering and is long. Instead of $w_c\tau \approx 1\div2$ we have $w_c\tau \gg 1$.

In this case instead of a monotonic increase of the near-resonance luminescence intensity with B we have observed magnetic oscillations of the intensity and polarization. These oscillations are shown in Fig. 5.

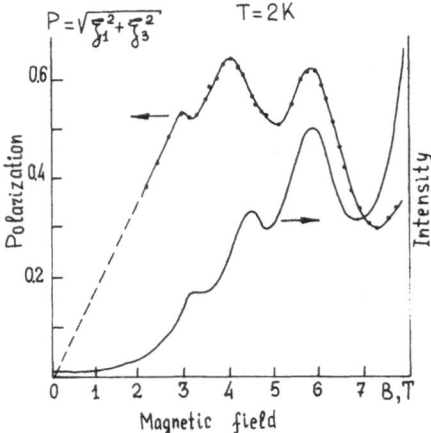

Fig. 5. Magnetic field dependence of near-resonance luminescence intensity at $\epsilon = 1$ meV and its polarization P in the case of electron energy $E_o < h\omega_{LO}$ (50/100 MQW structure). The energy of the initial electrons $E_0 = 32$ meV, ($h\omega_{LO} = 36.6$ meV in GaAs.)

The positions of the maxima are apparently related to the resonance of the excitation energy with the transitions between the Landau levels. There are now sharp features in the density of states that correspond to Landau levels.

The spatial localization of 2D e-h pairs in magnetic fields is favourable for geminate recombination.

In the case of overbarrier excitation of electron-hole pairs (that is, when the energy of the excitation exceeds the barrier height) all observed effects of magnetic fields on luminescence and Raman scattering are absent. So we believe that near-resonance luminescence is a peculiarity of 2D-systems.

The authors would like to thank A. G. Aronov, E. L. Ivchenko, V. I. Perel', V. B. Timofeev and B. P. Zakharchenya for helpful discussions in the course of this work.

REFERENCES

1. B. P. Zakharchenya, P. S. Kop'ev, D. N. Mirlin, D. G. Polyakov, I. I. Reshina, V. F. Sapega and A. A. Sirenko. Solid State Commun. 69, 203 (1989).
2. D. Gammon, R. Merlin, and H. Morcos, Phys. Rev. B35, 2552 (1987).
3. D. von der Linde, J. Kuhl, and E. Rosengart, Journal of Luminescence 24/25, 675 (1981).

INVESTIGATION OF TWO-ELECTRON-HOLE PAIR RESONANCES

IN SEMICONDUCTOR QUANTUM DOTS

N. Peyghambarian,[1] B. P. McGinnis,[1] K. I. Kang,[1]
Sandalphon,[1] S. W. Koch,[1,2] Y. Z. Hu,[2] M. Lindberg,[1]
B. Fluegel,[1] G. Kojoian,[3] L. C. Liu,[4] and S. Risbud[4]

[1]Optical Sciences Center, University of Arizona, Tucson, AZ 85721
[2]Physics Department, University of Arizona, Tucson, AZ 85721
[3]On leave from Department of Physics & Astronomy,
University of Wisconsin-Eau Claire, Eau Claire, Wisconsin 54701
[4]Department of Mechanical Engineering, University of California,
Davis, Bainer Hall, Davis, CA 95616

ABSTRACT

Our combined experimental and theoretical effort provide the first evidence for the presence of two-electron-hole pair resonances (or the biexciton states) in semiconductor quantum dots. Differential absorption measurements reveal the presence of an induced absorption feature on the high energy side of the bleached one-pair (exciton) resonance. The theory confirms that this induced absorption is the result of generation of excited biexciton states.

The recent interest in zero and one-dimensional confinement of semiconductors has inspired promising work in nanometer-scale spectroscopy.[1-19] At present, however, the most versatile structures are semiconductor microcrystallites embedded in glass matrices. In this paper we consider semiconductor microcrystallites precipitated from glass solution (doped glasses). These quantum dots are relatively inexpensive to produce, extremely robust, and have sufficient absorption for transmission studies.

The different dielectric constants between the semiconductor and glass keep carriers confined to the crystal. A typical semiconductor has a lattice constant of 4-6 Å, while its electron-hole Bohr radius is much larger, 20-150 Å. The crystallites investigated here have radii much larger than the lattice spacing, but comparable to the Bohr radius. Because the quantum dot covers many lattice sites, the properties associated with bulk semiconductors (bandgap, effective mass, etc.) still hold. However, an optically created electron-hole pair is confined by the boundary of the quantum dot. This quantum confinement effect causes the energy spectrum of the pair to become discrete. The absorption spectrum of an ideal quantum dot sample consists of a series of sharp lines. In practice, the samples are subject to a great deal of homogeneous and inhomogeneous broadening, due to the imperfections and the size distribution of the microcrystallites. Thus, the linear absorption spectrum exhibits broadened one-electron-hole pair resonances, which may be referred to as exciton resonances.

Figure 1 shows a typical linear absorption spectra of two CdS quantum dot samples with different sizes. These samples were prepared under two heat treatment conditions: 640^{0}C for six hours, and 700^{0}C for 1 hour. The noted temperatures and the times indicate the temperature and duration of the heat treatment. As is well-known, the larger the temperature and the longer the duration of the heat treatment, the larger are the semiconductor radii. The 640^{0}C/6 hr. sample has the smaller dot size compared with the 700^{0}C/1 hr. sample. As is clear from Fig. 1, the absorption spectra are blue shifted as a result of quantum confinement. The blue shift is larger for the smaller quantum dots as expected since quantum confinement is stronger for smaller sizes.

The measured linear absorption spectrum of the sample with small dots consists of a series of resonances. These resonances are considerably broadened for the larger dots. No distinct resonances are apparent in the absorption spectrum of the 700^{0}C sample. The transitions in the smaller dots are assigned to the quantum confined transitions between electron and hole confined states. The 640^{0}C samples clearly display two resonances which we assign to $e_{1s}h_{1s}$ and $e_{1p}h_{1p}$ transitions. The e and h refer to electron and hole states, and 1s and 1p refer to the quantum numbers for the particles, e.g. $e_{1s}h_{1s}$ means that both the electron and the hole are in the 1s-state. These quantum numbers are obtained from the solution of the Schrödinger equation for an electron and a hole inside a spherical crystal in the absence of Coulomb interaction, and with boundary conditions that the wavefunction vanishes at the surface of the spherical dot. The lowest energy transition is the $e_{1s}h_{1s}$ resonance, the next higher energy transition is $e_{1p}h_{1p}$, and so on. These allowed transitions are shown as vertical arrows in the inset of Fig. 1. The inclusion of Coulomb interaction only slightly modifies the linear absorption spectrum.

The nonlinear absorption of one of our CdS quantum dot samples at 15K was measured using a pump-probe technique. A tunable nanosecond laser was used as a pump pulse and excited the lowest energy transition. A weak broadband pulse was used as a probe, and its transmission was measured with and without the pump. The spectral position of the pump pulse, together with the linear and nonlinear absorption spectra, are shown in Fig. 2. The nonlinear absorption is plotted in the form of the negative change in absorption ($-\Delta\alpha L$). The positive peak in the ($-\Delta\alpha L$) spectrum around the pump frequency specifies bleaching of the 1s-1s transition, while the negative region on the high energy side of the bleached component is a result of an induced absorption. In our recent paper[16] we report on the observation of similar behavior in the CdSe quantum dots under femtosecond laser excitation.

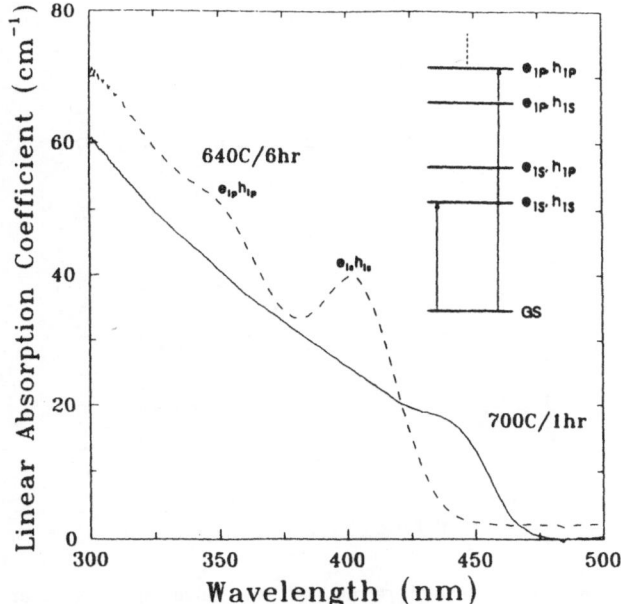

Fig. 1. The linear absorption spectrum for two samples of CdS quantum dots in glass. The crystallite sizes are smaller for the 640°C/6 hr. sample than the 700°C/1 hr. sample. The inset shows the schematics of the dipole allowed one-pair (exciton) transitions.

To explain these features, we have employed a theory for the nonlinear response of the quantum dots. Details of these calculations are presented in Ref. 19. The theory considers the one-pair states discussed above, and also the two-pair states that consist of two electrons and two holes in the quantum dot with the attractive and repulsive Coulomb terms. The quantum dots were modeled as dielectric spheres with perfect confinement of the optically generated electron-hole pairs. The wavefunctions of the exciton and biexciton states were expanded in terms of the eigenfunctions of the noninteracting electron-hole system. The expansion was truncated after M terms, and M was increased until satisfactory convergence was ensured. The kinetic energy terms were diagonal in the chosen basis and the Coulomb matrix elements were evaluated numerically. The resulting matrices for the one-and two-pair Hamiltonians were diagonalized numerically to yield the eigenfunctions and eigenvalues of the system. To improve the convergence of the procedure, we used the fact that the exciton and biexciton states were eigenfunctions of the total angular momentum operator and its z component. We started with a relatively small number, $M \cong 40$, of basis functions and successively increased M until the lowest energy eigenvalues converged to within a few percent. Typically, we used a few hundred eigenfunctions for the final results.

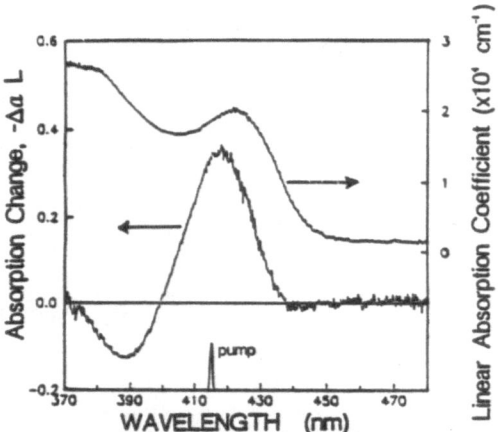

Fig. 2. Measured experimental results for the linear and nonlinear optical
properties of a sample of CdS quantum dots in glass. The linear
absorption spectrum, the absorption changes, $-\Delta\alpha L$, and the spectral
position of the pump are plotted.

The matrix-diagonalization procedure yielded the energies and wavefunctions not only for the exciton and biexciton ground state, but also for all the excited states included in the basic sets. Using these wavefunctions, we evaluated the various dipole matrix elements for transitions between the quantum dot ground state and the one-and two-electron-hole pair states. Inserting these results into the equation for the two-beam third-order susceptibility, we calculated the changes in absorption $\Delta\alpha$, which should be seen in a pump-probe experiment.

An example of the computed normalized transmission changes, $-\Delta\alpha$, is shown in Fig. 3 for quantum dots with $R/a_0 = 1.0$ where R and a_0 are the radius of the dot and the exciton Bohr radius, respectively, and for a broadening of $\hbar\gamma/E_R = 3$ with E_R being the bulk exciton binding energy. This spectrum was calculated for the case of resonant excitation of the energetically lowest exciton resonance, which in this case amounts to $(\hbar\omega-E_g)/E_R = 6.2$, where E_g is the bulk bandgap energy. The dominant feature in Fig. 3 is the bleaching of the exciton state, causing the positive peak in the spectra. In addition to this exciton saturation, we see a region of increasing absorption (negative $-\Delta\alpha$) in the region around $(\hbar\omega-E_g)/E_R \cong 12$, which is due to the transition to excited biexciton states. The dominant contribution to these transitions comes from biexcitons where the two electrons are in the 1s-state and the two holes are either both in the 1p-state, or one hole is in the 1s- and the other in the 2s-state, respectively. Such transitions would be dipole forbidden in the case without Coulomb interaction, and they are dipole allowed only because the Coulomb interaction breaks the symmetry of the quantum dots.

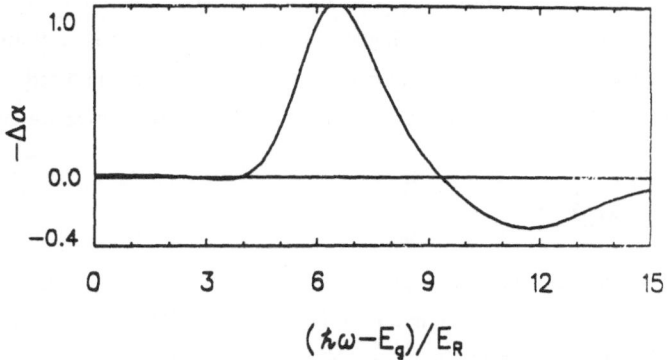

Fig. 3. Computed absorption change, $-\Delta\alpha$, for pump-probe excitation assuming pumping into the energetically lowest one-pair state. The parameters are $m_e/m_h = 0.24$, $R/a_0 = 1$ and $\hbar\gamma/E_R = 3$.

The good agreement between the experiment of Fig. 2 and theory of Fig. 3 strongly suggests the observation of excited biexciton states in quantum dots. The biexciton ground state is not seen in Fig. 2. Such a resonance is expected to be observed just below the exciton resonance as a result of the positive biexciton binding energy. However, the bleaching of the lowest exciton ($e_{1s}h_{1s}$ state) and the resulting increased transmission overwhelms the biexciton ground state absorption. In other words, the induced absorption just below the exciton due to the biexciton ground state is completely hidden by the exciton bleaching signal. The ground state of biexciton should be observable for systems with narrower exciton transitions. For example, in Fig. 4 we have plotted the calculated

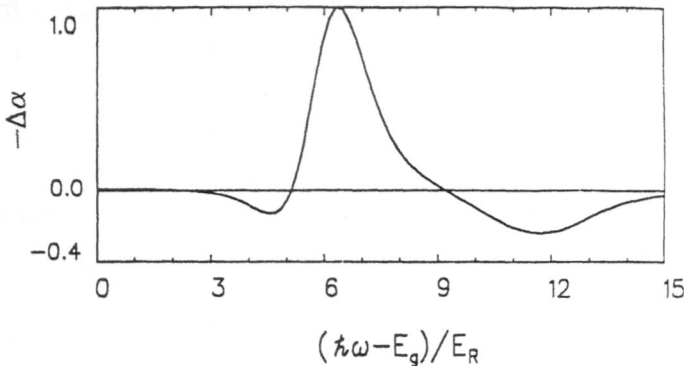

Fig. 4. Computed absorption changes, $-\Delta\alpha$ for pump-probe excitation assuming pumping into the energetically lowest exciton state. The parameters are the same as Fig. 3 except the broadening here is $\hbar\gamma/E_R = 2$.

absorption changes for a smaller broadening of $\hbar\gamma/E_R = 2$. The presence of induced absorption on the low energy side of the exciton bleaching is clearly demonstrated.

In conclusion, we have performed pump-probe experiments on quantum-confined CdS microcrystallites at low temperatures. We have measured bleaching of the lowest transition and an induced absorption at higher energy. By comparison with theory we attribute the bleaching of the one-pair states to state filling and the induced absorption to the generation of excited biexciton states as a result of the presence of Coulomb interaction.

ACKNOWLEDGEMENTS

We would like to acknowledge support from NSF (grant numbers ECS8909913 and INT8713068, travel grant), ARO (grant numbers DAAL-03-90-G-0006 and DAAL-03-89-K-0100), NATO, ONR/SDIO (grant number N00014-86-K-0719), the Optical Circuitry Cooperative of the University of Arizona, and CPU time from the John von Neumann Computer Center.

REFERENCES

1. A. L. Efros and A. L. Efros, *Sov. Phys.-Semicond.*, 16:772 (1982).

2. L. E. Brus, *J. Chem. Phys.*, 80:4403 (1984) and *IEEE J. Quantum Electron.*, QE-22:1909 (1986).

3. A. I. Ekimov and A. A. Onuschenko, *Sov. Phys.-Semicond.*, 16:775 (1982).

4. L. Banyai, M. Lindberg, and S. W. Koch, *Opt. Lett.*, 13:212 (1988; see also L. Banyai, Y. Z. Hu, M. Lindberg, and S. W. Koch, *Phys. Rev. B*, 38:8142 (1988).

5. S. Schmitt-Rink, D. A. B. Miller, and D. S. Chemla, *Phys. Rev. B*, 35:8113 (1987).

6. E. Hanamura, *Phys. Rev. B*, 37:1273 (1988).

7. U. Woggon and F. Henneberger, *J. De Physique C*, 2:225 (1988).

8. Y. Masumoto, H. Sugawara, and M. Yamazaki, *Proc. IQEC'88*, Tokyo, Japan, 1988, paper ThB3, pp. 618-619.

9. T. Takagahara, *Proc. IQEC'88*, Tokyo, Japan, 1988, paper ThB4, pp. 620-621.

10. N. F. Borrelli, D. W. Hall, H. J. Holland, and D. W. Smith, *J. Appl. Phys.*, 61:5399 (1987); D. W. Hall and N. F. Borrelli, *J. Opt. Soc. Amer. B*, 5:1650 (1988).

11. P. Roussignol, D. Ricard, C. Flytzanis, and N. Neuroth, *Phys. Rev. Lett.*, 62:312 (1989).

12. D. W. Hall, R. A. Haas, W. F. Krupke, and M. J. Weber, *IEEE J. Quantum Electron.*, QE-19:1704 (1983).

13. E. Hilinski, P. Lucas, and Y. Wang, *J. Chem. Phys.*, 89:3435 (1988).

14. L. J. Sham, *Superlattices and Microstructures*, 5:335 (1989).

15. N. Peyghambarian, S. H. Park, R. A. Morgan, B. Fluegel, Y. Z. Hu, M. Lindberg, S. W. Koch, D. Hulin, A. Migus, I. Etchepare, M. Joffre, G. Grillon, D. W. Hall, and N. F. Borrelli, in "Optical Switching in Low-Dimensional Systems," H. Haug and L. Banyai, Eds. New York: Academic, 1989, pp. 191-201.

16. N. Peyghambarian, B. Fluegel, D. Hulin, A. Migus, M. Joffre, A. Antonetti, S. W. Koch, and M. Lindberg, *IEEE J. Quant. Elect.*, 25:2516 (1989).

17. Y. Z. Hu, S. W. Koch, M. Lindberg, N. Peyghambarian, E. L. Pollock, and F. F. Abraham, *Phys. Rev. Lett.*, 64:1805 (1990).

18. S. H. Park, R. A. Morgan, Y. Z. Hu, M. Lindberg, S. W. Koch, and N. Peyghambarian, *Journ. Opt. Soc. Am. B*, accepted for publication.

19. Y. Z. Hu, M. Lindberg, S. W. Koch, and N. Peyghambarian, in proceedings of the SPIE conference, vol. 1216, ed. N. Peyghambarian, Jan. 16-17, 1990, Los Angeles, CA; see also Y. Z. Hu, M. Lindberg, and S. W. Koch, to be published in Phys. Rev.

MANY BODY EFFECTS IN HOMOGENEOUS QUASI 2D ELECTRON-HOLE PLASMA IN UNDOPED

AND MODULATION DOPED INGaAs SINGLE QUANTUM WELLS

L. V. Butov, V. D. Kulakovskii
ISSP of Academy of Sciences, Chernogolovka, 142432, USSR

T. G. Andersson, Z. G. Chen
Chalmers University of Technology, Gothenburg, SWEDEN

E. Lach, A. Forchel
Stuttgart University, Stuttgart, WEST GERMANY

D. Grutzmacher, RWTH, Aachen, WEST GERMANY

1. INTRODUCTION

During the past few years a number of studies have addressed the properties of an electron-hole plasma (EHP) in three-dimensional (3D) and two-dimensional (2D) semiconductor structures. The many body effects in the dense quasi-2D electron-hole (e-h) system were discussed in[1-6]. Interparticle interactions in a dense e-h system in a semiconductor lead to a renormalization both of the band gap and of the electron and hole dispersion laws. In many-particle theory, these changes are described by a self-energy Σ, which depends on the energy ϵ and quasi momentum k of the quasi particles[7,8]. ReΣ describes the renormalization of the dispersion laws for noninteracting electrons and holes.

$$\epsilon_{e,h}(k) = \epsilon^{o}_{e,h}(k) + Re\Sigma_{e,h}(k,\epsilon), \tag{1}$$

and Im$\Sigma_{e,h}(k,\epsilon)$ represents the damping of the one-particle states in the dense e-h system which is connected with Auger processes resulting from the three-particle collisions[9].

In 3D systems, the quantity $\Sigma(k,\epsilon)$ depends only weakly on k and ϵ and even on the particular species of charged particles because the screened interaction in the dense e-h system is of a short range nature[7]. The electron and hole bands in highly excited semiconductors thus undergo a basically rigid shift[7]. The rigid shift approximation has been used as well for a description of the emission spectra of the quasi 2D EHP in QW's[2]. The most reliable information on the density dependence both of a band gap shrinkage and of the carrier dispersion law may be obtained from magnetooptical measurements. The magnetic field perpendicular to the QW plane leads to a quantization of the motion of quasi-2D carriers in the plane of the QW and, as a consequence, to a discrete energetic spectrum of electrons and holes in QWs. The effective mass renormalization is easily detected as a change in the energy gaps between Landau levels.

We have investigated the influence of the multi-particle interactions on the dispersion law of the quasi-2D carriers in the dense ($r_s < 1$) EHP confined in QWs both in undoped and selectively doped heterostructures. In such a dense plasma the excitonic effects can be neglected. Here the dimensionless parameter $r_s = (\pi\, n_{eh} a_{ex}^2)^{-1/2}$ where n_{eh} is the EHP density and a_{ex} is the excitonic Bohr radius.

To obtain reliable results, great care was taken to prepare a photoexcited nonequilibrium e-h system of high homogeneity. An analysis of the photoluminescence spectra of a dense plasma in a wide region of plasma densities and magnetic fields has allowed us to measure with a high accuracy i) the damping of the one-particle states in the plasma (Sec. 4a), ii) the band gap renormalization for the lowest and higher index subbands (Sec. 4b), and iii) the renormalization of the reduced dispersion law for electrons and holes (Sec. 4c).

2. EXPERIMENTAL

To investigate the properties of a dense quasi-2D EHP we have used lattice matched undoped $In_{0.53}Ga_{0.47}As/InP$ SQW heterostructures (the QW width L_z = 15 nm) grown by the low pressure metalorganic vapor-phase epitaxy technique, and strained lattice, mismatched, selectively doped n-$Al_{0.3}Ga_{0.7}As$-$In_{0.18}Ga_{0.82}As$-GaAs SQW structures (L_z = 12 nm) grown by the molecular beam epitaxy method. In selectively doped structures an Si-doped 500 Å $Al_{0.3}Ga_{0.7}As$ layer ($N_{Si} \sim 10^{18}$ cm^{-3}) was separated from the QW by an undoped 100 Å $Al_{.3}Ga_{.7}As$ spacer. The dark electron density in the QW was $\sim 10^{12}$ cm^{-2} and remained essentially constant at excitation densities W < 10 W/cm^2. The maximum EHP density in a QW is determined by its depth and width. In our structures this maximum density was (5-6) × 10^{12}cm^{-2} ($r_s \sim 0.25$).

The photoluminescence measurements were carried out with the use of a cw Ar$^+$-ion laser with λ = 5145Å or a pulsed Cu-vapor laser with λ = 5105Å, a pulse duration of 20 ns and a repetition frequency of 10^4 Hz. The pulse length was sufficient to achieve quasisteady conditions. The emission was dispersed by a double monochromator with a dispersion of 10Å/mm and detected by a cooled photomultiplier with an S-1 cathode or a cooled Ge detector. The pulsed measurements were carried out only with the photomultiplier and a boxcar integrator (a strobe width of 5 ns) to realize a quasisteady regime.

The EHP of high homogeneity was created using an excitation laser beam spot on the sample larger than the entire area of the QW. To realize high densities of EHP under the cw Ar$^+$-ion laser excitation, we have defined mesa structures in the plane of the QW with dimensions down to 20 × 20 μm. The mesas were prepared by optical lithography and dry etching. The lateral confinement in the small mesa makes it possible to reach very high densities of about 5 × 10^{12} cm^{-2} for laser powers of 1W and to use a cw rather than a pulsed laser.

3. EHP IN QW WITHOUT MAGNETIC FIELD

Figure 1 represents the emission spectra of the QWs in the structures with and without a selective doping, recorded at 2K under various quasisteady excitation densities. The spectra for these two structures are strongly different only at low excitation densities. In this case the emission of localized excitons prevails in the spectra of the undoped heterostructure. The line halfwidth is about 5 meV and depends only on the quality of the QW. The free exciton transition energies for the n_z = 1-3 subbands determined by excitation spectroscopy are indicated by the arrows.

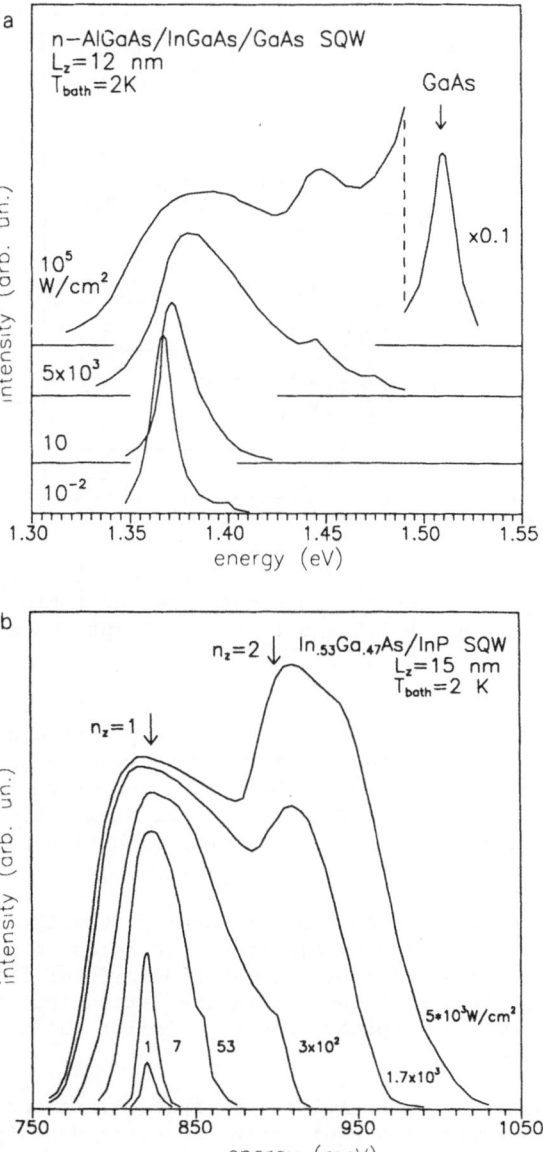

Fig. 1a. Photoluminescence spectra of a strained 120Å $In_{0.18}Ga_{0.82}As$ QW with an electron concentration due to the modulation doping of $1.1 \times 10^{12} cm^{-2}$ at the bath temperature $T_b = 2K$.

Fig. 1b. Photoluminescence spectra of an instrained 150 Å $In_{0.53}Ga_{0.47}As$ QW in the undoped InGaAs/InP structure at $T_b = 2K$ under the various excitation densities. The arrows indicate the $n_z = 1$ and 2 band edges at $n_{eh} = 0$ as found from the photoexcitation spectroscopy measurements.

For the excitation levels below 5 W/cm^2, the intensity of the emission line increases approximately proportionally to the excitation power, and the line width remains nearly constant. For higher excitation powers the line intensity saturates and we observe a broadening on its high energy side. The saturation of the maximum intensity corresponds to the complete filling of the conduction and valence $n_z = 1$ subband edges and is a direct consequence of the Pauli principle. At the higher excitations a new well-pronounced step appears in the emission spectra

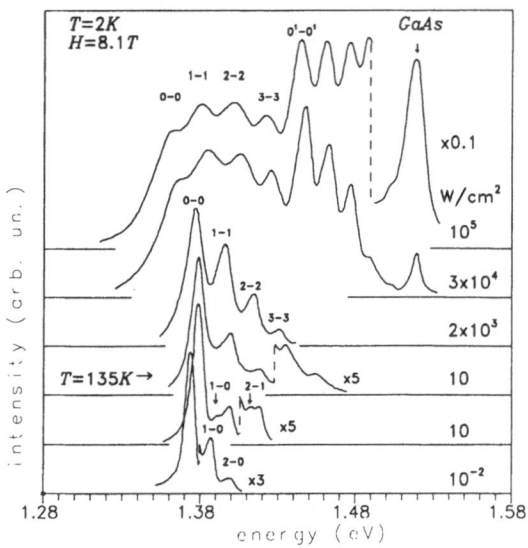

Fig. 2. Photoluminescence spectra of a strained 120Å $In_{0.18}Ga_{0.82}As$ QW with an electron concentration due to the modulation doping of 1.10×10^{12} cm^{-2} at H = 8.1 T and T_b = 2K (except one marked T_b = 135 K).

due to filling of the n_z = 2 subbands. This step corresponds to the allowed transitions between the electron and heavy-hole subbands (Δn_z=0). Note that a saturation of the maximum emission intensity was never observed in samples without mesas even at excitation levels as high as $10^6 W/cm^2$. This indicates a strong lateral expansion of EHP in large area samples.

Besides the strong emission line broadening associated with the increase in the kinetic energy of electrons and holes in a dense EHP, the emission spectra show as well a pronounced red shift of the low energy edge of the emission line with increasing excitation intensity. The second step in the emission spectra shows a similar shift, but with a markedly smaller rate.

Under low excitation intensities, the spectrum of the selectively doped heterostructure with the equilibrium electron density n_{2D}^o = 10^{12} cm^{-2} is dominated by an emission resulting from interband recombination of electrons and holes, and the line width of about 40 meV is approximately equal to the electron Fermi energy. Many particle effects have significant manifestations only at $W > 10^3 W/cm^2$, when the nonequilibrium photoexcited e-h pair density becomes comparable to n_{2D}^o (Fig. 1). Under the higher excitation densities, the behavior of the emission spectra is qualitatively similar to that in the undoped samples. We observe a strong broadening of the emission line and the appearance of the second step in the spectrum. These are due to the increase in the electron and hole Fermi energies and the filling of the n_z = 2 subbands. In addition, the emission line shifts to lower energies because of the renormalization of the band gap.

Unfortunately, the blurring of the steps, caused primarily by the pronounced decay of the one-particle states far from the Fermi level, is quite significant both in the undoped and selectively doped hetero-structures. It leads to large errors in the determination of the EHP

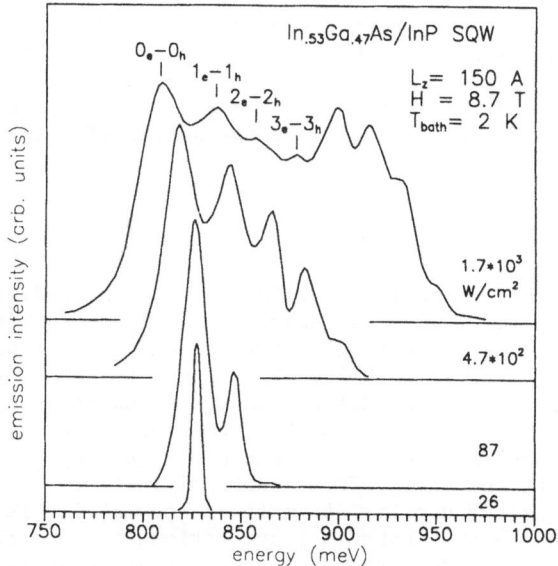

Fig. 3 Photoluminescence spectra of an unstrained 150Å In$_{0.53}$Ga$_{0.47}$As QW
in the undoped InGaAs/InP structure at H = 8.7 T and T$_b$ = 2K
under the various excitation densities.

parameters. To improve the situation we resort to measurements in a
magnetic field.

4. EHP IN THE QW UNDER A MAGNETIC FIELD

In the emission spectra of an EHP under a magnetic field
perpendicular to the QW plane, the effective masses, transition
probabilities, and magnitudes of the one-particle damping in EHP are
responsible for different spectrum parameters, namely, for the emission
line spacings, relative line intensities, and their half-widths,
respectively. Figures 2 and 3 illustrate the change in the emission
spectra of the EHP in a magnetic field with an increasing density. Only
the allowed j_e-j_h = 0 transitions between the electron (j_e) and hole (j_h)
Landau levels are dominated in the spectra of the undoped structures with
an equal filling of Landau levels. In the structures with selective
doping, at the low excitation levels there are three filled Landau levels
in the conduction band (due to selective doping) and only one partly
occupied level in the valence band. Therefore three lines in the spectrum correspond to the one allowed (0_e-0_h) and two forbidden (1_e-0_h and
2_e-0_h) transitions. Forbidden transitions appear due to an imperfectness
of the QW.

Figures 2 and 3 show that an increase in the EHP density gives rise
to i) appearance of new lines corresponding to the allowed transitions
between the higher Landau levels (due to a filling of these levels), ii)
broadening of the all lines (due to an increase of the one-particle
damping), and iii) a shift of all lines to lower energies (due to renormalization of the band gap). In the case of structures with selective
doping, there is a small contribution to the line shift from a change in
the electric field in the QW.

a) The damping of the one-particle states

The damping of the one-particle states in a dense e-h system is an important parameter in a many-particle theory. The damping increases for the one-particle states on going below the Fermi level because of an increase of the scattering probability[8]. In the 2D case, the halfwidths of the lines corresponding to transitions between discrete Landau levels directly reflects the reduced damping of the states (for the high quality QWs)

$$\Gamma^2 = \Gamma_e^2 + \Gamma_h^2 . \qquad (2)$$

The pronounced structure in the emission spectra in Figs. 2 and 3 indicates that, both in the undoped and selectively doped structures, the damping is small as compared with the particle energy (calculated from the Fermi level) for every $r_s > 0.25$. This means that the one-particle states in the quasi-2D EHP are well defined, and one can use the approximation (1) for the dispersion law. Figures 2 and 3 show that the halfwidth of the lines and, hence, the one-particle damping, increase sublinearly both with the energy from the Fermi level (at a fixed EHP density) and with the EHP density (at a fixed energy). This is in qualitative agreement with theoretical predictions[8], although more detailed calculations are desirable for a quantitative analysis.

b) Band gap renormalization

To determine the band gap renormalization we have measured the shift of the lowest EHP emission line 0_e-0_h. For each $n_z = 1$ and 2 subband, the energy gap at zero density was found from the spectral position of the excitonic lines in the photoexcitation spectra (with a correction for the excitonic Rydberg). We consider only EHP with two or more occupied Landau levels, and neglect the excitonic effects. Under these conditions the line 0_e-0_h moves monotonically to lower energies. A comparison of Figs. 1 and 2 demonstrates the advantage of a magnetic field. First, it gives a significantly higher accuracy to the determination of the band gap, because the band gap in EHP emission spectra recorded without a magnetic field is strongly obscured because of the broadening effects. Second, we have the possibility to obtain the EHP densities directly from the emission spectra by calculating the number of filled Landau levels, because the number of states in the Landau levels, determined only by the magnitude of the magnetic field and does not depend on the EHP density.

The dependence of the band gap renormalization on the EHP density for the $n_z = 1$ and 2 subbands is shown in Fig. 4 for a 150Å InGaAs QW in an undoped InGaAs/InP structure. It shows that the band gap shrinkage for the $n_z = 1$ subband is significantly larger than for the $n_z = 2$ subband. This result is opposite to that observed for a 3D EHP distributed over a few different bands. It is well established that in the case of the 3D plasma the deficiency in the exchange energy contribution for a weakly occupied band is strongly compensated by an increase in the correlation energy, their sum being a function of the total plasma density[10]. Figure 4 shows that the correlation energy contribution to the self-energy of the quasi-2D EHP for a weakly occupied second subband is too small to compensate the deficiency in the exchange energy arising from its weak filling. This experimental result is in qualitative agreement with recent calculations of the self-energy for an EHP in QWs[4,11].

The result of the available theoretical calculations for the band gap renormalization in the full RPA approximation is shown at Fig. 4 by the solid line. The gap shrinkage measured is in agreement with theoretical calculations only for $n_{eh} < 3 \times 10^{12}$ cm^{-2} ($r_s > 0.4$). The disagreement at the higher densities is striking because the RPA approximation should be more accurate the higher the EHP density. A

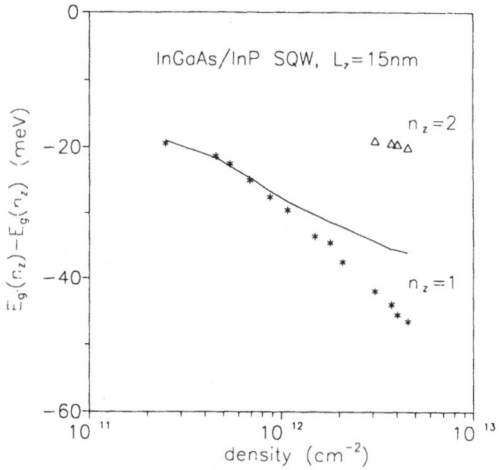

Fig. 4. Measured density dependences of the band gap renormalization for the n_z = 1 and 2 subbands in the 150Å $In_{0.53}Ga_{0.47}As$ SQW at T_{bath} = 2 K. The renormalization for each subband is plotted relatively to the corresponding subband edge at zero density. The solid line represents the calculated band gap renormalization for a 2D EHP in the full RPA approximation[12].

possible reason may be in an additional contribution to the band gap shrinkage from the electron-LO-phonon interaction. In particular, estimates of nonequilibrium LO-phonon occupation numbers in the QW show them to be rather high, unlike the acoustic ones. This is due to the high generation rate of LO phonons (the electron-hole recombination in a dense EHP is mainly nonradiative), and to the small velocity of these phonons which prevents them from leaving the QW (the lifetime of LO phonons is ~ 10 ps). As to the acoustic phonons, due to their high velocity of 10^5 cm/s they leave the 100Å QW before decaying into long wavelength phonons.

c) Renormalization of the dispersion law

As follows from Eq. (1), information about the dispersion of the one-particle states in a dense EHP can be obtained by the same methods as for free carriers in the empty band. Therefore, we have used the measurements of the energy gaps Δ_{ij} between the emission lines i_e-i_h and j_e-j_h corresponding to the allowed transitions between Landau levels. In the frameworks of the rigid band shift model, the gaps Δ_{ij} should remain unchanged, since the number of states in every Landau level is independent of the plasma density. Figures 2 and 3 show that this is not the case. The shifts of different lines are significantly different in both the neutral and charged EHP. Thus, we find that the approximation of a rigid band shift for the quasi-2D EHP is too crude.

Some important details of the changes in the density of states for carriers in the EHP are seen from the density dependencies of the gaps between the four low Landau levels (Fig. 5). At high densities of n_{eh} > 1.5 × 10^{12} cm^{-2} all the gaps Δ_{ij} show a similar behavior. They slightly increase both in the undoped and modulation doped structures. Taking into account that the density of states $\rho(\epsilon) \sim \Delta^{-1}$, one can conclude that below the Fermi level ρ decreases weakly and monotonically with increasing EHP density. At smaller densities a similar behavior of ρ is observed only for energies far from the bottom of the band. The

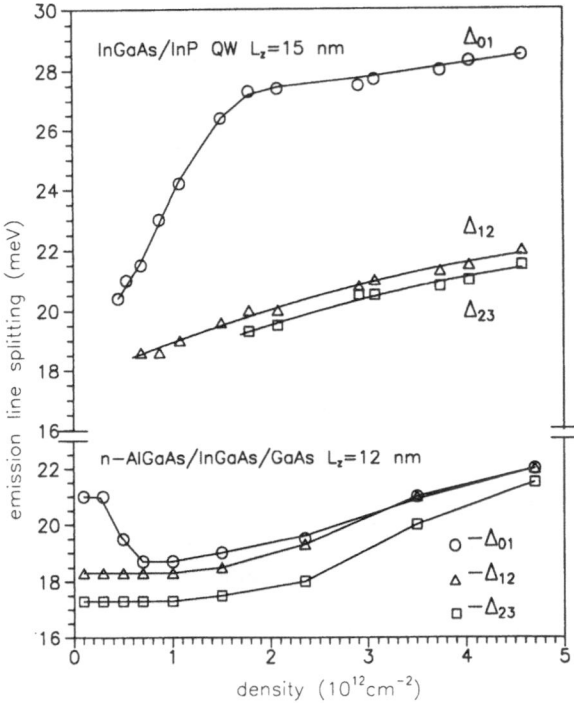

Fig. 5. Measured dependences for the energy gaps between the four low Landau levels in the quasi 2D EHP in the undoped unstrained $In_{0.53}Ga_{0.47}As$ QW (H = 8.7 T) and in the strained $In_{0.18}Ga_{0.82}As$ QW with a 10^{12} cm^{-2} equilibrium electron density due to modulation doping (H = 8.1 T) as functions of the photoexcited carrier density.

density dependence of Δ_{01} show, however, that the change in the density of states near the bottom of the band is strong and, in addition, has opposite signs for the neutral EHP in the unstressed QW in InGaAs/InP structures and charged EHP in the stressed QW in n-AlGaAs-InGaAs structures.

In the case of the charged ($n_{2D}^{o} = 10^{12}$ cm^{-2}) EHP in the strained QW we observe a noticeable increase of Δ_{01} with a decrease of the non-equilibrium hole concentration below 10^{12} cm^{-2}. This effect should be attributed to changes in the valence band, since the concentration of the photoexcited pairs is small compared to the equilibrium electronic concentration. This conclusion is supported by additional studies of the density of states in the purely electronic system in InGaAs QWs (n-AlGaAs-InGaAs-GaAs) with different equilibrium electronic concentrations. The measurements of the energy gap between the 0_e-0_h and 1_e-0_h emission lines at very low excitation densities have revealed a small change in m_e but of opposite sign, namely a decrease of approximately 5-7% with increasing density from 0.5 to 1×10^{12} cm^{-2}.

An increase of Δ_{01} with decreasing hole density is, in principle, expected at low n_h because the localization of holes in the InGaAs band gap fluctuations in the QW plane increases with decreasing hole density[12]. To exclude this effect we have increased the lattice temperature up to 135 K. However, the increase in Δ_{01} was found to be only very weakly dependent on temperature. This indicates that it cannot be explained only by hole localization effects.

The decrease in the density of a neutral EHP from 1.5 down to 0.4×10^{12} cm^{-2} leads to a 30% increase of the density of states near the band edge. Such a strong dependence is also unexpected. As was mentioned above, studies of the electronic system in the selectively doped InGaAs QWs have revealed only a weak increase of m_e in this density region. This is in agreement with the recent measurements m_e for low density quasi-2D electron system in n-AlGaAs-GaAs heterojunctions[13].

The decrease of the reduced density of states in the neutral EHP in an InGaAs QW with an increase of the plasma density may be to some extent enhanced by the changes in the valence band splitting due to the appearing of holes in the lowest hole subband. In particular, a possible increase of the gap between the light and heavy hole subbands should lead to some decrease of m_h and the density of states near the bottom of the band[14]. In the lattice mismatched AlGaAs-InGaAs QWs, this effect should be negligible because of a high strain induced splitting of the valence band. In this case the opposite behavior of Δ_{01} in the strained and unstrained QWs (Fig. 5) can be qualitatively explained.

As a crude estimate of the expected increase in the hole cyclotron frequency, we have used the results for the empty valence band[14]. The estimates result in a value too small to explain the experimental results even with as large an increase in the light-heavy hole splitting as 20 meV. This means that many-body effects should be taken into account.

5. CONCLUSION

Magnetooptical measurements have allowed to determine the main properties of quasi-2D EHP in QWs prepared with great care to have high homogeneity in space and time. Their properties are found to be different from those discussed earlier[6]. No saturation effects were found in the band gap and effective mass renormalization. In contrast with 3D EHP, the band gap renormalization for subbands with different occupations are significantly different. A very strong renormalization is found as well for the density of states in the bands. So, a change in the density of states as large as 30% is observed for a neutral EHP in unstrained InGaAs QWs at energies near the band edge. The effect decreases quickly with on approaching the Fermi level. Note, that recent RPA calculations[15] predict a strong change in the density of states near the renormalized band edge for EHP in QWs, but they do not explain the observed density dependence in a wide region of EHP densities. Additional calculations of the self-energy are necessary for quasi-2D EHP in QWs to clarify the situation.

ACKNOWLEDGEMENTS

We want to thank C. Ell, S. E. Esipov, I. V. Kukushkin, S. V. Meshkov, M. Pilkuhn, V. B. Timofeev, and R. Zimmermann for useful discussions. The financial support of our work by the Deutsche Forschungsgemeinschaft, the Swedish National Board for Technical Development (STU), and the Swedish Natural Science Research Council (NRF) is acknowledged.

REFERENCES

1. G. Trankle, E. Lach, A. Forchel, P. Scholz, C. Ell, H. Haug, G. Weimann, G. Griffiths, H. Kreemer, and S. Subbanna, Phys. Rev. B36, 6712 (1987).
2. G. Trankle, H. Leier, A. Forchel, H. Haug, C. Ell, and G. Weigmann, Phys. Rev. Lett. 58, 419 (1987).

3. E. Zielinski, F. Keppler, K. Streubel, F. Scholz, R. Sauer, and W. T. Tsang , "Superlattices and Microstructures, 5, 555 (1989).

4. C. Klingshirn, C. Weber, B. S. Chemla, D. A. B. Miller, G. E. Gunningham, C. Ell, H. Haug, NATO Workshop on "Optical switching in low dimensional systems," Marbella, Spain.

5. M. Potemski, J. C. Maan, K. Ploog, G. Weimann, Proc. of 19th Intern. Conf. on Phys. of Semicond., Warsaw, p. 119 (1988).

6. M. Potemski, J. C. Maan, K. Ploog, G. Weimann, Proc. of 8th Intern. Conf. EP2DS8, Grenoble, p. 265 (1989).

7. T. M. Rice, Solid State Physics, vol. 32, p. 1 (1977).

8. R. Zimmerman, Many particle theory of highly excited semiconductors, Teubner-Texte, Leipzig (1988).

9. P. T. Landsberg, Sol. State Electr. 10, 513 (1977).

10. Electron-hole droplets in semiconductors, eds. C. D. Jeffries and L. V. Keldysh, North-Holland, Amsterdam, N.Y., p. 95 (1983).

11. C. Ell, H. Haug, Phys. Stat. Sol., to be published.

12. T. Ando, A. B. Fouler, F. Stern, Rev. Mod. Phys., 54, 437 (1982).

13. I. V. Kukushkin, A. S. Plaut, K. von Klitzing, K. Ploog, Phys. Rev., submitted.

14. G. L. Bir, G. I. Pikus, Deformation Effects in Semiconductors, Science, Moscow, (1964).

15. R. Jalabet, S. Das Sarma, Proc. of the 8th Inter. Conf. EP2DS8, Grenoble, p. 369 (1989).

PULSED DIFFUSING-WAVE SPECTROSCOPY IN DENSE COLLOIDS

Arjun G. Yodh*, Peter. D. Kaplan*, and David J. Pine**

*Department of Physics, University of Pennsylvania
Philaadelphia, PA 19104
**Exxon Research and Engineering
Annandale, NJ 08801

There is a rich variety of systems in nature which are essentially dense aggregations of particle-like structures whose positions vary randomly in space and time. Some common examples include fog, smog, microemulsions, suspensions of glass or polymer microspheres, red blood cells, and ocean particles. In order to understand better and ultimately to gain control over these suspensions it is highly desirable to characterize the structure and dynamics of these systems.

Light scattering has been widely used and is particularly well-suited for structural and dynamical studies of colloids because the important length scales, such as particle size and interparticle spacing, are comparable to the wavelength of light. Unfortunately, this advantage is often offset by the fact that colloidal particles scatter light so strongly that multiple scattering becomes a significant problem at all but the lowest particle concentrations. As a consequence, progress in the study of dense colloids has lagged that of colloids at very low particle concentrations.

Recently a new spectroscopy has been developed and applied to study the properties of colloidal suspensions which multiply scatter light [1,2,3]. The technique, called diffusing-wave spectroscopy (DWS), exploits the diffusive nature of light transport in strongly scattering media to relate temporal intensity fluctuations of the scattered light to average particle motion. In contrast to more traditional dynamic light scattering methods [4,5], DWS probes particle motion over length scales much shorter than the light wavelength, and offers the possibility of studying the strongly correlated particle motions often present in dense media.

In this contribution we introduce a new diffusing-wave spectroscopic probe which utilizes light pulses and exploits the phase fluctuations of optically gated photons to eliminate the usual average over photon pathways. The principles are discussed and experimentally demonstrated. We have used the method to test several critical assumptions of the original multiple light scattering theories, and we have studied particle diffusion in the high volume fraction limit where hydrodynamic interactions are important. Since the general field is quite new we will review the important aspects of conventional DWS before discussing the pulsed ideas and our experiments.

Review of Diffusing-Wave Spectroscopy

The advances of DWS are best appreciated against the backdrop of the widely successful spectroscopy of quasielastic (or dynamic) light scattering (QELS) [4,5]. In QELS the low frequency noise spectrum of the scattered light is analyzed to obtain dynamical information about the various mechanical degrees of freedom of the scatterers.

An experimental set-up for a typical QELS experiment is shown in Figure 1. Here a volume of solvent containing N identical macromolecular scatterers is illuminated by a plane wave with a frequency ω and a wavevector $k_0 = 2\pi/\lambda$. Scattered light is detected far from the sample at an angle θ with respect to the incident propagation direction. For dilute samples we can ignore multiple scattering effects, and the electric field at the detector is *superposition* of the fields radiated by each of the individual particles. Since the position of each particle is random and varies randomly in time, the field at the detector, E(t) will fluctuate in time. The phase of the electric field at the detector due to the jth particle depends on its position r_j, its field strength A_j, and the scattering wavevector q, where $q = |q| = 2k_0\sin\theta/2$. The physical content in these fluctuation measurements is contained in the autocorrelation function, $g_1(\tau) = \langle E^*(t)E(t+\tau)\rangle/\langle|E(t)|^2\rangle$, of the scattered electric field. For Brownian motion in dilute systems we typically have $A_j = A$ and uncorrelated particle motion. In this case $g_1(\tau)$ reduces to an exponential function with a decay rate that is inversely proportional to the time it takes a particle to move a length $\sim 1/q$. Thus by varying the scattering angle θ it is possible to measure the relaxation of particle density fluctuations over length scales of $\lambda/2$ and larger. In practice the intensity autocorrelation function, $g_2(\tau) = \langle I(t)I(t+\tau)\rangle/\langle I(t)\rangle^2$, is often the quantity measured and the Bloch-Siegert [6] relation, $g_2(\tau) = 1+|g_1(\tau)|^2$, is used to derive $g_1(\tau)$ from the data.

QELS is a "single scattering spectroscopy" in the sense that the scattering problem is well posed only within the Born approximation. In more dense systems incident photons experience many scattering events before emerging from the medium and it becomes essential to incorporate the multiple scattering process directly into our interpretation of the fluctuation spectra. Indeed it is precisely because of this limitation that a great variety of systems remain unstudied.

At high concentration (i.e., volume fractions, $\phi > 0.2$), the simple system of strong macromolecular scatterers undergoing Brownian motion provides a good example of a class of dense, fluctuating random media that we wish to observe. Consider the experimental geometry depicted in Figure 2. Here a dense sample of Brownian particles is illuminated by a plane wave on the front face, and a portion of the light that has propagated through to the other face is collected through a small aperture at the output plane.

Microscopically one can envision each photon traveling ballistically between particles, and experiencing changes in propagation direction after each scattering event.

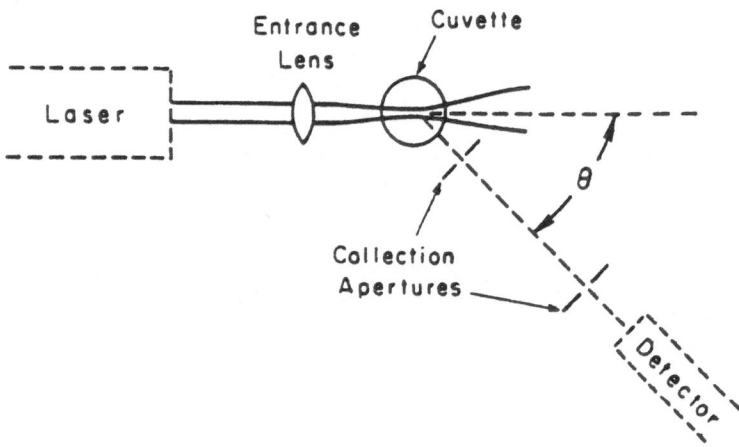

Figure 1. Experimental set-up for a typical QELS experiment. Scattered light is detected at an angle θ with respect to the incident propagation direction.

Three length scales characterize photon transport in the media: (1) s, the total distance traveled by a photon, (2) l, the mean distance traveled by the photon between particle encounters, and (3) l*, the transport mean free path of the photon. Physically, l* corresponds to the mean distance traveled by a photon before its propagation direction is completely randomized. Thus l* is the random walk step size for the "diffusing photons". Typically if s >> l* then the diffusion approximation is quite good. We assume this to be true, and we also assume that the disorder in the sample is uncorrelated.

Consider a single photon path of length s through the media. A typical photon will experience n=s/l scattering events. We can write the phase of the electric field for the light emerging along this pathway in terms of the position of the jth particle r_j, and the momentum transfer q_j for the jth scattering event [7,8]

$$E_s(t) \sim e^{-i\omega t} \prod_{j=1}^{N} e^{iq_j \cdot r_j}(t) \tag{1}$$

In contrast to QELS, we note that the phase shifts due to each scattering event enter *multiplicatively* rather than in an additive way, and that the momentum transfer q_j are different for each scattering event.

The time-averaged autocorrelation function $g_1^s(\tau)$ of the scattered field takes on a particularly simple form when the particles move independently and the particle displacement is a random Gaussian variable. Then, if we assume that the photon momentum transfer is independent of the particle displacement, we can show that [1,7,8],

$$g_1^s(\tau) = \exp[-k_0^2 \langle \Delta r^2(\tau) \rangle (s/l^*)/3] \tag{2}$$

where $\langle \Delta r^2(\tau) \rangle$ is the mean square displacement of a particle in time τ. Equation 2 is the primary result of the simplest DWS treatment.

Information on the dynamics of particle motion is contained in the decay of $g_1^s(\tau)$. In contrast to QELS, we see that the field correlation function is sensitive to particle motion

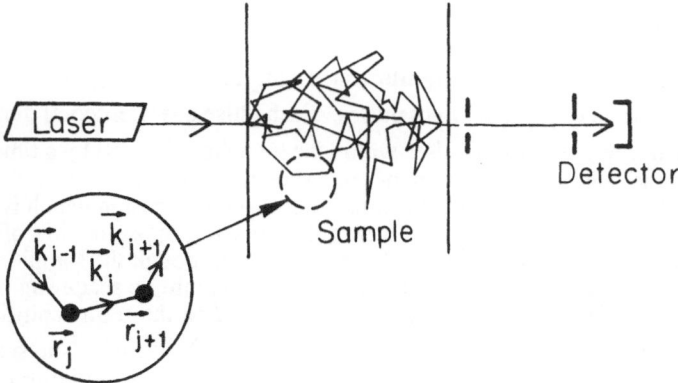

Figure 2. *Top:* Schematic of a typical DWS measurement. Light from laser is directed onto a dense colloidal suspension. Each photon travels through the sample along a complicated path. At the ouput face we collect a portion of the emerging light and direct it onto a photomultiplier tube. *Inset:* Magnification of a portion of a photon path. Here the photon encounters two particles at locations r_j and r_{j+1}. Thephoton travels ballistically between particles and experiences a momentum transfer in the jth collision given by $q_j = k_j - k_{j+1}$.

over length scales $\lambda/\sqrt{s/l^*}$, which, since $s \gg l^*$, is generally much less than λ. Thus, in addition to being able to probe the dynamics of optically turbid samples, DWS is sensitive to motion on a substantially different range of length and time scales than QELS.

Qualitatively we expect all of these dynamic light scattering correlation functions to decay in the time it takes the phase of the scattered field to change by π. In single scattering experiments this occurs when a typical particle position changes by $\sim 1/q$ $(> \lambda/2)$ along the direction parallel to q. In the multiple light scattering experiments this occurs when the *total* particle displacement projected along the direction of the output speckle wavevector changes by $\sim \lambda$. Since the scattered photons encounter many particles en route to emerging from the sample, the distance that *each* particle must move is much less than λ. Loosely speaking we can associate a phase shift of $\sim k_0(\Delta r(\tau))$ with each step in the photon random walk. Since the direction of $\Delta r(\tau)$ is random, the total phase shift along any particular direction will scale as the square root of the number of photon random walk steps. Equation 2 is a quantitative statement reflecting this simple idea.

The first experiments demonstrating these ideas were carried out with cw lasers [1,3,7] In this case the total electric field autocorrelation, $G_1(\tau)$, function is computed by incoherently summing the contributions of each path-dependent $g_1^s(\tau)$ weighted by $P(s)$, the probability that a photon will travel a distance s through the medium, i.e.

$$ G_1(\tau) = \int_0^\infty g_1^s(\tau) \, P(s) \, ds. \qquad (3) $$

For purely Brownian motion, $\langle \Delta r^2(\tau) \rangle = 6D_0\tau$, and $G_1(\tau)$ is the Laplace transform of $P(s)$ with scale factors that are simply related to D_0. The pulsed-DWS technique (PDWS) which we describe shortly, enables us to isolate the contributions of specific photon pathlengths and thereby directly measure the path-dependent autocorrelation function g_1^s (τ). This is particularly useful when the mean square particle displacement *does not* depend linearly on time, and in regimes where it is not apparent that the diffusion approximation is valid. In addition, with the same apparatus, we can directly measure l^*.

Pulsed Diffusing-Wave Spectroscopy

The basic ideas of PDWS are illustrated with the use of Figure 3. We employ a laser that emits a train of identical light pulses. Each pulse has a temporal duration Δt, and a carrier frequency ω_0. Adjacent pulses within the train are separated by a time T. A beam splitter divides the pulse train into reference and sample pulse trains. The reference train is optically delayed, and the sample train is directed into the suspension which is contained in a rectangular glass cell. Light pulses emerging from the opposite side of the cell are "stretched" due to the distribution of photon path lengths through the sample. In order for PDWS to be most effective, the pulse broadening due to multiple scattering must be large compared to the input pulse width and small compared to the train repitition rate, i.e., $\Delta t \ll \Delta s/c \ll T$, where Δs represents the characteristic width of $P(s)$. The scattered pulse train is then recombined with the reference train in a frequency doubling crystal, and a second harmonic (SH) pulse train is produced when the two input fields nonlinearly mix [9,10]. If s' is the *difference* in path length between the reference and sample arms when the sample is *removed*, then each pulse within the SH train will have a field $E_{s'}(2\omega_0,t)$, proportional to the reference field, $E_R(t)$, and the *path-dependent scattered field* $E_S(t,s')$. When fluctuations in the reference field are negligible, $E_R(t) = E_R$, and we have

$$ E_{s'}(2\omega_0,t) \sim E_R \, E_S(t,s') . \qquad (4) $$

In most cases of interest, the time scale of the fluctuation in the phase of $E_S(t,s')$ is much longer than T, and the autocorrelation function of the SH photons is given by

$$g_1(2\omega_0,\tau) \sim |E_S(t,s')|^2 \langle (E_S(t,s'))^* E_S(t+\tau,s') \rangle \sim P(s') g_1^S(\tau). \qquad (5)$$

Thus we see that the SH electric field will experience the *same fluctuations* due to particle motion as the scattered electric field for a single pathlength. By varying the path length difference, s', between the sample and reference arms, the reference pulse "gates" the total sample electric field, so that only a very narrow range of photon paths centered about s=s' contribute to the fluctuations of the upconverted field. The autocorrelation function of the SH field is simply the integrand of Equation (3) evaluated at the appropriate s,

$$g_1(2\omega_0,\tau) \sim \exp[-k_0^2 \langle \Delta r^2(\tau) \rangle (s/l^*)/3]. \qquad (6)$$

Note that the temporal behavior of the autocorrelation function no longer depends on the shape of $P(s)$, and for fixed s, a plot of $\ln[g_1(2\omega_0,\tau)]$ *vs* τ directly yields the time dependence of $\langle \Delta r^2(\tau) \rangle$. In DWS any process that affects $P(s)$, such as sample geometry or absorption, modifies the temporal decay of the measured autocorrelation function. Thus even processes that *do not* affect particle diffusion must be properly accounted for when analyzing DWS data. Since PDWS is insensitive to $P(s)$ these types of problems are eliminated.

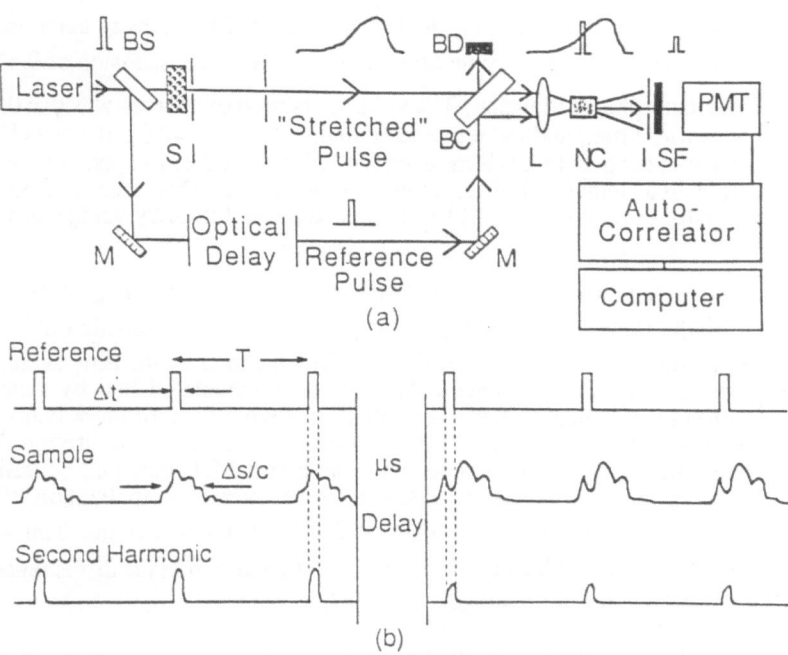

Figure 3. (a) Schematic of the PDWS experimental set-up: BC, beam combiner; BD, beam dump; BS, beam splitter; I, iris; L, lens; M,Mirror; NC, nonlinear crystal; PMT, photomultiplier tube; S, sample; SF, second harmonic spectral filter. (b) Sketch of the reference, sample, and second harmonic pulse intensities during two intervals separated by several microseconds. Temporal fluctuations in the sample pulse intensity arise from particle motion on the microsecond time scale.

An important additional feature of this scheme is that the dependence of the average SH intensity on the reference delay s, is proportional to P(s). Thus we can directly measure P(s) for any geometry. By fitting the results to predictions of photon diffusion theory, we can experimentally determine l* and the photon absorption length l_a.

Optical gating has been used with considerable success in time-resolved luminescience studies of semiconductors [10], and more recently in coherent backscattering measurements [11,12]. In contrast to these cases, the present application is the first to exploit the *phase fluctuations* of the gated photons. By eliminating the average over photon pathlengths, the new technique is essentially a very high resolution version of DWS. We expect that just as a number of high resolution laser spectroscopic techniques such as saturation spectroscopy for example, have revealed microscopic perturbations in the frequency domain, the spatial precision of PDWS will yield new information on problems concerning particle motion on different length scales.

To illustrate the basic features of PDWS we have carried out a number of measurements on the Brownian dynamics of dense colloidal suspensions. The light source was a mode-locked Nd:YAG laser ($\lambda = 1.06$ microns) that produced a 100 MHz train of 90-psec pulses. The average ouptut power of 5 Watts, was split evenly between the reference and sample beams (see Figure 3). A portion of the transmitted output from the sample was collected by two irises and imaged along with the reference beam into a KTP doubling crystal (5x5x5 mm, Type II). The second harmonic photons were spatially and spectrally filtered from the fundamental photons, and collected. Typical photon count rates were between 80 and 600 KHz, allowing normalized intensity autocorrelation functions, $g_2(\tau)$, to be obtained with 100 nsec time resolution in 2 to 20 minutes. The intensity correlation functions are related to the field correlation functions through the Siegert relation.

In the inset of Figure 4a we plot the log of a typical SH intensity autocorrelation function $[g_1(2\omega_0,\tau)]^2$ *vs* τ. The sample used in this case was a suspension of 0.460 μm-diameter polystyrene spheres in water. The volume fraction of spheres was $\phi = 0.30$, the sample thickness was 2 mm, and the reference arm delays were s=7.0 cm and s=13.0 cm. We emphasize that in contrast to DWS measurments the curves decay exponentially. Some of our runs exhibited a slight upward curvature at longer times. This effect was due to our relatively long pulse durations and will be eliminated in the future by using shorter laser pulses.

Using this sample we have performed measurments at different optical delays s. In figure 4a we plot the slope, Γ_1, of the $\ln[g_1(2\omega_0,\tau)]$ *vs* τ curve as a function of s. Below this plot, in figure 4b we show our measurment of P(s) obtained by the time averaged SH photon yield as a function of reference delay. We also calculated P(s) by solving the diffusion equation [13] subject to boundary conditions which insure there is no flux of diffusing photons into the medium [14] (dashed curve). The solid line through the data represents the best fit of theory to experiment after we account for finite pulse duration and the small absorption of light by water. Aside from an overall normalization, the only adjustable parameter is l*. From our P(s) data we deduce that l*=31.6 μm. This value of l*, coupled with the measured slope of the Γ_1 *vs* s curve yields a particle diffusion constant of D=6.20x10-9 cm^2/sec for this sample.

Within the limitations of the current apparatus, these measurements corraborate the primary result of DWS. That is, the electric field autocorrelation of a photons that have diffused s/l* steps, decays by exp(-τ 2k_0^2 D) per step. This is explicitly demonstrated over path lengths ranging from 2200 to 4100 steps. Experiments are currently underway to test this hypothesis in the more complicated backscattering geometry where it is expected to breakdown.

Table 1. Summary of diffusion and l* data for various volume fractions of polystyrene in water. Also shown are the theoretical extimates for the self diffusion coefficient, $D=D_0(1-1.83\phi)$, [15], where D_0 is the Stokes-Einstein diffusion coefficient ($k_BT/6\pi\eta a$) and ϕ is the volume fraction of spheres.

Concentration (volume fraction)	l*(μm) (exp.)	D(cm²/s) (exp.)	D(cm²/s) (theo.)
0.05	148±16	$1.06\pm0.10\times10^{-8}$	1.14×10^{-8}
0.10	61.8±4.0	$7.34\pm0.8\times10^{-9}$	1.03×10^{-8}
0.20	41.7±3.4	$7.85\pm0.78\times10^{-9}$	7.97×10^{-9}
0.30	31.6±2.2	$6.20\pm0.47\times10^{-9}$	5.67×10^{-9}

It addition to the measurements described above we report some preliminary results of a concentration dependent study undertaken with this technique in the high volume fraction limit. In Table 1 we have tabulated our measured values of l*, and D for various particle volume fractions. We also indicate theoretical estimates of D based on hydrodynamic corrections to the motion [15]. There clearly exists a substantial deviation between theory and experiment as we approach the highest densities. At present we are not sure of the origin of these deviations, but quantitative studies are underway to try and understand these differences.

In conclusion we have introduced a pulsed DWS technique which substantially improves on the spatial resolution and the interpretation of conventional DWS experiments. With shorter light pulses it will be possible to observe ballistic particle motions, and more complicated systems in greater detail. The technique has enabled us to explicitly test a fundemental assumption of DWS theory, and it can be applied in the important backscattering geometry where some of the diffusive assumptions are known to break down. We note also that this general idea of performing an autocorrelation measurement on optical gated photons can be extended to conventional QELS measurements. In this

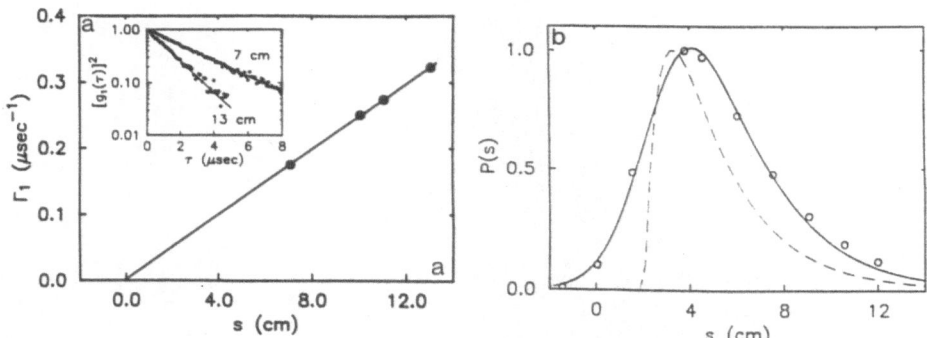

Figure 4. (a) Plot of the decay rate, Γ_1, of the SH temporal field correlation function *vs* reference arm delay s (same scale as (b)). Solid line is a least squares fit to the data. Inset: Plot of the log of the SH intensity autocorrelation function $[g_1(2\omega_0,\tau)]^2$ *vs* τ for s=7.0, 13.0 cm. Both curves exhibit the expected single-exponential decay. (b) P(s): measured (open circles) and calculated (solid line). Dashed curve represents P(s) for delta function input pulses.

case the gating feature can be used to do optical ranging since there is no multiple light scattering. Projects along these lines are underway in our laboratories.

This work has been supported by the National Science Foundation MRL Program through Grant No. DMR-8519059 and through equipment loans from the MRL Laser Central Facility. We also acknowledge support from the Petroleum Research Fund, the PEW Foundation and start-up equipment funds provided by the University of Pennsylvania.

References

1.   G. Maret and P.E. Wolf, Z. Phys. B **65**, 409 (1987).
2.   M.J. Stephen, Phys. Rev. B **37**, 1 (1988).
3.   D.J. Pine, D.A. Weitz, P.M. Chaikin, and E. Herbolzheimer, Phys. Rev. Lett. **60**, 1134 (1988).
4.   B.J. Berne and R. Pecora, *Dynamic Light Scattering* (Wiley, New York, 1976).
5.   N.A. Clarke, J.H. Lunacek, and G.B. Benedek, Am. J. Phys. **38**, 575 (1970).
6.   E.O. Schulz-Dubois, in *Photon Correlation Techniques in Fluid Mechanics,* ed. E.O. Schulz-DuBois, p.6 (Springer-Verlag, Berlin, 1983).
7.   D.J. Pine, D.A. Weitz, G. Maret, P.E. Wolf, E. Herbolzheimer, and P.M. Chaikin, in *Scattering and Localization of Classical Waves in Random Media*, ed. P. Sheng, (World-Scientific, to appear 1990).
8.   F.C. MacKintosh and S. John, Phys. Rev. B **40**, 2383 (1989).
9.   Y.R. Shen, *The Principles of Nonlinear Optics* (Wiley, New York, 1984).
10.   J.Shah, IEEE J. Quantum Electronics **24**, 276 (1988).
11.   R. Vreeker, M.P. Van Albada, R. Sprik, and A. Lagendijk, Phys. Lett **A 132**, 51 (1988).
12.   K.M. Yoo, Y. Takiguchi, and R.R. Alfano, Applied Optics **28**, 2343 (1989).
13.   H.S. Carslaw, and J.C. Jaeger, *Conduction of Heat in Solids, 2nd Edition* (Clarendon Press, Oxford, 1959).
14.   A. Ishimaru, *Wave Propagation and Scattering in Random Media, Vol. 1* (Academic Press, New York, 1978).
15.   P.N. Pusey and W. van Megan, J. Phys. (France) **44**, 285 (1983); P. Mazur, Faraday Discuss. Chem. Soc. **83**, 33 (1987).

WAVES ON CORRUGATED SURFACES: K-GAPS AND ENHANCED BACKSCATTERING

V. Celli and P. Tran

Department of Physics
University of Virginia
Charlottesville, VA 22901 USA

A. A. Maradudin, Jun Lu, and T. Michel

Department of Physics
and Institute for Surface and Interface Science
University of California
Irvine, CA 92717 USA

Zu-Han Gu

Surface Optics Corporation
P. O. Box 261602
San Diego, CA 92126 USA

INTRODUCTION

Several unexpected phenomena involving the propagation of electromagnetic waves on corrugated surfaces arise from high-order processes in the interaction of a photon (or a surface polariton) with the surface corrugations. Although these phenomena are expected to occur for waves of any type, they have been observed and discussed more extensively for visible light.

The dispersion curves of surface polaritons on classical gratings display in some cases a range of forbidden wave vectors, a k-gap, instead of the more usual energy gap at a zone boundary. This anomaly is seen most prominently in the angular distribution of light emitted by surface polaritons excited by electrons. It is due to the mixing of surface polaritons through the emission and reabsorption of photons, and depends sensitively on the dielectric properties of the surface and on the surface profile.[1-4]

An enhanced scattering of light from a moderately rough random surface into the retroreflection direction results primarily from the coherent interference of each doubly-reflected optical path with its time-reversed partner. Numerical simulations of this effect, which are feasible for one- and two-dimensional surface profiles, reproduce the salient features of experimentally determined backscattering peaks. In the presence of surface polaritons enhanced backscattering occurs even for weak corrugations, and is connected with the weak localization of the surface polaritons.[5-8]

In this contribution we present several new results bearing on both of these properties of electromagnetic waves on corrugated surfaces, and on related phenomena.

K-GAPS

The dispersion relation of any wave in a weakly periodic structure of period a displays a frequency gap (ω gap) at $k=n\pi/a$, n integer, of magnitude $2|V(2n\pi/a)|$, if the potential V is hermitian. For surface waves, however, the effective matrix element can be predominantly anti-hermitian, as illustrated in Fig. 1 for a gap at $2\pi/a$ in the surface polariton spectrum of a metal grating. If the grating profile $\zeta(x_1)$ has glide-reflection symmetry, i.e., if $\zeta(x_1+a)=-\zeta(x_1)$, $V(4\pi/a)$ vanishes. The dominant coupling between the states at $k=\pm2\pi/a$ is then due to a second order process, passing through resonant continuum states near $k=0$, as shown. The anti-hermitian part of this coupling is the same as the radiative width W of the states near $\pm2\pi/a$, while its hermitian part is comparatively negligible. With this coupling, the two standing waves have frequencies $\omega_0-i\Gamma$ and $\omega_0-i(\Gamma+2W)$, where Γ is the ohmic width. Thus one branch has no radiative width, the other twice the usual. There is neither an ω-gap nor a k-gap; in particular the simple prediction of a k-gap equal to 2W is wrong in this case[9].

Nevertheless, the EM response of a grating shows peaks that trace out a k-gap near $\pm2\pi/a$, and the theory reproduces these peaks. Three types of experiments have been analyzed; reflectivity[1,2], light emission excited by 80 keV electron bombardment[3,10] and light emission from corrugated metal-oxide-metal (MOM) tunnel junctions[4,11]. In each case the measured quantity can be written as $|N/D|^2$, where D has poles at the complex frequencies described above. The k-gap is seen when one plots the peaks of $|N/D|^2$ for real k and real ω. As an example, in Fig. 2 we show the computed light emission intensity from a sinusoidal Ag grating with period 555 nm and peak-to-valley height 30 nm. It is supposed that a white-noise spectrum of EM waves is present in the grating, which represents the top layer of a MOM structure. The mechanism by which these waves are excited by tunneling electrons is not specified in detail. There is clearly a k-gap at the crossing $\hbar\omega=\hbar\omega_0= 2.155$ eV, and its magnitude is given approximately by $c\Delta k=2[\Gamma(2W-\Gamma)]^{1/2}$. It is interesting that ohmic damping must be present to produce the k-gap, but it must not exceed 2W. When $2W\gg\Gamma$, Δk is linear in the grating height h, although it arises from a second order process (W is quadratic in h). This behavior is seen experimentally (Fig. 3); however, the theoretical W appears to be too small by a factor of ~ 2.

ENHANCED BACKSCATTERING OF LIGHT FROM A RANDOMLY ROUGH METALLIC SURFACE

One of the most interesting phenomena associated with the scattering of light from a randomly rough surface is that of enhanced backscattering. This is the presence of a well-defined peak in the retroreflection direction in the angular distribution of the intensity of the incoherent component of the light scattered from such a surface.[5,8,12-14]

It is generally believed that enhanced backscattering is a multiple scattering effect. In the case of a large-amplitude random surface the light ray striking the surface undergoes $n(n \geq 1)$ deflections, at the last of which it is scattered back into the vacuum away from the surface. All such scattering sequences are assumed to be uncorrelated due to the random nature of the surface. However, any such sequence and its time-reversed partner, in which the light is scattered from the same points on the surface, but in the reverse order, interfere constructively if the

316

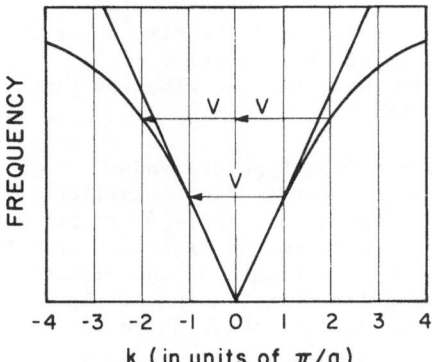

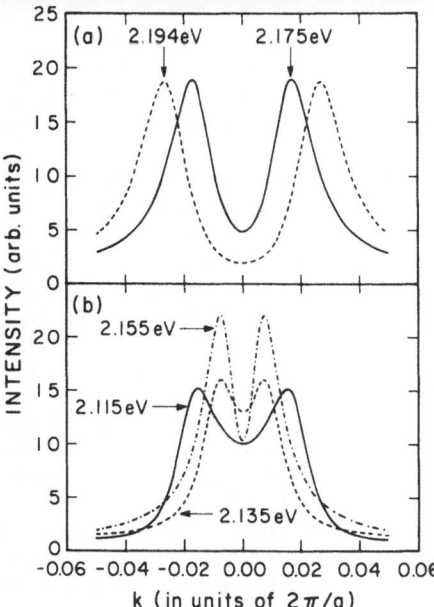

Fig. 1. Plot of the surface polariton dispersion curve, light line, and the second order process coupling surface polaritons at k = ± 2π/a. V is V(2π/a).

Fig. 2. Constant frequency scans for the intensity of light emitted in the plane perpendicular to the grooves by a sinusoidal silver grating with a = 555 nm and h = 30 nm.

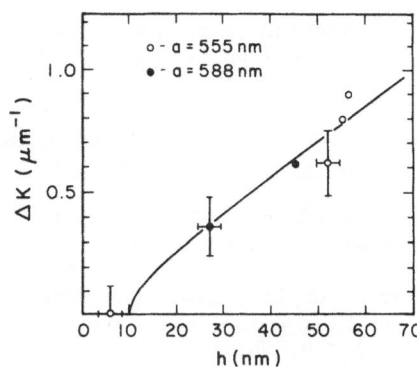

Fig. 3. Dependence of the k-gap at 2π/a on the peak to valley height of the surface corrugation. The data points, from Ref. 11, are for emission from the silver surface of a MOM junction with trapezoidal corrugation. The solid line is $2[\Gamma(2W-\Gamma)]^{1/2}$ with $\Gamma = 0.075$ and $W = 3.4 \times 10^{-4} h^2$. With $\epsilon = -7.5+0.5i$ theory would give $\Gamma = 0.058\ 2\pi/a$ and $W = 1.41 \times 10^{-4} h^2$ for a sinusoidal grating with period a = 555 nm.

wave vectors of the incident and final waves are oppositely directed. These two waves have the same amplitude and phase and add coherently in forming the intensity of the scattered light. For scattering into directions other than the retroreflection direction the different partial waves have a nonzero phase difference and very rapidly become incoherent, so that only their intensities add. Thus, the intensity of the scattering into the retroreflection direction is a factor of two larger than the intensity of scattering into those other directions, because of the cross-terms that appear in the expression for the intensity in the former case. The contribution of the single-scattering processes must be subtracted off in obtaining this factor of two enhancement, because it is not subject to coherent backscattering.

The enhanced backscattering of light from a weakly rough metal surface of the kind studied in Refs. 5-6 arises from the excitation of a surface polariton, its multiple scattering through the surface roughness, and its conversion back into light propagating away from the surface. The coherent addition of such a scattering sequence and its time-reversed partner again leads to a two-fold enhancement of the intensity of scattering into the retroreflection direction when the contribution from single-scattering processes is subtracted off.

For scattering at normal incidence from either type of random surface the phase difference ϕ between a given light/surface polariton path and its time-reversed partner is proportional to $(2\pi/\lambda)\theta_s d$, where θ_s is the scattering angle, d is the distance between the first and last scattering points on the surface, and λ is the wavelength of the light. One thus expects subsidiary maxima in the angular distribution of the intensity of the scattered light whenever the average phase shift $<\phi>$ is a multiple of 2π, i.e. at angles of observation given by $\theta_s = n\lambda/<d>$, where n is the order of interference. From this result we expect that the reflected intensity increases by up to the above-mentioned factor of 2 inside a region of angular width of order $\lambda/<d>$ about the retroreflection direction. Since the average value of d for the shortest scattering sequence (n=2) is the mean distance between two scattering events, i.e. the elastic mean free path ℓ of the light interacting with the surface, the angular width of the enhanced backscattering peak is expected to be of the order of λ/ℓ.

In the remainder of this section we present some new results bearing on enhanced backscattering of light from randomly rough, one- and two-dimensional metallic surfaces.

Enhanced Backscattering From a One-Dimensional Random Surface and its Dependence on the Surface Height Correlation Function

A one-dimensional random surface is defined by the equation $x_3 = \zeta(x_1)$, so that its generators are parallel to the x_2-axis. We assume that the region $x_3 > \zeta(x_1)$ is vacuum, while the region $x_3 < \zeta(x_1)$ is a metal, characterized by a complex, isotropic, frequency-dependent dielectric constant $\epsilon(\omega)=\epsilon_1(\omega) + i\,\epsilon_2(\omega)$. We will be interested in the frequency range in which $\epsilon_1(\omega) < 0$, while $0 < \epsilon_2(\omega) << |\epsilon_1(\omega)|$, i.e. in which the metal is strongly reflecting. The surface profile function $\zeta(x_1)$ is assumed to be a single-valued function of x_1, and to constitute a stationary, Gaussian, stochastic process, defined by the equations $<\zeta(x_1)> = 0$ and $<\zeta(x_1)\zeta(x_1')> = \delta^2 W(|x_1-x_1'|)$. Here, the angle brackets denote an average over the ensemble of realizations of the surface profile, while $\delta^2 = <\zeta^2(x_1)>$ is the mean-square departure of the surface from planarity.

If enhanced backscattering is a multiple-scattering phenomenon, we should expect to see it from any random surface that can multiply scatter light or surface polaritons. Here we present results of numerical simulations of the scattering of p-polarized light from random metallic gratings, for four different forms of the two-point correlation function $W(|x_1|)$. The computational method used is the one described in Refs. 7-8. The plane of incidence (the x_1x_3-plane) is perpendicular to the generators of these gratings. The four forms for $W(|x_1|)$ that we consider are

$$W(|x_1|) = \exp(-x_1^2/a^2); \tag{1}$$

$$W(|x_1|) = a^2/(x_1^2+a^2) \tag{2}$$

$$W(|x_1|) = \sin(\pi x_1/a)/(\pi x_1/a) \tag{3}$$

$$W(|x_1|) = \frac{a}{\tan^{-1}aQ_o} \int_o^{Q_o} dQ \frac{\cos Q x_1}{1+a^2Q^2} . \tag{4}$$

The surface defined by Eq. (4) can be termed a band-limited fractal surface, since in the limit as $Q_o \to \infty$ it has the fractal dimensionality D = 1.5.[15] The characteristic length a is called the transverse correlation length of the surface roughness. A good approximation to the mean distance between consecutive peaks and valleys on the random surface is given by a for Eqs. (1)-(3)[16] and by $\sqrt{3}\pi a/(aQ_o)$ for Eq. (4) in the limit of large Q_o.[17]

In Fig. 4 we present the incoherent contribution to the differential reflection coefficient, as a function of the scattering angle θ_s, for p-

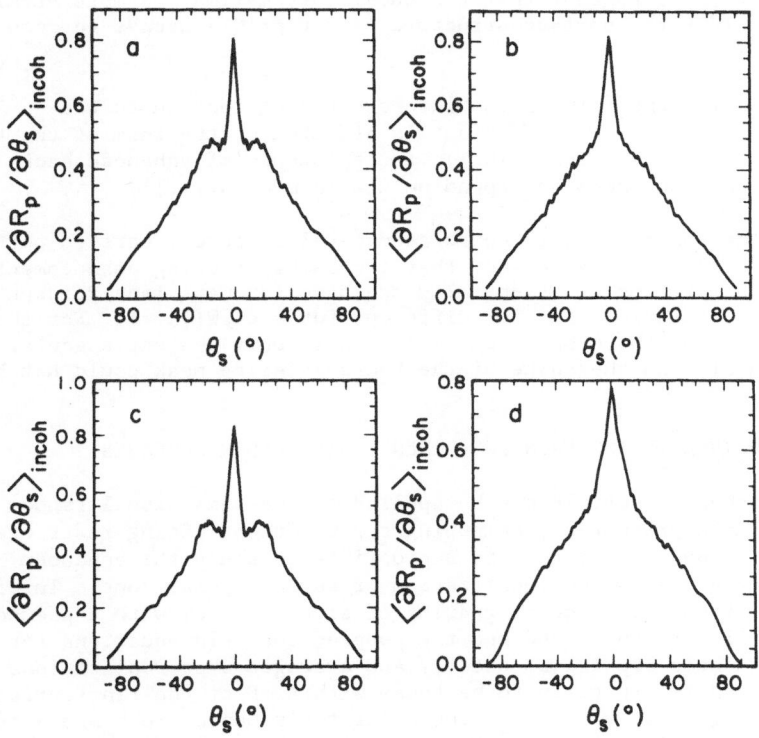

Fig. 4. The incoherent contribution to the mean differential reflection coefficient for the scattering of a beam of p-polarized light of wave length $\lambda = 6127\text{Å}$ incident normally on a random silver grating. $\epsilon(\omega) = 17.2 + i0.498$, $a = 2\mu m$. (a) Eq. (1), $\delta = 1.2\mu m$; (b) Eq. (2), $\delta = 1.2\mu m$; (c) Eq. (3), $\delta = 1.2\mu m$; (d) Eq. (4), $\delta = 0.3893\mu m$, $Q_o = 15.382 \ \mu m^{-1}$.

polarized light of wavelength $\lambda = 6127\text{Å}$ incident normally on a random grating ruled on a silver surface, when the roughness is characterized by the four correlation functions (1)-(4). In each case the root-mean-square slope of the surface has been set equal to 0.8485. For each form of $W(|x_1|)$ a well-defined peak in the retroreflection direction ($\theta_s = 0^\circ$) is present. The angular width of the peak for the cases defined by Eqs. (1)-(3) is given very closely by λ/ℓ, if for ℓ we take the mean distance between consecutive peaks and valleys on the surface. In the case defined by the correlation function (4) the angular width of the peak is given closely by λ/ℓ, if for ℓ we take the mean distance between consecutive peaks (or between consecutive valleys) on the surface. In the case of scattering from the surfaces defined by the correlation functions (1) and (3) well-defined subsidiary maxima are observed at scattering angles θ_s that are quite close to $\theta_s = \pm\lambda/\ell$. The differential reflection coefficients calculated with the use of the correlation functions (2) and (4) show no such subsidiary maxima. The explanation for these results lies in the statistics of the random quantity d, the distance between the first and last scattering points on the surface, in terms of which we have argued the positions of subsidiary maxima are expected at scattering angles $\theta_s \simeq n\lambda/\langle d\rangle$. The standard deviation σ_ϕ of the phase difference ϕ between time-reversed paths at the angle of observation $\theta_s \simeq n\lambda/\langle d\rangle$ is proportional to $n\,\sigma_d/\langle d\rangle$, where σ_d is the standard deviation of d[8]. Thus, the results presented in Fig. 4 suggest that for n=2, σ_ϕ is large enough to destroy all interference effects in scattering from the surfaces defined by Eqs. (1) and (3), while these effects are already washed out for n=1 in scattering from surfaces defined by Eqs. (2) and (4). These results also suggest that the subsidiary maxima are more prominent the more rapidly the surface structure factor $g(|Q|)$ decays to zero with increasing $|Q|$.

Thus, while the <u>form</u> of the incoherent contribution to the differential reflection coefficient is affected by the form of the two-point correlation function $W(|x_1|)$, the <u>existence</u> of enhanced back-scattering does not seem to depend on the form of $W(|x_1|)$.

Calculations for weakly corrugated metal surfaces, carried out by the method of Ref. 18, also show that the backscattering enhancement due to surface polaritons is essentially the same for Gaussian, Poisson, or exponential statistics, and for different forms of $W(|x_1|)$. For these weak corrugations the enhancement was seen in only the antispecular output channel, and the shape of the backscattering peak could not be studied.

ENHANCED BACKSCATTERING FROM TWO-DIMENSIONAL RANDOM SURFACES

The method of Ref. 18 can be applied to two-dimensional random surfaces within present-day computing capabilities. Being restricted to weak corrugations, the method is appropriate to study the enhanced backscattering due to the localization of surface polaritons. In brief, many realizations of a square grating of side $L \simeq 10^4$ nm with a pseudo-random profile are generated and the coupled channels equations for the resonant channels are solved exactly and averaged over realizations. The resonant channels are taken to be those within $\alpha\Gamma$ of the flat-surface polariton frequency, where Γ is the polariton's ohmic width and α is allowed to range from 1 to 3. For a given α the number of resonant channels, and thus the result of the calculation, fluctuates with L. After averaging over these fluctuations, the result is extrapolated to $\alpha=\infty$. Figure 5 shows typical results obtained with 500 realizations for each of 45 L values from 9140 to 9700 nm. The calculations show a backscattering enhancement that is as pronounced for two-dimensional roughness as it is for a one-dimensional random profile. The results are in general agreement with those of the diagrammatic theory of McGurn and Maradudin[19] which are shown by dashed lines.

ENHANCED TRANSMISSION

Not all of the manifestations of weak localization in the interaction of light with random metallic grating are in reflection. It was shown recently[20] that the angular distribution of the intensity of the incoherent component of p-polarized light transmitted through a thin metal film surrounded by vacuum, whose illuminated surface is randomly rough while the back surface is planar, displays a well-defined peak in the direction of transmission that is directly opposite to the direction of specular reflection of the incident light (the antispecular direction). The physical origin of this effect is believed to be the scattering, by the surface roughness, of the surface polaritons excited in the film by the incident light. The coherent interference of a doubly-scattered light/surface polariton path with its time-reversed partner gives the dominant contribution to enhanced transmission.

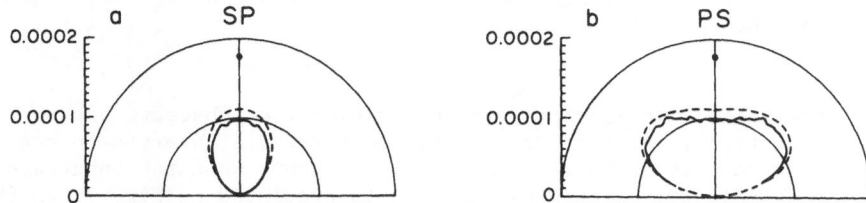

Fig. 5. Polar plots of the computed cross-polarized scattering intensity at normal incidence from a silver surface with two-dimensional roughness (δ = 5 nm, a = 120 nm). The intensity in the backscattered channel is shown by a heavy dot. The results of the theory of Ref. 19 are shown by a dashed line.

The calculations described in Ref. 20 were perturbation-theoretic in nature, and consequently limited to films with small amplitude random grating surfaces. The restriction to weakly rough surfaces can be lifted if the calculation of the intensity of the incoherent component of the light transmitted through a thin metallic film with a random grating surface is carried out by an extension to this structure of the kinds of numerical simulations described in Refs. 7-8. In Fig. 6 we present results obtained in this way for the incoherent contribution to the differential reflection and transmission coefficients for p-polarized light of wavelength λ = 4579Å incident on an 800Å thick silver film surrounded by vacuum. The angle of incidence of the light is θ_o = 20°.

The rms height and transverse correlation length of the roughness are δ = 282.84Å and a = 1500Å, respectively. A total of N_p = 1500 different surface profiles was used in these calculations. A well defined peak in the retroreflection direction (θ_s =-20°) is seen in Fig. 6a, and represents enhanced backscattering from this structure. Similarly, a well-defined peak in the antispecular direction (θ_t = -20°) is observed in Fig. 6b, and represents enhanced transmission. The difference between the magnitudes of these two coefficients reflects the attenuation of the light in traversing the film.

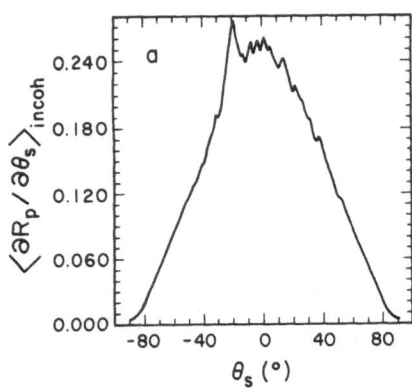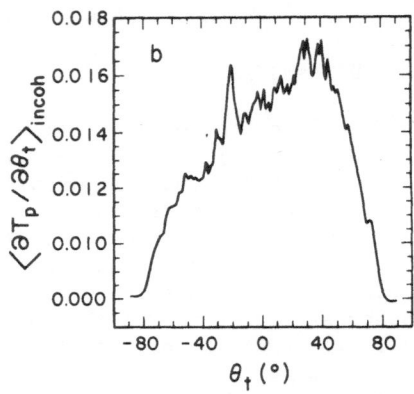

Fig. 6. The incoherent differential reflection (a) and transmission
(b) coefficients as functions of θ_s and θ_t, respectively, for light of
wavelength $\lambda = 4579\text{Å}$ incident at an angle $v_o = 20°$ on a silver film
surrounded by vacuum. The dielectric constant of silver at this wavelength is $\epsilon(\omega) = -7.5 + i0.24$. The roughness of the upper surface is
characterized by $\delta = 282.84\text{Å}$ and a = 1500Å. The mean thickness of the
film is 800Å. At total of $N_p = 1500$ different surface profiles were used
in obtaining these results.

When the metal film is deposited on a dielectric substrate whose
index of refraction is greater than unity, features are observed in the
differential transmission coefficient that are absent when the substrate
is vacuum. This is illustrated by the results presented in Fig. 7, which
correspond to the same film and experimental conditions that were used in
obtaining Fig. 6, with the sole exception that the film now rests on a
semi-infinite dielectric medium whose index of refraction is $n_d \approx 1.5$.
The incoherent contribution to the differential reflection coefficient
(Fig. 7a) displays a well-defined peak in the retroreflection direction
$(\theta_s = -20°)$, and both qualitatively and quantitatively is very close to
the curve depicted in Fig. 6a. The situation is quite different for the
incoherent contribution to the differential transmission coefficient
depicted in Fig. 7b. We see, first, that the magnitude of the enhanced
transmission peak relative to the value of the background at its position
is smaller than in the case that the substrate is vacuum (Fig. 6b). The
presence of a dielectric substrate tends to reduce the enhanced transmission. Second, we see that the enhanced transmission now occurs at an
angle θ_t which is not equal to $-\theta_o$. This result is easily understood.
Enhanced backscattering and enhanced transmission both occur when the
component q of the wavevector of the scattered and transmitted light that
is parallel to the mean surface equals the negative of the component of
the wavevector of the incident light parallel to the surface, -k. In the
vacuum, $k = (\omega/c) \sin\theta_o$ and $q = (\omega/c) \sin\theta_s$, so that enhanced back-
scattering occurs when $\theta_s = -\theta_o$. In the dielectric, however, $q = n_d(\omega/c)\sin\theta_t$, where n_d is the index of refraction of the dielectric substrate, so that enhanced transmission occurs when $\theta_t = -\sin^{-1}(\sin\theta_o/n_d)$.
For an angle of incidence $\theta_o = 20°$ and a value of $n_d = 1.5$, we have
$\theta_t = -13.18°$, which is very close to the value of θ_t at which enhanced
transmission is predicted in the result of our computer simulation
studies presented in Fig. 7b. Finally, we see two prominent peaks in the
differential transmission coefficient at values of $\theta_t \approx \pm 50°$. These are
due to the roughness-induced resonant excitation of a leaky surface
polariton whose wave number lies in the interval $(\omega/c) \lesssim q < n_d(\omega/c)$.

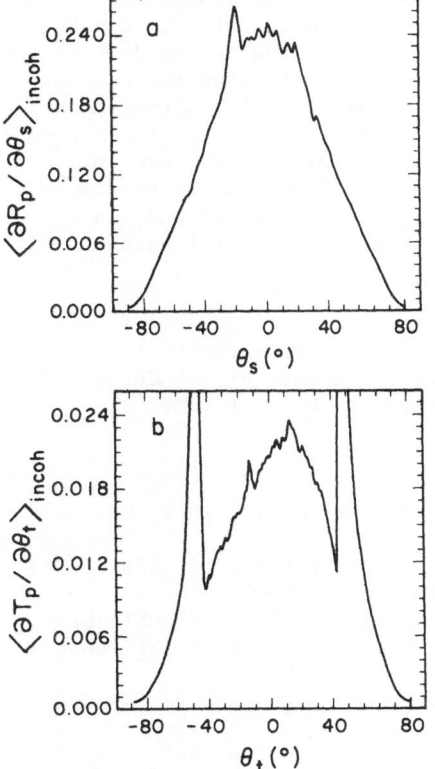

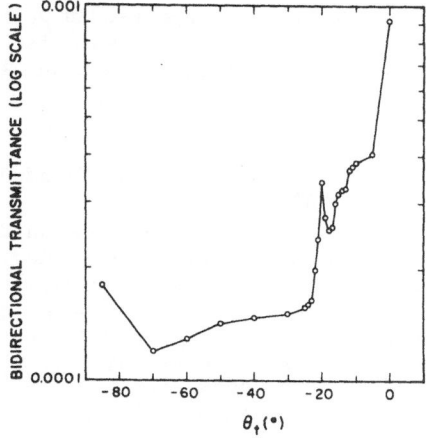

Fig. 7. The same as Fig. 6, except that the silver film rests on a semi-infinite dielectric medium whose index of refraction is $n_d = 1.5$.

Fig. 8. The bidirectional transmittance of p-polarized light through a silver film with a randomly rough illuminated surface, as a function of the angle of transmission θ_t. The wavelength of the incident light is $\lambda = 0.6328 \mu m$ and the angle of incidence is $\theta_o = 20°$. The thickness of the film is $d = 850$ Å. The roughness of the illuminated surface is characterized by $\delta = 118$Å and $a = 1200$Å. The film is deposited on a glass plate whose index of refraction is $n_d = 1.51$ and whose thickness is 5 mm.

The phenomenon of enhanced transmission has now been observed experimentally. Figure 8 is a plot of the bidirectional transmittance distribution function of p-polarized light through a silver film deposited on a glass substrate whose index of refraction is $n_d = 1.51$. The illuminating source was a He-Ne laser with wavelength $\lambda = 0.6328 \mu m$. The illuminated surface of the silver film is a two-dimensional randomly rough surface, rather than a one-dimensional random surface, as was the case in calculating the results depicted in Fig. 7. The plane of transmittance coincides with the plane of incidence, and it is the p-polarized component of the transmitted light which has been measured. The angle of incidence is $\theta_o = 20°$. A sharp peak in the bidirectional transmittance is observed at a transmission angle $\theta_t = -20°$, which represents enhanced transmission. The occurrence of the enhanced transmission peak at

$\theta_t = -20^{\circ}$ rather than at $\theta_t = -13.18^{\circ}$ as in Fig. 7, is due to the fact that the glass substrate in the experiment has a finite thickness rather than being semi-infinite as was assumed in calculating Fig. 7, and that the two plane surfaces of the glass are parallel. The use of Snell's law twice (at the entrance and exit of the substrate), returns the peak of the enhanced transmission to $\theta_t = -20^{\circ}$. The absence of a strong peak at $\theta_t \cong -50^{\circ}$ in the present case has the same origin. Thus it is seen that although the surface of the silver film studied in the present work is a two-dimensional randomly rough surface, rather than the random grating for which the theoretical calculations of Refs. 20 and the present paper were carried out, enhanced transmission still occurs.

Acknowledgements

This work was supported in part by U.S. Army Research Office Grant No. DAAL-03-89-C-0036. It was also supported in part by the University of California, Irvine, through an allocation of computer time.

References

1. V. Celli, P. Tran, A. A. Maradudin, and D. L. Mills, Phys. Rev. B37, 9089 (1988).
2. P. Tran, V. Celli, and A. A. Maradudin, Optics Lett. 13, 530 (1988).
3. P. Tran and V. Celli, Phys. Rev. B (in press, 1990).
4. P. Tran, Ph.D. thesis, University of Virginia (1990) (unpublished).
5. A. R. McGurn, A. A. Maradudin, and V. Celli, Phys. Rev. B31, 4866 (1985).
6. V. Celli, A. A. Maradudin, A. M. Marvin, and A. R. McGurn, J. Opt. Soc. Am. A2, 2225 (1985).
7. A. A. Maradudin, E. R. Méndez, and T. Michel, Optics. Lett. 14, 151 (1989).
8. A. A. Maradudin, E. R. Méndez, and T. Michel, in Scattering in Volumes and Surfaces, eds. M. Nieto-Vesperinas and J. C. Dainty (North-Holland, Amsterdam, 1990), p. 157.
9. In an optically active medium, Γ can be negative and k-gaps are known to occur. In particular, if Γ + W = 0, there is a k-gap equal to 2W, as noted by H. Kogelnik and C. V. Shank, J. Appl. Phys. 43, 2327 (1972). A general classification of gap shapes when Γ + W = 0 is given by P. Halevi and O. Mata-Mendez, Phys. Rev. B39, 5694 (1989).
10. D. Heitmann, N. Kroo, C. Schultz, and Zs. Szentirmay, Phys. Rev. B35, 2660 (1987).
11. N. Kroo, Zs. Szentirmay, J. Felszerfarvi, Phys. Lett. 86A, 445 (1981).
12. E. R. Méndez and K. A. O'Donnell, Optics. Commun. 61, 91 (1987).
13. J. C. Dainty, M.-J. Kim, and A. J. Sant, In Scattering in Volumes and Surfaces, eds. M. Nieto-Vesperinas and J. C. Dainty (North-Holland, Amsterdam, 1990), p. 143.
14. Zu-Han Gu, R. S. Dummer, A. A. Maradudin, and A. R. McGurn, Appl. Optics 28, 537 (1989).
15. M. V. Berry, J. Phys. A12, 781 (1979).
16. A. A. Maradudin and T. Michel, J. Stat. Phys. 58, 485 (1990).
17. A. A. Maradudin and T. Michel (unpublished work).
18. P. Tran and V. Celli, J. Opt. Soc. Am. A5, 1635 (1988).
19. A. R. McGurn and A. A. Maradudin, J. Opt. Soc. Am. B4, 910 (1987).
20. A. R. McGurn and A. A. Maradudin, Optics. Commun. 72, 279 (1989)

BLACK HOLE RADIATION: CAN VIRTUAL PHOTOCONDUCTIVITY

PRODUCE A SIMILAR EFFECT IN SEMICONDUCTORS?

E. Yablonovitch

Bell Communications Research
Navesink Research Center
Red Bank, N.J. 07701-7040

ABSTRACT

Shortly after Hawking's prediction of thermal radiation from Black Holes, it became apparent that there were other contexts in which such radiation could appear. For example, it was predicted that accelerating observers are bathed in thermal radiation (Unruh radiation). Even a stationary observer who is looking at an accelerating mirror should see such radiant energy. The effect is very weak, however. An acceleration $g = 980$ cm/sec^2 produces a radiation temperature of only $\sim 4 \times 10^{-20}$ °K, making its detection a major experimental challenge. A nonlinear optical window, whose refractive index is changing rapidly with time, appears, to an observer, to be a window into an accelerating world. The sudden injection of a virtual electron-hole plasma into a semiconductor window can change its refractive index on a sub-picosecond time scale, and can produce an apparent acceleration $\sim 10^{20}$ g.

INTRODUCTION

It is sometimes mistakenly believed that the black hole concept is a product of 20th century physics. Actually the black hole was first introduced 200 years ago[1,2], and it relied purely on the precepts presented by Isaac Newton, fully 300 years ago. A good starting point is the Newton's Law of the gravitational force between two objects of mass m and M:

$$\text{Force} = \frac{GmM}{r^2} \tag{1}$$

where G is the gravitational constant and r is the distance between the centers of the two objects. As is well-known to college students, the Force Law (1) leads directly to the concept of the "escape velocity" from massive bodies. The gravitational potential energy is simply made equal to the kinetic energy:

$$\frac{GmM}{r} = \frac{mv^2}{2} \tag{2}$$

where v is now the escape velocity.

Two hundred years ago, the Newtonian corpuscular model of light was firmly entrenched, and in addition there was some approximate knowledge of the speed of light, c. It made sense then to think of light as a particle and to ask whether it could escape from a gravitational potential. Taking v= c in eq'n. (2), the mass m cancels from both sides, and we are led to a simple expression for the radius r of a compact body whose gravitational field is strong enough to prevent light from escaping:

$$r = \frac{2GM}{c^2} \tag{3}$$

The Schwarzschild radius, r, was already known 130 years before General Relativity. In 1784, Michell[1], a leading English astrophysicist came up with the idea during his study of double stars. Independently in France, Laplace[2] writing in 1796 introduced the same concept. The elementary mathematics and physics leading to the Schwarzschild radius, r, is accessible even to a college freshman. By sheer coincidence, the factor 2 originating from the non-relativistic kinetic energy in eq'n. (2), yields the correct relativistic expression in eq'n. (3).

The Schwarzschild radius defines the event horizon of the black hole, the edge beyond which it is impossible to see. By the same elementary approach we can determine the acceleration, a, at the event horizon, which we will find useful later on:

$$a = \frac{GM}{r^2} = \frac{c^4}{4GM} \tag{4}$$

All that we have done so far is based on the physics of 200 years ago. An excellent review of the intellectual history of black holes is given by Werner Israel[3].

We will now skip ahead in time, past Einstein's General Relativity, past the discussions of collapsed stars, right into the 1960's. During that decade, the famous theoretician John Wheeler at Princeton was conducting a serious campaign to improve the understanding of black holes. Uniqueness theorems had recently been proven for the solutions of the gravitational field equations. It had become apparent that the most general black hole could be described by its mass, charge and angular momentum, and nothing more. This led Wheeler and others to the "Black Holes Have No Hair" conjecture. This is whimsical way of saying that there is no fine structure on the surface of black hole which could be used to distinguish one black hole from another. Indeed an uncharged, non-rotating black hole should be fully and uniquely specified by one scalar number, its mass M.

The idea that "Black Holes Have No Hair" led to a paradox, which is illustrated in Fig. 1. A garbage pail can be tossed into the black hole. The garbage pail could contain, for example, old issues of **Physical Review** which would carry a substantial amount of entropy, i.e. it would require an enormous amount of information to specify the physical state of those pages. However the black hole state can be fully specified by a single scalar number, (its mass M). Upon falling into the black hole, the garbage pail entropy would disappear. This "Wheeler's demon" would violate the Second Law of Thermodynamics. As a solution to waste disposal problems, it is too good to be true.

In 1971, Wheeler addressed this problem to his graduate student, Bekenstein[4]. To resolve the paradox Bekenstein had to attribute some entropy to the black hole, but the only variable at his disposal was the mass M, (for the uncharged, non-rotating case). Since the irreversible combination of two black holes should have more entropy than the sum of the two, Bekenstein postulated that the entropy should be proportional to mass, M, squared. For the entropy to have units of Boltzmann's constant, K, dimensional analysis required the prefactor to consist of certain

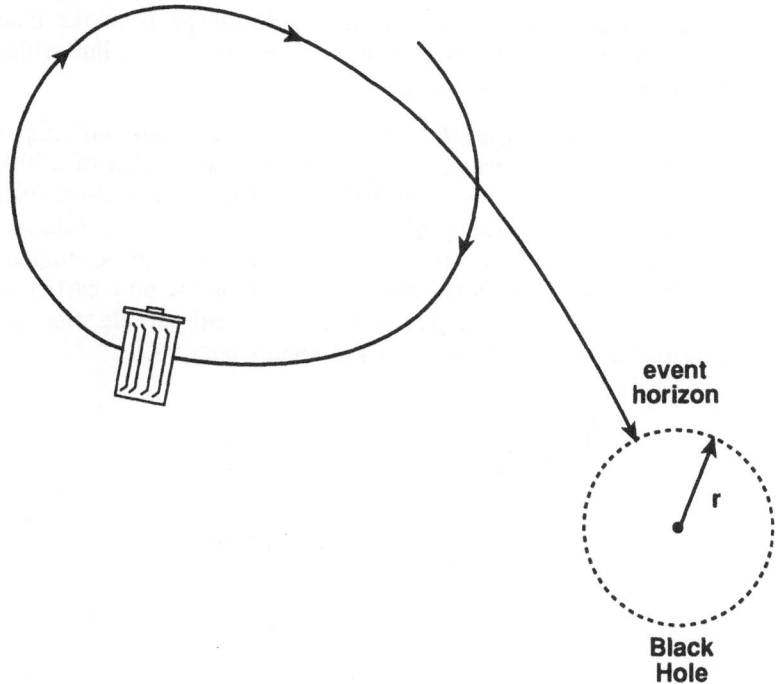

Figure 1. An illustration of the paradox associated with the "Black Holes Have No Hair" conjecture. The garbage pail and its entropy are tossed into the black hole, violating the 2nd Law of Thermodynamics.

combinations of fundamental constants, leading him to the following[5] "ansatz" for the entropy:

$$\frac{S}{K} = \frac{4\pi G}{\hbar c} M^2 \qquad (5)$$

where $\hbar \equiv$ Planck's constant$/2\pi$. Thus Bekenstein was able to guess all but the numerical prefactor 4π, which had to await Hawking's work. Since we also happen to know the rest energy of the black hole, $U = Mc^2$, Bekenstein's entropy "ansatz", eq'n. (5), allows us to construct the temperature of a black hole, $T \equiv \partial U / \partial S$:

$$KT \equiv K \frac{\partial U}{\partial S} = \frac{\hbar}{8\pi} \frac{c^3}{GM} \qquad (6)$$

With this in hand, Bekenstein showed that the paradox illustrated in Fig. 1 is resolved, and the 2nd Law of Thermodynamics is saved. Amazingly, a small amount of guesswork led all the way to a black hole temperature.

Bekenstein's gravitational physics colleagues[6] were not amused. They regarded the identification of eq'n. (5) with entropy as incorrect, merely a mathematical coincidence. They were content to abandon thermodynamics and retain the black holes. I find their attitude somewhat surprising since the 2nd Law of Thermodynamics is one of the pillars of physics, while the black hole is a hypothetical object which has never been detected.

Nevertheless, Bekenstein still had one problem which he could not resolve. The black hole could never achieve radiation equilibrium since thermal radiation could fall in, but by·definition no electromagnetic radiation could escape from the black hole. Hawking was particularly skeptical of Bekenstein's entropy idea, but within three years, in 1974, he would be the one to rescue it.

Hawking showed[7] that a black hole was also a black body, emitting thermal radiation. This astounding result was contrary to the very definition of a black hole. Hawking's picture, illustrated in Fig. 2, included particles and anti-particles being pulled out of the vacuum in the vicinity of the event horizon. The anti-particle could fall in, while the particle would be emitted. For us the picture can be simpler: Since the temperature will be low in our discussion, the photon is the only particle which is energetically accessible, and since the photon is its own anti-particle, this can all be summarized by a simple, thermal spectrum of low energy photons.

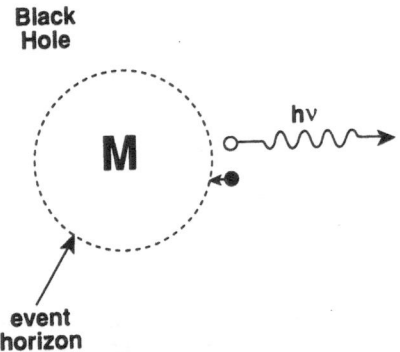

Figure 2. Particle and anti-particle emission from black holes. We need only consider the photon, which is its own anti-particle.

Hawking derived eq'n. (6) from quantum electrodynamics[7], supplying the numerical factor that had been missing from Bekenstein's equations. For a black hole of mass equal to our sun, the temperature is only $\approx 6 \times 10^{-8}$ °K. A flurry of checking and re-checking was triggered by this prediction. Other theorists wanted to find alternate ways of deriving the same result, and to find more examples of the same type of effect. Within months, it became apparent that there was a more general phenomenology. To see this let us combine the acceleration at the event horizon, eq'n. (4), based on eighteenth century physics, and the temperature, eq'n. (6), based on Bekenstein's "ansatz":

$$KT = \frac{\hbar}{2\pi} \frac{a}{c} \tag{7}$$

Re-interpreted, eq'n. (7) means that the thermal radiation is associated with the acceleration, a, and does not require black holes. Indeed, exactly this was proven by Unruh[8] and Davies[9] shortly after Hawking's work. Observers of the electromagnetic field in an *accelerating reference frame* should see thermal radiation at a temperature T. Black holes are not required! Today, the black hole temperature eq'n. (6), is regarded as a special case of the more general and elegant-looking Unruh radiation temperature, eq'n. (7). The effect as seen by an accelerating observer on a rocket ship is illustrated in Fig. 3.

The basic physical mechanism for this radiation is that zero point quantum fluctuations in an inertial frame become transformed into real thermal photons when Fourier analyzed in terms of the photon modes of the accelerating frame. In a frame experiencing linear acceleration at $g = 980$ cm/sec^2, equivalent to the acceleration

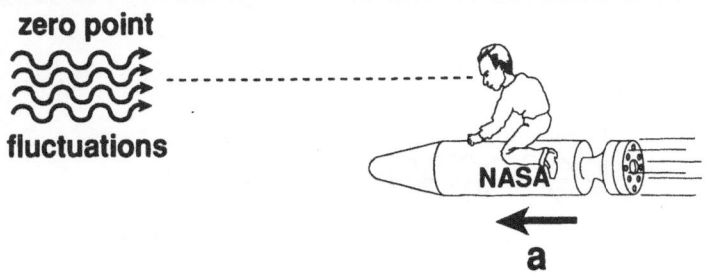

Figure 3. An observer on an accelerating rocket looks out at the zero point electromagnetic field, but sees a finite thermal excitation above the zero point as given by eq'n. (7).

experienced at the earth's surface, this thermal radiation is at a temperature of only 4×10^{-20} °K.

RADIATION FROM MOVING MIRRORS

Thus emerges the pattern of our discussion. What started out as an esoteric effect in black holes, now merely requires an impossibly large acceleration of the observer. Presently, we will find that the observer can be stationary and that large accelerations are required only for a mirror. Following that, we will see that an accelerating mirror is actually not needed, a rapidly changing refractive index will do. At each step we get closer to an experiment that can actually be performed with present-day technology!

Now we will see that the observer need not experience the acceleration himself. DeWitt[10], Fulling[11] and Davies[12] argued that the zero point field fluctuations near an accelerating mirror are already being subjected to an accelerating motion. In effect, the moving mirror forces a relative accelerating motion between the electromagnetic quantum fluctuations and a stationary observer. Remarkably, the conversion of zero point quantum fluctuations into real thermal photons is equally effective whether the mirror is advancing toward the observer or receding away from him. This is illustrated in Fig. 4:

One of Davies' papers[12] had an interesting title, "Radiation From Moving Mirrors and Black Holes". The origin of the acceleration effects is similar in both cases as illustrated in Fig. 5. In front of the mirror is a zero point photon standing wave. As the mirror accelerates, the node of the standing wave remains tied to the its surface, and the standing wave pattern is accelerated along. Likewise, zero point fluctuations are constantly being pulled into the black hole. In both instances zero point fluctuations are being accelerated relative to an observer, giving rise to a finite excitation above the zero point.

How to create a moving mirror? Couldn't nonlinear optics be defined as doing "tricks" with moving mirrors? During the mid-1970's, when all these wonderful theoretical ideas were being derived, I was specializing in nonlinear optics and doing[13] laser induced plasma experiments. For example, a CO_2 laser would be focused in air, transforming it into a plasma. Before the ionization event the refractive index of the air was unity. Afterward, the overdense plasma would have a refractive index of zero. The phase modulation associated with such an index change was shown[13] to produce nearly a 10% blue shift in the transmitted laser photon energy, a very substantial effect. Likewise, in a semiconductor, the sudden creation of electron-hole pairs can reduce the refractive index from ~3.5 to 0 in a very brief time period.

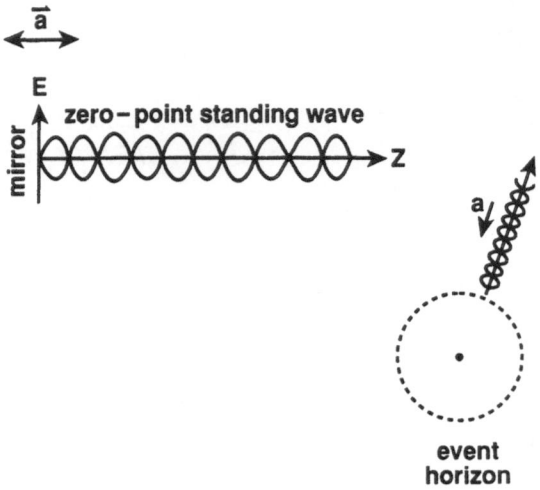

Figure 4. An observer looking at the reflection of the zero point electromagnetic field in an accelerating mirror. The relative motion of the field and the observer is the same as if the observer were on the rocket ship of Fig. 3. The stationary observer sees a finite thermal excitation above the zero point.

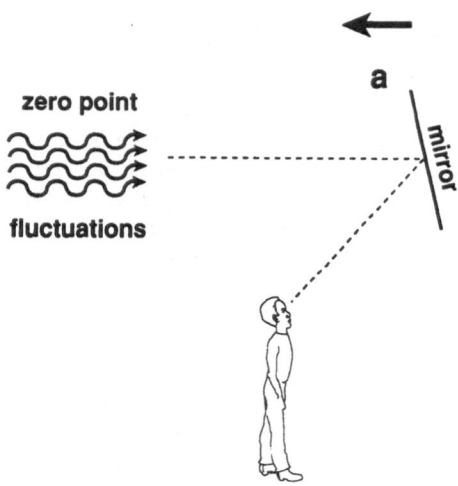

Figure 5. Two ways of accelerating zero point fluctuations, by gravitational attraction, and by means of an accelerating mirror.

Now let us imagine looking through one of these windows, of thickness z, whose refractive index is changing. The optical path is n× z, the phase shift is nzω/c and the frequency shift is ṅzω/c where n(t) is the time-dependent window refractive index and ω, is the incident angular frequency. To the observer looking at a laser in transmission through the window, its frequency appears to be Doppler shifted by the velocity ṅz. This window sees a fast moving world, a Doppler-shifted world. The equivalent velocity is given by:

$$\omega + \Delta\omega = \omega(1 + \frac{v}{c}) \qquad (8)$$

Suppose further, at the next order of approximation, that the frequency shift $\Delta\omega$

becomes a frequency chirp $\dot{\omega}$. Then,

$$\omega + \dot{\omega}t = \omega(1 + \frac{at}{c}) \qquad (9)$$

The apparent Doppler shift is now changing with time. It appears to the observer that the image in the window is accelerating.

Therefore we can regard such a gas plasma or a semiconductor slab as an observational window on an accelerating world. The main point of our discussion[14] is that *frequency chirp is equivalent to acceleration*. Chirp is ubiquitous in nonlinear optics. The relative motion of the observer and the electromagnetic field is controlled by the rate of refractive index change or alternatively by the time varying plasma density as illustrated in Fig. 6:

Figure 6. An observer looking through a semiconductor slab and seeing the frequency chirped image. For the observer this seems like a window into an accelerating world.

We may speak alternately, in acceleration units, or in chirp rate units. Equation (9) shows that:

$$a = c\frac{\dot{\omega}}{\omega} = \frac{c}{\tau} \qquad (10)$$

where τ is a characteristic chirp time. For $\tau = 10^{-13}$ sec, well within current sub-picosecond technology, the acceleration c/τ is 3×10^{23} cm/sec^2 or $\approx 3 \times 10^{20}$g. According to eq'n. (7) this is enough to produce an Unruh radiation temperature of a few degrees Kelvin, with frequencies extending into the millimeter wave band.

MATHEMATICAL DIGRESSION

Let us now make a brief digression towards quantum field theory. The operative mechanism is that frequency chirp is produced by a changing refractive index. The normal modes of the electromagnetic field which exist at the outset of the experiment, are different from the new normal modes which have adapted to the changed refractive index. A zero point excitation in the original normal modes might have to be expanded in terms of zero *and* finite photon-number wave-functions of the new modes. The mathematical recipe for this transformation is called the Bogolyubov Transformation. Let $A(z,t)$ be a quantum field operator which can be written as a linear combination of the original normal modes $A_k(z,t)$:

$$A(z,t) = \sum_{all\ k} [a_k A_k(z,t) + a_k^+ A_k^*(z,t)] \qquad (11)$$

where a_k and a_k^+ are the annihilation and creation operators appropriate to these normal modes which might be, for example, the standard plane waves:

$$A_k(z,t) = c[2\pi\hbar/\omega]^{1/2}\exp\{i(kz - \omega t)\} \qquad (12)$$

As the refractive index changes, the quantum field operator $A(z,t)$ can be re-written in terms of the new normal modes $A_k(z,t)$ which solve Maxwell's eq'ns.

$$A(z,t) = \sum_{\text{all } k} [b_k \overline{A}_k(z,t) + b_k^+ \overline{A}_k^{\hspace{0.1em}*}(z,t)] \tag{13}$$

where b_k and b_k^+ are the new annihilation and creation operators. The Bogolyubov transformation consists of writing the new operators as a linear combination of the original operators:

$$b_k = \sum_{k'} (\alpha_{kk'}^* a_{k'} - \beta_{kk'}^* a_{k'}^+) \tag{14}$$

A key role is played by the coefficients $\beta_{kk'}$ which measure the admixture of annihilation and creation operators, or equivalently the admixture of positive and negative frequency modes. The expectation value of the accelerated mode k photon-number operator, in the original zero point state, $|0>$, is:

$$<0| b_k^+ b_k |0> = \sum_{k'} \left| \beta_{kk'} \right|^2 \tag{15}$$

For case of a step-function change of refractive index from n_0 to n at time $t = 0$, the transformation is straightforward:

$$\alpha_{kk'} = \delta_{kk'}(n + n_0)/2\sqrt{n_0 n} \quad \text{and} \quad \beta_{kk'} = \delta_{kk'}(n - n_0)/2\sqrt{n_0 n} \tag{16}$$

where $\delta_{kk'}$ is the Kronecker delta. We see that in this example, the induced photon occupation number can be of order unity, and will have a frequency independent spectrum. In a finite duration pulsed laser experiment, as in this case, the spectrum will generally be non-thermal. An *exact* thermal spectrum evolves only asymptotically[14] at infinitely long times. In that respect we fail to exactly mimic a black hole. This concludes our digression into quantum electrodynamics.

VIRTUAL PHOTOCONDUCTIVITY

In the remainder of this article let us discuss how to implement these ideas in a real experiment. Basically, we require an optical nonlinearity in a real material, capable of a substantial refractive index change. The index change associated with an electron-hole plasma in a semiconductor satisfies this requirement. A very small density of carriers results in a very large index change in the microwave spectral region. This is due to the zero-frequency resonant character of the plasma dielectric response. Such a low frequency dielectric response is well-matched to eq'n. (7), which restricts us to low "effective Unruh temperatures" at the moderate accelerations which are experimentally achievable.

Since laser produced plasmas have already been shown to generate a substantial frequency chirp, this nonlinearity would seem promising. Furthermore, a semiconductor crystal is readily cooled to 1°K, reducing background thermal radiation in the microwave region. There remains one problem however. The free electrons and holes in the optically injected semiconductor plasma will experience mutual collisions. These collisions will generate bremstrahlung radiation in the microwave region, a background effect that could mask the Unruh radiation. We would prefer to find a purely reactive nonlinearity in which dissipative absorption of the incident laser beam is absent. Instead of creating real electrons and holes, one could tune the incident laser beam in the transparent region just below the semiconductor band edge, creating[15] virtual electron-hole pairs. The distinction is illustrated in Fig. 7.

The change of zero-frequency electric susceptibility induced by optical photons of energy $h\nu = \hbar\omega$, conventionally a third-order nonlinear susceptibility $\chi^{(3)}(0, 0, -\omega, \omega)$, has been given various names in the past, including virtual photoconductivity, the inverse quadratic electro-optic effect, and in the case of quantum wells, the AC-DC Stark effect[16]. Since this particular susceptibility is resonant at the band edge, for small detunings it will dominate other nonlinearities such as two-photon absorption which is parity-forbidden at zone center.

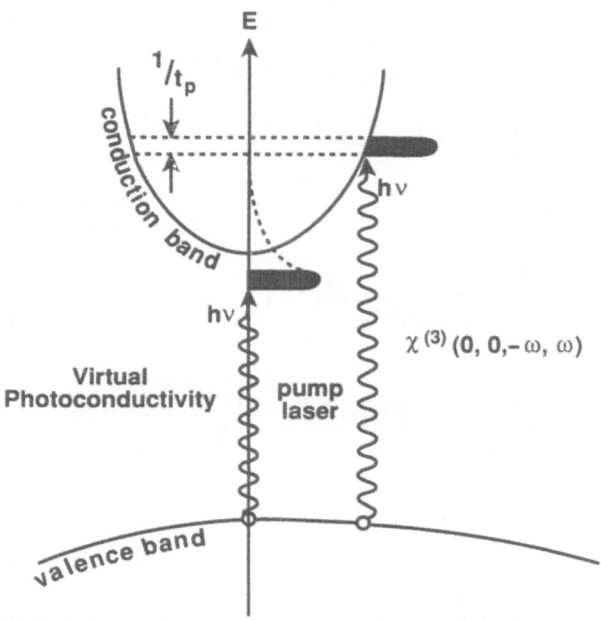

Figure 7. A comparison of transitions from the valence band to the conduction band in ordinary photoconductivity and in virtual photoconductivity. The electron population distribution for the two cases is illustrated by the dashed lines. The frequency width, $1/t_p$, corresponds to the real electron distribution induced by ordinary photoconductivity. Virtual photoconductivity, due to non-absorbed photons, has the advantage of being purely reactive and non-dissipative.

The effect of virtual electron-hole pairs is large because they behave as though bound by the small detuning energy and are therefore highly polarizable. Calculations of virtual photoconductivity have recently been extended[17] to the limit of very strong optical fields. The change of low frequency dielectric constant saturates at ~0.5 unit for an optical intensity of ~10 MW/cm^2, for reasonable detunings. Different experimental geometries suggest themselves for the excitation of the semiconductor slab, as illustrated in Fig. 8.

There is yet another way of thinking about this effect, by making contact with the famous Casimir[18] Force between metal plates, as illustrated in Fig. 9. This Force can be thought of as arising from zero-point energy of the electromagnetic field. It is therefore regarded as one of the most direct experimental manifestations of the reality of the quantum field. Between two closely spaced metallic plates, long wavelength electromagnetic field modes are unable to fit. Outside the plates these modes exist, and their zero-point mean-square field exerts an unbalanced pressure on the plates.

In effect, when the metal plates are brought together, the empty space between them becomes filled with the electron plasma of the metal. If this is done very quickly, the zero point electromagnetic field has no time to re-adjust adiabatically and suffers non-adiabatic excitation in the course of trying to escape from the forbidden region between the plates. Likewise, a semiconductor slab may be filled with zero point electromagnetic fields. If suddenly the semiconductor were converted to a plasma, by electron-hole excitation, the zero-point electromagnetic field would be abruptly forced to escape from that forbidden volume. In doing so, there would be some probability of non-adiabatic excitation, leading to real photons.

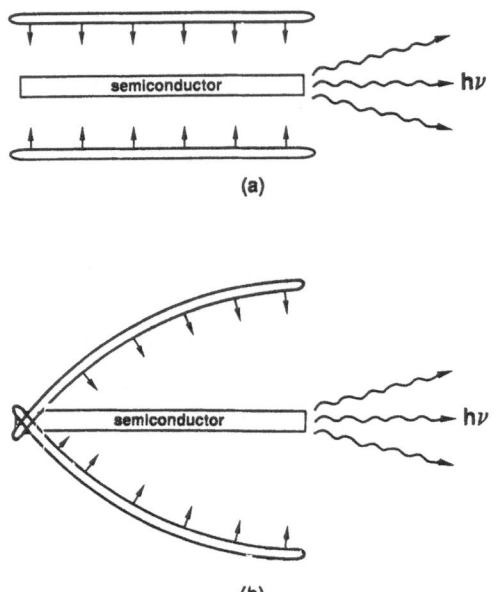

Figure 8. Two possible experimental geometries for exciting a semiconductor slab dielectric waveguide for millimeter waves. The pancake-shaped objects are the incident sub-picosecond optical pulses. In (a), the pancake-shaped sub-picosecond pulses bombard the semiconductor slab simultaneously throughout, mimicking the step function model we solved earlier. In (b) there is a traveling wave excitation, which, due to the incident angle of optical pulses, sweeps down the slab faster than the speed of light. The emitted microwave photons are written as hν. This not a Cerenkov effect since polarizability, not polarization is moving faster than the speed of light.

The design of an experimental configuration for generating and detecting this radiation owes much to Heinrich Hertz, who first made radio waves 100 years ago, and to the more modern photoconductive switches[18] which have been used with mode-locked lasers to generate millimeter waves. H. Hertz used the switching action of a spark gap to generate radio waves. Likewise, a fast photoconductor, when illuminated by a sub-picosecond light pulse, will conduct a burst of current which radiates in the millimeter wave spectral region. Such a configuration is shown in Fig. 10. Cleverly, a photoconductor can be used both as a generator and coherent detector of the radio waves. The detector switch is turned on by a synchronous optical pulse from the same mode-locked laser. If the detector photoconductor is in the on-state for one-half cycle of the millimeter wave electric field, a small dc current will be pushed through. If this repeats for every pulse in the mode-locked train, the average current can be monitored by a pico-ammeter.

Suppose that the bias voltage + V were disconnected. It might seem that there would be no millimeter wave signal at all. But the generator photoconductor is always experiencing a fluctuating bias field from zero-point electromagnetic waves. In an octave bandwidth around 500 GHz, the zero-point field strength magnitude averages a few tens of millivolts per cm, small but not negligible. This assures a signal even in the absence of an overt bias voltage.

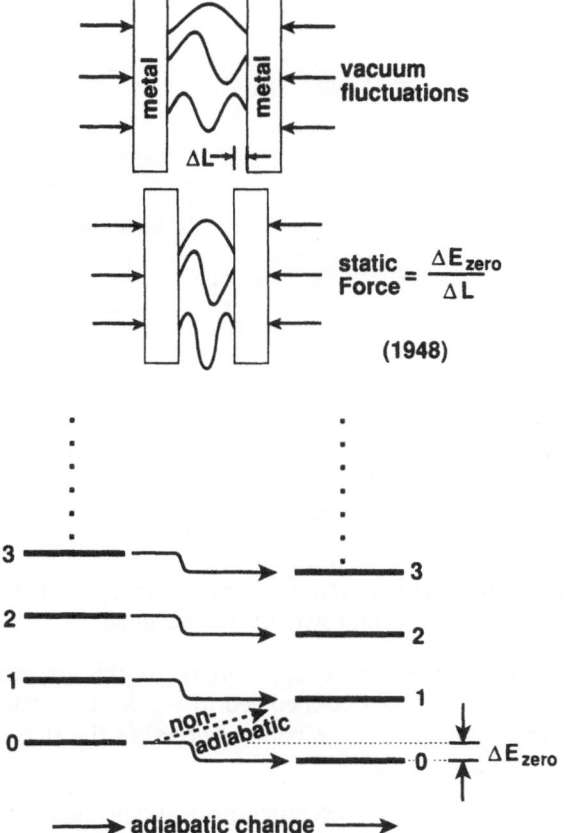

Figure 9. The Unruh radiation, re-interpreted as a non-adiabatic Casimir effect. The Casimir force is calculated by making a slow, adiabatic, differential displacement ΔL of the spacing between the plates. This changes the entire ladder of harmonic oscillator levels representing the electromagnetic field. Adiabatically, the zero-photon energy level goes into a new zero-photon level, the one-photon level goes into a new one-photon energy level, etc. The adiabatic shift in zero point energy is ΔE_{zero}. The Casimir Force is then the differential energy divided by differential displacement, $\Delta E_{zero}/\Delta L$. This is one possible way of calculating the Force. Now what if the displacement change was large, very sudden, and too fast for the electromagnetic field to respond adiabatically. Then there would a small probability of a non-adiabatic transition, for example from the old zero-photon level to the new one-photon level, as illustrated. Such one-photon and multiphoton states are what we would hope to detect.

In our case, of course, we would use a virtual photoconductor which changes its dielectric constant rather than its conductivity. Radiation produced this way would be equivalent to Unruh radiation. It would be detected by conventional photoconductivity in the detector switch, acting as a radio mixer. Employing a similar arrangement as Fig. 10, the photoconductive switch has been shown[19] to have extremely low noise, such that signals as weak as quantum noise can be detected readily. Such radio receivers are finding increasing[18,19] use in opto-electronic technology. The noise temperature is not so low, but fortunately the detector switch

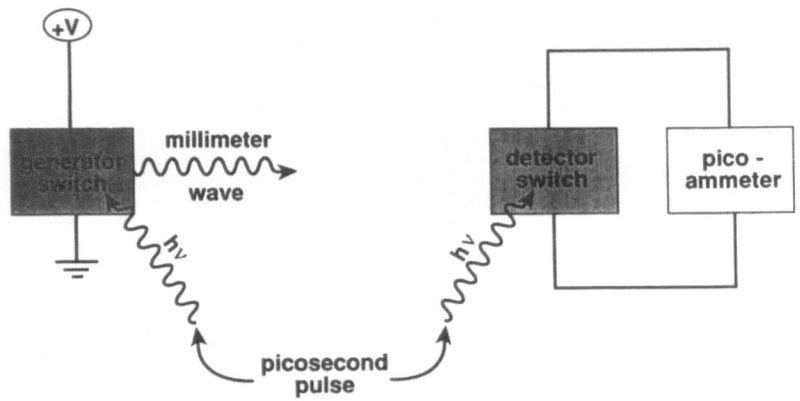

Figure 10. Photoconductive switches are turned "on" by a picosecond optical pulse. The generator switch, biased at voltage + V, produces a millimeter wave which drives a very tiny electric current through the detector switch.

has little or no noise in the waiting period between successive picosecond pulses. The noise is present during the picosecond switching pulse only.

An experimental geometry more appropriate to this experiment is shown in Fig. 11. The generator is a slab of GaAs cooled to $\sim$1°K and thick enough to support waveguide modes in the millimeter wave band. Therefore the slab would contain zero point electromagnetic waves.

The experimental signature of accelerated zero-point fluctuations is very clear in this experiment, since there is no other mechanism which could cause a reduction of noise below normal quantum fluctuations. In that respect it resembles the noise reduction seen in optical squeezing[20] experiments. The increase of noise on the trailing edge of the laser pulse is much less definitive, since inadvertent bremsstrahlung or other noise mechanisms could be responsible.

Since a reduction in mean-square noise power is the experimental signature, discrimination against coherent signals is quite easy. Nevertheless, it is best to minimize coherent signals by diminishing any stray electric fields inside the GaAs generator slab. Accordingly the slab is doped n-i-n to minimize doping-induced fields, and the laser beam should avoid the periphery of the slab where there are depletion fields.

CONCLUSION

In this paper I chose to introduce the excitation of the zero-point electromagnetic field by means of black holes, Hawking radiation and Unruh radiation. It should be obvious by now that this approach is only one of many possible choices. In Fig. 9, we gave an alternate interpretation in terms the non-adiabatic Casimir effect. On the other hand, those who are experts in optical squeezing[20] may insist that this effect is merely a case of single-cycle millimeter wave squeezing. Other observers will call this the parametric excitation of the vacuum, still others, the inverse quadratic electroptic effect by virtual photoconductivity with zero-point-photon input waves. There are many choices.

But there is one final explanation which I prefer the most; the hand waving explanation. Consider the human hand. Being mainly composed of water, no doubt its dielectric constant differs greatly from unity. As we wave our hand, indeed as we walk through life, unbeknownst to us, we are forcing the zero-point electromagnetic field to constantly change and adapt to us. This all takes place automatically, adiabatically and almost imperceptibly since the Casimir forces are so weak. What if we were to make a sudden, abrupt, motion of our hand that is so quick, that the zero-point electromagnetic field could not respond adiabatically? The quantum field would make itself felt. Our hand-waving would generate a few detectable microwave photons.

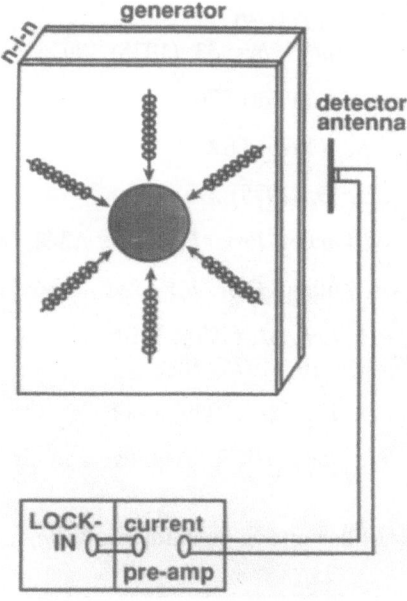

Figure 11. A picosecond laser pulse at normal incidence is transmitted through the center of the slab, generating a local population of virtual carriers, indicated by the shaded region in the center of the slab. Virtual carriers, unlike real carriers, cause an increase rather than a decrease in the local dielectric constant. This would pull the waveguided zero-point electromagnetic field fluctuations (illustrated as standing waves) toward the laser spot, as shown. Since virtual photoconductivity is purely reactive, the dielectric constant would return to normal on the trailing edge of the laser pulse. The excess concentration of field fluctuations at the center of the laser spot would then be expelled. The detector photoconductive switch, designed as an antenna and located at the slab periphery, would initially see a reduction in zero-point noise in synchronism with the leading edge of the laser pulse and then an increase in noise as the field fluctuations were expelled on the trailing edge of the pulse.

REFERENCES

1. John Michell, *Phil. Trans. Royal. Soc.* **74**, (1784) 35-57. Reprinted in *Black Holes: Selected Reprints,* ed. S. Detweiler, (American Assoc. of Physics Teachers, Stony Brook, NY, 1982).

2. P. S. Laplace, *Exposition du Systeme du Monde, vol. 2,* (J. B. M. Duprat, Paris, 1796), p. 305. See also S. W. Hawking and G. F. R. Ellis, *The Large Scale Structure of Space-Time,* (Cambridge Univ. Press, Cambridge, 1974), pp. 365-8.

3. W. Israel, in *Three Hundred Years of Gravitation,* ed. S. W. Hawking and W. Israel, (Cambridge Univ. Press, Cambridge, 1987).

4. J. D. Bekenstein, *Physics Today,* **33,** no. 1, (Jan. 1980) 24.

5. J. D. Bekenstein, *Nuovo Cimento Lett.* **4** (1972) 737.
   J. D. Bekenstein, *Phys. Rev.* **D7** (1973) 2333.

6. S. W. Hawking, *A Brief History of Time,* (Bantam Books, NY, 1987).

7. S. W. Hawking, *Nature* **248,** (1974) 30.
   S. W. Hawking, *Commun. Math. Phys.* **43,** (1975) 199.

8. W. G. Unruh, *Phys. Rev.* **D14,** (1976) 870.

9. P. C. W. Davies, *J. Phys.* **A8,** (1975) 609.

10. B. S. DeWitt, *Phys. Reports* **19,** (1975) 295.

11. S. A. Fulling and P. C. W. Davies, *Proc. Roy. Soc.* **A348,** (1976) 393.

12. P. C. W. Davies and S. A. Fulling, *Proc. Roy. Soc.* **A356,** (1977) 237.

13. E. Yablonovitch, *Phys. Rev. Lett.* **32,** (1974) 1101.
    E. Yablonovitch, *Phys. Rev.* **A10,** (1974) 1888.

14. E. Yablonovitch, *Phys. Rev. Lett.* **62,** (1989) 1742.

15. E. Yablonovitch, J. P. Heritage, D. E. Aspnes and Y. Yafet, *Phys. Rev.* **63,** (1989) 976.

16. D. S. Chemla, D. A. B. Miller and S. Schmidt-Rink, *Phys. Rev. Lett.* **59,** (1987) 1018.

17. Y. Yafet and E. Yablonovitch, *Phys. Rev. B* to be published.

18. H. B. G. Casimir, *Proc. Kon. Ned. Akad. Wet.* **51,** (1948) 793.

19. B. B. Hu, J. T. Darrow, X.-C. Zhang, D. H. Auston and P. R. Smith, *Appl. Phys. Lett.* **56,** (1990) 886.
    C. H. Lee, *Appl. Phys. Lett.* **30,** (1977) 84.

20. M. van Exter and D. Grischkowsky, *IEEE Trans. Microwave Th. & Tech.* **MTT-38,** (1990) to be published.

21. R. E. Slusher, L. W. Hollberg, B. Yurke, J. C. Mertz and J. F. Valley, *Phys. Rev. Lett.* **55,** (1985) 2409.
    R. M. Shelby, M. D. Levenson, S. H. Perlmutter, R. G. DeVoe, and D. F. Walls, *Phys. Rev. Lett.* **57,** (1986) 691.

COUPLING BETWEEN CHARGE TRANSFER AND INTRAMOLECULAR

VIBRATIONAL MODES IN $(TMTSF)_2X$, X = PF_6 or ReO_4

M. Krauzman, J. Breitenstein and R.M. Pick

Département de Recherches Physiques (URA 71)
Université P. et M. Curie, 4, place Jussieu
75252 Paris cedex 05, France

1. INTRODUCTION

The Bechgaard's salts [1], of general formula $(TMTSF)_2X$ represent one of the most interesting family of quasi 1-d organic conductors. They have been extensively studied during the past decade by a variety of techniques [2,3,4,5]. Nevertheless, among the various light scattering experiments [6,7,8,9] performed, only Resonant Raman Scattering (R.R.S.) enabled some of us [8,9] to record more than a few modes. This was first reported[8] for X = PF_6 which remains metallic down to OK. Only some of these modes, namely those which are the totally symmetric modes of the isolated TMTSF molecule, were then identified. More recent experiments[9] on X = PF_6, and also on X = ReO_4 which undergoes a metal insulator transition at 182 K have enabled us to interpret some of the extra modes as resulting from a dynamical charge transfer (C.T.) mechanism, a phenomenon typical of these organic conductors.

2. RESONANT RAMAN SCATTERING IN THE METALLIC PHASE

In their metallic phase, the Bechgaard's salts crystallize in a centered cell with one formula unit per cell. Each TMTSF molecule (Fig. 1) is

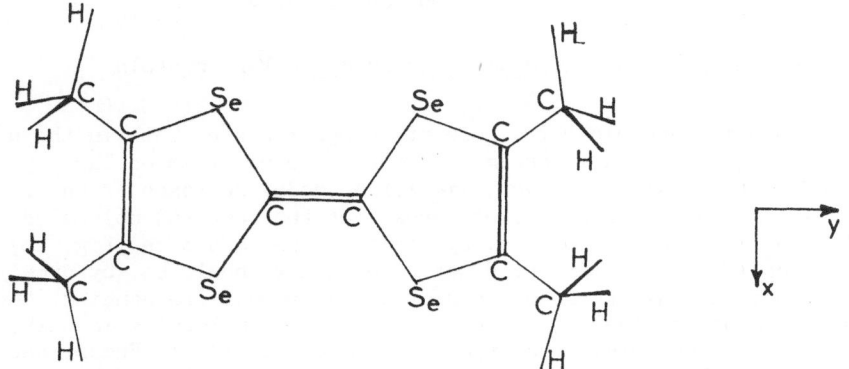

Fig. 1. Schematic drawing of an isolated TMTSF molecule (D_{2h} symmetry). The irreducible representations of this group are labelled according to the axes represented on the r.h.s. of the figure.

nearly flat and approximately keeps the D_{2h} symmetry it possesses as an individual molecule. In the crystal, these molecules stack one on top of the other along $\vec{c}$, and for each stack, there is a $\vec{c}$ axis, which forms approximately a twofold screw axis, and also contains the centers of inversion between the molecules of the stack. The X molecules (acceptors) are located between the stacks, and are large enough to suppress nearly any overlap between neighbouring stacks, leading to quasi 1-d conduction along $\vec{c}$.

The TMTSF molecule contains two types of double bonds, the vibrations of which we shall focus on in this paper. Both types of bonds are the sources of the electronic donnor character of TMTSF, the central bond electron being more easily transferred than the electron on the external bond. Because of this donnor character, the stretching motion of these bonds is strongly Raman active. Conversly, as light is absorbed by the conduction electrons, any metal is a very poor Raman scatterer, which explains why only the central bond stretching mode was clearly identify in the first[6,7] Raman experiments.

Fortunately it exists an optical absorption, involving an electronic transition between the conduction band and the first valence band[10], which is polarized perpendicular to $\vec{c}$, and takes places around 15 000 cm^{-1}. Consequently, using incident and scattered light polarized perpendicular to $\vec{c}$, with $\nu = 15\ 453$ cm^{-1}, R.R.S. involving only this transition turned out to be possible[8]. As each of these two electronic bands generates from a linear combination of only one type of molecular orbital of the TMTSF molecule, an elementary theory[8,9], based on this model shows that only the *modes* which are *totally symmetric* in the pseudo monoclinic(C_{2h}) crystal factor group will resonnate through this mechanism. Focusing on the TMTSF intramolecular modes one expects to detect in this process all the modes with a_{1g} symmetry (irreducible representations of the D_{2h} molecular point group are labelled with lower case symbols), and indeed, the 12 a_{1g} modes detected or computed by Meneghetti et al.[11] for the isolated molecule were recorded in our PF$_6$ experiments [8], with frequencies in agreement with a charge of half an electron per molecule.[7,8] We shall discuss, here, the modes, detected both with PF$_6$ in our first experiments[8], and with ReO$_4$, in its metallic, and, with important modifications, in its insulating phase[9], which can be attributed to a dynamical C.T. mechanism between the double bonds.

3. CHARGE TRANSFER IN THE METALLIC BECHGAARD'S SALTS

1. Dynamical coupling and Raman activity in molecular crystals

In order to make clear the role of charge transfer both in the dynamics and in the detection mechanim of internal modes of molecular crystals, it is usefull to consider the very simplified model represented on Fig. 2. It consists of a molecular dimer formed of two centered molecules. The monomolecular odd mode generates, by symmetry, an odd mode (Fig. 2b) and an even mode (Fig. 2a) for the dimer. The latter should be, by symmetry, Raman active, and the two modes should have different frequencies. Nevertheless, as long as the coupling between the two molecules is weak, the modes remain degenerate, and the even mode practically Raman inactive because it can be considered as the sum of two monomolecular Raman inactive modes. *Coupling* between the two modes simultaneously *lifts* this *degeneracy* and provides *Raman activity* to the mode with *even symmetry*.

Following elementary chemistry, the length of a C-C bond depends on the charge on this bond. If the corresponding electronic molecular orbitals of two bonds overlap, in an antisymmetric stretch of the two bonds, charge will flow from one bond to the other to minimize the energy. C.T., between the bonds is thus a very efficient coupling mechanism between double bond stretching motions and a source of strong Raman activity.

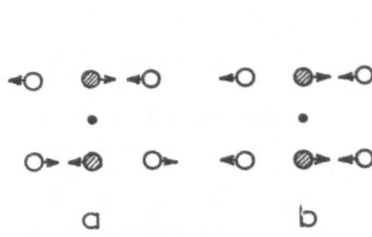

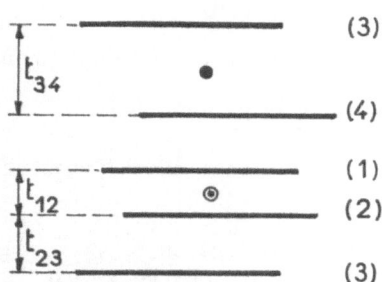

Fig. 2. Even (a) and odd (b) molecular I.R. active modes for a dimer.

Fig. 3. Schematic stacking of TMTSF molecules in the insulating phase of $(TMTSF)_2ReO_4$.

In the Bechgaard's salts, as the TMTSF molecules along a stack are rather far apart, one expects that all the internal modes with b_{2u}, b_{3u} and b_{1g} symmetry, which, in principle, could lead to crystalline modes with A_{1g} symmetry, will be invisible, even under resonant conditions. Nevertheless, this will not be true for the stretching modes of the donor bonds and their A_{1g} combination will become Raman active, due to C.T. coupling. The basic argument that C.T. lifts the degeneracy between degenerate modes has already been given, using a localized electron picture, by M.J. Rice et al. [12] and applied to I.R. absorption experiments. We shall give here a straight forward description of the mechanism based on a quadratic vibrational free energy which includes a coupling to electronic variables treated in the adiabatic approximation. As we shall see, this method turns out to be easily extended to complex situations.

2. Vibrational energy and Charge Transfer for the central bond

Let us discuss the case of the central bond of the TMTSF molecules. For the isolated molecule, its vibrational energy may be written as :

$$F_i = \frac{1}{2} \omega_c^2 Q_i^2 + \Delta\omega_c Q_i R_i + \frac{1}{2} a_c R_i^2 \qquad (1)$$

where i (= 1 or 2) labels the two molecules of the unit cell,
ω_c is the frequency of the central bond stretching mode with normal coordinate Q_i,
R_i is the extra charge which can be located on this bond,
a_c is a positive constant which mimics the interaction of each molecule with the counter ions and tends to localize half an electron on each TMTSF molecule at high temperature. Finally $\Delta\omega_c$ is a coupling constant between Q_i and R_i and allows for the change of the molecule equilibrium confi-

guration under the influence of a charge modification on the molecule.

The charge transfer between the two molecules of the unit cell originates from the overlap between the corresponding electronic molecular orbitals which generate the conduction band. Within the approximation that the stacks have a twofold screw axis, there is only one overlap integral, v, between neighbouring molecules. In the spirit of the harmonic approximation of Eq. 1., we mimic the influence of this overlap on the charge transfer mechanism by

$$F_c = 2t \, R_1 \, R_2 \quad . \tag{2}$$

Writing the total vibrational energy as

$$F = F_1 + F_2 + F_c \quad , \tag{3}$$

the two vibrational eigen frequencies can be easily obtained from F, noting that

a) the total charge on the two molecules is constant whence $R_1 + R_2 = 0$,
b) the electronic charge is always in equilibrium with the molecular deformations : $\dfrac{\partial F}{\partial R_i} = 0$.

F is then trivially diagonalized by group theory method yielding

$$\omega_+^2 = \omega_c^2 \quad ; \qquad Q_+ = \frac{1}{\sqrt{2}} \, (Q_1 + Q_2) \tag{4a}$$

$$\omega_-^2 = \omega_c^2 - \frac{\Delta\omega_c^2}{a-2t} \quad ; \qquad Q_- = \frac{1}{\sqrt{2}} \, (Q_1 - Q_2) \tag{4b}$$

In the present case where Q_1 and Q_2 have a_{1g} molecular symmetry, Q_-, whose frequency is affected by the charge transfer mechanism has B_{1u} symmetry and is thus I.R. but not Raman active. The only Raman active mode is Q_+, with A_{1g} symmetry ; this mode does not involve any charge transfer so that its frequency is the same as for an isolated molecule with half an electron on the bond, and it has been found[7] that ω_c is a linear function of the total charge on the molecule.

Note that the parameter t must be related to the overlap integral v. In order for the model to be consistent, one can assume that

$$t \simeq - \left(\frac{1}{v}\right)^n \tag{5}$$

where n is some positive number which ensures that ω_- tends towards ω_c if v tends to zero.

3. Vibrational energy and Charge Transfer for the external bonds

The previous model can easily be extended to the case of the external double bond stretching modes. For the isolated molecule the situation is now very reminiscent of the model of Fig. 2 : due to the inversion center existing on the molecule, the stretching of the external bonds generates two internal vibrations, one with a_{1g} symmetry and the other with b_{1u} symmetry respectively. Furthermore, the electronic molecular orbital which generates the crystalline conduction band provides an efficient coupling mechanism between the stretching modes of the two halves of the TMTSF molecule so that for the individual molecules imbeded in the crystal one can write in agreement with Eqs 1 - 3

$$F_0 = \sum_{i=1}^{2} \left[\sum_{\lambda=1}^{2} \left(\frac{1}{2} \omega_e^2 Q_{i\lambda}^2 + \Delta\omega_e Q_{i\lambda} R_{i\lambda} + \frac{1}{2} a_e R_{i\lambda}^2 \right) + \frac{1}{2} \sum_{\lambda,\lambda'=1}^{2}{}' d\ R_{i\lambda} R_{i\lambda'} \right]. \quad (6)$$

Here λ labels the two halves of the molecule, ω_e is the external bond "bare" frequency, $\Delta\omega_e$ and a_e having the same meaning as in Eq. 1 while d characterizes an intramolecular C.T. between the two external bond stretching modes, ensuring that the b_{1u} mode has a lower frequency than the a_{1g} mode. The intermolecular C.T. acts between one bond of one molecule and the other bond of the second molecule so that the corresponding coupling term may be written as

$$F_c = \sum_{\lambda,\lambda'=1}^{2}{}' 2t\ R_{1\lambda} R_{2\lambda'}. \quad (7)$$

$F = F_0 + F_c$ thus represents the total vibrational energy of the our coupled external double bonds. Using again the charge neutrality condition as well as the adiabatic condition, F leads now to two Raman active modes. One is the totally symmetric mode, the four bonds vibrating in phase, with frequency ω_e. The second is the exact analog of the mode described at the beginning of this section and one finds its frequency to be

$$\omega_-^R = \left[\omega_e^2 - \frac{\Delta\omega_e^2}{a_e - 2t - d} \right]^{\frac{1}{2}} \quad (8)$$

while the two I.R. active mode frequencies are given by

$$\omega_\pm^{IR} = \left[\omega_e^2 - \frac{\Delta\omega_e^2}{a_e \pm (2t-d)} \right]^{\frac{1}{2}}. \quad (9)$$

Eq. 8 gives an explanation for the appearance, of a second Raman active mode, at a frequency $\omega_-^R < \omega_e$, nearly as intense as ω_e under resonance

Table 1. Measured frequencies in cm^{-1}
The data of column 2, 3 and 4 are pertinent to the insulating phase. Column 2 refers to the central bond ; columns 3 and 4 to the external bonds. Raman data are from [8], IR data from [15] (column 2) and [14] (column 4).

ω_c	1.462	ω_{c+}^R	1.469	$\omega_e^R\lambda$	1.610	$\omega_e^{IR}1$	1.601
$\omega_c^0 - \omega_c^+$	140	ω_{c-}^R	1.330	$\omega_e^R\lambda$	1.595	$\omega_e^{IR}2$	1.591
ω_e	1.599	ω_{c+}^{IR}	1.380	$\omega_e^R\lambda$	1.568	$\omega_e^{IR}3$	1.560
$\omega_e^0 - \omega_e^+$	55	ω_{c-}^{IR}	1.250	$\omega_e^R\lambda$	1.559	$\omega_e^{IR}4$	1.538
ω_-^R	1.581						

condition [8,9]. This agrees with our experiments on $X^- = PF_6^-$,[8] the difference between ω_e and ω_-^R being equal to 18 cm^{-1} .

4. INSULATING PHASE OF $(TMTSF)_2 ReO_4$

Below $T_c = 182$ K, $(TMTSF)_2 ReO_4$ undergoes a weakly first order
metal-insulator phase transition[13]. For the present purpose, it is sufficient to know that, below T_c, the cell doubles along the $\vec{c}$ axis. The low
temperature phase has a P$\bar{1}$ structure, so that one center of inversion out
of two is lost. Schematically, the conducting chain dimerizes as shown on
Fig. 3. Molecules 1 and 4 are no longer equivalent, even if their relative
distance and orientation are similar to what they were in the metallic
phase, because the two centers of inversion of Fig. 3 are different, the
distance between 1 and 2 being shorter than between 3 and 4. This dimerization alters the vibrational properties in three different ways.

a) The number of molecules per cell doubles, yielding now two Raman
active modes for the central bond, and four for the external bonds. These
effects are quite visible in our spectra, and in particular the central
bond activity is so large that, even under a non resonant condition, the
second mode is still detectable [9]. Because the latter couples with C.T.,
its frequency is lowered with respect to the fully symmetric mode, and
this lowering turns out to be very large (~ 135 cm^{-1}).

b) The unique bond electron is shared unevenly between molecules 1
and 4, the charges being respectively $\left(\frac{1}{2} \pm \delta q\right)e$ instead of $\frac{e}{2}$ in the metallic case. As the "bare" stretching mode frequencies appear to be linear
functions of the net charge on the molecule one must replace for the central double bond ω_c by

$$\omega_c \implies \omega_c \left[1 \pm \lambda_c \delta q\right] \qquad \text{with} \qquad \lambda_c \omega_c = \omega_c^0 - \omega_c^+ \qquad (10)$$

ω_c^0 and ω_c^+ being the frequencies of the central double bond for the isolated neutral (ω_c^0) and for the fully ionized (ω_c^+) molecule. A similar expression holds for the external bond

$$\omega_e \implies \omega_e \left[1 \pm \lambda_e \delta q\right] \quad . \qquad (11)$$

This static C.T. will result in a small but perfectly detectable increase of the totally symmetric mode frequencies both for the central and
for the external double bonds.

c) Finally, due to the dissymmetry between the different molecular
distances along the $\vec{c}$ axis, the overlap integrals, and the resulting C.T.
coefficients between the three couples of neighbouring molecules differ one
from another.

Due to these three effects, the vibrational energy of the problem is substantially modified and, for instance, for the central bond case, using the
notations of Fig. 3, one obtains

$$F = \frac{1}{2} \omega_c^2 (1+\lambda_c \delta q)^2 (Q_1^2 + Q_2^2) + \frac{1}{2} \omega_c^2 (1-\lambda_c \delta q)^2 (Q_3^2 + Q_4^2) +$$

$$+ \sum_{i=1}^{4} (\Delta\omega_c Q_i R_i + \frac{1}{2} a_c R_i^2) + t_{12} R_1 R_2 + t_{23} (R_2 R_3 + R_1 R_4) + t_{34} R_3 R_4 \quad . \qquad (12)$$

Making use again of the charge neutrality condition and of the adiabatic approximation, the 2x2 determinants leading to the frequencies of the two Raman active modes on the one hand, of the two I.R. active modes on the other hand are easily computed. They show that the totally symmetric Raman active mode has a frequency slightly higher than $\omega_c \left[1 + \frac{1}{2} (\lambda_c \delta q)^2 \right]$, an effect totally due to the dissymmetry between molecules 1 and 4, which turns to be of the order of 7 cm^{-1}. Conversely, the main effect on the low frequency Raman mode is governed by the charge transfer mechanism, yielding approximately

$$\omega^R \simeq \omega_c - \frac{\Delta\omega_c^2}{\omega_c \ (2a_c + t_{12} + 2t_{23} + t_{34})} . \tag{13}$$

Finally the two I.R. active modes are also strongly affected by C.T. but have different frequencies, due to the difference in the coupling integrals, an effect which has been measured by Bozio et al.[15]. As shown by Breitenstein[9] a similar treatment can be made for the case of the external bonds from which one can deduce the frequencies of the four Raman active modes, and of the four I.R. active modes. Though some more unexpected Raman lines have been detected in the corresponding spectral region (1540 cm^{-1} - 1610 cm^{-1}) , using various polarizations of the incident and scattered light, we have been able to identify the four modes involved here, while the four I.R. modes have been measured by Homes et al.[14].

The entire set of input data, obtained in our Raman experiments, on the metallic (X = PF$_6$) and insulating (X = ReO$_4$) salts or derived from the corresponding I.R. experiments are given in Table 1. We have fit these data with the various expressions obtained for both cases . Only the ratios of a, t or d with the corresponding $\Delta\omega^2$ are observable quantities, and we have found that, in the metallic case, a correct fit could be obtained using the same a and $\Delta\omega$ for the central and the external bonds. Using the same rule in the insulating case, the fitting procedure turned out to be more difficult. In view of the obvious oversimplifications represented by the fitting model (see below) we did not try to look for the best possible adjustement using all its flexibilities. Reasonnable agreement with the data of Table 1 were obtained for the changes of parameters reported in the second column of Table 2. The important points to be noticed are :

a) the static charge transfer between the two molecules in the insulating state is quite large ; there is approximately .66e on one type of molecules and .33e on the other in the dimerized chain instead of .5e on each molecule in the metallic case, in quite reasonable agreement with the values derived from the I.R. data[15] on the central bond mode (.7e and .3e).

b) The intramolecular charge transfer coefficient is more important than the intermolecular one.

c) The low temperature dimerization process strongly modifies the C.T. mechanism between the various molecules and *increases* its mean value with respect to the metallic phase.

Let us finally note that the basis of our model is a quadratic expansion of the free energy in term of charge distribution but the microscopic meaning of $\Delta\omega$, a and t (or d) will depend on the model used to describe the electronic interactions in the 1-d chain ; without specifying it, we have found useless to use all the flexibilities of the model to get the best possible agreement.

Table 2. Values of the different fitting parameters.
First column : metallic case, the coefficients being in 10^{-6} cm^{-1}, having arbitrarily set $\Delta\omega_e = \Delta\omega_c = 1$ cm^{-1}.
Second column : changes of the coefficients in the insulating case.

$$
\begin{array}{|c|c|}
\hline
a_c = a_e = 43 & \delta q \quad = \quad 0.33 \\
d = 13 & \dfrac{t_{12} + 2\,t_{23} + t_{34}}{4t} = 0.85 \\
t = 6.4 & \dfrac{t_{12} - t_{34}}{2t} \quad = \quad 0.36 \\
\hline
\end{array}
$$

REFERENCES

1. K. Bechgaard, Sol. State. Physics, 16:3535 (1980)
2. N. Thorup, G. Rinsdorf, H. Soling and K. Bechgaard, Acta Crystallogr., Sect. B 37:1236 (1981)
3. K. Bechgaard, C.S. Jacobsen, K. Mortensen, H.J. Pedersen and N. Thorup, Solid State Commun., 33:1119 (1980)
4. R. Bozio, C. Pecile, J.C. Scott and E.M. Engler, Mol. Cryst. Liq. Cryst. 119:211 (1985)
5. P. Bernier, M. Audenaert, R.J. Schweizer, P.C. Stein, D. Jerome, K. Bechgaard and A. Moradpour, J. Physique Lett. 46:L675 (1985)
6. H. Kuzmany, L. Iwahana, F. Wudl and E. Aharon Shalom, Mol. Cryst. Liq. Cryst. 79:39 (1982)
7. P.V. Huong, C. Garrigou-Lagrange, J.M. Fabre and G. Giral in "Raman Spectroscopy" ; Lascombe ed., J. Wiley, New York, 447 (1982)
8. M. Krauzman, H. Poulet and R.M. Pick, Phys. Rev. B 33:99 (1986)
9. J. Breitenstein, Thesis, Université Paris 7, unpublished (1988) ; J. Breitenstein, M. Krauzman and R.M. Pick, in "Raman Spectroscopy", R.J.H. Clark and D.A. Long ed., J. Wiley, New York, 545 (1988)
10. C.S. Jacobsen, D.B. Tanner and K. Bechgaard, Phys. Rev B 28:7019 (1983)
11. M. Meneghetti, R. Bozio, I. Zanon, C. Pecile, C. Ricotta, and M. Zanetti, J. Chem. Phys. 80:6210 (1984)
12. M.J. Rice, V.M. Yartsev, and C.S. Jacobsen, Phys. Rev. B 21:3437 (1980)
13. R. Moret, J.P. Pouget, R. Comes, and K. Bechgaard, Phys. Rev. Lett. 49:1008 (1982)
14. C.C. Homes, and J.E. Eldrige, Synth. Metals 27:B49 (1988).
15. R. Bozio, M. Meneghetti, D. Pedron, and C. Pecile, Synth. Metals 27:B109 (1988).

A THEORY OF DIFFRACTION OF WAVEGUIDE POLARITONS

T. A. Leskova, I. Merkhasin, V. M. Agranovich

Institute of Spectroscopy, USSR Academy of Sciences
Troitsk, Moscow Region 142092, USSR

ABSTRACT

   An analytical theory of diffraction phenomena in a wave guide which
is partly coated with a thin dielectric film is developed. With the use
of the modified Wiener-Hopf method transformation coefficients for waveguide polaritons are calculated, in particular, in the vicinity of the
resonance of waveguide modes with vibrations in a thin film. Possible
nonlinear optical phenomena for an interferometer for waveguide modes are
discussed.

FAST PHOTOREFRACTIVE EFFECT IN HIGH-DOPED SEMICONDUCTORS:
APPLICATIONS TO OPTICAL BISTABLE INTERFEROMETRIC DEVICES

F.V. Karpushko, S.A. Bystrimovich, V.P. Morozov,
S.A. Porukevich, G.V. Sinitsyn, and I.A. Utkin

B. I. Stepanov Institute of Physics
BSSR Academy of Sciences
Leninskii prospekt 70, Minsk, 220602, USSR

INTRODUCTION

Resonant photoinduced electrooptical refractive index
changes of a semiconductor close to its band structure singular
points have been proposed for large optical nonlinearities and
to realize fast all-optical bistability in thin-film
interferometric devices [1,2,3]. In the case of a thin-film
interferometer strong electrooptical Franz-Keldysh effects can
arise from photoinduced electric field changes in the space
charge region (SCR) of heterojunctions, taking place at the
boundaries of the interferometer structure layers. In general,
this suggestion follows from the analysis of the semiconductor
photoreflectance spectra. The photoreflectance method to
investigate fine structure spectrum of the refractive index
uses a probe light beam to observe the reflection changes which
are caused by the absorption of another light beam within the
band-band transition spectrum (see, for instance [4,5]). In other
words, in the photoreflectance method "linear" (relative to
probe beam) spectrum of the refractive index is measured which
depends on the modulation, caused by absorption of a pump beam.
However, to observe a "nonlinear" (relative to measuring beam)
response it is important to understand the self influence, i.e.
both the changes of the refractive index as a result of the
photoinduced electric field and the measurement of these
effects involve the same light beam. Perhaps, in the context
of the photoreflectance method, such nonlinear problem can be
called as a nonlinear photoreflectance (photorefraction).

Just below the intrinsic absorption edge of a
semiconductor, i.e., in spectrum region where these
semiconductor layers are transparent enough to exhibit bistable
behavior in a thin-film interferometer, the singularities of
the semiconductor energy band structure are represented by: 1)
the transition between the top of the valence band and the
bottom of the conduction band, 2) the exciton transition, and
3) the transition between the top (bottom) of the valence
(conduction) band and the donor (acceptor) levels, for a doped
semiconductor. The intensity of the exciton transition
decreases as the temperature increases in contrast with the
band-impurity transition which is high temperature in nature.

Below we are reporting the experimental display of the doped semiconductor nonlinear photorefractive spectra in the impurity absorption spectrum area obtained, as far as we know, for the first time [6]. We discuss also in detail the fast and strong intrinsic optical nonlinearity and bistability of the same Franz-Keldysh effect nature which can be realized in a homogeneous doped semiconductor layer with a single type of conductivity under conditions of nonuniform sinusoidal lighting [7].

SEMICONDUCTOR OPTICAL CONSTANTS DEPENDENCE ON THE ELECTRIC FIELD

According to the Franz-Keldysh effect in the spectrum areas corresponding to the direct band-to-band and band-to-impurity transitions of a semiconductor the refractive index dependence on the electric field strength E can be given as[8,9]:

$$n(E) = n_0 + \frac{\lambda}{4\pi}*B_{g,i}*G(\eta_{g,i})*\left(h\Delta\omega_{g,i}\right)^{1/2}$$

(1)

Here n_0 is the linear refractive index, λ is the wavelength, $G(\eta_{g,i})$ is the second kind of electrooptical function, $\eta_{g,i} = (\mathcal{E}_{g,i} - h\omega)/h\Delta\omega_{g,i}$, $\mathcal{E}_g$ is the energy gap; $\mathcal{E}_i = \mathcal{E}_g - \Delta\mathcal{E}$, $\Delta\mathcal{E}$ is the impurity ionization energy, $h\Delta\omega_{g,i} = (e^2E^2h^2/2m_{g,i})^{1/3}$ is the Franz-Keldysh shift energy, e is the electron charge; h is the Plank's constant; $m_g = (m_n^{-1} + m_p^{-1})^{-1}$ is reduced effective carrier mass and m_n and m_p are the effective masses of the c-band electrons and v-band holes. The subscript g corresponds to the band-band, and i to the band-impurity transitions. The corresponding absorption coefficient dependence on E also can be written as

$$\alpha(E) = \alpha_0+B_g*\left(h\omega-\mathcal{E}_g\right)^{1/2}*\upsilon(h\omega-\mathcal{E}_g)+B_{g,i}*F(\eta_{g,i})*\left(h\Delta\omega_{g,i}\right)^{1/2}$$

(2)

where $F(\eta_{g,i})$ is the first kind of electrooptical function and $\upsilon(h\omega - \mathcal{E}_g)$ is the step function: $\upsilon = 1$ at $h\omega > \mathcal{E}_g$ and $\upsilon = 0$ at $h\omega < \mathcal{E}_g$. α_0 in Eq. 2 takes into account the background absorption caused by impurities and, also, Urbach's tail of the intrinsic absorption edge. Obviously, α_0 is the initial reason photocarriers are generated at $h\omega < \mathcal{E}_g$.

As it is seen from Eq. 2 the constant B_g characterizes the parabolic absorption edge of the undoped semiconductor and is determined by the following expression[8]

$$B_g = \frac{2e^2}{m^2cn_0\omega}*|eP_{cv}(0)|^2*\left(\frac{2m_g}{h^2}\right)^{3/2}$$

(3)

Here m is the free electron mass, c the velocity of light and $|eP_{cv}(0)|^2$ the square of the inter-band matrix element of the momentum operator. For the v-band- to - donor impurity transition, according to the near acting potential model of a impurity center the constant B_i is[8,10]:

$$B_i = \frac{16\pi N_i e^2 |eP_{cv}(0)|^2}{m^2 c n_0 \omega (\Delta \mathcal{E})^{3/2}} * \left(1 + \frac{m_p}{m_n} \frac{h\omega - \mathcal{E}_i}{\Delta \mathcal{E}}\right)^{-2} \qquad (4)$$

where N_i is the donor impurity density. It should be noted, also, that for the v-band - to - donor impurity transition Franz-Keldysh shift energy $h\Delta\omega_i$ is determined only by the hole effective mass m_p since for the donor level the electron is localized. As it is seen from Eqs. 1 and 2 the dependence of the optical constants of a semiconductor on the electric field are completely known through the special integral function $G(\eta_{g,i})$ and $F(\eta_{g,i})$. However, in this paper we are interested mainly in estimating the maximum values of the nonlinear refractive index changes Δn which occur at $h\omega = \mathcal{E}_{g,i}$. Just close to these frequencies, which represent two singularities of the doped semiconductor band structure, the broadened electrooptical function $G(\eta_{g,i})$ has an extreme, and the broadened electrooptical function $F(\eta_{g,i})$ crosses zero (see, for example Ref. 9). Thus, for the spectrum areas corresponding to the semiconductor energy band structure singular points discussed above we can assume $\alpha(E) = \alpha_0$ and

$$\Delta n(E) = n(E) - n_0 = \beta |E|^{1/3} \qquad (5)$$

where β is a constant, independent of E. This follows from Eq. 1 with $G(\eta_{g,i}) = G_0$. Below we will use Eq. 5 to calculate the maximum values of the nonlinear response of a semiconductor layer interacting with light.

THE MEASUREMENT METHOD AND EXPERIMENTAL DATA

To experimentally investigate the photoinduced electrooptical refractive index nonlinearity in n-GaAs samples we use the self diffraction of light beams by dynamic light induced gratings[11]. In our case such nonlinear diffraction gratings occurring only in the SCR near the surface of the semiconductor plate because there the photogenerated carriers were separated by the SCR electric field. Fig.1 shows schematically the nonlinear interaction geometry. The plane light beams of initial intensities I_1 and I_2 propagate inside the sample at an angle of $2\Theta_0$ and form an interference light field within the semiconductor sample:

$$I(x,y) = \frac{I_m(y)}{1 + A} \left(1 - A\cos\frac{2\pi x}{\Lambda}\right), \qquad (6)$$

where $I_m(y) = \mathbb{T}_F^0 * (\sqrt{I_1} + \sqrt{I_2})^2 * \exp(-\alpha y/\cos\Theta_0)$ is the intensity of the interference pattern maxima at a distance y from the front surface of the sample; $A = 2\sqrt{I_1 * I_2}/(I_1 + I_2)$; $\mathbb{T}_F^0$ is the transmission

(Fresnel) of the sample surface; $\Lambda = \lambda/2n_0\sin\Theta_0$ is the period of the interference pattern; and x is the coordinate across the

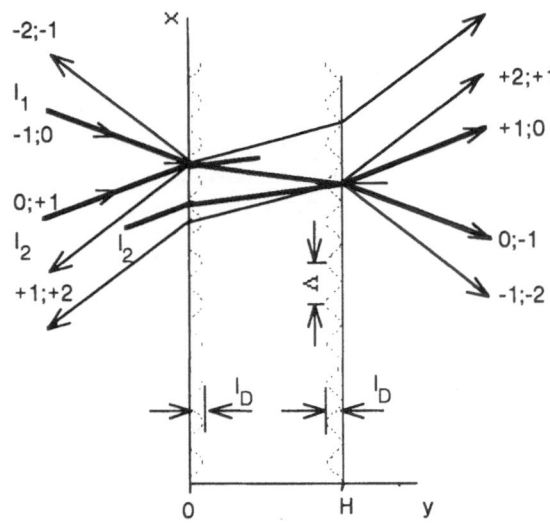

Fig. 1. The experimental method for measurement of the nonlinear refractive index changes in the semiconductor SCR. The figure schematically indicates the self diffraction orders.

sample. Note, in this part of the paper we are considering the case when the angle Θ_0 is small enough, so that the interference pattern period is greater than the SCR thickness (well known as the Debye length). This allows us to neglect the nonuniform distribution and consequently the division of the photocarriers along the x-direction and to concentrate solely on the photocarrier separation processes along y. According to such an interaction model, the nonlinear response is determined only by light absorbed within the SCR. Later we will also consider the case of small Λ when the nonlinear interaction spreads to whole volume of the semiconductor layer. When a potential barrier at a semiconductor surface is substantially more than k_0T, where k_0 is Boltzman's constant and T is the temperature, the electric field strength E within the SCR thickness is practically a linear function of the y-coordinate[12]. For our case one can write

$$E^{(0)}(x,y) = E_0 * \{ 1 - y/l_D(x,0) \}$$

and
$$E^{(H)}(x,y) = E_0 * \{ 1 - (H - y)/l_D(x,H) \} \tag{7}$$

where E_0 is the barrier electric field strength without lighting. At the boundary conditions when the semiconductor surface states are fully charged the Debye length is determined as $l_D = N_s/(N_i + \Delta N)$ where N_s is the surface state density and $\Delta N(x,y=0;H)$, in our case, is the nonequilibrium carrier density, under illumination, for the front $(y=0)$ and back $(y=H)$ SCRs of the semiconductor plate. For a highly doped semiconductor far from saturation conditions the nonequilibrium carrier density is much less than N_i and proportional to the light intensity. Thus, assuming $\alpha_0 l_D \ll 1$ in the order to neglect the intensity dependence on y within the SCRs, and using Eq. 5,6 and 7 we can write:

$$n^{(0)}(x,y) = n_0 + \beta |E_0|^{1/3} * \{1 - y/l_D(x,0)\}^{1/3}$$

and

$$n^{(H)}(x,y) = n_0 + \beta |E_0|^{1/3} * \{1 - (H-y)/l_D(x,H)\}^{1/3}, \qquad (8)$$

Eqs. 8 describes the periodical structures (gratings) in x direction with nonuniform optical thickness along y for any x cross section. At small values of $2\Theta_0$ the considered gratings are substantially thin: $n_0 l_D \ll \Lambda$. Then, the phase change of a light wave for one pass of the SCR along any x cross section is determined by the total optical SCR thickness which is

$$L^{(0,H)}(x) = \begin{cases} \displaystyle\int_0^{l_{D0}} n^{(0)}(x,y)dy \\[6pt] \displaystyle\int_{H-l_{D0}}^{H} n^{(H)}(x,y)dy \end{cases} = n_0 l_{D0} + l_{D0} \Delta n^{(0,H)}(x) \qquad (9)$$

where $\Delta n^{(0,H)}(x) = \dfrac{3}{4}\beta E_0|^{1/3} * \left\{1 - a\dfrac{I_m^{(0,H)}}{1 + A}\left[1 - A\cos\dfrac{2\pi x}{L}\right]\right\}^{1/3} \qquad (10)$

with a proportionality constant a and $l_{D0} = N_s/N_i$.

Taking in account the above assumptions, and also the modulation of the Fresnel reflection and transmission coefficients along x one can easy obtain the expressions for the self diffraction efficiency[6]. In the case of light polarization perpendicular to the plane of incidence for the first diffraction orders they are:

$$D_{1\tau}^{(0,H)} = \frac{n_0}{4} * \frac{|\delta n^{(0,H)}|^2}{(n_0 + 1)^4} * \left\{1 + \left[\frac{3\pi(n_0 + 1)\,l_D}{2\lambda}\right]^2\right\}, \qquad (11)$$

$$D_{1\rho}^{(0)} = \frac{|\delta n^{(0)}|^2}{4(n_0 + 1)^4}, \qquad (12)$$

$$D_{1\rho}^{(H)} = \frac{|\delta n^{(H)}|^2}{4(n_0 + 1)^4} * \left\{1 + \left[\frac{3\pi(n_0^2 - 1)\,l_D}{2\lambda}\right]^2\right\} \qquad (13)$$

Here $\quad |\delta n^{(0,H)}| = \dfrac{3}{2}\beta\,|E_0|^{1/3}\,\dfrac{a\,I_m^{(0,H)}\,A}{1 + A} \qquad (14)$

is the total magnitude of the first order refractive index gratings and subscripts τ and ρ indicate forward and backward diffraction efficiencies. Eqs. 12-14 determine the diffraction efficiencies of the gratings formed in the semiconductor SCRs. To analyze the experimental data it is necessary also to take into account the intensity losses of the incident and diffracted beams caused by both the sample absorption $(\exp[-\alpha y/\cos\Theta_0])$ and Fresnel losses at the surfaces. It should be noted when the sample is a plane parallel plate, spectra of the registered intensities of both forward and backward diffraction orders will be modulated as a result of interference between beams going out both front and back side gratings.

To realize the above measurement method we used "MALSAN-201", LiF color center laser[13], as a source of tunable radiation in the GaAs absorption edge area. The tunable laser line parameters were: a linewidth of 0.1 nm; pulse energy ~0.1-0.3 mJ; pulse duration of 4 - 10 ns; and repetition frequency of 10 Hz. The beam splitter we used made two coherent beams with intensity ratio 1:4 which were brought together at the angle of ~ 10^0 , to form the interference field. The samples were placed in the interference pattern area so that the light field intensity was modulated along the surfaces of the sample. The ratio of the sum of the intensities of the (+1) order beams diffracted from both the front and back SCR gratings to the incident beam intensity was measured. Measurements were made at room temperature for two n-GaAs:Te samples with different donor impurity concentrations. The sample thicknesses were 40 μm (#1) and 350 μm (#2).The #2 sample was a slightly wedge that is why we could observe the diffracted beams from both its front and back SCR gratings propagating separately in space. Fig. 2 demonstrates the obtained experimental spectra of self diffraction and its time dependence. The relatively narrow resonant curves of the diffraction efficiencies and the

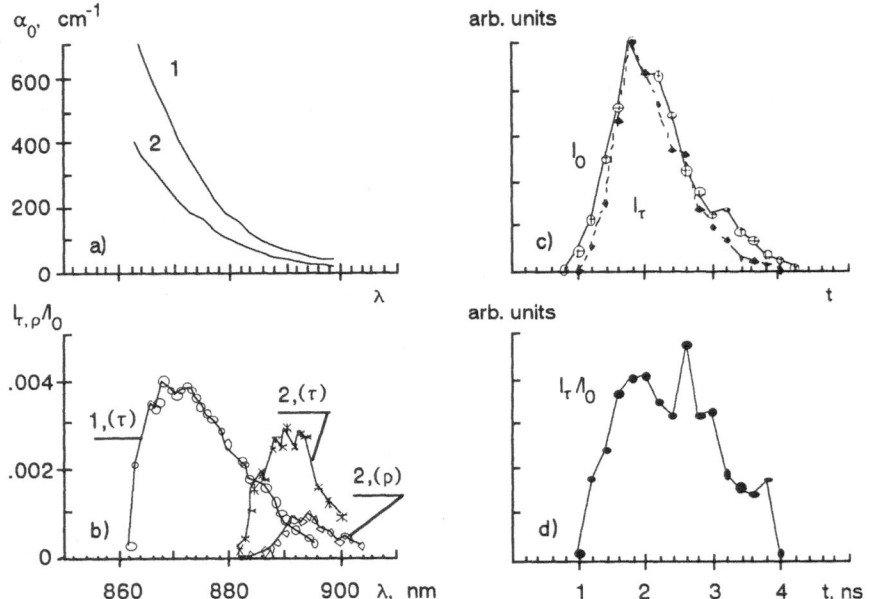

Fig.2. a) - the absorption spectra α_0, and b) - the self diffraction efficiency spectra $I_{\tau, \rho}/I_0$ for GaAs:Te samples with (1) donor density 2.10^{18} cm^{-3} and 40 μm of thickness and with (2) donor density 5.10^{18} cm^{-3} and 350 μm of thickness.c) - the time dependence of both the incident I_0 and (+1) order diffracted I_τ intensities, and d) - the time dependence of the self diffraction efficiency I_τ/I_0 for the sample #2 at λ = 892 nm. τ and ρ indicate forward and backward diffraction, respectively.

spectrum positions indicate the impurity nature of the observed nonlinear response. The previous analysis of these experimental data based on the above formulas (with $N_s = 10^{14}$ cm^{-2}) gives the photoinduced refractive index changes $1.1*10^{-2}$ and $1.6*10^{-2}$, respectively, for the samples #1 and #2. Obviously, to calculate more exactly the values of $|\delta n|$ it is necessary to take into account the possible energy transfer between interacting beams because of their intensities were not equal in our first experiment. It also should be noted the lifetime of the observed refractive index nonlinearity seems to be smaller or order of 10^{-9} s. This follows from Fig. 2c,d which were taken with 10^{-10} s time resolution by using the boxcar integration technique.

THE PHOTOVOLTAGE AND ELECTROOPTICAL EFFECTS IN A HOMOGENEOUS SEMICONDUCTOR LAYER IN A STANDING WAVE FIELD

If a semiconductor is placed in a light field with a sinusoidal (nonuniform) intensity distribution one of the mechanisms of arising built-in electric field required for Franz-Keldysh effect can be the Dember's mechanism which takes in account the spatial separation of the nonuniformly generated photoelectrons and photoholes as a result of their different mobilities in a homogeneous semiconductor. For the doped homogeneous semiconductor layer refractive index nonlinearity caused by nonuniform standing wave field absorption, the general nonlinear problem wording was described earlier[7]. Below we will analyze in detail the important case of highly doped n-GaAs. Really, for bulk GaAs the mobility and diffusion coefficients of the p-subsystem can be neglected as compared to the similar parameters of the n-subsystem. Then, from the well known diffusion, continuity and Poisson's equations (see, for instance, Ref. 14) using the quasi-state approach it is easy to obtain the following Eqs. for the electron subsystem:

$$E(x) = -\frac{k_0 T}{e} * \frac{1}{N_n} * \frac{dN_n}{dx} \, , \qquad (15)$$

$$\frac{d^2(\ln N_n)}{dx^2} - \frac{\varepsilon^2}{\varepsilon_0 \varepsilon k_0 T} \left\{ N_n - N_i - \frac{\alpha_0 I_s(x)}{h\omega\gamma N_n} \right\} = 0 \qquad (16)$$

Here N_n is the basic carrier (electron) density; ε_0 is the free space permittivity; ε is the dielectric constant; γ is the recombination constant; $I_s(x)$ is the light intensity distribution inside Fabry-Perot interferometer; the others are the same which were used before. Note, also, that the abscissa x of the nonuniform lighting direction is now directed across the internal semiconductor interferometer layer of the thickness H. For the values $n_0 H$ close to $r\lambda/2$, where r is the interference order, the shape of the standing wave field intensity distribution $I_s(x)$ inside the interferometer depends insignificantly on the nonlinear part of $n(E)$ in Eq. 1, because the last is small as compared to the linear refractive index n_0. But, on the other hand, the magnitude of this distribution can change very much according to the strong interferometer transmission dependence on $\Delta n = n(E) - n_0$ in the area of the interferometer transmission peak. Below we are interested to estimate the maximum values for Δn, based on Eq. 5. According to the given intensity distribution shape approach[7], the inside interferometer intensity can be written by the above Eq. 6 with

$$I_m = \mathbb{T} \, I_0 \frac{1 + \sqrt{R}}{1 - \sqrt{R}} \; ; \quad \Lambda = \frac{\lambda}{2 n_0 \cos\Theta} \; ; \quad A = \frac{2\sqrt{R}}{1 + R} \; ; \quad (17)$$

where I_0 is the input intensity, Θ is the propagating angle inside interferometer, R is the interferometer mirrors reflection coefficient, and

$$\mathbb{T} = \frac{(1 - R)^2 \, \exp(-\frac{\alpha_0 H}{\cos\Theta})}{[1 - R\exp(-\frac{\alpha_0 H}{\cos\Theta})]^2 + 4R\exp(-\frac{\alpha_0 H}{\cos\Theta}) \, \sin^2[\pi\frac{H}{\Lambda}(1 + \frac{\Delta n}{n_0})]} \quad (18)$$

is the transmission function of the nonlinear Fabry-Perot interferometer which depends on $\Delta n = \frac{1}{H} \int_0^H n[E(x)]dx - n_0$. It should be noted, there is one more component in the expression for $I_s(x)$, determined by the traveling component of light field inside the interferometer. However for the interferometer with high finesse its magnitude is small: $\frac{\alpha_0 H}{\cos\Theta} \frac{1 - R}{1 + R} \ll 1$ (see Ref. 7). In the highly doped semiconductor case when the nonequilibrium carrier concentration is much less than the equilibrium one ($N_n - N_i \ll N_i$), Eq.6 for $I_s(x)$ with Eqs.17,18 allows us to integrate Eq.16, keeping the parametric dependence on the nonlinear refractive index in the expression for $N_n(x)$. By using $Y = \frac{N_n - N_i}{N_i} \ll 1$, Eq.16 can be easily linearized and then we obtain the following solution

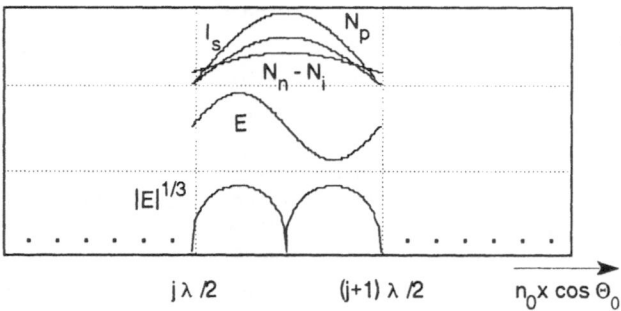

Fig. 3. The schematic steady-state distributions of the light intensity I_s, the nonequilibrium electron ($N_n - N_i$) and hole N_p concentrations, the photoinduced electric field E, and the cube root of its magnitude $|E|^{1/3}$ along j-th period of the standing light wave.

$$Y = \frac{\alpha_0 I_m}{1 + A} * \frac{1}{h\omega N_i} * (\gamma N_i)^{-1} - \frac{A}{1 + A} * \frac{\alpha_0 I_m}{h\omega N_i} * (\gamma N_i)^{-1} * \frac{\cos 2\pi \frac{x}{\Lambda}}{1 + (2\pi \frac{L_d}{\Lambda})^2} \quad (19)$$

where $L_d = \{\varepsilon_0 \varepsilon k_0 T / e^2 N_i\}^{1/2}$ is the free path of the conductive band electrons. Consequently for $E(x)$ from Eqs. 15 and 19 it follows:

$$E(x) = - E_D \sin 2\pi \frac{x}{\Lambda} \,, \quad (20)$$

with

$$E_D = \frac{k_0 T}{e} * \frac{2\pi}{\Lambda} * \frac{A}{1 + A} * \frac{\alpha_0 I_m}{h\omega N_i} * \frac{(\gamma N_i)^{-1}}{1 + (2\pi \frac{L_d}{\Lambda})^2} \quad (21)$$

To illustrate the Dember's mechanism of the electrooptical processes arising in a homogeneous semiconductor layer in the case of a sinusoidal intensity distribution lighting, Fig. 3 schematically represents some parameters of a semiconductor in such a case.

OPTICAL BISTABILITY

To consider optical bistability in a Fabry-Perot interferometer with the above properties of the internal layer, we should use now the electric field dependence of the refractive index $\Delta n(E)$, which provides the feedback to the interferometer transmission through the internal layer nonlinearity. Using Eqs. 5, 20, and 21 and averaging through the standing wave period we obtain

$$\Delta n(E) = \frac{1}{\Lambda} \int_{x}^{x+\Lambda} n[E(x)]dx - n_0 = \beta_I I_T^{1/3} \quad (22)$$

with

$$\beta_I^{(g,i)} = \lambda G_0 B_{g,i} \Gamma^{-3}(1/3) \left\{ k_0 T \frac{4\pi h}{\Lambda \sqrt{2m_{gi}}} \frac{1}{\Delta\phi_0 \sqrt{\mathbb{T}_0}} \frac{\alpha_0}{h\omega N_i} \frac{(\gamma N_i)^{-1}}{1 + (2\pi \frac{L_d}{\Lambda})^2} \right\}^{1/3} \quad (23)$$

Here $I_T = \mathbb{T} I_0$ is the transmitted intensity;

$$\Delta\phi_0 = \frac{1 - R\exp(-\frac{\alpha_0 H}{\cos\Theta})}{2\sqrt{R}\exp(-\frac{\alpha_0 H}{2\cos\Theta})} \quad \text{and} \quad \mathbb{T}_0 = \frac{(1 - R)^2 \exp(-\frac{\alpha_0 H}{\cos\Theta})}{\{1 - R\exp(-\frac{\alpha_0 H}{\cos\Theta})\}^2}$$

are the phase halfwidth of the transmission peak and its maximum value. Also it is the interesting thing to note that the term $4\pi h/\Lambda\sqrt{2m_{gi}}$ in brackets in Eq. 23 is, in essence, the square root of the addition carrier energy represented by means of its momentum within the potential wells, which are the induced by the standing wave light field superlattice with the wave vector of $4\pi/\Lambda$.

Let $\pi H/\Lambda = r\pi + \Delta\phi$ were $\Delta\phi$ is the initial (at $I_0 = 0$) phase detuning of the interferometer and $|\Delta\phi| \ll \pi$. Then, replacing the sine in Eq. 18 by its small argument and taking into account

Eq. 22 one can obtain the algebraic equation connecting the transmitted intensity I_T with the incident one I_0:

$$I_0 = \frac{I_T}{\mathbb{T}_0} \left\{ 1 + (\Delta\phi_0)^{-2} \left[\Delta\phi + \frac{r\pi + \Delta\phi}{n_0} \beta_1 I_T^{1/3} \right]^2 \right\} . \qquad (24)$$

According to Eq. 24 the dependence I_T on I_0 is a typical s-like curve which characterizes any bistable system. It is easy to

Table 1. The calculated parameters of a bistable thin-film n-GaAs interferometer with the electrooptical Franz-Keldysh nonlinearity induced by nonuniform absorption in a standing wave field

α_0, cm^{-1}	$|\Delta n_{thr}|$	$h\omega = \mathcal{E}_g$			$h\omega = \mathcal{E}_g - \Delta\mathcal{E}$		
		I_0^{thr}, kW/cm^2	W_{thr}, fJ/μm^2	E_0, kV/cm	I_0^{thr}, kW/cm^2	W_{thr}, fJ/μm^2	E_0, kV/cm
500	0.0205	9,130	9,130	1,410	24,000	24,000	3,750
100	0.0041	73	73	11.3	195	195	30
50	0.0021	9.13	9.13	1.41	24	24	3,75
20	0.0008	0.58	0.58	0.09	1.56	1.56	0,24

show that the initial phase detuning of the considered interferometer corresponding to the optical bistability threshold is

$$|\Delta\phi_{thr}| = \sqrt{15} \, |\Delta\phi_0| , \qquad (25)$$

and, consequently, the threshold input intensity is

$$I_0^{thr} = 1.6 \, \mathbb{T}_0^{-1} \left(\frac{4}{5} \frac{n_0 |\Delta\phi_{thr}|}{|\beta_1| r\pi} \right)^3 . \qquad (26)$$

From an analysis of Eq. 26 with $R \sim \exp(-\frac{3}{4} \frac{\alpha_0 H}{\cos\Theta})$ and with the corresponding interferometer transmission value $\mathbb{T}_0(R) \sim 0.18$ the threshold value of the nonlinear refractive index change is

$$|\Delta n_{thr}| \sim 5.5 \frac{\alpha_0 \lambda}{4\pi} . \qquad (27)$$

As an example, in Table 1 there are given some estimates of $|\Delta n_{thr}|, I_0^{thr}, E_0$, and also the bistability threshold switching

energies $W_{thr} = I_0^{thr} * (\gamma N_i)^{-1}$ at the different background

absorption coefficients α_0 for both g and i singularities, calculated from the above formulas at $H = \lambda/2n_0$ and $G_0 = -0.4$ with the parameters: $N_i = 10^{18}$ cm^{-3}; $\lambda = 0.82$ μm ($h\omega = 2.4 * 10^{-19}$ J); $n_0 = 3.6$; $\Theta = 0$; $\gamma = 8 * 10^{-9}$ cm^{3}s^{-1} [14]; $T = 300$ K; $\varepsilon = 10$; $m_g = 0.07$, and $m_i = m_p = 0.5$ of free electron mass which are close to those of the n-GaAs samples. For $B_{g,i}$ we use the value of $B_g = 5*10^{13}$ J$^{-1/2}$cm^{-1} from Ref. 14. It has been shown experimentally[5] that these constants are approximately equal for both v-band - to - c-band and v-band - to - donor impurity transitions.

CONCLUSION

In this paper the strong optical nonlinearities of the Franz-Keldysh effect nature in highly doped semiconductor layers has been described. With sinusoidal intensity distribution of lighting in an impurity absorption spectral region the strong Franz-Keldysh effect can be caused by the photoinduced electric field changes due to photocarrier separation either by a barrier potential in the space charge region or by diffusion processes with different carrier mobilities with spatially nonuniform photocarrier generation. The experimental method and data have been given for n-GaAs samples. Refractive index changes have been estimated to be about 10^{-2} in magnitude.

The calculated data show the possibility of the all-optical bistable semiconductor interferometer devices with the small switching energies near the quantum statistical limit (in the femtojoule range) and with high operating speed (in the picosecond range), at room temperature. However, it should be noted, that the schemes of the photovoltage and electrooptical processes in a homogeneous semiconductor layer considered above (by using band - to - band and band - to - impurity transitions) do not exhaust yet all opportunities to reach the limiting switching speeds of such a system. The next step to increase the switching on/off speeds is to use the photoinduced refractive index changes within the spectral areas corresponding to the transitions between the band structure singular points within the same band. As an example of such transitions there are the transitions between the light and heavy hole (electron) subbands and the spin-orbit-split off subband within the valence (conduction) band of a semiconductor. In such a case the dynamics of the nonlinear refractive index changes will be determined by hot carrier thermalization with the corresponding time constants in the subpicosecond range.

ACKNOWLEDGEMENT

The authors would like to acknowledge A. Kost for his help in editing the English version of this paper.

REFERENCES

1. F. V. Karpushko, Nonlinearity of thin-film semiconductor interferometers due to interlayer boundary photoEMF and electrooptic processes, Journal de Physique, Colloque C2, Suppl. au #6, 49:C2-87 (1988).
2. F. V. Karpushko, Fast and strong nonlinearity of thin-film semiconductor interferometers due to intrinsic photo- and electrooptical effects of the semiconductor interlayer, Phys.Stat. Sol.(b) 150:791 (1988).

3. F. V. Karpushko, Photoinduced electrooptical mechanism of nonlinearity and bistabilitv in thin-film semiconductor interferometers, Vesti AN BSSR, seria phys.-math. nauk, #1:13(1989), in Russian.

4. M. Cardona, K. L. Shaklee, and F. H. Pollak, Electroreflectance at a semiconductor-electrolyte interface, Phys. Rev. 154:696(1967).

5. Yu. E. Maronchuk, A. P. Shestryakov, and A. S. Tokarev, Photoreflectance of Gallium Arsenide, Phys. Tekh. Poluprov.7:552 (1973), in Russian, - Sov. Phys. Semicond. 7:386 (1973).

6. S. A. Bystrimovich, F. V. Karpushko, V. P. Morozov, S. A. Porukevich, G. V. Sinitsyn, and I. A. Utkin, "Experimental measurement of the photoinduced refractive index nonlinearity corresponded to the impurity absorption of a doped semiconductor", Preprint # 565, B. I. Stepanov Institute of Physics, Minsk (1989), in Russian.

7. F. V. Karpushko, "Semiconductor interferometer optical nonlinearity induced by photovoltage effects in a standing wave light field," Preprint # 562, B. I. Stepanov Institute of Physics, Minsk (1989), in Russian.

8. V. A. Tyagai, O. V. Snitko, "Light electroreflection in semiconductors," Naukova Dumka, Kiev (1980), in Russian.

9. M. Cardona, "Modulation spectroscopy," Academic Press, N.Y. (1969).

10. V. S. Vinogradov, "Effect of electric fields on the absorption of light by deep impurity centers, Phys. Tverdova Tela 15:285(1973), in Russian, Sov. Phys. Solid State 15:395 (1973).

11. A. S. Rubanov, Dynamic holography, in: "Problems of novel optics and spectroscopy," B. I. Stepanov, ed. , Nauka i tekhnika, Minsk (1980), in Russian.

12. M. E. Levinstein, and G. S. Simin, "The barriers," Nauka, Moscow (1987), in Russian.

13. T. T. Basiev, F. V. Karpushko, S. M. Kulashchik at al. Zh. Pricl.Spektroskopii 47:682 (1987), in Russian.

14. Ja. I. Pankove, "Optical processes in semiconductors," Dover Publications, N.Y. (1971).

BALLISTIC SIGNALS FROM A HOT PHONON SPOT

Y.B. Levinson

Department of Physics
Institute of Microelectronics Technology
and High Purity Materials
USSR Academy of Sciences
142432 Chernogolovka, Moscow, USSR

The aim of the paper is to discuss the experiment made by the group of J. P. Wolfe[1] and to show that this experiment demonstrates phonon hot spot explosion.

1. Primary Phonons and Ballistic Phonons

The paper deals with ballistic phonon pulses excited in a semiconductor by laser pulses absorbed in band-to-band transitions. Let us recall what processes occur in this situation.

A laser pulse (typical pulse duration 10ps-10ns) is absorbed near the front surface of the crystal. When a phonon is absorbed, an electron-hole pair is created, the electron and the hole being hot. The carriers diffuse into the bulk of the crystal and relax via emission of optical and intervalley high-energy phonons. The energy relaxation time is about 10ps, of the order of or shorter than the laser pulse duration. As a result, during the laser pulse about one-half of the pulse energy is imparted to the phonon system. (The rest of the laser pulse energy is transferred to the phonons in the course of electron-hole recombination). The high-energy relaxation phonons are created in a near-surface layer the thickness of which is of the order of a few microns. A bolometer of a tunnel-junction on the rear surface of the crystal is used to detect the nonequilibrium phonons. The distance L between the detector and the focal spot of the laser is about 1 cm.

Let us emphasize that the frequency of the primary phonons created during the hot carrier relaxation is of the order of the Debye frequency ω_D irrespective of the excitation intensity. Therefore, the primary phonons are strongly scattered by isotopic impurities (elastic scattering) and due to the phonon anharmonicity (inelastic phonon-phonon scattering). In all crystals the elastic scattering time $\tau'(\omega)$ is shorter than the inelastic scattering time $\tau(\omega)$. As a result, during the time interval between two successive inelastic scattering events the motion of the phonons is diffusive. Due to the strong phonon scattering the crystal domain occupied by the nonequilibrium phonons spreads slowly.

Nevertheless in some experiments with Ge[2], Si[1] and GaAs[3] sharp ballistic signals are detected (Fig. 1).

These signals are due to phonons arriving onto the detector without scattering. The absence of scattering is confirmed by three facts: 1) the arrival time of the signal is exactly L/s, where s is the sound velocity in the given direction; 2) the signals are short; (3) the signals are very sensitive to phonon focusing. The ballistic signals are detected because a small number of low frequency phonons with mean free path l > L are created in the excited crystal domain as a result of the decay of the high-frequency phonons. In other words, the nonequilibrium domain of the crystal is a source of ballistic phonons. The ballistic signal from this source contains information about the time evolution of the excited domain. It is this point that is discussed in what follows.

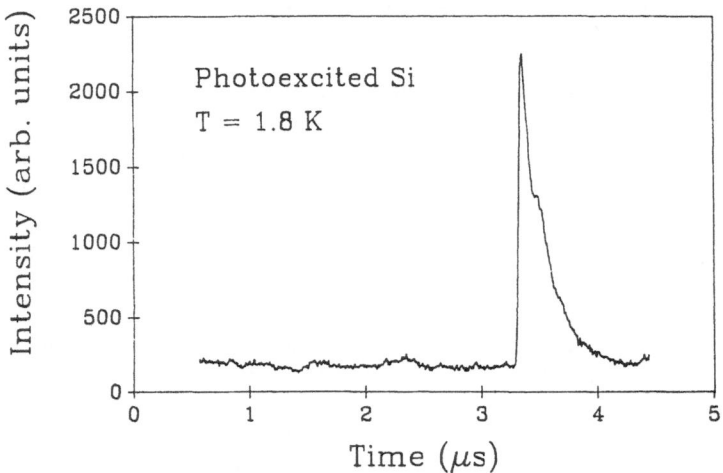

Fig. 1. Ballistic phonon signal from photoexcited silicon. The propagation is near [001]. The onset at t = 3.2 μs corresponds to the slow-transverse sound velocity (from [1]).

2. Hot Spot and Decay Spot

Two models are known to describe the time evolution of the excited crystal domain occupied by high-frequency phonons, namely the hot spot model (Hensel and Dynes[4]) and the phonon generation model (Kazakovtsev and Levinson[5]). These two models can be discriminated from the properties of the ballistic signal.

For simplicity we assume the primary phonons to be generated in a hemisphere with radius R_o and the laser pulse to be very short.

The phonon generation model is valid for low excitation levels. In this case the phonon occupation numbers are small, and spontaneous phonon decay is the dominant phonon-phonon scattering process. Since in spontaneous decay the phonon energy is divided approximately in half, the primary phonons with frequency $w_o \approx w_D$ give rise to a chain of phonon generation with frequencies w_o, $w_1 = w_o/2$, $w_2 = w_1/2 = w_o/2^2$, ... The life-time of the generation with frequency w_k is $\tau(w_k)$, where $\tau(w)$ is the phonon life-time against spontaneous decay. Each generation spreads by diffusion and the spreading length of the generation is of the order of the diffusion length $l(w_k)$ where

$$l(w_k) = [D(w)\tau(w)]^{1/2}, \quad D(w) = 1/3 \, s^2 \tau'(w). \tag{1}$$

For order-of-magnitude estimates one can use the long-wave approximation

$$\tau(w) = \tau x^{-5}, \quad \tau'_1(w) = \tau' x^{-4}, \quad l(w) = 1 x^{-9/2}, \quad x = w/w_D. \tag{2}$$

In Si the branch-averaged nominal values are: $\tau = 320ps$, $\tau' = 13ps$, $1 = 0.22 \, \mu m$ and $w_D/2\pi = 13.4$ THz[6]. Let $R_o = 5 \, \mu m$ and assume the frequency of the primary acoustic phonons to be $w_o = w_D/2$. From (2) one calculates $l(w_o) = 5 \, \mu m$ and $l(w_1) = 110 \, \mu m$. This means that the second generation w_1 escapes from the hemisphere R_o. The radius of the hemisphere, occupied by the nonequilibrium phonons grows with time. Let us find how this radius $R(t)$ depends on time.

The characteristic phonon frequency w_t at the time t can be found from the relation $\tau(w_t) = t$. If $l(w_t) < R_o$, then at the time t the phonons do not escape from the hemisphere R_o, where the primary phonons were excited, and hence $R(t) = R_o$. If $l(w_t) > R_o$ then the radius of the excited hemisphere is of the order of the diffusion length of the last generation, i.e. $R(t) = l(w_t)$. As a result, we obtain

$$R(t) = R_o, \qquad\qquad t < t_o$$
$$= R_o(t/t_o)^{9/10}, \qquad t > t_o, \qquad t_o = \tau(R_o/1)^{10/9}, \tag{3}$$

where t_o is the time of escape from the initial hemisphere R_o. In the above metioned numerical example $t_o = 10ns$. Two important conclusions follow from the above consideration of the phonon generation model. 1) The size of the excited domain grows smoothly and unlimitedly. 2) This size and its growth rate do not depend on the laser power (Fig. 2).

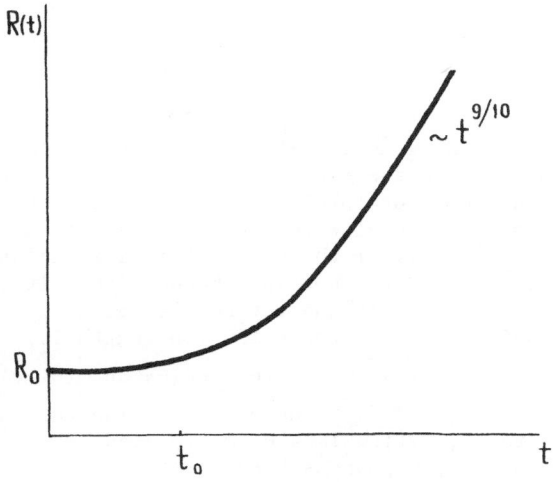

Fig. 2. Spreading of the decay spot.

At higher excitation levels not only spontaneous decay but also phonon coalescence and induced phonon decay are important. When the number of phonons is so high that the coalescence rate and the decay rate are of the same order of magnitude, a Planckian phonon distribution is established due to phonon-phonon scattering. In this situation the excited domain of the crystal can be specified by a nonequilibrium temperature T which is higher than the bath temperture T_B. This is the phonon hot spot model.

The initial nonequilibrium phonon temperature T_0 can be calculated from energy balance

$$E/V_{R_0} = \epsilon_{T_0} , \qquad (4)$$

where E is the energy absorbed from the laser pulse, V_{R_0} is volume of the hemisphere R_0, and ϵ_T is the phonon energy per 1cm^3 at temperature T. The Planckian with temperature T_0 is established, if $R_0 > l_{T_0}$, where l_{T_0} is the diffusion length $l(\omega)$ of thermal phonons with $\omega = 2.82T$. From the condition $R_0 = l_T$, combined with (4) one can find the cross-over energy E^*. If $E > E^*$, the hot spot model is appropriate. If $E < E^*$, the decay model is appropriate. Let us adopt the long-wave approximation

$$\epsilon_T = \epsilon(T/T_D)^4, \qquad T_D = h\omega_D \qquad (5)$$

where $\epsilon = 2.5 \times 10^4$ J/cm^9 for Si. Then for a focal spot with $R_0 = 5$ μm the cross-over energy $E^* = 10$ nJ.

Now let's discuss the time evolution of the hot spot size and temperature. This evolution is governed by the equations

$$\epsilon_T V_R = E, \qquad dR^2/dt = D_T , \qquad (6)$$

the first of which is the energy conservation law and the second is the diffusion law. V_R is the volume of a hemisphere with radius R and D_T is $D(\omega)$ for thermal phonons with $\omega = 2.82T$. It follows from Eqs. (6) in the case of the long-wave approximation, that

$$R(t) = R_0[1-(E^*/E)(t/t_0)]^{-1},$$

$$T(t) = T_0[R(t)/R_0]^{-3/4}, \qquad (7)$$

Where t_0 has the same meaning as in (3). One can see from (7), that the hot spot dynamics depends on the imparted energy E. The higher this energy, the slower the hot spot spreading. This is in contrast to the decay spot, the spreading rate of which does not depend on the imparted energy E. Another peculiarity of the hot spot is its explosion. In a finite time interval $(E/E^*)t_0$ the hot spot becomes infinitely large, and its temperature becomes zero. To understand what actually happens, compare the hot spot radius R(t) and the diffusion length $l_{T(t)}$ at various times. As the temperature T(t) falls, the radius and the diffusion length grow: $R \approx T^{-4/3}$, $l_T \approx T^{-9/2}$, the diffusion length growing faster. Therefore, the condition $l_T < R$ is violated at some moment t^+ (Fig. 3). At this moment the phonon-phonon scattering becomes too weak to support the Planckian distribution and the hot spot is destroyed.

364

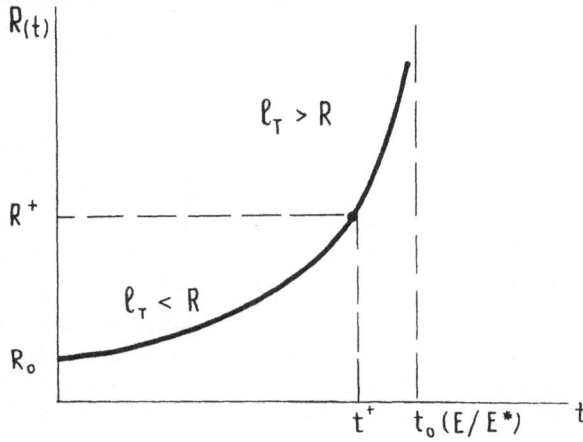

Fig. 3. Spreading and explosion of the hot spot.

In other words, Eqs. (6) and (7) are valid only for $t < t^+$.

The radius and temperature of the hot spot at the moment of its destruction R^+ and T^+ can be obtained from the condition $R^+ = l_{T^+}$ giving

$$T^+ = T_o(l_{T_o}/R_o)^{6/19} \qquad R^+ = R_o(l_{T_o}/R_o)^{-8/19}. \qquad (8)$$

Combining Eqs. (8) with Eqs. (7), we find the destruction moment

$$t^+ = t_o(E/E^*)[1-(l_{T_o}/R_o)^{8/19}] \approx t_o(E/E^*). \qquad (9)$$

At the moment t^+ the hot spot is transfored to a decay spot and accordingly the slow spreading is transformed into fast spreading.

In the case of the above considered example in Si for $E = 100$ nJ one obtains for the initial temperature of the hot spot $T_o = 230K$. The thermalization length $l_T = 0.2$ μm is short compared to $R_o = 5$ μm. The hot spot is destroyed at $t^+ = 67$ ns, having at the moment of destruction the temperature $T^+ = 83K$ and the radius $R^+ = 19$ μm.

3. Ballistic Signals from a Hot Spot and a Decay Spot

Now we compare the theoretical predictions about the ballistic signal emitted from a hot spot and from a decay spot. The shape of a ballistic signal emitted from a decay spot does not depend on the laser pulse energy, the intensity of the signal being proportional to this energy. The decay spot can be considered as a localized source of ballistic phonons until the nonequilibrium high-frequency phonons are in the excitation domain determined by the laser focal spot. When these phonons escape from the excitation domain the size of the source grows rapidly. As a result the decay spot is to be considered as a source of ballistic phonons with size R_o and lifetime t_o. When ballistic phonons are emitted from a phonon hot spot not only the intensity of the ballistic signal depends on the laser pulse energy, but also the shape of the signal. The hot spot spreads slowly to destruction. Hence, the hot spot is to be considered as a source of ballistic phonons with size R^+ and life-time t^+. Both these source properties depend on the laser energy: $R^+ \approx E^{9/19}$, $t^+ \approx E$, in contrast to size R and lifetime T_o of the ballistic phonon source in the case of a decay spot.

Now we are going to compare theory and experiment. First we consider the experiments in Si performed with high resolution[1]. The width of the caustic Δx was measured in a phonon focusing experiment. This width is of the order of the size of the ballistic phonon source. The dependence of Δx on E is shown in Fig. 4. This dependence is strongly pronounced in the region E > 10 nj, where $\Delta x \approx E^{0.36}$. This high-energy region corresponds to phonon emission from a hot spot. The theory predicts the size of the phonon source to be $R^+ \approx E^{0.47}$, which is in fairly good agreement with the experiment. In the region E < 10 nJ the dependence of Δx on E is weak. This region corresponds to phonon emission from a decay spot. The experimental cross-over energy E = 10 nJ agrees with the calculated one.

It is also possible to discuss the source size. For E = 100 nJ the theory predicts the size $2R^+$ = 40 μm. The measured value Δx = 100 μm. For E = 1 nJ the theory predicts the size to be the focal spot diameter $2R_o$ = 10 μm. The measured value Δx = 25 μm. Bearing in mind that the long-wave approximation of the scattering rate (2) and of the phonon energy density (5) is crude for high frequencies and temperatures, we find that the theory agrees with the experiment quite satisfactorily.

Experiments with Ge[2] measure the life-time of the ballistic phonon source. This life-time depends on pulse energy $\approx E^{0.7}$. The theory predicts the life-time $t^+ \approx E$ for ballistic emission from a hot spot. This looks like a fairly good agreement, but it should be remembered that much higher energy excitation levels are used in the experiments with Ge, and the Debye temperature for Ge crystal is lower, so the long-wave approximations for Ge are much less appropriate than for Si.

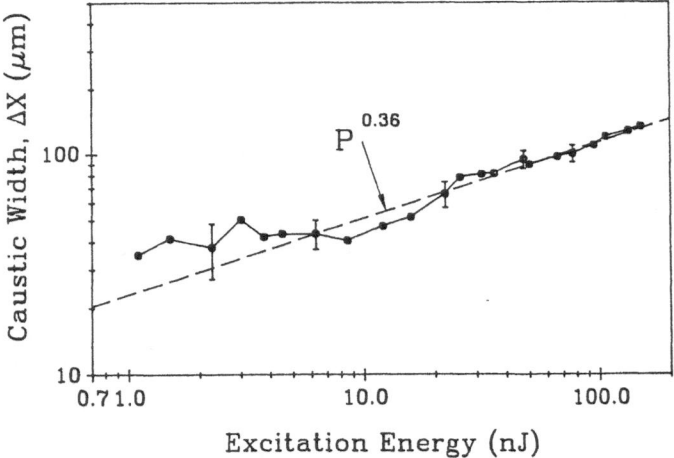

Fig. 4. Power dependence of the caustic width in silicon (STA mode [001]) (from [1]).

REFERENCES

1. J. A. Shields and J. P. Wolfe, Preprint (1988).
2. M. Greenstein, M. A. Tamor, and J. P. Wolfe, Phys. Rev. B26, 5604 (1982).
3. B. A. Danilchenko, V. N. Poroshin, M. I. Slutskii, and M. Asche, phys. stat. sol. (b) 136, 63 (1986); B. A. Danilchenko, D. V. Kazakovtsev, and M. I. Slutskii, Solid State Comm. Phys. Lett. A 138, 77 (1988).
4. J. C. Hensel and R. C. Dynes, Phys. Rev. Lett. 39, 969 (1977).
5. D. V. Kazakovtsev and Y. B. Levinson, Sov. Phys. JETP 61, 1318 (1985).
6. S. Tamura, Phys. Rev. B31, 2574 (1985).

OPTICAL STARK EFFECT OF EXCITONS IN SEMICONDUCTORS

Dietmar Fröhlich

Institut für Physik
Universität Dortmund
4600 Dortmund 50
Federal Republic of Germany

INTRODUCTION

The first observation of the Stark effect in "rapidly varying fields" was reported by Autler and Townes.[1] They applied a typical pump and probe technique in the microwave and radiofrequency region. They observed a splitting in the microwave spectrum (probe) of OSC molecules (carbonyl sulfide) at about 24 GHz if the molecules were simultaneously pumped close to a radio-frequency resonance either in the groundstate (about 13 MHz) or in the excited state (about 38 MHz). A field strength of several $V \cdot cm^{-1}$ was sufficient to clearly demonstrate this splitting which is now refered to as the "Autler-Townes effect". For a review on the first observation of shifts and splitting of atomic levels in the radiation field of optical frequency (ruby laser) we refer to Bonch-Bruevich and Khodovoi. High resolution experiments on sodium were reported by Liao and Bjorkholm[3] and Gray and Stroud.[4] Laser intensities of less than $1 \, W \cdot cm^{-2}$ were sufficient to clearly resolve the Autler-Townes splitting. The observation of such an effect in solids, however, seems to be very difficult, because the linewidths of electronic transitions in solids are in general much larger than in atoms and molecules. The first observation of the optical Stark effect in a semiconductor was reported by Fröhlich, Nöthe and Reimann[5] in the classical exciton spectrum of Cu_2O. The authors observed the dynamical coupling of the 1S and 2P exciton by an intense CO_2 laser. The possibility of the observation of a "stimulated Stark effect" in Cu_2O was already mentioned in a review article by Agekyan.[6]

The interest in the field grew considerably after the first observation of the optical Stark effect in multiple-quantum-wells (MQW) by Mysyrowicz et al.[7] and von Lehmen et al.[8] In these experiments the optical Stark effect results from a coupling of the groundstate and e. g. the first exciton by an intense pump beam. The transient spectrum of the same first exciton, which exhibits a characteristic blue-shift, is then measured by a weak probe beam. As shown in the schematic diagram (Fig. 1) for light-induced effects we call this a "two-level Stark effect" as compared to a coupling of two excited states which is refered to as a "three-level Stark effect". Further experiments concerning the two-level Stark effect in bulk semiconductors were reported recently by Fluegel et al.[9] and Peyghambarian et al.[10] In these experiments coherent transients are resolved with the use of femtosecond spectroscopy. For the theory and further literature we refer to publications by Schmitt-Rink, Chemla and Haug[11], Lindberg and Koch[12], Balslev, Zimmermann and Stahl[13], and Chemla et al.[14]

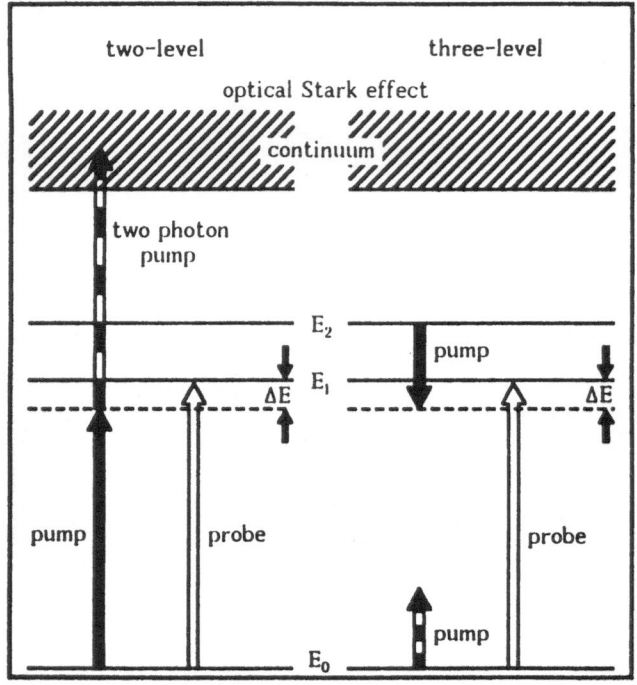

Fig. 1. Schematic level diagram to demonstrate light-induced effects.

As discussed in detail by Knox et al.[15] and indicated in Fig. 1 real carrier excitation by two-photon absorption leads to drastic effects in the transient spectra. The authors were able to measure the photocurrent which showed a quadratic dependence on the pump intensity. In the case of the three-level Stark effect as sketched on the right side of Fig. 1 real carrier or exciton population can be neglected, because the high intensity laser is far off resonance for a transition from the groundstate even if one considers many-photon processes. Another example for the three-level Stark effect is reported in a MQW by Fröhlich et al.[16] They give experimental evidence of an optical quantum-confined Stark effect, where two conduction band sublevels are coupled by an intense CO_2 laser pulse.

In this contribution new results of the optical Stark effect in Cu_2O are reported. In Cu_2O one has the unique possibility to monitor the dynamical coupling of the 1 S and 2 P exciton either as a change of transmission of the 2 P spectrum (dipole transition) or the 1 S spectrum (quadrupole transition). We will first brievely describe the experimental setup and then report the experimental results which are analyzed in terms of a nonlinear optical susceptibility.

EXPERIMENTAL SETUP

The experiments on the quadrupole transition to the 1 S exciton in Cu_2O were performed with a picosecond-setup as shown in Fig. 2. The central part is an AML YAG laser system, which is described in detail in Ref. 17. The high power CO_2 laser is synchronized to the dye-laser by plasma switching in a germanium plate, as first proposed by Alcock, Corkum and James.[18] The CO_2 laser pulse from a single mode CO_2 laser (Laser Science PRF-150 S) is incident at Brewster angle with perfect horizontal polarization on a germanium plate. The YAG pulse, which is synchronous with the dye pulse, excites a plasma in the germanium plate which leads to an instantaneous rise of reflectivity for the CO_2 pulse. The finite lifetime of the plasma yields rather long CO_2 pulses (250 ps). The short pulse was further amplified in a TEA amplifier. The dye laser beam was focussed to a waist of about 60 μm to be in perfect overlap with the CO_2 laser (waist about 150 μm). The high purity Cu_2O crystal was mounted in a closed-cycle refrigerator and kept at a temperature of about 20 K.

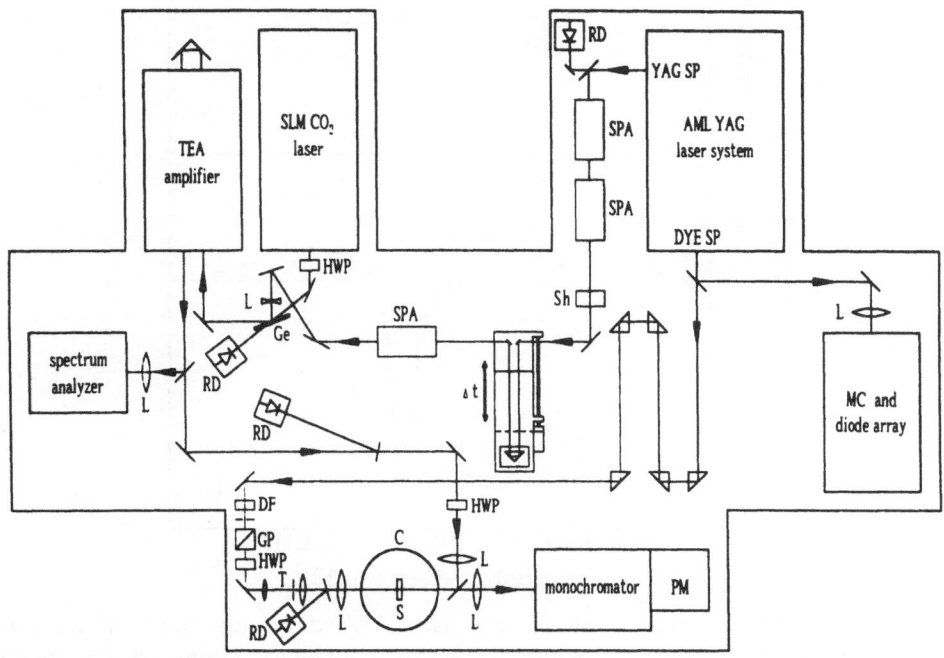

Fig. 2. Block diagram of the experimental setup. AML YAG laser system, active-mode-locked YAG pumped dye laser system; C, cryostat; DF, neutral density filter; Δt, optical delay; Ge, germanium plate; GP, Glan polarizer; HWP, half-wave plate; L, lens; MC monochromator; PM, photomultiplier; RD, reference diode; S, sample; Sh, shutter; SLM CO_2 laser, single longitudinal mode CO_2 laser; SP, single pulse; SPA, single pulse amplifier; T, telescope; TEA amplifier, transversely excited atmospheric CO_2 amplifier.

EXPERIMENTAL RESULTS AND DISCUSSION

In Fig. 3 experimental results of the optical Stark effect on the 1S exciton (Γ_5^+ symmetry) are presented. The spectra are measured at a temperature of 20 K for a k-vector along a [1$\bar{1}$0] direction. From the polarization dependence for a quadrupole transition ($\Gamma_1^+ \rightarrow \Gamma_5^+$), as given by eq. 1, we expect a signal for a polarization vector $\vec{\varepsilon} = (0,0,1)$ and no signal for $\vec{\varepsilon} = 1/\sqrt{2}\,(1,1,0)$

$$|q_{01}|^2 \sim (\varepsilon_y k_z + \varepsilon_z k_y)^2 + (\varepsilon_z k_x + \varepsilon_x k_z)^2 + (\varepsilon_x k_y + \varepsilon_y k_x)^2 . \tag{1}$$

As shown by the T-spectrum in Fig. 3a there is a drastic change of transmission induced by the CO_2 laser beam. The T-spectrum shows the same quadrupole polarization dependence as the T_0-spectrum. No dependence on the polarization of the CO_2 laser was detected. This is expected, since the dynamical coupling of the 1S exciton to the 2P exciton (3-fold P-envelope) is independent of the CO_2 laser polarization. As discussed in detail in Ref. 17 there is a pronounced polarization dependence, if the change of transmission of the 2P exciton (dipole allowed) is measured.

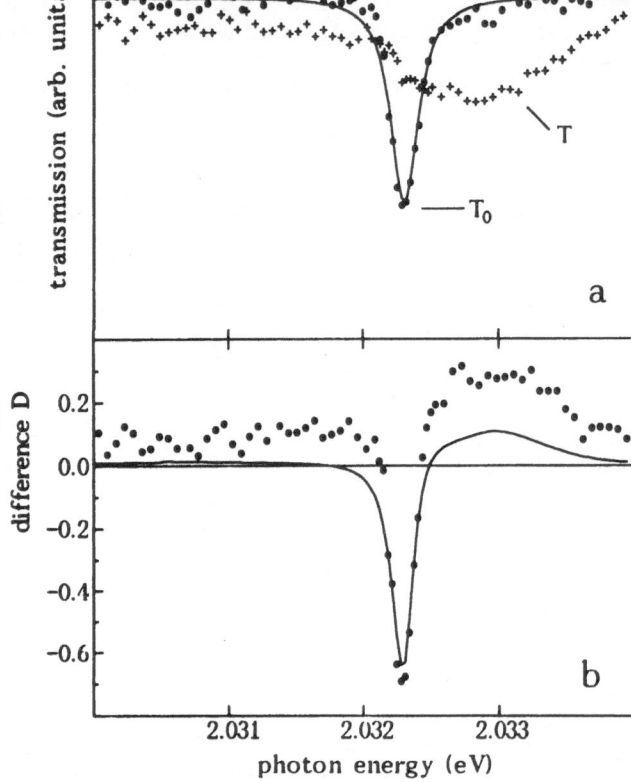

Fig. 3. Transmission spectra of dye laser; filled circles: transmission T_0 without CO_2 pulse; crosses: transmission T with CO_2 pulse, resonance parameter $\Delta E = 1.8$ meV; solid line: fit as described in text. b) Difference spectrum calculated from the experimental results (filled circles) and fit, as described in text (solid line).

The difference spectrum, which is defined by $D = \ln(T_0/T)$ is shown in Fig. 3 b. From a density matrix approach for a three-level system as shown in Fig. 5 one can derive a quadrupolar nonlinear susceptibility χ^{NL}:

$$\chi^{NL}(E, V_{12}) = \frac{N \,|q_{01}|^2}{\varepsilon_0} \cdot \frac{E_{2P} - (E + \hbar\omega) - i\Gamma_{2P}}{(E_{1S} - E - i\Gamma_{1S})[E_{2P} - (E + \hbar\omega) - i\Gamma_{2P}] - |V_{12}|^2}$$

$$= \frac{N \,|q_{01}|^2}{\varepsilon_0} \cdot \frac{1}{E_{1S} - E - i\Gamma_{1S}} \cdot \frac{1}{1 - \dfrac{|V_{12}|^2}{(E_{1S} - E - i\Gamma_{1S})[E_{2P} - (E + \hbar\omega) - i\Gamma_{2P}]}} \quad . \quad (2)$$

N is the density of oscillators and q_{01} the quadrupole matrixelement for a transition from the groundstate to the 1S exciton (E_{1S}). V_{12} is the transition dipole moment between the 1S and 2P exciton (E_{2P}) multiplied by the electric field strength of the CO_2 laser. Γ_{1S} and Γ_{2P} represent the damping constants of the corresponding excitons, which are derived from their absorption linewidth. An equivalent formula was derived by Saikan et al.[19] for a three-level system coupled by dipole transitions.

The change in absorption (D spectrum in Fig. 3 b) is proportional to the imaginary part of

$$\Delta\chi(E, V_{12}) = \chi^{NL}(E, V_{12}) - \chi^{NL}(E, 0) \tag{3}$$

Our fit (solid line Fig. 3 b) to eq. 3 (imaginary part) takes into account the laser linewidth (0.07 meV) and the attenuation of the CO_2 laser beam by the sample of about 3 mm. For the fit we used the following parameters: $E_{1S} = 2.0323$ eV, $E_{2P} = 2.1473$ eV, $\Gamma_{1S} = 0.1$ meV, $\Gamma_{2P} = 1$ meV for $(E + \hbar\omega) > E_{2P}$ and $\Gamma_{2P} = 2.5$ meV for $(E + \hbar\omega) < E_{2P}$. The resonance parameter, which is defined as $\Delta E = \hbar\omega - (E_{2P} - E_{1S})$, was 1.8 meV. The scaling factor is determined by a fit of $\chi^{NL}(E, 0)$ to the T_0-spectrum (solid line in Fig. 3 a). For the matrixelement V_{12} we get 3 meV. There is no quantitative agreement between the fit and the experimental data. (Fig. 3 b). One has to keep in mind, that there is only one free parameter for the fit (matrixelement V_{12}), since the scaling factor was chosen to reproduce the T_0-spectrum (Fig. 3 a). The spectrum shows qualitatively the expected behaviour. As expected there is a blue shift ($\Delta E > 0$) and an induced absorption on the low energy side.

In Fig. 4 we present first results for different resonance parameters. The spectra show again the right behaviour, a quantitative analysis, however, has to await more experimental data. As seen in Fig. 4 d, there is an induced background absorption, whose origin is not clear at the moment.

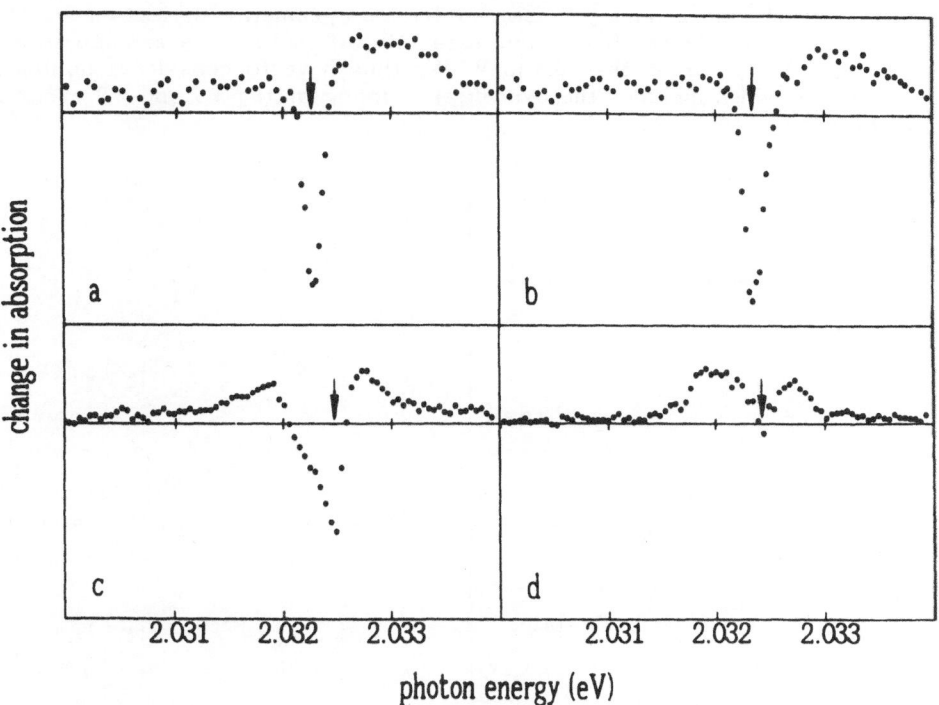

Fig. 4. Induced change in absorption as a function of dye laser energy. The corresponding resonance parameters $\Delta E = \hbar\omega - (E_{2P} - E_{1S})$ are a) 1.8 meV, b) 1.3 meV, c) 0.8 meV, d) 0.4 meV. The arrows indicate energy values E_{1S}, which depend on sample temperature.

From eq. 2 one can easily derive the odd terms of the susceptibility $\chi^{(1)}, \chi^{(3)}, \chi^{(5)} \ldots$ as indicated schematically in Fig. 5.

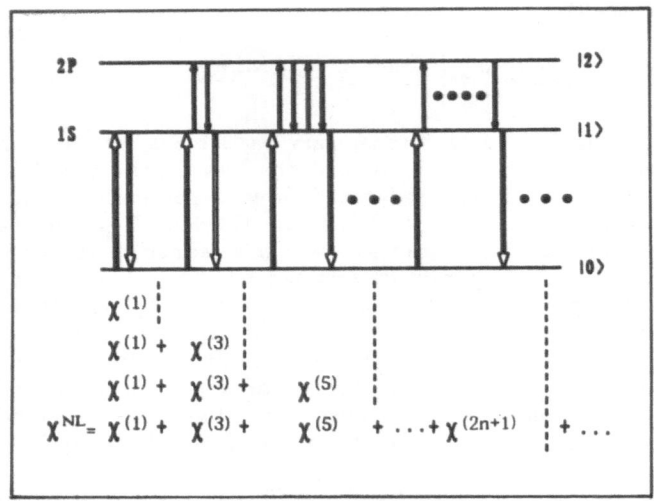

Fig. 5. Schematic three-level diagram to demonstrate nonlinear optical interaction in different orders

In Fig. 6 we compare the imaginary part of $\Delta\chi$ (eq. 3) for the total nonlinear susceptibility χ^{NL} (eq. 2) and $\chi^{(3)}$ (second term of geometric series, eq. 2). The comparison clearly shows that in our case ($\Delta E = 1.8$ meV) there are appreciable deviations if V_{12} is larger than 0.5 meV. We thus have to consider a nonlinear susceptibility, which includes the dynamical coupling of the 1S and 2P excitons to all orders.

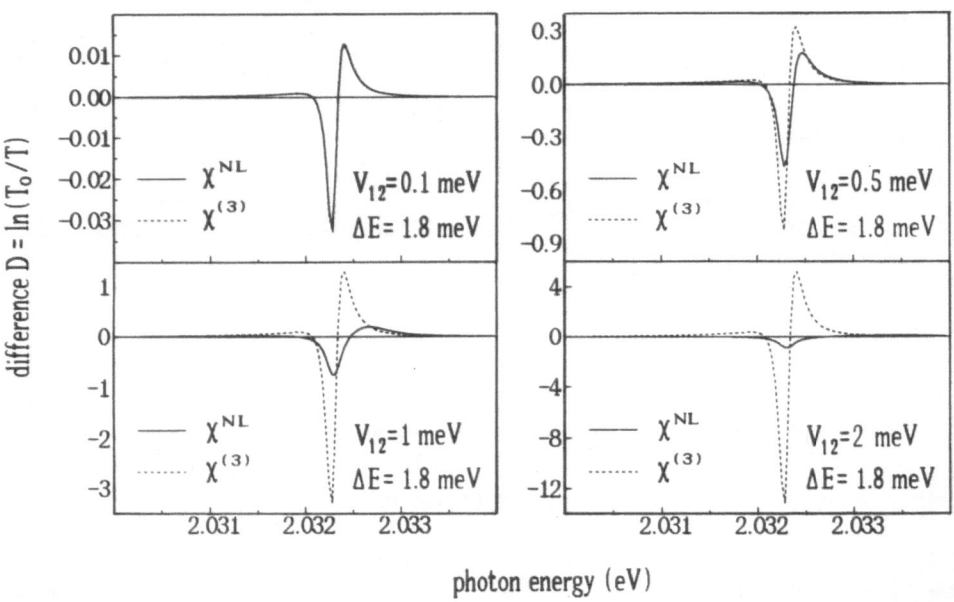

Fig. 6. Imaginary part of nonlinear susceptibility to all orders (χ^{NL}) and to third order ($\chi^{(3)}$) for different values of matrixelement. The parameters correspond to data of Cu_2O.

For a quantitative explanation of the experimental results, the theoretical analysis has to be improved. Higashimura, Iida and Komatsu[20] have done calculations on the optical Stark effect in Cu_2O taking into account the exciton-phonon interaction. They consider explicitly the Fano effect, which is caused by the interference between the 2 P exciton absorption and the phonon background. They apply their theory to the experimental results of Fröhlich. Nöthe and Reimann.[5]

In a further refinement of the analysis the degeneracy of the exciton levels should be considered. Shmiglyuk et al.[21] have discussed interexciton transitions in Cu_2O. They consider all twelve 2 P excitons, which can be constructed from a $\Gamma_7^+(2)$ valence band, a $\Gamma_6^+(2)$ conduction band and a P-envelope ($\Gamma_4^-(3)$ symmetry) $\Gamma_7^+ \otimes \Gamma_6^+ \otimes \Gamma_4^- = \Gamma_2^-(1) + \Gamma_3^-(2) + \Gamma_4^-(3) + 2\Gamma_5^-(3)$. The degeneracy of the levels is given in parentheses. If the 2 P exciton of $\Gamma_4^-(3)$ symmetry is coupled to the 1 S exciton ($\Gamma_5^+(3)$ symmetry) by the appropriate polarization of the CO_2 laser. one expects sum-frequency generation. The polariton dispersion of the dipole-allowed 2 P exciton should allow phase-matching. Because of reabsorption one expects only weak signals.

For a further analysis of our data propagation effects should be taken into account. For the Stark effect on the 1 S exciton as considered in this contribution the pulse width of our probe pulse (about 60 ps) is certainly shorter than the lifetime.[22] The pulse width is expected to be longer than the coherence relaxation time. Bigot and Hönerlage[23] have investigated this case for a three-level system. Although they apply their model to the exciton-biexciton system in CuCl. similar considerations should be applicable in our case.

ACKNOWLEDGEMENT

We acknowledge the financial support of this project by the Deutsche Forschungsgemeinschaft.

REFERENCES

1. S. H. Autler and C. H. Townes, "Stark effect in rapidly varying fields", Phys. Rev. **100**, 703 (1955).
2. A. M. Bonch-Bruevich and V. A. Khodovoi, "Current methods for the study of the Stark effect in atoms", Sov. Phys. Usp. **10**, 637 (1968).
3. P. F. Liao and J. E. Bjorkholm, "Direct observation of atomic energy level shifts in two-photon absorption", Phys. Rev. Lett. **34**, 1 (1975); J. E. Bjorkholm and P. F. Liao, "AC Stark splitting of two-photon spectra", Opt. Commun. **21**, 132 (1977).
4. H. R. Gray and C. R. Stroud, Jr., "Autler-Townes effect in double optical resonance", Opt. Commun. **25**, 359 (1978).
5. D. Fröhlich, A. Nöthe, and K. Reimann, "Observation of the resonant optical Stark effect in a semiconductor", Phys. Rev. Lett. **55**, 1335 (1985).
6. V. T. Agekyan, "Spectroscopic properties of semiconductor crystals with direct forbidden energy gap", phys. stat. sol. (a) **43**, 11 (1977).
7. A. Mysyrowicz, D. Hulin, A. Antonetti, A. Migus, W. T. Masselink, and H. Morkoc, "Dressed Excitons in a Multiple-Quantum-Well Structure: Evidence for an optical Stark effect with femtosecond response time", Phys. Rev. Lett. **56**, 2748 (1986).
8. A. von Lehmen, D. S. Chemla, J. E. Zucker, and J. P. Heritage. "Optical Stark effect on excitons in GaAs quantum wells", Opt. Lett. **11**, 609 (1986).
9. B. Fluegel, N. Peyghambarian, G. Olbright, M. Lindberg, S. W. Koch, M. Joffre, D. Hulin, A. Migus, and A. Antonetti, "Femtosecond studies of coherent transients in semiconductors", Phys. Rev. Lett. **59**, 2588 (1987).

10. N. Peyghambarian, S. W. Koch, M. Lindberg, B. Fluegel, and M. Joffre. "Dynamic Stark effect of exciton and continuum states in CdS", Phys. Rev. Lett. **62**, 1185 (1989).

11. S. Schmitt-Rink, D. S. Chemla, and H. Haug, "Nonequilibrium theory of the optical Stark effect and spectral hole burning in semiconductors", Phys. Rev. B **37**, 941 (1988).

12. M. Lindberg and S. W. Koch, "Transient oscillations and dynamic Stark effect in semiconductors", Phys. Rev. B **38**, 7607 (1988).

13. I. Balslev, R. Zimmermann, and A. Stahl, "Two-band density-matrix approach to nonlinear optics of excitons", Phys. Rev. B **40**, 4095 (1989).

14. D. S. Chemla, W. H. Knox, S. Schmitt-Rink, J. B. Stark, and R. Zimmermann, "The excitonic optical Stark effect in semiconductor quantum wells probed with femtosecond optical pulses", J. Lumin. **44**, 233 (1989).

15. W. H. Knox, D. S. Chemla, D. A. Miller, J. B. Stark, and S. Schmitt-Rink, "Femtosecond ac Stark effect in semiconductor quantum wells: Extreme low- and high-intensity limits", Phys. Rev. Lett. **62**, 1189 (1989).

16. D. Fröhlich, R. Wille, W. Schlapp, and G. Weimann, "Optical quantum-confined Stark effect in GaAs quantum wells", Phys. Rev. Lett. **59**. 1748 (1987).

17. D. Fröhlich, Ch. Neumann, B. Uebbing, and R. Wille, "Experimental investigation of three-level optical Stark effect in semiconductors", phys. stat. sol. (b) to be published.

18. A. J. Alcock, P. B. Corkum, and D. J. James, "A fast scalable switching technique for high-power CO_2 laser radiation", Appl. Phys. Lett. **27**, 680 (1975).

19. S. Saikan, N. Hashimoto, T. Kushida, and K. Namba, "Variation of inverse Raman spectrum near resonance", J. Chem. Phys. **82**, 5409 (1985).

20. T. Higashimura, T. Iida, and T. Komatsu, "Optical Stark effect in an exciton-phonon system", phys. stat. sol. (b) **150**, 431 (1988).

21. M. I. Shmiglyuk, P. I. Bardetskii, and S. D. Tiron, "Nonlinear optical nutation on interexciton transitions", phys. stat. sol. (b) **154**, 209 (1989); M. I. Shmiglyuk, P. I. Bardetskii, and E. V. Vitiu, "Exciton spectrum in Cu_2O crystal: An investigation by a double resonance method", Opt. Spectrosc. **50**, 436 (1981).

22. D. Fröhlich, K. Reimann, and R. Wille, "Time-resolved two-photon emission in Cu_2O", Europhys. Lett. **3**, 853 (1987).

23. J. Y. Bigot and B. Hönerlage, "Nonlinear propagation of picosecond pulses interacting with a three-level system", Phys. Rev. A **36**, 715 (1987).

FOUR-WAVE MIXING OF SHORT PULSES: SUBPICOSECOND VIBRATIONAL DYNAMICS OF

ELECTRONIC EXCITATIONS

V. V. Hizhnyakov

Institute of Physics of Estonian Academy of Sciences
Riia 142, 202400 Tartu
Estonian SSR, USSR

Pulse interference methods for resonantly exciting the nonlinear polarization of a medium enable the dynamics of the temporal development of excitations in the system to be investigated in the subpico- and femto-second region. Four-wave mixing of short pulses[1] belongs to these methods. It is usually realized in the form of a stimulated echo (SE)[1,2], i.e. in the form of a fourth pulse excited by three short pulses in an inhomogeneously-broadened absorption band. If the pulses act at the times $t_1 = 0$, $t_2 > 0$ and $t_3 > t_2$, then the SE occurs as a pulse at the time $t = t_2 + t_3$. The physical nature of the SE lies in the filtration (modulation) of the spectrum of the third pulse by the first two pulses; this spectral modulation is manifested in the time dependence of the transmitted signal in the form of the SE pulse.

SE is a coherent phenomenon: its intensity (proportional to N^2, where N is the number of excited centers) depends on the phase relaxation processes disturbing coherence. Therefore, SE is used for the investigation of the processes indicated (see, e.g. Refs. 2-8). The advantage of the SE method lies in the possibility of studying slow as well as fast relaxation processes. Moreover, when an interferometer (beam splitter and delay line) is used to obtain mutually coherent excitation pulses, the SE method also makes possible the investigation of subpicosecond and femtosecond processes by the use of significantly longer pulses[5-9]. In the case of such an experimental arrangement the duration of the pro-cesses under study is restricted from below only by the autocorrelation time of the excitation radiation τ_o, which is determined by its inverse spectral width; in the case of coherent pulses it is shorter than their duration. Thanks to this feature, the SE method can be used to investigate ultrafast dynamical processes, such as the dynamics of the wave packets of the continuum levels of molecules and crystals (including the dynamics of quasiparticles in crystals), intramolecular dynamics of strongly fluctuating systems, e.g. biological objects in vivo, etc. In such investigations the fluctuations which in the time $t_2 + t_3$ change the frequencies of optical transitions by less than t_2^{-1}, are manifested as static. For $t_2, t_3 < \bar{\omega}^{-1} \sim 10^{-13}$ s the role of "static" fluctuations can be played by the lattice vibrations of the crystal (ω is the mean vibrational frequency.)

In this work, the influence of vibrations on SE in the pico- and subpicosecond regions is considered. The investigation is based on the method of correlation functions. In this method vibrations (including vibrational relaxation) are described dynamically within the framework of the Hamiltonian formalism. As the reasons for vibrational relaxation, the dephasing of the phonons of the quasi-continuous spectrum (the smearing of a phonon packet) and their anharmonic decay are considered. Two different contributions to SE are elucidated. The first one is caused by absorption, the second one by the amplification of the third excitation pulse. A theoretical description of both processes and the effect of their interference is given. It is also demonstrated that inhomogeneous broadening of absorption bands is not necessary for observing SE in the subpicosecond region ($t_2 \sim 0.1\omega^{-1} \sim 10^{-14}$ sec). Here the model of a frozen lattice turns out to be valid, in which the role the of inhomogeneous field is played by the vibrations. Successful experiments have been carried out which have allowed such an SE signal to be observed for the first time: SE has been detected for F centers in KBr, which have a broad homogeneous absorption band.

The influence of fast relaxation processes on the four-wave mixing of pulses has been also considered in Ref. 11, where a stochastic description of the processes in the medium was used. Some general problems of the theory of four-wave mixing of pulses of multilevel dynamic systems have been considered in Ref. 12.

1. SE after vibrational relaxation

Let us consider a crystal (solution) which contains impurity centers (molecules) differing from one another only by the frequencies of the purely-electronic transition ω_o. The crystal is excited in the impurity absorption band by two similar short light pulses delayed by an interval t_2 (such pulses can be obtained from a light source by means of an interferometer, i.e. a beam splitter and a delay line). The spectrum of such excitations contain a grating with a period $2\pi t_2^{-1}$:

$$i(\omega) = | \int_{-\infty}^{\infty} dt\, \mathcal{E}_1(t)\exp(-i\omega t)\left[1+\exp(-i\omega t_2)\right]|^2 =$$

$$= 2|\mathcal{E}_1(\omega)|^2(1+\cos\omega t_2), \qquad (1)$$

$\mathcal{E}_1(t)$ is the field strength of one pulse, and

$$\mathcal{E}_1(\omega) = \int_{-\infty}^{\infty} d\omega\, \exp(-i\omega t)\mathcal{E}_1(t).$$

By this excitation some centers are excited and, as a result, the absorption spectrum is changed; this change, measured after vibrational relaxation in the excited state, equals

$$\Delta\kappa(\omega) \sim \int d\omega_o \left[K(\omega-\omega_o)+J(\omega-\omega_o)\right] \Delta N(\omega_o), \qquad (2)$$

where $K(\omega-\omega_o)$ and $J(\omega-\omega_o)$ are the normalized absorption and luminescence spectra of a single center, the frequency of whose purely-electronic transition is ω_o;

$$\Delta N(\omega_o) = -N(\omega_o) \int d\omega'\, K(\omega'-\omega_o)i(\omega') \qquad (3)$$

is the change of the distribution $N(\omega_o)$ of the centers with the frequency ω_o under radiation (1) (it is assumed that $\Delta N(\omega_o)/N(\omega_o) \ll 1$). In Eq. (2), the term $\sim K(\omega-\omega_o)$ allows for the decrease of light absorption due to depopulation of the lower electronic state, while the term $\sim J(\omega-\omega_o)$, takes into account the amplification of the light by stimulated radiative

378

transitions from the relaxed excited electronic state.

In the case of large widths of the excitation spectrum $|\mathcal{E}_1(\omega)|^2$ and inhomogeneous broadening $N(\omega_o)$ (in comparison with the widths of the homogeneous spectra $K(\omega-\omega_o)$ and $J(\omega-\omega_o)$), Eq. (2) takes the form

$$\Delta\kappa(\omega) \sim -N|\mathcal{E}_1|^2\{2+\mathrm{Re}\ \exp(-i\omega t_2)F(-t_2)[F(t_2)+L(t_2)]\}, \qquad (4)$$

where $F(t)$ and $L(t)$ are the Fourier transforms of $K(\omega-\omega_o)$ and $L(\omega-\omega_o)$, respectively. These Fourier transforms are determined by the vibrational dynamics of the system in both the ground and excited electronic states.

One can see that excitation with two similar pulses induces a grating in the absorption spectrum of the system whose contrast is determined by the Fourier transforms of the absorption and stimulated emission spectra.

Let us consider now the effect of this grating on the response of the system to the third short pulse $\mathcal{E}_3(t)$. This effect is described by the equation

$$\Delta\mathcal{E}_3(t) \sim \int_{-\infty}^{\infty} dt'\, \mathcal{E}_3(t-t')\, \Delta X(t'),$$

where

$$\Delta X(t) = \theta(t) \int_{-\infty}^{\infty} d\omega\, e^{i\omega t}\, \Delta\kappa(\omega)$$

is the change of the response function. Taking $\mathcal{E}_3(t) = \mathcal{E}_3\delta(t-t_3)$, we obtain the response at the time $t > t_3$ in the form of an SE:

$$\Delta\mathcal{E}_3(t) \sim \mathcal{E}_4(t_2)\delta(t-t_2-t_3), \qquad (5)$$

where

$$\mathcal{E}_4(t_2) \sim N|\mathcal{E}_1|^2 \mathcal{E}_3\mathrm{Re}\ F(-t_2)[F(t_2)+L(t_2)] \qquad (6)$$

is the SE amplitude. Thus, in the region of small $t_2 \lesssim t_r$, the SE amplitude depends significantly on the vibrational dynamics (t_r is the characteristic time of vibrational relaxation). In the case under consideration, $t_3 \gg t_r$, this dependence is determined by the Fourier transform of the homogeneous vibronic spectra of absorption and luminescence.

In the preceding discussion the spectrum of the third pulse was assumed to be overlapped by both the absorption and the luminescence band (the spectrum of the first two pulses may be overlapped only by the absorption band). If it is overlapped only by the absorption (or only by the luminescence) band, then in Eq. (6) the second (or the first) term in brackets should be omitted. In case of a partial overlapping of the spectrum of the third pulses, e.g. by the luminescence band, the second term in Eq. (6), which has into account the stimulated emission, is weakened.

2. SE in the course of vibrational relaxation

Let us consider the case of arbitrary times t_2, t_3, and t_r. We take into account that the amplitude of the SE signal is determined by the macroscopic nonlinear polarization induced by the three exciting pulses[1]. In the activated crystal (solution) excited in the absorption band of the impurity centers, during the time $\tau_o \ll t \ll \gamma^{-1}$ this polarization equals[13]

$$P_\alpha^{(3)}(t) \sim \int_o^\infty \int \int d\mu d\tau d\tau' \sum_{\alpha_1 \alpha_2 \alpha_3} X_{\alpha\alpha_1\alpha_2\alpha_3}^{(3)}(\mu,\tau,\tau') \times$$

$$\times E_{\alpha_1}(t-\mu-\tau')E_{\alpha_2}(t-\mu)E_{\alpha_3}(t-\tau). \tag{7}$$

Here

$$X_{\alpha\alpha_1\alpha_2\alpha_3}^{(3)}(\mu,\tau,\tau') \sim \mathrm{Re} \ll A_{\alpha_1\alpha_2\alpha\alpha_3}(\mu,\tau,\tau')\gg, \tag{8}$$

$\ll...\gg$ denotes an ensemble average,

$$A_{\alpha_1\alpha_2\alpha\alpha_3}(\mu,\tau,\tau') = \langle M_{\alpha_1}\exp(i\tau'H_2-\gamma\tau')M_{\alpha_2}\exp(i\mu H_1) \ M_\alpha \times$$

$$\times \exp(-i\tau H_2-\gamma\tau)M_{\alpha_3}\exp\left[-i(\mu+\tau'-\tau)H_1\right]\rangle \tag{9}$$

is the correlation function of the secondary emission of a single center[10]; H_1 and H_2 are vibrational Hamiltonians of the center in the ground and excited electronic states; M_α is the electronic matrix element of the dipole transition in the polarization $\alpha = x,y,z$; and γ denotes the radiative decay constant of the excited state.

Let us examine the excitation by three consecutive short pulses:

$$E(t) = \mathcal{E}_1\delta(t) + \mathcal{E}_2\delta(t-t_2) + \mathcal{E}_3\delta(t-t_3).$$

We take into account that the coherent response under consideration satisfies the law of conservation of the wave vector[2] $\vec{k} = \vec{k}_2+\vec{k}_3-\vec{k}_1$ ($\vec{k}_i$ is the wave vector of the field, and $\vec{k}$ is the wave vector of the induced polarization and the response). In the case of noncoinciding $\vec{k}_1$ and $\vec{k}_{2,3}$ the signal in the direction $\vec{k} \neq \vec{k}_i$ (it is nonzero at $t > t_3$) is determined by the terms proportional to $\mathcal{E}_1\mathcal{E}_2\mathcal{E}_3$. There are two such terms:

1) $E_{\alpha_1} \sim \mathcal{E}_1$, $E_{\alpha_2} \sim \mathcal{E}_2$, $E_{\alpha_3} \sim \mathcal{E}_3$; $\mu = t-t_2$, $\tau = t-t_3$, $\tau' = t_2$;

2) $E_{\alpha_1} \sim \mathcal{E}_1$, $E_{\alpha_2} \sim \mathcal{E}_3$, $E_{\alpha_3} \sim \mathcal{E}_2$; $\mu = t-t_3$, $\tau = t-t_2$,, $\tau' = t_3$;

the remaining terms $\sim \mathcal{E}_1\mathcal{E}_2\mathcal{E}_3$ vanish for $t_1 < t_2 < t_3 = 0$. The following contributions to $P^{(3)}(t)$ correspond to these two terms:

$$P_a(t) \sim \mathrm{Re} \ll A(t-t_2,t-t_3,t_2)\gg;$$

$$P_e(t) \sim \mathrm{Re} \ll A(t-t_3,t-t_2,t_3)\gg \tag{10}$$

(the polarization indices have been omitted). $P_a(t)$ takes into account the change in the absorption of the third pulse under the action of the first two pulses; $P_e(t)$ allows for the amplification of the third pulse by stimulated radiative transitions.

In systems with a large inhomogeneous broadening of the absorption band, the averaging $\ll...\gg$ (averaging over the frequency of the purely electronic transition and space) gives

$$P_{a,e}(t) \sim \int d\omega_0 \, \exp\left[i\omega_0(t-t_2-t_3)\right] N(\omega_0) \sim$$

$$\sim \delta(t-t_2-t_3)\delta(\vec{k}+\vec{k}_1-\vec{k}_2-\vec{k}_3).$$

This is the SE signal.

In the case under consideration, $t_3 < t_r$, when the SE signal is emitted during vibrational relaxation, the vibrational dynamics of the system is revealed in the dependence of SE on t_2 as well as on t_3. Let us find this dependence in a model that takes into account an arbitrary linear and a weak quadratic vibronic interaction in the Condon approximation. In this model[10]

$$A(\mu,\tau,\tau') = \exp\left[-\gamma_1(|\mu+\tau'-\tau|-|\mu-\tau|+\tau) + g(\mu)+g(\mu+\tau'-\tau) + \right.$$

$$\left. + g(\tau') + g(-\tau)-g(\mu+\tau')-g(\mu-\tau)\right] \tag{11}$$

$(\mu,\tau,\tau' \geq 0)$, while $F(t)$ and $L(t)$ are given by

$$F(t) = L(-t) = \exp\left[(-\gamma+\gamma_1)|t|+g(t)\right], \tag{12}$$

where γ_1 is the modulational broadening of the zero-phonon line (for $T \to 0$, $\gamma_1 \to 0$), $g(t) = f(t) - f(0)$,

$$f(t) = \sum_s \xi_s^2 \exp(-\Gamma_s|t|)\left[(n_s+1)\exp(i\omega_s t)+n_s\exp(-i\omega_s t)\right]$$

is the Fourier transform of the one-phonon absorption sideband, ω_s is the frequency, $\Gamma_s \ll \omega_s$ is the decay constant, and ξ_s, is the parameter of the interaction of the electronic transition with the vibrational modes, $n_s = \left[\exp(\omega_s/kT)-1\right]^{-1}$.

On substituting (11) into (10), one obtains the following SE amplitude for t_2, $t_3 \ll \gamma^{-1}$:

$$\mathcal{E}_4 = \mathcal{E}_a + \mathcal{E}_e \sim \exp(-2\gamma_1 t_2)\left[D(t_2,t_3) + D(t_3,t_2)\right], \tag{13}$$

where

$$D(t_2,t_3) = \exp\left[2\operatorname{Re} g(t_2)\right] \operatorname{Re} \exp\left[2g(t_3)-g(t_2+t_3)-g(t_3-t_2)\right]. \tag{14}$$

Consequently, the dependence of the SE intensity $I \sim \mathcal{E}_4^2$ on t_2 and t_3 is determined by the vibrational dynamics, in this case described by the phonon function $g(t)$.

If the vibronic interaction is strong, the SE has a significant intensity only in the region of small t_2, $t_3 \ll \omega^{-1}$. In this region, the SE amplitude is equal to

$$\mathcal{E}_4(t_2,t_3) \sim \exp(-m_4 t_2^2 t_3^2)\left[\cos(m_3 t_2^2 t_3)+\cos(m_3 t_2 t_3^2)\right],$$

where m_ℓ is the ℓth moment of the absorption band of a single center. It can be seen that the SE oscillates and decays; the characteristic rates of oscillation ($\sim \bar{\omega}S^{1/3}$) and decay ($\geq \bar{\omega}S^{1/4}$) exceed by many times the average phonon frequency $\bar{\omega}$ ($S = \sum_s \xi_s^2$ is the dimensionless Stokes losses parameter).

3. Excitation by incoherent pulses

If the experiment is carried through by the method of (5)-(8) with the use of an interferometer and incoherent exciting pulses whose duration is $t_o > \bar{\omega}^{-1}$ and whose spectral width is $\sigma > m_2^{1/2}$, then the intensity of the SE signal in the direction $\vec{k} = 2\vec{k}_2 - \vec{k}_1$ is determined by the following integral[13]:

$$I(t_2) \sim \left[1 + \Phi(t_2\sigma/2)\right] \int_0^{t_o} d\tau \; \tau \, \bar{\mathcal{E}}_4^2(t_2, t_2 + \tau), \tag{15}$$

and $\Phi(x)$ is the probability integral,

$$\bar{\mathcal{E}}_4(t_2, t) \sim \exp\left[-t_2^2 m_2(t)\right]\left\{\cos\left[t_2^2 m_2(t)\right] + \exp\left[-4m_1^2(t)/3\sigma^2\right] \times \right.$$

$$\left. \times \left[\cos 2t_2 m_1(t)\right]\right\}, \tag{16}$$

where $m_\ell(t) = (-i)^\ell d^\ell g(t)/dt^\ell$.

Figures 1a and 1b depict the dependence of the intensity of the SE signal on the delay time t_2 in the case of a vibronic interaction with one decaying pseudolocal vibration at low temperature, described by the function $g(t) = S[\exp(i\omega_1 t - \Gamma t) - 1]$, where ω_1 and Γ are the frequency and the decay rate of the pseudolocal vibration. The decrease in the SE intensity in the region of small and negative t_2 is connected with the

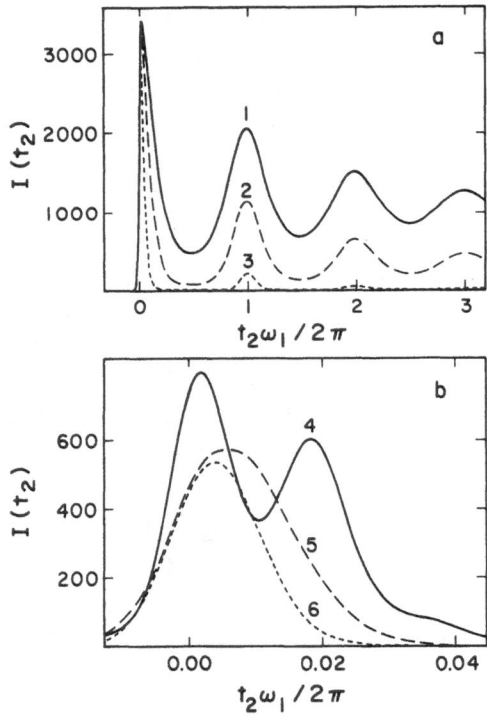

Fig. 1. The dependence of the stimulated echo intensity on the delay time t_2 of the exciting pulses, caused by the vibrational dynamics of a pseudolocal mode with the frequency $\omega_1 \sim 10^{13}$ sec^{-1}; $\Gamma = 0.1\omega_1$, $t = 30\omega_1^{-1}$; $\sigma = 30\omega_1$; S equals: (a) 0.3(1), 0.6(2), 1.5(3); (b) 25(4), 50(5), 100(6).

finite duration of the exciting pulses, while its oscillations and decrease in the region of positive $t_2 > \sigma^{-1}$ is due to the phase relaxation of the electronic excitation which is caused by the vibrations. The dependence of the signal in the direction $\vec{k}' = 2\vec{k}_1 - \vec{k}_2$ on t_2 is equal to $I(-t_2)$, i.e. it is mirror-symmetric with respect to the signal in the direction $\vec{k} = 2\vec{k}_2 - \vec{k}_1$.

4. SE of F centers

In the case of a strong vibronic interaction, the SE with the characteristics found above can be observed not only on inhomogeneously broadened bands, but also on homogeneous absorption bands. Indeed, for $S \gg 1$ and small t_2, t_3, $P^{(3)}(t) \sim \exp[-m_2(t-t_2-t_3)^2/2]$, i.e. the SE is of a finite but short (compared to t_r) duration $\sim \omega^{-1} S^{-1/2}$. Thus, in the region of small t_2, $t_3 < \omega^{-1}$, SE can be observed in centers with broad absorption bands, e.g. F centers in alkali halide crystals. The reason is the following. In the initial stage of vibrational relaxation, which determines the absorption spectrum when $S \gg 1$, the electron energy V changes according to the law[14] $V(t) \simeq -m_3 t^2/2 \sim S\omega^3 t^2/2$. During the dephasing of the coherent polarization $t \approx m_2^{-1/2}$ this change (it is of the order $\leq \omega^{-1}$) is considerably smaller than the vibration induced width of the absorption band $m_2^{1/2}$. Therefore, in the case that $S \gg 1$, in the time region actual for dephasing, vibrations appear as nearly static fluctuations, while the vibrational broadening of the absorption band is quasi-inhomogeneous. This allows one to observe SE in these systems for t_2, $t_3 < S^{1/2} \omega^{-1}$. Such an SE signal was indeed observed experimentally recently[15].

As a source of broadband radiation an XeCl excimer-laser-pumped dye laser was used. To get a maximally broad generation spectrum in the DL resonator the diffraction grating was replaced by a mirror with a high reflection coefficient. By using Rhodamine C, a spectral width of 11 nm at the wavelength of 604 nm was obtained. The peak-width $\tau_0 \sim 100$ fs of the autocorrelation function of the exciting pulses corresponds to such a spectrum. The DL pulses were channelled into a Michelson interferometer, where a pair of pulses with the wave vectors $\vec{k}_1$ and $\vec{k}_2$ was formed which were directed on the sample at an angle of 2° with respect to each other. The interval between pulses t_2 was changed with the aid of a monitoring delay line. As the duration of the pulses ($\tau_p \sim 10$ ns) was much longer than t_2 and t_r, then for the third probe pulse the tail of the second pulse could be used. The signal SE pulse was observed in the direction $\vec{k} = 2\vec{k}_2 - \vec{k}_1$ (for $t_2 > 0$) or in the direction $\vec{k}' = 2\vec{k}_1 - \vec{k}_2$ (for $t_2 < 0$). The one-sidedness of the signal enabled one to distinguish the SE from the self-diffraction phenomenon. The sample, a single crystal of KBr with F centers, had a width ~ 0.7 mm and an optical density $D = 0.9$ at $\lambda = 605$ nm. The sample was immersed in a helium cryostat with the temperature kept near 2K. Pulses of frequency 10 Hz and peak intensity 1 kW were focused on an ~ 1 mm^2 area of the sample. The SE signals in the directions $\vec{k}$ and $\vec{k}'$ were separated by a spatial shutter and directed to a multiplier. The intensities of the signals as functions of the delay t_2 were recorded by a boxcar and a microcomputer.

In Fig. 2, the SE signals which were recorded in the directions $\vec{k} = 2\vec{k}_2 - \vec{k}_1$ and $\vec{k}' = 2\vec{k}_1 - \vec{k}_2$ are shown as functions of the delay time t_2

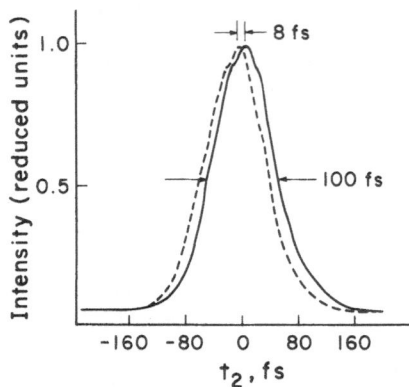

Fig. 2. Stimulated echo intensity observed on F centers in KBr in the directions $\vec{k} = 2\vec{k}_2 - \vec{k}_1$ (solid line) and $\vec{k}' = 2\vec{k}_1 - \vec{k}_2$ (dashed line) as a function of the delay time t_2 between the $\vec{k}_1$ and $\vec{k}_2$ pulses.

between the pulses. In the case of positive values of t_2 the pulse in the direction $\vec{k}_2$ reached the sample after the pulse in the direction $\vec{k}_1$, but the order of their arrival was reversed on scanning through the point $t_2 = 0$ (which corresponds to the negative delay in Fig. 2). Signals in both directions were shifted with respect to each other by 8±4 fs. The width of the SE curves is $\sigma \sim 100$ fs, which coincides within the limits of registration accuracy with the width of the autocorrelation function of the laser pulses measured by the method of noncollinear generation of second harmonics in the KPD crystal. The shape of the registered signal is almost symmetrical, which is also a result of the very short dephasing time, < 100 fs, in the system under the examination. Note that the response signal in this experiment can, in principle, contain for $t_2 < \tau_0$, besides the SE, a certain contribution from the simple self-diffraction by the pulse-induced spatial interference grating. However, this contribution should be insignificant as no self-diffraction signal in the resonance medium mentioned has been observed at room temperatures. The distinct shift of the intensity maxima of signals in the directions $\vec{k}$ and $\vec{k}'$ with respect to the zero value of the delay t_2 is also an indication of the dominating contribution of SE in the response signal. The registered value of the shift, 8 fs, gives, in accordance with the theory, an estimate of the phase relaxation time, 10 fs, for F centers in KBr.

REFERENCES

1. P. Ye and Y. R. Shen, Phys. Rev. A25, 2183 (1982).
2. T. Mosberg, A. Flushberg, R. Kachru and S. R. Hartmann, Phys. Rev. Lett. 42, 974 (1979).
3. T. Yajima and Y. Taira, J. Phys. Soc. Japan 47, 1620 (1979).
4. H. Fujita, H. Nakatsuka, H. Nakanishi, and M. Matsuoka, Phys. Rev. Lett. 42, 1665 (1979).
5. A. J. Taylor, D. J. Eskine, and C. L. Tang, Appl. Phys. Lett. 43, 989 (1979).
6. A. M. Weiner, E. P. Ippen, Opt. Lett. 9, 53 (1984).
7. H. Asaka, H. Nakatsuka, M. Fujiwara, and M.Matsuoka, Phys. Rev. A29, 2286 (1984).
8. R. Beach and S. R. Hartmann, Phys. Rev. A53, 663 (1984).
9. N. Morita, T. Yajima, Phys. Rev. A30, 2525 (1984).
10. V. Hizhnyakov and I. Tehver, phys. stat. sol. 21, 755 (1967); K. Rebane, V. Hizhnyakov and I. Tehver, Proc. Acad. Sci. of Estonian SSR, Phys.* Math. 25, 207 (1967).

11. B. D. Fainberg, Optika i Spektroskopiya 60, 120 (1986) (in Russian).
12. S. Mukamel and R. F. Loring, J. Opt. Soc. America B3, 595 (1986).
13. V. Hizhnyakov, phys. stat. sol. (in press).
14. I. Yu. Tehver, V. V. Hizhnyakov, Sov. Phys.-JETP 42, 305 (1975).
15. M. Rätsep, Proc. Estonian Acad. Sci., Phys.* Math. 39, 449 (1989).

FEMTOSECOND REAL AND VIRTUAL EXCITATIONS IS GaAs QUANTUM WELLS:

PHYSICS AND APPLICATIONS

Wayne H. Knox

AT&T Bell Laboratories
Holmdel, NJ 07733

Introduction

   The ultrafast dynamical response of semiconductors is of
interest for a number of potential applications such as high-
speed optical information processing. Such applications will
require a complex interplay between light and semiconductors,
as some tasks are better accomplished with light, and some
with electronics. In the case of quantum-confined structures,
large optical nonlinearity and electro-optical properties are
obtained in room-temperature conditions. We discuss in the
present text one recent application of short pulse optical
techniques to the generation, propagation and detection of
ultrafast electrical transients using quantum wells.

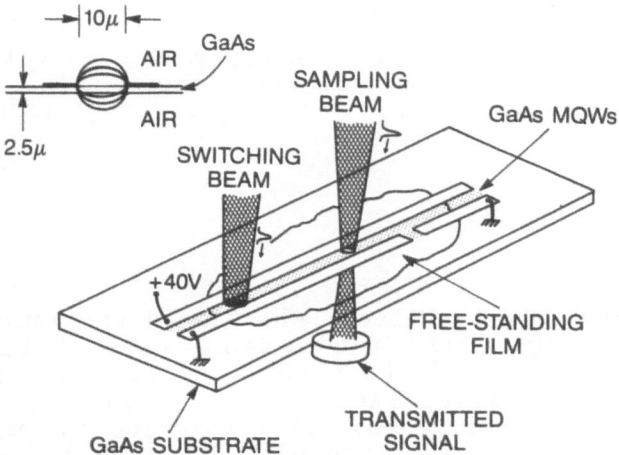

Fig. 1. Structure for femtosecond electric field
        transient generation in GaAs quantum wells.

Excitonic Electroabsorption in Quantum Wells

The bandedge absorption properties of semiconductors are strongly affected by quantum confinement which is possible today with epitaxial growth techniques. In particular, the appearance of a sharp excitonic feature which persists at room temperature in quantum wells makes possible a number of new approaches to optoelectronics. Two field configurations are possible: applied field perpendicular to the layers, and parallel to the layers. In the former case the exciton is

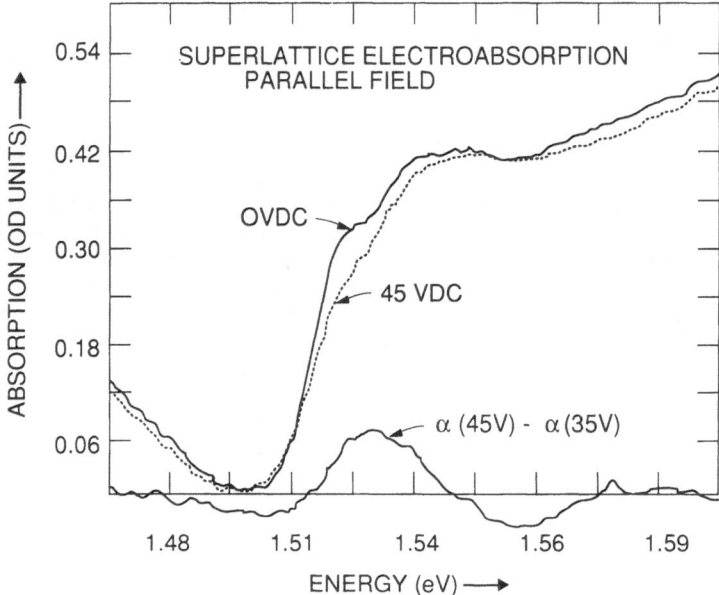

Fig. 2. Transmission change when applying a 45 volt bias to the 10 micron coplanar stripline.

quantum-confined stark-shifted, whereas in the latter case it is broadened by field-ionization [1]. Although the perpendicular case leads to higher contrast and lower loss modulators, the parallel-field case is readily adaptable to very high speed electrical signal generation and propagation. We show in Figure 1 a structure with which we can explore the ultrafast dynamical response of the quantum well excitons to a rapidly changing field transient [2]. This structure is made with a 0.75 micron thick film of quantum wells grown with alternating layers of GaAs and AlGaAs with 30 % Al fraction. A 10 micron gold coplanar stripline is evaporated onto the sample, and the back is etched away with a selective chemical etch, leaving a very thin film. When a DC voltage is applied across the contacts, a strong electric field is established in the plane of the quantum wells, leading to parallel-field electroabsorption. Figure 2 shows the change in absorption spectrum which results. In the narrow-barrier limit (superlattice-excitons) the spectral region of positive transmission change is wide enough to modulate a broadband femtosecond light pulse; thus we have a technique of electrical pulse detection in the same structure as the generator. The experiment is, then, to focus a short laser pulse onto the stripline at the generation point, propagate

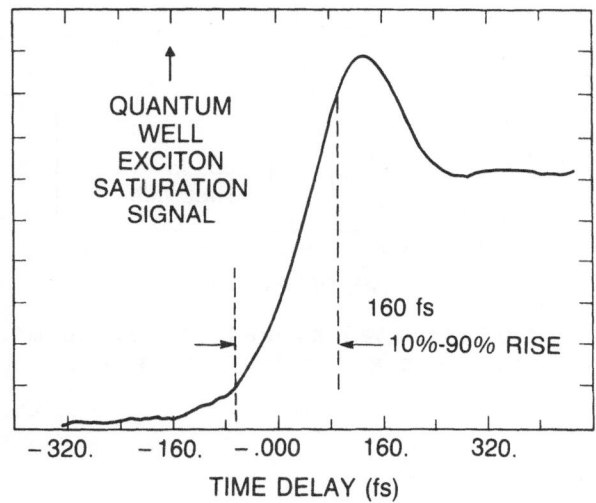

Fig. 3. Absorption saturation signal with pump and probe overlapped on the sample. This yields the system impulse response.

the resulting ultrafast electrical transient to the desired point, and then detect it using the excitonic response to electric fields by passing a second delayed short pulse at the exciton wavelength through the sample at the point where the electrical transient is to be measured. In the small-signal limit, the transmission change resulting from the passage of the electrical transient will be proportional to the change in applied voltage. Interestingly, the use of excitons as both the generator and detector yields some advantages compared to conventional optoelectronics. First, the exciton is a sensitive nonlinear optical detector which can be used as a time marker and represents the optical impulse excitation and

time response of the measurement system. Figure 3 shows the
excitonic nonlinear response when the pump and probe beams are
overlapped on the sample. Figure 4 shows the electrical signal
which propagates away from the excitation region, with a rise
time of 380 fs. Our sample structure provides not only a means
to observe the nonlinear optical properties of the thin film,
but the coplanar stripline on a very thin dielectric substrate
is an interesting new way to transport ultrafast electrical

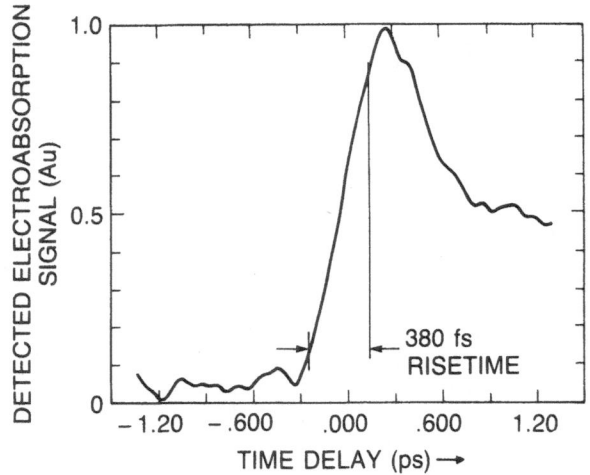

Fig. 4. Detected electrical signal after 200 microns of
propagation in the coplanar stripline.

pulses. When a very thin substrate is used, the electric field
lines propagate in air on the top and the bottom of the
structure, resulting in greatly reduced modal dispersion,
phonon loss and radiation losses [3]. One consequence of this
is a propagation speed of nearly the speed of light in vacuum.
Fig. 5 shows the propagation delay measured for the 0.75
micron thick sample. If the stripline were embedded in a GaAs
slab, the speed would be only 28% of the speed of light in
vacuum.

Summary

Optical probes of electrical transients provide a number
of interesting features. First, the inherent speed of optics
can be directly used to obtain ultrafast electrical time
response. Second, optical probes can be relatively non-
invasive compared to electrical probes which have to be placed
in contact with signals to be measured. In the present case,

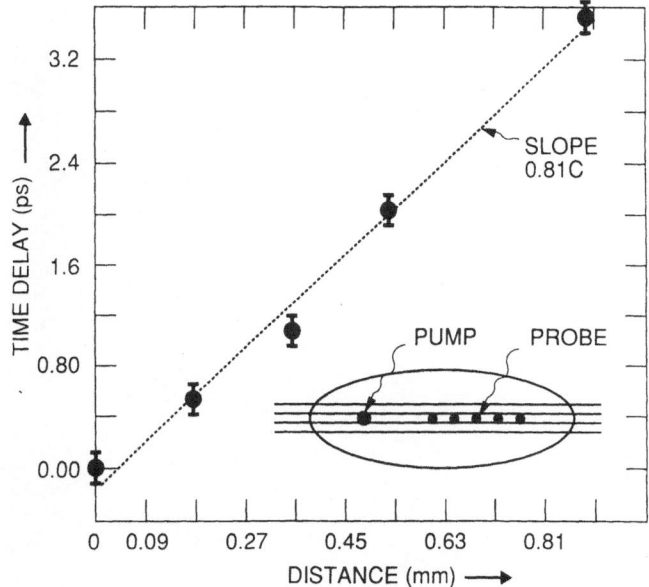

Fig. 5. Propagation delay in coplanar quantum well stripline. The speed of propagation is nearly that of light in vacuum because the substrate is very thin.

the opportunity to simultaneously utilize the nonlinear optical response as well as the electroabsorption response of quantum well excitons provides new capabilities in ultrafast optoelectronics.

REFERENCES

[1] D.A.B. Miller, D.S. Chemla, T.C. Damen, A.C. Gossard, W. Weigmann, T.H. Wood and C.A. Burrus, Phys. Rev. B32, 1043 (1985).
[2] W.H. Knox, J.E. Henry, K.W. Goossen, K.D. Li, B.J. Tell, D.A.B. Miller, D.S. Chemla, A.C. Gossard, J. English and S. Schmitt-Rink, IEEE J. Quantum Electron., 25, 2586 (1989).

[3] G. Hasnain, K.W. Goossen and W.H. Knox, Appl. Phys. Lett. 56, 515 (1990).

QUANTUM THEORY OF LASER FIELD EFFECTS ON IMPURITY ION RELAXATION IN
CRYSTALS: APPLICATIONS TO TRANSIENT SPECTROSCOPY

P. A. Apanasevich, S. Ya. Kilin, A. P. Nizovtsev and N. S.
Onishchenko

Institute of Physics of the BSSR Academy of Sciences, Leninsky
Prospekt 70, 220602 Minsk, USSR

INTRODUCTION

Theoretical[1-12] and experimental[13-18] studies of optical
relaxation in a strong laser field unambiguously demonstrate i) a
restricted applicability of the Bloch equations for the description of
relaxation, and ii) the necessity of considering for this purpose
numerous details of the dephasing peruturbations which are out of the
Bloch scheme.

According to the classification proposed in Ref. 3 the whole diver-
sity of relaxing quantum systems can be divided into 3 types (Fig. 1)
differing in the manifestation of their relaxation in nonlinear coherent
optical processes:

$\underline{\alpha\text{-systems}}\left(T_1 < T_2 = (K_0 + \frac{1}{2} T_1)^{-1}, K_0 = \text{Re}\int_0^\infty dt <U^t U^o>, U^t$ - is the quantum

system - environment interaction energy); $\underline{\beta\text{-system}}$ $(T_1 > T_2 > \tau_c, \tau_c$ -
is the characteristic damping time of the correlation function $<U^t U^o>$;
and $\underline{\gamma\text{-systems}}$ $(T_2 < \tau_c)$. Theoretical analysis of the experimental data
of [13] on the non-Bloch behavior of free induction decay (FID) in an
impurity ion crystal $Pr^{3+}:LaF_3$ based on different models for the Pr^{3+}
transition frequency fluctuations (models of random telegraph (RT),
Gauss-Markov, and generalized RT stochastic processes)[3-12] permits
classifying $Pr^{3+}:LaF_3$ among the β-systems[3,6]. However, in the framework of β-systems there has been no consistent explanation of the recent
experimental results on the dependence on the laser intensity of
transient hole burning (HB)[17], FID[16], and rotary echo (RE)[15]
in ruby.

In this report we demonstrate the features of the dephasing
manifested in HB, FID and RE in the case of γ-systems and conclude that a
consistent explanation of the recent experiments[15-17] is possible in
the framework of γ-systems. The physical reason for the difference of
relaxation in ruby from that in $Pr^{3+}:LaF_3$ is that unlike the dephasing of
the Pr^{3+} optical transition resulting from the stochastic modulation of
its frequency due to the paired flip-flops of fluorine nuclei spins

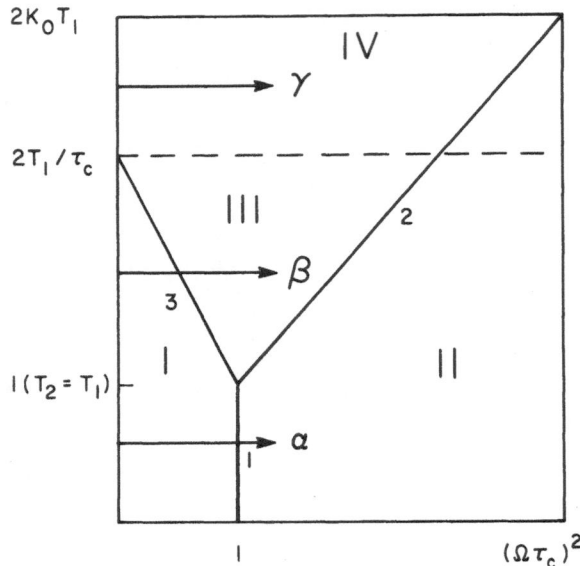

Fig. 1. Classification of relaxing quantum systems in a strong laser field. I - the region of applicability of the Bloch description of relaxation (Bloch region). II - The region of the "suppression" of the dephasing perturbations (Redfield region). III, IV - the regions where effects associated with the non-Markovian nature and nonlinearity of relaxation are the most important; they correspond to strongly (III) and weakly (IV) stochastic perturbations.

connected with the nuclear spin of Pr^{3+} by a weak dipole-dipole interaction, the dephasing of the Cr^{3+} optical transition is induced by a stronger interaction of its nonzero electron spin with the surrounding Al nuclei.

THEORY

As the model of stochastic fluctuations of the resonance optical transition frequency, we use the random telegraph process. The choice of the model is due to the possibility of both exactly describing the fluctuations and preserving the main features of the γ-systems from the point of view this simple model. The starting point of the analysis is the Liouville equation describing the interaction with the high power radiation of a quantum system whose resonance transition frequency fluctuates under the action of the environment: $\Omega_{21}(t) = \omega_{21} + \epsilon_t$, where ϵ_t is a RT process characterized by the frequency of jumps $v(\tau_c = v^{-1})$ and variance σ. On averaging the Liouville equation over realizations of the process ϵ_t we obtain[3] the following equations for the density matrix p_{ij}^t of the quantum system

$$\dot{p}_{22}^t = -\dot{p}_{11}^t = iV(p_{21}^t - p_{12}^t) \tag{1a}$$

$$\dot{\rho}_{21}^t = (\dot{\rho}_{12}^t)^* = -i\epsilon\rho_{21}^t + iV(\rho_{22}^t - \rho_{11}^t) - \int_0^t dx\, \gamma^{t-x}\rho_{21}^x - \int_0^t dx\, \beta^{t-x}\rho_{12}^x, \tag{1b}$$

where

$$\gamma^x = \sigma^2 e^{-\nu x}(2s^2 + c_+^2 e^{-i\Omega x} + c_-^2 e^{i\Omega x})/4, \quad \beta^2 = \sigma^2 e^{-\nu x} s^2 (\cos(\Omega x) - 1)/2 , \quad (2)$$

(2)

and $s = 2V/\Omega$, $c = \epsilon/\Omega$, $\Omega^2 = 4V^2 + \epsilon^2$, $c_+ = 1 \pm c$. Equations (1) do not take into account the spontaneous transition 2-1 since we assume that $t \ll T_1$. The transient solution of Eq. (1) is of the form

$$\rho_{ij}^t = \sum_\alpha e^{p_\alpha} \lim_{p \to p_\alpha} (p - p_\alpha) \rho_{ij}^{[p]} , \quad (3)$$

and is expressed in terms of Laplace transforms of ρ_{ij}^t and the roots p_α of the characteristic equation of Eq. (1), which in this case is reduced to

$$x^3 + 2(\Omega^2 + \sigma^2)x^2 + \left[(\Omega^2 + \sigma^2)^2 + (\Omega\nu)^2 - (2\epsilon\sigma)^2\right]x + (2V\nu\sigma)^2 = 0, \quad (4)$$

where $x = p(p + \nu)$. The six roots p_α break up into pairs

$$p_{\beta\pm} = -\frac{\nu}{2} \pm \left[x_\beta + \left(\frac{\nu}{2}\right)^2\right]^{1/2}$$

($\beta = 0, 1, 2$). Two real roots p_{0+} and $p_{0-} = -\nu - p_{0+}$ are negative and bounded $0 \leq |p_{0+}| \leq \nu$, and the rest are complex-conjugate pairs $p_{1+} = (p_{2+})^*$, $p_{1-} = (p_{2-})^* = -\nu - p_{1+}$. In the majority of the cases being considered a good approximation for the root x_0 of Eq. (4) is given by

$$x_0 \simeq -(2V\nu\sigma)^2 / \left[(\Omega^2 + \sigma^2)^2 + (\Omega\nu)^2 - (2\epsilon\sigma)^2\right] ,$$

which is valid for $x_0 \ll \Omega^2 + \sigma^2$ or $x_0 \ll \nu^2$. The weights of the contributions associated with different roots differ among various transient optical phenomena. Calculations show that the contribution due to the root p_{0+} is dominant in describing the HB and FID, whereas the roots p_{1+} and p_{2+} determine the RE-signal.

COMPARATIVE ANALYSIS OF TRANSIENT OPTICAL EFFECTS

Due to the consideration of the γ-systems we had to reinterpret the information given by different optical transients. First of all we illustrate the relaxation of γ-systems using two simple examples of field independent transients.

The amplitude of the echo signal arising at the time 2τ is proportional to the characteristic functional

$$\Psi(\epsilon_\tau) = \left\langle \exp\left\{-i\int_0^\tau dx \epsilon_x + i\int_\tau^{2\tau} dx \epsilon_x\right\}\right\rangle$$

of the dephasing perturbations ϵ_τ. In the case of the RT, it is possible to calculate this functional exactly and to obtain the amplitude of the echo at time 2τ

$$A^{2\tau}_{echo} = A_o e^{-\nu\tau}\left[\left(2\frac{\sigma}{\eta}\right)^2 - \frac{\nu}{\eta}\left(sh(\eta\tau) + \frac{\nu}{\eta} ch(\eta\tau)\right)\right], \qquad (5)$$

where $\eta = (\nu^2 - 4\sigma^2)^{1/2}$. Within the Bloch limit $\sigma\to\infty$, $\nu\to\infty$, $\lim\sigma^2/\nu = K_o$) expression (5) yields a standard result: $A^{2\tau}_{echo} = A_o\exp(-K_o 2\tau)$. But if $\sigma^2/\nu^2 = k \geq 1/4$ (for the involved samples of Cr^{3+}: Al_2O_3 $k = 20$, see below) $A^{2\tau}_{echo}$ becomes oscillating (see also [18]) and the decay occurs with the rate $\nu/2$.

Another example illustrating the relaxation of γ-systems is polarization decay. Calculations show that for the RT model the solution of Eqs. (1) for $V = 0$ is given by

$$\rho^t_{21} = \rho^0_{21} e^{-\nu t/2}\left[ch(\eta t/2) + \frac{\nu}{\eta} sh(\eta t/2)\right]. \qquad (6)$$

In the Bloch limit we again have the standard result $\rho^t_{21} = \rho^0_{21}\exp(-K_o t)$, while for $k \geq 1/4$ the decay rate is $\nu/2$. Thus in the case of a γ-system, the role of the time $T_2(K_o^{-1})$ from the point of view of the RT model is played by the time $2\tau_c = (\nu/2)^{-1}$. Therefore, in analyzing the experiments of Refs. (15)-(17) according to the echosignal decay time 15 μs [18] we assumed $\tau_c = 7.5$ μs. The second parameter of the RT, its variance σ, remains unknown and is to be determined by fitting the theoretical dependences to experimental data.

1. Transient Hole Burning

The shape of a hole burned in an inhomogeneously broadened line is determined by the excited state population $\rho^T_{22}(\epsilon)$ induced by the laser field at time T. In the case of the RT model the hole shape for a relatively weak laser field $2V \ll \sigma$ and for $\nu \ll \sigma$ has a two peaked-form with the peaks at frequencies $w_{21}\pm\sigma$. As the radiation intensity increases, these two spectral components are field broadened and at $2V \stackrel{\sim}{=} \sigma$ they begin to overlap. These general considerations are confirmed by the results of numerical calculations of the shape of the holes $\rho^T_{22}(\epsilon)$ by Eqs. (1)-(4)[*]. Figure 2 shows the dependence of the hole halfwidth Γ_{HB} (HWHM) vs Rabi frequency $(2\pi)2V$ calculated at $T = 400$ μs and $\tau_c = 7.5$ μs for different values of $k = \sigma^2/\nu^2$. It is seen that in the region of large V the ratio $k = \sigma^2/\nu^2 = 20$ gives a satisfactory

---

[*] It should be noted that the exact (for the RT model) non-autonomous Eqs. (1) lead to a modulation of the hole shape as opposed to the consideration based on the autonomous (t >> τ_c) equations used in Refs. (15-17).

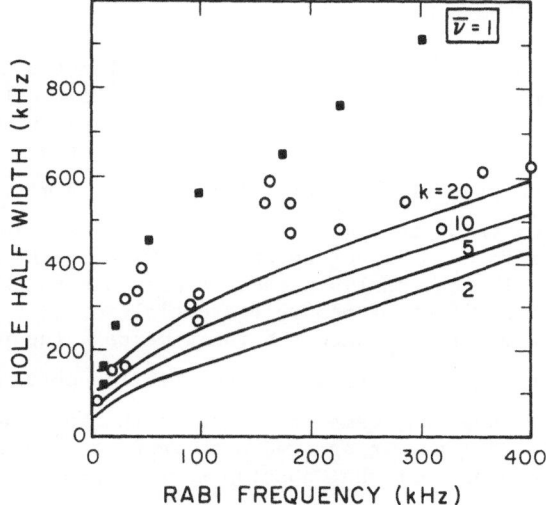

Fig. 2. Optical linewidth of the prepared hole vs Rabi frequency. Theoretical curves are plotted for the random telegraph model assuming $\nu x(7.5\mu s) = 1$, and $k = 2, 5, 10, 20$. Experimental points are taken from Ref. 17.

coincidence with the experimental data of (17). Note the asymptotic limit $\Gamma_{HB} \approx 2V$ for large V. In the region of relatively small V, however, there are disagreements with the experiment. These disagreements are associated with the RT model which does not take into account the real spectrum of the dephasing interactions in ruby. Indeed, the Cr^{3+} ion in ruby is connected by superhyperfine (SHF) interaction with the surrounding Al nuclei owing to which its resonance levels consist of sublevels i with a dense distribution of their energies $E_1+\epsilon_i^{(1)}$ and $E_2+\epsilon_i^{(2)}$. Transitions between these sublevels analogous to the jumps in the RT can take place owing to the part of the SHF interaction Hamiltonian nondiagonal in the quantum states of the Al nuclear spins, or they can be induced by the interaction of the Al nuclei closest to Cr^{3+} forming its frozen core with the Al nuclei in the bulk of the Al_2O_3 crystal. This two-band model under the assumption of the absence of jumps within the bands yields

$$\rho_{22}^T(\epsilon) = 4V^2 \sum_i W_i(1-\cos(\Omega_i T))/\Omega_i^2 , \qquad (7)$$

where $\Omega_i = \left[(\epsilon+\epsilon_i)^2 + 4V^2\right]^{1/2}$, $\epsilon_i=\epsilon_i^{(2)}-\epsilon_i^{(1)}$, W_i- is the distribution of the energy ϵ_i in the bands. Proceeding in (7) from summation to integration and regarding the distribution W_i as the Lorentzian $W_i = (\Gamma_o/\pi)/(\Gamma_o^2+(\epsilon_i)^2)$ we can show that the hole shape is approximately (for $2VT \gg 1$) of the form $\rho_{22}^T(\epsilon) = 2V(2V+\Gamma_o)/\left[(2V+\Gamma_o)^2 + (\epsilon)^2\right]$ with the width $2V + \Gamma_o$, which is consistent with the RT model assuming $\sigma \approx \Gamma_o$ in the case of a γ-system. The presence of jumps within the distribution W_i will obviously lead to a narrowing of the hole due to the effect of motional narrowing for the closest sublevels, and as a result to deviations from the RT model. For large values of V when the hole broadening is caused mainly by the effect of saturation, the specific structure of the spectrum of dephasing interactions in ruby becomes unessential and the RT model predicts correct results with a proper choice of the parameters σ and ν.

2. Free induction decay

Usually the FID is studied by means of observing the decay beat signal of the polarization field and of the laser field after switching its frequency out of resonance. The theoretical analysis of the FID signal at time $t + T$, where T is the laser pulse duration, was carried out in the following way. The polarization ρ_{21}^{t+T} for time $t > 0$ was calculated on the basis of the Eqs. (1)-(4) with a step field $V(x)$ different from zero for time x belonging to $[0,T]$. Unlike the analysis carried out in Ref. 16, our analysis made it possible to take into account correctly the effect of memory (essential for γ-systems) in dephasing for the times $t < \tau_c$ when the ion remembers the previous action of the already switched-off radiation. The solution obtained for ρ_{21}^{t+T} was averaged over ϵ (over the Gaussian shape of the inhomogeneously broadened transition 21) and integrated with respect to V from zero to V_0 in order to take into account the Gaussian spatial profile of the laser beam.

Analysis of the general expression for the FID signal amplitude has shown that for $\sigma^2 \gg \nu^2$ it assumes the simple form

$$A_{FID}(t) = \frac{e^{-\nu t/2}}{At^2}\left[e^{-\nu t/2}(1+\nu t/2)-e^{-\mu t}(1+\mu t)\right], \qquad (8)$$

$$\mu = \left[4v_0^2+\nu^2\left(1+4v_0^2/(\sigma^2+4v_0^2)\right)/4\right]^{1/2} \approx \left[4v_0^2+\nu^2/4\right]^{1/2}, \quad A = (\mu^2-\nu^2/4)/2.$$

In addition in the general expression for $A_{FID}(t)$ there is a small (proportional to k^{-1}) oscillatory term describing the weak modulation of the FID signal given by (8) (see also Ref. 19). Note that according to (8) at a specified value of the Rabi frequency $2V_0$ the temporal behavior of the FID is determined by the jump frequency ν and is practically independent of σ. The latter result is associated with the fact that for $\sigma^2 \gg \nu^2$ the dephasing becomes quasireversible, i.e. associated with the quasi-inhomogeneous distribution of the quantum system over frequencies $\omega_{21} \pm \sigma$ which does not show up in the FID signal due to the averaging over a wide inhomogeneous distribution of frequencies ω_{21} in ruby crystal. A similar independence of the FID signal from the values of ϵ_i and on the distribution W_i is also obtained within the framework of the two-band model in the absence of jumps between sublevels. Figure 3 shows the results of calculations on the basis of (8) of the time behavior of the FID signal for three values of the Rabi frequency made at $\tau_c = 7.5\ \mu s$ ($\nu = 21.1$ kHz) and $k = \sigma^2/\nu^2 = 20$. It is obvious that Eq. (8) describes the experimental data of Ref. 16 quite satisfactorily. Especially good agreement exists in the case of large Rabi frequencies.

3. Rotary Echo

As opposed to HB and FID, the rotary echo signal arising at time $t \approx 2T$ as result of the change of the phase of the strong strong field by π at the time T is described by the roots $p_{1\pm}$ and $p_{2\pm}$ of the characteristic Eq. (4), and $p_{1-} \approx i\Omega-\nu/2-G$, $G = ((\nu/2)^2 + x_0 -(2c\sigma^2)^2)^{1/2}$, where x_0 is determined by the relation given after (4). The RE amplitude in the strong field limit ($\Omega \gg \nu,\sigma$) is of the form

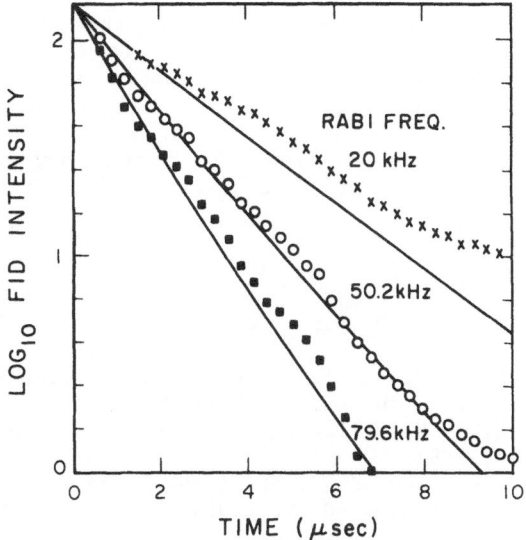

Fig. 3. Free induction decay intensity $(A_{FID}(t))^2$ vs time, where $A_{FID}(t)$ is obtained from Eq. (8) for different Rabi frequencies assuming $(\nu/2)^{-1} = 15\mu s$, k = 20. Experimental points are taken from Ref. 16.

$$A_{RE}(t) = \int d\epsilon \ \sin\left[\Omega(t-2T)\right]\bar{A}_{RE}(\epsilon,V,t,T), \qquad (9)$$

where $\bar{A}_{RE}(\epsilon,V,t,T)$ is the rotary echo "spectrum":

$$\bar{A}_{RE}(\epsilon,V,t,T) = -4i\left(\frac{V}{\Omega}\right)^3 \frac{1}{G^2}\left\{\left[(1+G)^2+4c^2k\right]e^{Gt}+\left[(1-G)^2+4c^2k\right]e^{-Gt} - \right.$$
$$\left. -2(1-G^2+4c^2k)\,\mathrm{sh}\left[G(t-2T)\right]\right\}e^{-\nu t/2} . \qquad (10)$$

It follows from (10) that depending on the sign of $g = (\nu/2)^2 + x_0 - (2c\sigma)^2$, the decay rate of the RE "spectrum" is different: for $g < 0$ ($\epsilon > \epsilon_0 \approx V\nu/\sigma$) it is equal to $\nu/2$, and for $g > 0$ it varies from $\nu/2$ at $\epsilon = \epsilon_0$ to $-2x_0/\nu \approx 2\nu\sigma^2(2V/(4V^2+\sigma^2))^2$ at $\epsilon = 0$. Consequently, the signal $A_{RE}(t)$ will be biexponential with two decay rates $\gamma_1 = \nu/2$ and $\gamma_2 \approx \nu\sigma^2/2V^2$. The weight of the first exponent will predominate, since the region of integration with respect to ϵ in (9) contributing to the second exponent is much smaller for $k \gg 1$ than the region corresponding to the first exponent. This conclusion is supported by numerical calculations. It follows from the foregoing that the rotary echo decay in the case of γ-systems is practically independent on the Rabi frequency, as was observed in Ref. 15 for ruby.

CONCLUSIONS

It is shown that the dephasing of quantum states in the presence of a slow stochastic modulation of the frequency of optical transitions similar in nature to the reversible dephasing in inhomogeneously broadened systems displays a number of characteristic features which manifest themselves in nonlinear optical transients: i) the width of the HB is determined by the width of the resonance optical transition frequency distribution formed during the slow stochastic modulation and by the Rabi frequency; ii) the time dependence of the FID is universal, it is not associated with the distribution of the possible values of the resonance optical transition frequency, and is determined by the Rabi frequency and the frequency of jumps between these values; iii) the RE signal does not depend on the Rabi frequency and is determined only by the jump frequency. The facts permit concluding that the features observed in the experiments of Refs. 15-17 in ruby can be explained on the basis of the idea about $Cr^{3+}:Al_2O_3$ as a system with a slow and strong modulation of the transition frequency.

We thank J. H. Eberly and P. Berman for turning our attention to the recent papers by A. Szabo. We are also grateful to A. Szabo for sending us his work [17].

REFERENCES

1. P. A. Apanasevich and A. P. Nizovtsev, Sov. J. Quant. Electr. <u>5</u>, 895 (1975).
2. P. A. Apanasevich, Fundamentals of the theory of light-matter interaction, Nauka i teckhnika, Minsk (1977).
3. P. A. Apanasevich, S. Ya. Kilin and A. P. Nizovtsev, J. of Appl. Spectr. <u>47</u>, 1213 (1988).
4. E. Hanamura, J. Phys. Soc. Jpn., <u>52</u>, 1269 (1983).
5. J. Javanainin, Opt. Commun. <u>50</u>, 26 (1984).
6. P. A. Apanasevich, S. Ya. Kilin, A. P. Nizovtsev, and N. S. Onishchenko, Opt. Commun. <u>52</u>, 279 (1984).
7. M. Yamanoi and J. H. Eberly, Phys. Rev. Lett. <u>52</u>, 1353 (1984).
8. A. Schenzle, M. Mitsunaga, R. G. DeVoe, and R. G. Brewer, Phys. Rev. A<u>30</u>, 325 (1984).
9. P. R. Berman and R. G. Brewer, Phys. Rev. A<u>32</u>, 2784 (1985).
10. K. Wodkiewicz and J. H. Eberly, Phys. Rev. A<u>32</u>, 2784 (1985).
11. P. R. Berman, J. Opt. Soc. Am. B<u>3</u>, 564 (1986); <u>3</u>, 572 (1986).
12. P. A. Apanasevich, S. Ya. Kilin, A. P. Nizovtsev and N. S. Onishchenko, J. Opt. Soc. Am. B<u>3</u>, 587 (1986).
13. R. G. De Voe and R. G. Brewer, Phys. Rev. Lett. <u>50</u>, 1269 (1983).
14. T. Endo, T. Muramoto, and T. Hashi, Opt. Commun. <u>51</u>, 163 (1984).
15. T. Muramoto and A. Szabo, Phys. Rev. A<u>38</u>, 5928 (1988).
16. A. Szabo and T. Muramoto, Phys. Rev. A<u>39</u>, 3992 (1989).
17. A. Szabo, T. Muramoto, and R. Kaarli (to be published).
18. A. Szabo, J. Opt. Soc. Am. B<u>3</u>, 514 (1986).
19. S. Nakanishi, O. Tamura, T. Muramoto, and T. Hashi, Opt. Commun. <u>31</u>, 344 (1979).

MODE LOCKED SEMICONDUCTOR LASERS

P. A. Morton, D. J. Derickson, R. J. Helkey, A. Mar and J. E. Bowers

Department of Electrical and Computer Engineering
University of California
Santa Barbara, CA 93106.

I   INTRODUCTION

Mode locked semiconductor laser diodes offer the possibility of producing small, cheap and reliable sources of stable subpicosecond pulses over wide wavelength ranges and with moderate peak powers. They can be used in telecommunication systems for time division multiplexing or for high bit rate systems using an external modulator. Semiconductor lasers are ideal candidates for use in practical commercial electro-optic sampling systems and optical analog to digital converters due to their small size, low cost, low noise and small timing jitter in comparison to the use of the more complex and less reliable sources such as pulse compressed YAG lasers.

Short pulses have been obtained by both active [1-4] and passive [5-7] mode locking of semiconductor lasers. The shortest pulses to date have been achieved from active mode locking. Using active mode locking and a high speed laser diode capable of being modulated efficiently at 16 GHz, the results in section II were found. They show a pulse width of 580 fs, assuming a $sech^2$ pulse shape. There are, however, multiple peaks in the autocorrelation trace, due to a multiple pulse output from the laser, which is seeded by reflections from the imperfect anti-reflection (AR) coating on the laser. The multiple pulse phenomena is due to a mechanism called 'dynamic detuning' which is described in the theoretical section III.

One way to overcome the problem of reflections from the anti-reflection (AR) coated facet, and at the same time remove the need for an external cavity is to utilize monolithic structures [7-9]. The integrated structure shown in section IV is constructed from a single continuous waveguide so that multiple pulses do not build up from internal reflections, and has produced pulses as short as 1.4 ps with no signs of a multiple pulse output. Monolithic devices could be fabricated to operate in a way analogous to the colliding pulse modelocked (CPM) dye laser (which has produced the shortest mode locked optical pulses to date), although in a linear configuration, unlike the more generally used form of CPM dye laser which uses a ring structure.

In some situations a low modelocking frequency is necessary, thus causing a major problem when using sinusoidal drive waveforms. As the frequency is reduced, so the effective applied electrical waveform increases in pulsewidth - thus giving much longer optical pulses. Step recovery diodes (SRD's) can be used, however, for lower frequencies the pulsewidth attained from a typical SRD is quite broad (~100 ps). 'Self Modelocking' is one way to overcome the need for a microwave signal source (synthesizer) [10]. This is a new modelocking technique in which the optical output of a modelocked laser is fed back and used to drive the modelocked laser. Experimental work carried out on self modelocking is described in section V.

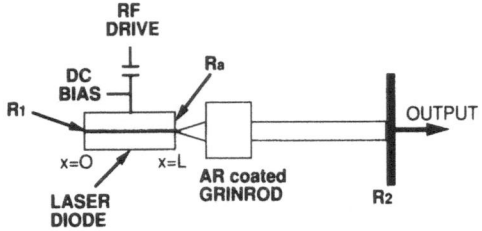

Fig. 1 Schematic of the linear cavity modelocking arrangement.

II   ACTIVE MODE LOCKING

This section describes results obtained from active mode locking a high speed laser diode in an external cavity [2]. Fig. 1 shows a schematic of the linear cavity arrangement. The laser output is coupled with an AR coated graded index lens to an external cavity in air which terminates in a 77% reflectivity dielectric mirror. The optical output is taken from this mirror. The semiconductor laser is AR coated on one facet and has a high reflectivity (70%) on the other. The laser is 260 µm long, with a round trip cavity time of 7 ps. The laser is biased around threshold and a high power, high frequency sinusoid is applied to the device.

Several features are necessary to achieve short pulses. The first is high coupling to the external cavity. The second feature is good light-current characteristics. The laser must be driven with a large rf drive to achieve short pulses. If the light-current curve rolls over due to turn-on of parasitic current blocking junctions, then high current through the active layer is not possible. Perhaps the most important factor is a large modulation bandwidth. If the laser has significant parasitics, it may still be possible to mode lock it at high frequencies, but it requires correspondingly more rf power, which may damage the laser, and will cause the chip temperature to rise. The final important feature is frequency stability of the microwave oscillator driving the laser. To obtain short pulses a low phase noise microwave synthesizer is needed. With these provisions, the pulse shapes shown in Fig. 2 are repeatably obtained. The main pulse has an autocorrelation full width at half maximum (FWHM) of 890 fs, and the different longitudinal modes of the original laser diode cavity are locked together, as can be seen in the optical spectrum. This corresponds to a 580 fs pulsewidth, assuming a $sech^2$ pulse shape. The shortest pulses were obtained with the laser biased near threshold, and the average output power was 0.5 mW, with a peak power of 30 mW. Much higher average output powers, as much as an order of magnitude higher, were obtained, but with much broader pulses, typically 5 ps.

Two features of these results are surprising. First, the pulse widths are an order of magnitude shorter than have previously been obtained with active mode locking. The main reason is the shortness of the current pulses passing through the laser. The period is 62.5 ps, and the time when the total cavity gain is within a factor of two of the peak gain is under 20 ps. Consequently, with a high frequency, low capacitance laser where significant modulation of the gain occurs at microwave frequencies, a large amount of pulse shaping occurs per pass. Another feature of Fig. 2 is surprising, namely, if the AR coating is really 0.5%, then why is the trailing pulse one-half as big as the first pulse? Extensive modeling has been carried out, discussed in the following section, of active mode locking in semiconductor lasers to answer this question.

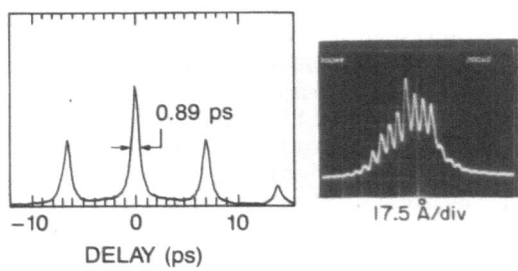

Fig. 2 Second harmonic autocorrelation trace and optical spectrum for mode locking at 16 GHz.

III   MODELING OF ACTIVE MODE LOCKING

All results published to date showing mode locking of semiconductor lasers with pulse widths below one picosecond have a second harmonic intensity autocorrelation trace with multiple peaks. These multiple peaks are always spaced in time by the round trip time of the laser diode which is typically 5-10 picoseconds. The extra peaks are therefore built up from reflections internal to the laser diode, so the small reflectivity of the AR coated facet still contributes to the output waveform. This kind of multiple pulse phenomena cannot be explained by previous theories for active mode locking of laser diodes, which make the approximations of a sinusoidal gain waveform and low modulation depth.

The following theoretical model [11] uses the traveling wave approach to include spatial variations of the electron and photon densities within the laser diode, together with a non-zero reflectivity for the AR coated facet. The theoretical model is based on the traveling wave rate equations for electron density $N(x,t)$ and forward and backward travelling photon fluxes $S^+(x,t)$ and $S^-(x,t)$:

$$\frac{\partial N}{\partial t} = \frac{J(t)}{ed} - \frac{N}{\tau_n} - g(N - N_t)(S^+ + S^-) \tag{1}$$

$$\frac{\partial S^{\pm}}{\partial t} \pm v_g \frac{\partial S^{\pm}}{\partial x} = \Gamma g(N - N_t) S^{\pm} - \alpha_i S^{\pm} + \frac{\Gamma \beta M N}{\tau_n} \tag{2}$$

where $J(t)$ is the applied current density waveform, e the electronic charge, d the active layer thickness, τ_n the electron lifetime, g the differential gain coefficient, N_t the electron transparency density, v_g the group velocity, Γ the confinement factor, α_i the internal loss, β the spontaneous emission coupling into each external cavity mode, and M the number of external cavity modes oscillating (given by M= Round trip time / Pulse width). These rate equations are integrated numerically using finite difference approximations, with boundary conditions at the laser facets of:

$$S^+(0,t) = R_1 S^-(0,t) \tag{3}$$

$$S^-(L,t) = R_A S^+(L,t) + R_2 C^2 S^+(L,t-\tau_{ext}) \tag{4}$$

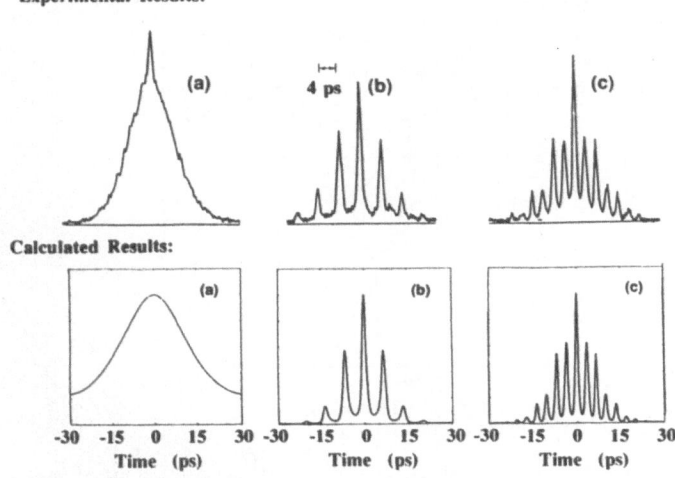

Experimental Results:

(a)   4 ps (b)   (c)

Calculated Results:

(a)   (b)   (c)

-30 -15   0   15   30 -30 -15   0   15   30 -30 -15   0   15   30
       Time (ps)   Time (ps)   Time (ps)

Fig. 3 Comparison of experimental and theoretical autocorrelation traces with increasing levels of RF current.

where R_1 and R_2 are the power reflectivities of the high reflectivity mirrors, R_A the power reflectivity of the AR coated facet, C the coupling from laser to external cavity, and τ_{ext} the external cavity round trip time. The applied current density waveform J(t) consists of a d.c. bias plus a large sinusoidal component at the resonance frequency of the combined cavity.

Experimental results and calculated autocorrelation results from the simulations are compared in Fig. 3, for three different levels of r.f. current. For a low r.f level, a broad peak is seen with a FWHM of over 10 ps. As the r.f level is increased, a subpicosecond multiple peak trace is found, with the spacing between peaks being the round trip time of the laser diode. If the r.f. level is increased further a second set of peaks is seen. The calculated autocorrelation traces show very good agreement with the experimental results at all three r.f. current levels, giving confidence to the theoretical model used in the simulations.

The build up of a mode locked pulse train over many round trips is shown in Fig. 4. The first trace shows the applied current density waveform during one modulation period, which is clipped in the negative direction by the laser diode. The two other traces show the electron density and optical output waveforms at the AR coated facet of the laser diode after 25, 75 and 1000 periods of modulation. Initially a gain switched pulse is produced, which travels around the external cavity and returns to seed the output of the next modulation period. After 25 periods the output pulse is still fairly broad (14 ps), and has a peak power of only 3 mW. As the pulse builds up in power the front edge starts to deplete the carriers (and therefore the gain) as it passes through the laser diode, and so the trailing edge sees less gain and reduces in power. After 75 round trips this process is occurring, the effect being to move the pulse to an earlier point in the modulation period. As this initial pulse moves earlier in the modulation period it moves away from the peak in the gain waveform, which would occur at the center of the modulation period. Therefore, the gain can rise up again after the initial pulse passes through the laser, so the small reflections from the AR coated facet can be amplified. These reflections build up after many passes through the laser diode until they become much larger than the initial reflection. After steady state conditions have been reached (1000 periods), a subpicosecond pulse train is seen. This has a powerful initial pulse followed by pulses seeded by the reflections off the AR coated facet, which are separated by the round trip time of the laser diode. The initial pulse has moved to an earlier point in the modulation period. A three dimensional plot

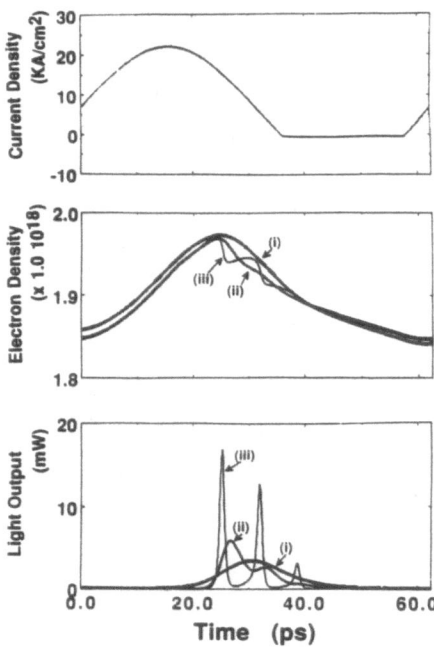

Fig. 4 Waveforms over one period of applied current density, electron density at the AR coated facet and light output at the AR coated facet after (i) 25 round-trips. (ii) 75 round-trips and (iii) 1000 round trips.

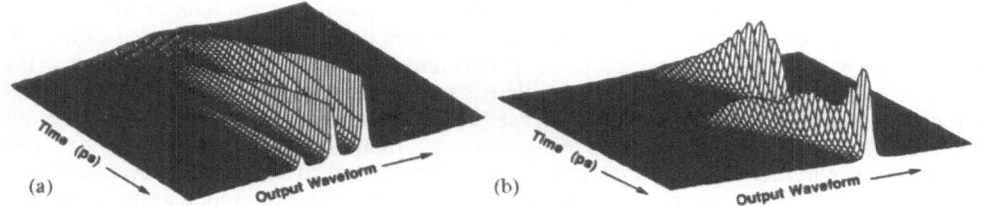

Fig. 5 Three dimensional plot showing the build up of the mode-locked output waveform
for (a) 0.5 % reflectivity AR coating and (b) perfect AR coating.

of the build up of the mode-locked output is shown in Fig. 5a. If the r.f. current is increased to higher levels than in Fig. 4, eventually the gain will rise up sufficiently between the initial and first reflected pulses for a separate mode locked pulse to be sustained between them. This new mode locked pulse will itself cause reflections from the AR coated facet and so a second set of pulses will build up in the output waveform. There is no particular time separation between the two sets of pulses, in fact the time difference changes as parameters such as the r.f. current are varied. For very high levels of r.f. current, many sets of pulses are observed in the output waveform.

We have called the pulse stabilization mechanism 'Dynamic Detuning' as it is a dynamic process which detunes the position of the pulse away from the original peak in the gain waveform. This allows small reflections to be amplified and build up. The dynamic detuning process has only one stable solution which defines the pulse width, peak power and shape of the pulse. This mechanism is a limiting factor on the minimum pulse width achievable from mode locked semiconductor lasers, independent of the finite gain bandwidth of the laser material and dispersion in the cavity. The dynamic detuning mechanism accounts for the long (580 fs) pulses seen from mode locked semiconductor laser diodes compared to the theoretical limits of about 50 fs for the gain linewidth and 100 - 200 fs for dispersion.

The effects of varying the r.f. current on the main pulse parameters are shown in Fig. 6. For a low r.f. current, broad pulses of 10 to 20 ps are seen. These pulses have a low peak power and occur near the center of the modulation period. As the r.f. is increased, the pulses become narrower with a corresponding increase in peak power. The pulses can be seen to move towards an earlier position in the modulation period. For an r.f. current of 40 mA, subpicosecond pulses are seen, with a peak power of about 20 mW. It should be recognized that for a modulation frequency of 16 GHz, as in this case, it is not a trivial problem to achieve 40 mA peak r.f. current through the active region of a laser diode.

The effect of having a perfect AR coated facet has been modeled, as it has been thought that such a device will provide the shortest pulses. If the reflectivity of the AR coating is zero, no reflected pulses occur. Fig. 5b shows a three dimensional plot of how the output waveform can build up in such a case. An initial mode locked pulse starts to build up, becoming shorter and more powerful and so moving to an earlier point in the modulation period. As it moves, the

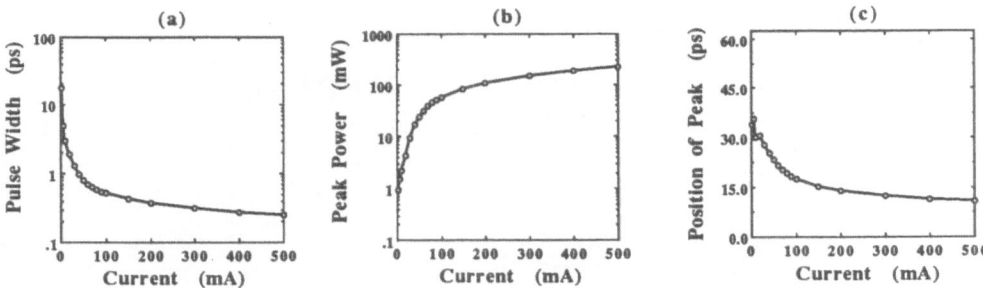

Fig. 6 (a) Pulsewidth and autocorrelation width, (b) peak power and (c) pulse position
in the modulation period, versus RF current.

gain starts to rise to higher levels later in the period and eventually a second mode locked pulse starts to build up. This second mode locked pulse increases in power at the expense of the first, so that it moves to an earlier position in the period and takes over from the first pulse. This whole process repeats itself as the pulses oscillate in position around the modulation period, and the output is unstable. However, in such a case the second harmonic intensity autocorrelation trace will show a stable single peak of a few picoseconds in duration. The timing jitter between these pulses is extremely large so that their usage in most practical systems is prevented. This instability may be overcome under certain conditions by detuning the modulation frequency slightly or by increasing the cavity round trip time to be much longer than the carrier recombination time. This decouples the effects of the optical output on subsequent carrier density waveforms. Simulations show that the unstable output seen for devices with perfect AR coatings can also occur for coatings with power reflectivities of less than 0.1%. The subpicosecond multiple pulse output is seen for AR coating reflectivities of between 0.1% and 3%. For coatings of 5% and more, much broader pulses (20 -30 ps) are predicted.

The secondary pulses can be considered to be very high frequency mode locked relaxation oscillations (>140 GHz). In the case of a stable output with perfect AR coatings, the time difference between the peaks will vary with operating parameters, however for finite AR reflectivities (0.1% - 3%) these relaxation oscillation peaks are seeded by the small reflections from the AR coated facet.

IV   MONOLITHIC MODE-LOCKED DEVICES

Monolithic modelocked structures overcome the need for an external cavity by the use of a long integrated waveguide [8,9]. The integrated structure described here [9] is constructed from a single continuous waveguide so that multiple pulses should not build up from internal reflections. The device has a gain region integrated with an active waveguide and saturable absorber to provide pulse shortening by both active and passive mode locking.

The monolithic modelocked lasers are fabricated from semi-insulating, channeled substrate, buried heterostructure 1.3 µm GaInAsP laser material. Each device is fabricated from a single continuous optical waveguide split into three functionally separate regions by separate contact pads. A schematic diagram and photograph of an array of these devices is shown in Fig. 7. The device consists of a 75 µm gain section, a 2300 µm active waveguide and a 75 µm saturable absorber. Each section is separated by a 50 µm spacer region which typically provides 5 kΩ of isolation between contacts. The threshold current of these devices is typically between 70 mA and 80 mA and is uniform across an array of devices. This low value of threshold current for such a long device can be attributed to an extremely high quality optical waveguide with low scattering losses.

The curves for light output versus current in the active waveguide are shown in Fig. 8 for various bias levels in the gain section. The saturable absorber section is left unconnected to give the required characteristics, that is, it provides a loss until it is optically pumped up to transparency and then the loss falls to a very small value. The curves show threshold currents of around 70 mA, depending on gain section bias, with hysteresis in the traces due to the saturable absorption. Low values of gain section bias vary the loss in this section, and change the optical output considerably more than for a similar current variation in the active waveguide.

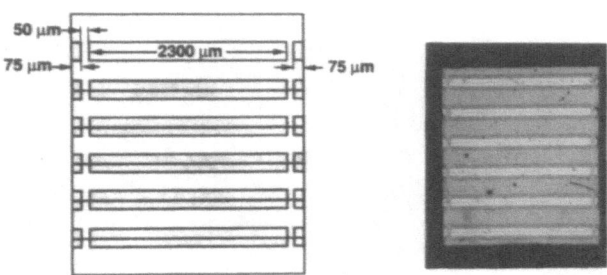

Fig 7. Schematic diagram and photograph of an array of hybrid mode-locked monolithic devices.

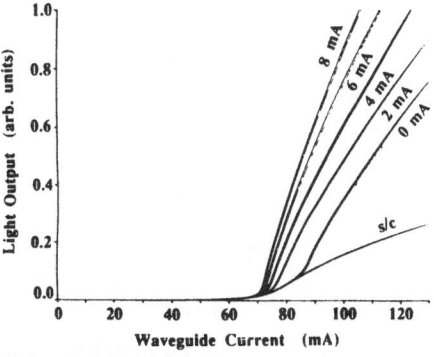

Fig. 8 Light output vs current in the active waveguide section for various gain section bias levels.

By biasing the device with a small or zero current in the gain section it is possible to use this mechanism to provide more efficient modulation of the device.

The device is biased with a d.c. current to the waveguide and a d.c. current plus r.f. modulation is applied to the gain section. The output is taken from the saturable absorber end of the device. The optical output is focussed onto a high speed GaInAs p-i-n detector which is connected to either a spectrum analyser (0 - 22 GHz) or a 20 GHz sampling oscilloscope. Small signal microwave measurements of the frequency response of these devices show a very large peak at 15 GHz which is the reciprocal of the round trip time of the device. This peak at 15 GHz was measured to be only 12 dB below the response at 50 MHz, while away from the peak the response falls down by over 27 dB. This indicates a strong feedback from the integrated cavity. Under d.c. bias conditions, the microwave spectrum shows a noise peak at the reciprocal of the overall cavity round trip time. This noise peak is shown in Fig. 9, together with a trace in which +10 dBm modulation is applied to the device at the frequency at which the noise trace peaks. The linewidth of the noise peak is around 10 MHz, compared to an instrument limited linewidth measured for the peak under modulation, of under 30 Hz. The noise peak is suppressed by the presence of the strong modulation signal so that the overall noise level becomes flat for high levels of r.f. power.

A noncolinear second order intensity autocorrelation technique is employed in order to measure the width of the optical pulses. Fig. 10 shows a second harmonic intensity trace versus the time delay in one arm of the measurement. This trace shows a FWHM of 2.17 ps, which corresponds to a pulsewidth of 1.4 ps assuming a $sech^2$ pulse shape. The average output power from the device is typically 0.5 mW, giving peak powers of about 20 mW. The optimum biasing conditions necessary to achieve short pulses were a zero bias current in the gain section, and a bias current well above transparency (110 mA) in the waveguide section. In devices with low isolation, the shortest pulses were achieved. This may be due to a traveling wave effect of the r.f. modulation in which the optical pulse is also partially shaped while passing under the active waveguide. In this case, lower values of r.f. power (+12 dBm) were necessary to produce short pulses than for devices with high values of isolation. Results for devices with high isolation between the waveguide and gain section showed pulsewidths down to 3.5 ps for 27 dBm applied r.f. power.

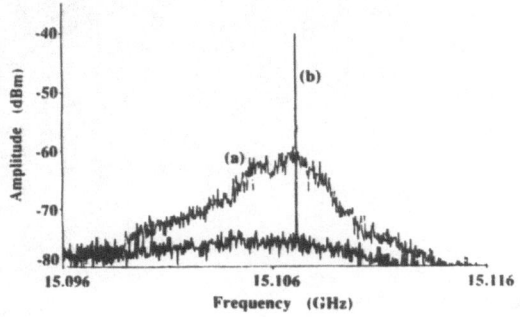

Fig. 9 Microwave noise spectrum around the cavity frequency for (a) no modulation and (b) 10 dBm modulation.

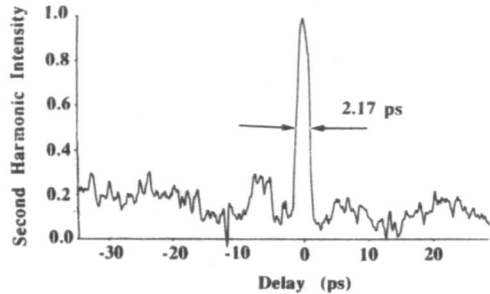

Fig. 10 Second Harmonic Autocorrelation trace for hybrid mode-locked monolithic device

V   SELF MODELOCKING

In this section a new mode-locking technique, self mode-locking, is described [10]. This uses the detected short optical pulses from the mode-locked laser as the active driving source. This technique forms narrow-width mode-locked optical pulses at low repetition rates without the use of a microwave synthesizer. Active modelocked semiconductor lasers have produced optical pulses as short as 580 fs (section II). This result was achieved using the positive portion of a high frequency sinusoid as an electrical drive signal. The sinusoid used in the experiment was at 16 GHz. For many applications of modelocked lasers it is more useful to have the mode-locked optical pulses at a much lower repetition rate. This requires an electrical drive source that produces short electrical pulses at this lower rate. Step recovery diodes can be used, but they typically have pulse widths up to 100 ps. Since the mode-locked optical pulses themselves are very short, they are good candidates to drive the mode-locking action. Self mode-locking requires a high speed optical to electrical (O/E) converter in a positive feedback configuration to convert the output optical pulses back into electrical drive signals in a regenerative process. Advances in high speed electrical components allow the short optical pulse to create a short electrical drive pulse.

A block diagram of the self mode-locked semiconductor laser is shown in Fig. 11. A semiconductor laser is placed in an external cavity (ring geometry used in this case) with a round trip delay time of 2 ns. The laser is a high speed 1300 nm SIPBH laser with anti-reflection coatings on both facets. The light from the laser is collimated by the use of two AR coated GRIN lenses. The optical to electrical converter consists of a high speed PIN photodetector and a 4 stage MESFET distributed amplifier. The O/E converter has an overall responsivity of 20 amps per watt. The impulse response of the O/E converter was measured to be a 50 ps FWHM pulse with a peak amplitude of 3.5 volts into a 50 Ω system.

The laser is biased above the CW lasing threshold so that counter-propagating optical signals start to build up in the ring cavity. One of the two output signals is fed into the O/E converter which converts it into an electrical drive signal which is applied to the direct modulation input of the laser. If the round trip delays of the optical cavity and the signal path

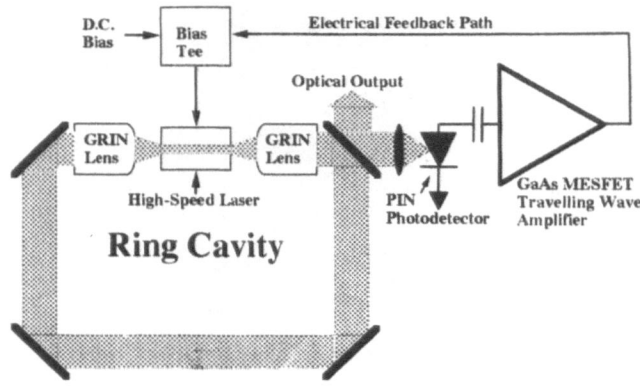

Fig. 11 Block diagram of self-mode-locking cavity geometry.

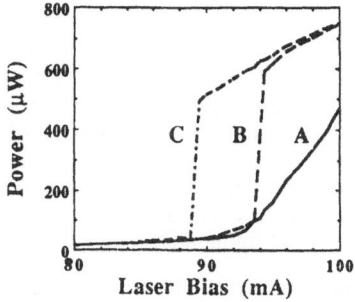

Fig. 12 Light output vs current for the self mode-locked laser (A) without feedback, (B,C) with feedback.

through the O/E converter are properly chosen, a regenerative process will build up a mode-locked pulse in the laser. In order for self mode-locking to occur, the loop gain of the electrical feedback path must be greater than one. The other condition necessary for self-oscillation is that the optical path time delay, T_O, must be related to the electrical path time delay, T_e, by $T_O/n = T_e/m$ (n and m are integers). This condition forces the electrical pulses and the optical pulses to arrive in the gain section of the laser at the same time. For stable operation, the electrical delay time must be shortened slightly to allow the carrier density to rise before the arrival of the optical pulse.

Fig. 12 shows the measured average optical power versus d.c. current drive to the self mode-locked laser for the case of n=3 and m=10. Part A shows the response without O/E converter feedback. When feedback is added, the laser breaks into pulsation just above the normal laser threshold. For this particular system, the laser amplitude limits when the peak feedback current available from the O/E converter is reached. Part B shows that the average power increases with laser bias but with a lower slope efficiency. Part C shows that once self-mode-locking has been established, the laser can be biased below the CW lasing threshold and it will continue to oscillate. The hysteresis shown is possible because the O/E converter supplies the extra current necessary to maintain the oscillation below d.c. threshold.

Pulse repetition frequencies from 500 MHz (n=1,m=3) to 6 GHz (n=12,m=37) have been measured. The pulse rate is adjusted by substituting various lengths of coaxial transmission line delays and varying the length of the optical ring cavity for fine adjustments. The width of the optical pulse is very sensitive to the relative lengths of the optical and electrical feedback loops. Fig. 13 shows how the optical pulse FWHM varies for changes in the optical path delay, with the electrical path delay held constant. The left side of the plot is for the electrical pulse arriving at the laser slightly earlier than the optical pulse. By shortening the optical cavity, the pulse width narrows until the optical pulse starts to arrive too late to achieve significant optical gain, and the optical pulse width rises dramatically. Fig. 13 also shows the theoretical detuning behavior which agrees qualitatively with the experimental data. The theoretical results are based on a partial integration solution of the rate equations [12]. The modeled current drive has a d.c. bias plus a component due to the O/E feedback. This O/E feedback current is digitally filtered and limited to match the measured impulse response and saturation characteristics of the O/E converter.

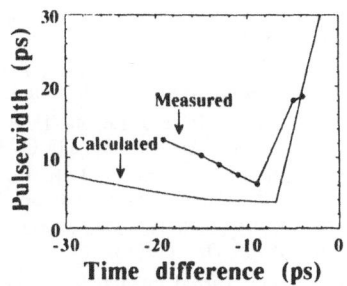

Fig. 13 Optical Pulse width vs arrival time difference between electrical and optical pulses at the gain region.

SUMMARY

The use of active mode locking to produce short optical pulses has been characterized extensively, with the shortest pulses ever observed from a semiconductor laser (580 fs) being measured. The output pulse train, however, contains multiple pulses and so the useable pulsewidth becomes the envelope of these pulses. The results of theoretical calculations have led to a new mechanism called dynamic detuning which causes multiple pulse output and is the limiting factor in achieving shorter optical pulses, not dispersion or the finite gain bandwidth of the optical medium, as previously believed. These multiple pulses are mode locked relaxation oscillations which have been observed at frequencies up to 300 GHz.

The first monolithic hybrid modelocked device has been described. This device overcomes the constraint of using an external cavity in modelocking of semiconductor lasers, by integrating a waveguide on the device. It also overcomes the problem of multiple pulse output due to the exclusion of the interface between the gain section and the external cavity, from which reflections occur, and by the inclusion of a saturable absorber which will inhibit multiple pulse build up. By the effects of both active and passive mode locking, single optical pulses as short as 1.4 ps have been obtained. Future devices using passive waveguide sections to reduce dispersion, and optimized geometries to capitalize on colliding pulse modelocking action, should produce subpicosecond single pulse output from monolithic devices.

A new technique in mode locking of semiconductor lasers called 'Self Modelocking' has been described. This technique permits mode locking of a semiconductor laser without a microwave synthesizer to drive the laser diode. Initial results show that stable modelocking occurs, and by varying either electrical or optical feedback paths, modelocking at various harmonics of the optical cavity length can be achieved. The use of electrical thresholding or nonlinearity may be interesting for improving results of this modelocking technique.

In the future, a hybrid mode locking approach (both active and passive mode locking) will produce subpicosecond, single pulses from semiconductor laser devices. Measurements of timing jitter from mode-locked semiconductor lasers show them to be orders of magnitude quieter than conventional larger mode-locked laser systems. These attributes, coupled with the inherent advantages of size, reliability and cost of these devices will then open up a vast array of new practical uses of short optical pulses in communications and measurements applications.

The authors would like to acknowledge the Office of Naval Research and the National Science Foundation for sponsoring this work.

REFERENCES

[1] J. P. van der Ziel, "Mode Locking of Semiconductor Lasers", in Semiconductors and Semimetals, vol 22B, edited by W. T. Tsang, Academic Press, Inc. (1985).

[2] S. W. Corzine, J. E. Bowers, G. Przybylek, U. Koren, B. I. Miller, and C. E. Soccolich, "Active Mode Locked GaInAsP Laser with Subpicosecond Output", Appl. Phys. Lett. **52**, 348 (1988).

[3] J. E. Bowers, P. A. Morton, A. Mar, S. W. Corzine, "Actively Mode Locked Semiconductor Lasers", IEEE J. Quantum Electron. **QE-25**, 1426 (1989).

[4] G. Eisenstein, R. S. Tucker, U. Koren, S. Korotky, "Active Mode Locking Characteristics of InGaAsP Single Mode Fiber Composite Cavity Lasers", IEEE J. Quantum Electron., **QE-22**, 142 (1986).

[5] E. P. Ippen, D. J. Eilenbert, and R. W. Dixon, "Picosecond Pulse Generation by Passive Mode Locking of a Diode Laser", Appl. Phys. Lett. **37**, 267 (1980).

[6] Y. Silberberg, and P. W. Smith, "Subpicosecond Pulses from a Mode Locked Semiconductor Laser", IEEE J. Quantum Electron. **QE-22**, 759 (1986).

[7] K. Y. Lau, and I. Ury, "Passive and Active Mode Locking of a Semiconductor Laser Without an External Cavity", Appl. Phys. Lett. **46**, 1117 (1985).

[8] R. S. Tucker, U. Koren, G. Raybon, C. A. Burrus, B. I. Miller, T. L. Koch, G. Eisenstein, '40-GHz Active Mode Locking in a Monolithic Long-Cavity Laser', Electron. Lett., **25**, 621 (1989).

[9] P. A. Morton, J. E. Bowers, L. A. Koszi, M. Soler, J. Lopata, D. P. Wilt, 'Monolithic Hybrid Mode Locked 1.3 µm Semiconductor Lasers', Appl. Phys. Lett., **56**, 111 (1990).

[10] D. J. Derickson, R. J. Helkey, A. Mar, P. A. Morton, J. E. Bowers, 'Self Mode-Locking of a Semiconductor Laser using Positive Feedback', Appl. Phys. Lett., **56**, 7 (1990).

[11] P. A. Morton, R. J. Helkey, J. E. Bowers, 'Dynamic Detuning in Actively Mode Locked Semiconductor Lasers', IEEE J. Quantum Electron., **QE-25**, Dec. (1989).

[12] R. J. Helkey, P. A. Morton, J. E. Bowers, 'Partial Integration Method for Analysis of Mode-Locked Semiconductor Lasers', Optics Lett., **15**, 112 (1990).

FEMTOSECOND PULSE SHAPING AND THE FUNDAMENTAL DARK SOLITON

J. P. Heritage, A. M. Weiner, and R. N. Thurston

Bell Communications Research, 331 Newman Springs Road
Red Bank, New Jersey 07701, U.S.A.

ABSTRACT

The technique of femtosecond pulse shaping provides unprecedented control of the amplitude and phase of utrashort optical pulses. One class of pulse shapes that can be fabricated by this technique is the so-called "dark pulse". Dark pulses are a rapid dip or hole in a slowly varying intensity envelope. This pulse is of considerable interest because it resembles the formal solitary wave solution to the nonlinear Schrödinger equation in the positive group velocity dispersion region of optical fibers. This is true provided there is an abrupt π phase jump in the center of the dark pulse. We experimentally verify the existence of the fundamental dark soliton by propagating a 185-fsec dark pulse at a wavelength of 0.62 microns in a 1.4 m length of single mode fiber. At the appropriate power level the dark pulses propagate without broadening. Our measurements are in quantitative agreement with numerical solutions to the nonlinear Schrödinger equation.

CLOSING REMARKS

Professor K. K. Rebane
Institute of Physics, Tartu

Ladies and Gentlemen!

The Fourth Binational US/USSR Symposium on "The Physics of Optical Phenomena and Their Use as Probes of Matter" is very near to be over.

We listened to a number of interesting lectures; our discussions were active and, I would say, mostly to the point. I am sure that the Irvine Symposium was not only just a place to exchange information, but that in the course of the symposium some new understanding and some new ideas were created that did not exist before. We can think, that every-one leaving this symposium takes with her or him at least one-two new ideas and pieces of knowledge useful for his or her work.

I would like also to say that our symposium certainly was a positive contribution to the traditionally friendly relationship between the Soviet and American physicists.

I was told and I felt it myself, that there was a common feeling among the participants that this series of symposia has to be continued.

On Thursday night there was a meeting of the organizers and it was proposed that the next, 5th symposium, take place in the U.S.S.R., in 1992, probably in spring or early summer. As to the site, then as the first choice Tallinn, Estonia was proposed and as a second choice Troitsk, the city near Moscow, where the Institute of Spectroscopy is located, was proposed. I am glad to invite you to the 5th symposium and to the Soviet Union in 1992!

On behalf of the Soviet participants I thank all the organizers of the Fourth binational symposium, most of all Professor Alexei Maradudin and his very efficient assistants Mrs. Peggy Maradudin and Ms. Jeannie Brown. On the Soviet side we are grateful to Dr. Ludmila Bureyva, whose precise and highly responsible work with all the papers, contacts and tickets made the travel of this considerably large body of 20 Soviet physicists to USA possible!

Thank you!

Professor A. A. Maradudin
University of California, Irvine

As the Fourth Binational US/USSR Symposium on "The Physics of Optical Phenomena and Their Use as Probes of Matter" draws to a close, I would like to follow up on Karl Rebane's remarks and say that I hope that as the participants leave they will take away with them not only new ideas useful for their work, but also the feeling of having made new friends, and possibly new colleagues, as a result of having taken part in the Symosium.

On behalf of the U.S. Organizing Committee for this Symposium, I wish to express our appreciation to Professor Karl K. Rebane and the Soviet Organizing Committee for their efforts in putting together such a large and strong group of physicists and bringing it to this Fourth Symposium in Irvine. Special thanks are due to Dr. Lyudmila A. Bureyva for her skillful handling of the logistical problems involved in accomplishing this.

I wish to express my thanks to the other members of the U.S. Steering Committee, to the Organizing and Program Committee, and to the Local Organizing Committee for the many hours of time and the effort they expended in creating a Symposium that not only reflected the state of the art in the field, but brought a group of excellent speakers to describe it as well, in settings that were well-equipped and comfortable.

We look forward to the Fifth Symposium, be it in Tallinn or in Troitsk. May there be as much growth and development in our field, and in the relations between our countries, between now and then as there has been since the last Symposium.

Tomorrow our Soviet colleagues depart for an intense week of visits to many universities and laboratories across the United States. I wish them safe, pleasant, and scientifically profitable journeys, as they meet old, and make new, friends in this country. And to our international and American participants - a safe trip home!

I now declare the Symposium closed.

PARTICIPANTS

Aaviksoo, J., Institute of Physics, Tartu, Estonian SSR
Agranovich, V. M., Institute of Spectroscopy, Troitsk, USSR
Akhmanov, S. A., Moscow State University, Moscow, USSR
Anderson, D. Z., University of Colorado
Apanasevich, P. A., Institute of Physics, Minsk, Byelorussian SSR

Bagaev, V. S., P. N. Lebedev Physical Institute, Moscow, USSR
Birman, J. L., City College of the City of New York
Bjorklund, G. C., IBM Almaden Research Center
Bloembergen, N., Harvard University
Bowers, J., University of California, Santa Barbara
Boyd, R. W., University of Rochester
Bron, W. E. University of California, Irvine
Bureyeva, L. A., Scientific Council on Spectroscopy, Moscow, USSR

Califano, S., European Laboratory of Nonlinear Spectroscopy, Florence,
   Italy
Celli, V., University of Virginia
Chebotaev, V. P. Institute of Thermal Physics, Novosibirsk, USSR

Falcone, R., University of California, Berkeley
Fayer, M. D., Stanford University
Flytzanis, Chr. Ecole Polytechnical, Palaiseau, France
Fröhlich, D., Universität Dortmund, Federal Republic of Germany
Fujimoto, J. G., Massachusetts Institute of Technology

Garmire, E., University of Southern California

Hanamura, E., University of Tokyo, Japan
Heinz, T. F., IBM T. J. Watson Research Center, Yorktown Heights
Heritage, J. P., Bell Communications Research, Red Bank
Hizhnyakov, V. V., Institute of Physics, Tartu, Estonian SSR, USSR

Ipatova, I. P., A. F. Ioffe Physico-Technical Institute, Leningrad, USSR

Jaccarino, V., University of California, Santa Barbara

Kaplyanskii, A., A. F. Ioffe Physico-Technical Institute, Leningrad, USSR
Karpushko, F. V., B. I. Stepanov Institute of Physics, Minsk,
   Byelorussian SSR, USSR
Keldysh, L. V., P. N. Lebedev Physical Institute, Moscow, USSR
Kimble, H. J., California Institute of Technology
Klein, M. V., University of Illinois, Urbana-Champaign
Knox, W. H., AT&T Bell Laboratories, Holmdel
Koroteev, N. I., Moscow State University, Moscow, USSR
Kulakovskii, V. D., Solid State Physics Institute, Chernogolovka, USSR

Lagendijk, A., Universiteit van Amsterdam, The Netherlands
Leskova, T. A., Institute of Spectroscopy, Troitsk, USSR
Levanyuk, A. P., Institute of Crystallography, Moscow, USSR
Levinson, Y. B., Institute of Microelectronics Technology, Chernogolovka, USSR
Lyons, K. B., AT&T Laboratories, Murray Hill

Mavrin, B. N., Institute of Spectroscopy, Troitsk, USSR
Mirlin, D. N., A. F. Ioffe Physico-Technical Institute, Leningrad, USSR
Österberg,
            U., U.S. Air Force Academy
Peyghambarian, N., University of Arizona
Pick, R. M., Université Pierre et Marie Curie, Paris, France
Pinczuk, A., AT&T Bell Laboratories, Murray Hill

Rebane, K. K., Institute of Physics, Tartu, Estonian SSR, USSR

Shen, Y. R., University of California, Berkeley
Singer, K. D., AT&T Bell Laboratories, Princeton
Steel, D. G., University of Michigan
Sulewski, P. E., AT&T Bell Laboratories, Murray Hill

Tartakovskii, I. I., Institute of Solid State Physics, Chernogolovka, USSR

van Driel, H. M., University of Toronto

Winful, H. G., University of Michigan

Yablonovitch, E., University of Pennsylvania
Yodh, A. G., University of Pennsylvania

Zakharchenya, B., A.F. Ioffe Physico-Technical Institute, Leningrad, USSR

AUTHOR INDEX

$CH_3(CH_2)_{17}(Me)_2N^+(CH_2)_3Si(OMe)_3$
Cl^-(DMOAP), 50, 52

Domain structure, 183
  nonequilibrium, 182
Domain wall, 184-185
  bending of, 185
Domain wall vacancies, 184
Donor, 161
Donor-donor transfer, 159
Doped glasses, 289
Doppler effect, 95
Doppler shift, 95, 330
  changing with time, 331
Double monochromator, 167
Double stars, 326
Dressed polariton, 19, 21, 23-24
  damping of, 24
Drift velocity
  of excitons, 40
Drude formula, 233
Dynamic charge transfer mechanism,
  340
Dynamic detuning, 402, 406, 411
Dynamic fluctuations
  in a dipolar glass, 187
Dynamic light scattering, 307
Dynamic slowing down, 192
Dynamic, light-induced, gratings,
  351
Dynamical charge transfer, 339
Dynamical instabilities, 99-100
Dynamical matrix
  third-order, 169, 176
Dynamics
  of molecular crystals, 168
  of natural Raman oscillations, 97
Dyson equation, 170

Echo, 396
Echo signal, 395
Echo signal decay time, 396
Effective diffusion coefficient, 155
Effective Hamiltonian, 126-127
Effective mass
  exciton, 62-63, 67
  for exciton center of mass motion,
    207
  of Coulomb exciton, 151, 154
  renormalization of, 297, 305
Effective Unruh temperature, 332
8CB, 50, 52
Elastic deformation, 193
Elastic light scattering
  at phase transitions, 181
  defect-induced, 181
  from KTN, 190
Elastic mean free path of light, 318
Elastic scattering time, 361
Electric dipole approximation, 238
Electric dipole interaction, 247,
  249
Electric field poling, 251, 253

Electric field poling (continued)
  under uniaxial stress, 251
Electric field potential, 253
Electric quadrupole interactions,
  243, 249
Electroabsorption response
  of quantum well excitons, 391
Electro-absorptive media, 223
Electro-absorptive modulator, 224
Electro-optic Cerenkov effect, 20
Electro-optic effect, 222
  in CdTe, 226
Electro-optic modulation, 228
Electro-optic photorefractive
  effect, 226-227
Electro-optic sampling systems, 401
Electro-optic switching, 228
Electro-optical properties, 387
Electro-refraction, 219, 226
Electro-refractive media, 222
Electro-refractive photo-refractive
  effects, 226, 227
Electrodynamic Green's function, 115
Electromagnetic field, 199
  nonclassical character of, 205
  quantum character of, 205
Electromagnetic quantum
  fluctuations, 329
Electromagnetic response of a
  grating
  light emission, 316
  reflectivity, 316
Electron density
  fluctuations of, 27-28
Electron density fluctuation, 30
Electron diffraction, 132, 237
Electron dynamics, 71
Electron energy bands
  nonparabolicity of, 32
Electron gas
  two-dimensional, 71-72, 80
Electron hole excitation, 334
Electron lifetime, 260
Electron mean free path, 29
Electron scattering
  from optical phonons, 262
Electron spin resonance, 245
Electron subband, 300
Electron traps, 268
Electron wave function, 77
Electron-electron interactions, 273,
  278-279
Electron-hole Bohr radius, 290
Electron-hole dynamics, 265
Electron-hole pairs, 38, 283, 329,
  361
  optically created, 290
  optically generated, 292
  overbarrier excitation of, 287
  virtual, 332